NASA SP-2009-4012

NASA HISTORICAL DATA BOOK
Volume VII

(Part 1)

NASA Launch Systems, Space Transportation/ Human Spaceflight, and Space Science 1989–1998

Judy A. Rumerman

The NASA History Series

National Aeronautics and Space Administration
NASA History Division
Office of External Relations
Washington, D.C.   2009

Library of Congress Cataloging-in-Publication Data
(Revised for vol. 7)

NASA historical data book.

(The NASA historical series) (NASA SP ; 2009-4012)
Vol. 1 is a republication of: NASA historical data book,
1958–1968./ Jane Van Nimmen and Leonard C. Bruno.
Vol. 7 in series: The NASA history series.
Includes bibliographical references and indexes.
Contents: v. 1 NASA resources, 1958–1968 / Jane Van
Nimmen and Leonard C. Bruno — v. 2. Programs and projects,
1958–1968 / Linda Neuman Ezell — v. 3. Programs and projects,
1969–1978 / Linda Neuman Ezell — v. 4. NASA resources, 1969–
1978 / Ihor Gawdiak with Helen Fedor — v. 5. NASA launch
systems, space transportation, human spaceflight, and space
science, 1979–1988 / Judy A. Rumerman — v. 6. NASA Space
Applications, Aeronautics and Space Research and Technology,
Tracking and Data Acquisition/Support Operations, Commercial
Programs, and Resources 1979–1988
    1.   United States. National Aeronautics and Space
Administration—History. I. Van Nimmen, Jane. II. Bruno,
Leonard C. III. Ezell, Linda Neuman. IV. Gawdiak, Ihor. V.
Rumerman, Judy A — v. VI.
    VI. Series. VII. Series. VIII. Series: NASA SP; 4012.

CONTENTS

LIST OF FIGURES AND TABLES

vi

Chapter Three: Human Spaceflight

Chapter Four: Space Science

xiv

PREFACE AND ACKNOWLEDGMENTS

This volume of the NASA Historical Data Book is the seventh in the series that describes NASA's programs and projects. Covering the years 1989 through 1998, it includes the areas of launch systems, human spaceflight, and space science, continuing the volumes that addressed these topics during NASA's previous decades. Each chapter presents information, much of it statistical, addressing funding, management, and details of programs and missions. This decade, which followed the Agency's return to flight after the *Challenger* accident, was especially productive. Upgraded expendable launch vehicles sent missions into Earth orbit and toward the outer reaches of space; 66 Space Shuttle missions were successfully launched; the Space Station received its first components; and 30 space science missions, most of which met their scientific goals, began returning scientific data to Earth. These events took place in an environment both of international cooperation and one in which NASA learned to make the best use possible of its resources.

A forthcoming companion volume will describe NASA's Earth science missions; aeronautics and space research activities; tracking and space operations; facilities; resources; and personnel areas.

A large group of people assisted in preparing this volume and should be recognized. Most valuable and essential was my research assistant, Tai Edwards, who gathered material, organized it superbly, entered data into tables, and proofed and edited draft chapters, all while attending graduate school and getting married. It would have been impossible to deal with the quantity of information I faced without her help. The NASA History Division archivists, Colin Fries, John Hargenrader, Liz Suckow, and chief archivist Jane Odom, helped gather information. Stephen Garber managed the project and dealt with contractual matters. Interns Matt Barrow and Clare Kim, also helped shepherd this project through the production cycle. Nadine Andreassen assisted in a myriad of ways.

On the production end, special thanks go to the NASA Headquarters Communications Support Services Center: Shelley Kilmer-Gaul carefully laid out this volume; Andrew Jarvis edited the layout; and Hanta Ralay oversaw the critical final step of printing. Many thanks are due to all these professionals.

Many people at NASA and in the NASA community gladly provided information and helped explain events and resolve discrepancies. Staff members at every NASA Center willingly offered their assistance and supplied material. Individuals in the program offices at Headquarters and in the various projects at Goddard Space Flight Center spent hours talking with me and finding documents. Graphics personnel both at Headquarters and at the Centers regularly filled my requests for "high resolution graphics."

A special thanks goes to the reviewers of the draft chapters. They took on the arduous job of reading drafts that often numbered in the hundreds of pages, finding errors, and making valuable suggestions.

Their work improved the quality of this document immensely.

I'd also like to thank my husband, Howard, who provided continual support, would listen to my concerns, and frequently resolved computer issues that could have proven disastrous.

ABOUT THE AUTHOR

Judith A. Rumerman is a professional technical writer who has written or contributed to numerous documents for the National Aeronautics and Space Administration. She has written documents describing various spaceflight programs, in-house procedures used at Goddard Space Flight Center, and various materials used for training. She was also the compiler of *U.S. Human Spaceflight: A Record of Achievement, 1961–1998*, a monograph for the NASA History Office detailing NASA's human spaceflight missions, and volumes five and six of the *NASA Historical Data Book, 1979–1988*. In the years preceding the 2003 Centennial of Flight, Ms. Rumerman served as technical lead and prime author of the series of essays written for the Centennial of Flight Commission describing all aspects of aviation and spaceflight aimed at young people of high-school age.

Ms. Rumerman has degrees from the University of Michigan and George Washington University. She grew up in Detroit and presently lives in Silver Spring, Maryland.

NOTES ON SOURCES

The bulk of sources used in preparing this volume are official NASA documents and references. Whenever possible, the author attempted to use primary sources prepared by the organizations or individuals most directly. involved in a program or mission. NASA Web sites were also used extensively. Secondary sources were most often used to provide perspective rather than data The following paragraphs describe major sources. Detailed footnotes are located in each chapter.

Annual Budget Estimates: These documents are issued each year by the NASA Office of the Chief Financial Officer when the annual budget request is presented to Congress. These lengthy documents, filling several loose-leaf binders each year, contain breakdowns of three fiscal years of budgets: the year just ending, the next fiscal year, and the fiscal year two years out. Budget figures are presented by appropriation, program office, installation, program, and in any other way that may be of interest to budget preparers. Toward the end of this decade, "full cost" accounting was adopted, and budget figures for major programs were presented in both the traditional way and in "full-cost" figures. The budget estimate documents also provided comprehensive narrative descriptions of programs and activities, describing both what had occurred during a prior fiscal year (and occasionally farther back) and what the Agency's plans were for the next two years. These descriptions provide a useful account of a program's evolution.

Press and Media Kits: NASA prepares press or media kits for every Space Shuttle mission and for a number of major robotic missions. They describe launch events, payloads, planned experiments, astronaut biographies, and other mission-unique information. Designed for non-technical audiences and the media, they provide a comprehensive description of NASA missions. All Shuttle press kits and most other press kits are available online.

Mission Operation Reports: Every NASA mission is required to prepare a pre-launch and post-launch mission operation report. These reports are designed for the use of senior management and, while they are part of the NASA Historical Reference Collection, they may not always be available to the public. They provide material similar to that found in the press kits but may also include more technical information, and the post-launch reports may include assessments of the success of various mission elements.

Aeronautics and Space Reports of the President: These annual reports describe the aeronautics and space activities of all government agencies that engage in these types of activities. They provide a good overview and an excellent starting point for research.

Press Releases: NASA Headquarters and each NASA Center regularly issue press releases describing newsworthy events.They provide the current status on various events including scientific missions, management and organizational changes, contract awards, and changing Agency priorities. They are often the only source of current, detailed information about a mission. Headquarters press releases have been posted on the NASA Web site since the early 1990s. The Centers began posting their press releases in the mid-1990s.

Exploring the Unknown, Selected Documents in the History of the U.S. Civil Space Program, Volume V: Exploring the Cosmos and *Volume VI: Space and Earth Science*, edited by John Logsdon: Particularly in the space science area, the introductory essays preceding the documents in these two volumes, written by eminent individuals in their fields, provide outstanding descriptions of the major events in the history of the space program.

International Reference Guide to Space Launch Systems, Third Edition and Fourth Edition, by Steven J. Isakowitz, Joseph P. Hopkins, Jr., and Joshua B. Hopkins: Published by the AIAA, the two editions of this reference contain thorough descriptions of every launch vehicle used during this decade, as well as information related to performance, cost, flight history, vehicle design, payload accommodations, production and launch operations, and vehicle history.

Faster, Better, Cheaper: Low-Cost Innovation in the U.S. Space Program, by Howard McCurdy: This book offers an excellent introduction to NASA's management approach, describing what very likely was the dominant philosophy at NASA during this decade.

Web Sites: The past few years have seen an explosion of material posted on the Internet. Every NASA program has a Web site (too many to list here) and posts a wide variety of information about a project. This has had both positive and negative consequences. On the positive side, official documents such as legislation, policies, Agency reports, and directives are readily available. NASA programs post huge amounts of material describing all phases of missions including: mission parameters and specifications, instrument descriptions, scientific results, implications, etc. This information enables researchers to acquire a great deal of information without the need to cull through files or archives. However, it is also very easy for errors to be perpetuated, even when information is located on NASA Web sites. Information is easily copied from one Web site to the next, often without question, and errors are inadvertently introduced when material is not carefully edited. It is necessary for the researcher to verify information carefully before using it. Another issue is the removing of information from Web sites because of storage considerations without archiving the information. Information "disappears" or is moved to another location on the internet. This happens especially when information becomes "out-of-date" without concern for the historical value of the material. Broken links, due both to technical difficulties and the removal or moving of Web pages without revising the referring link, are also a problem. Web material has been used extensively in this volume, but care has been taken to ensure its reliability. An "access" date is always included, and a printed copy of all Web pages used has been provided to the NASA History Division.

Space Shuttle Mission Chronologies: These short mission descriptions provide launch and landing information, a crew list, and mission highlights. They originally existed as individual pages for each Space Shuttle mission available from the main Human Spaceflight Web page at Kennedy Space Center. Half way through preparing this volume, these disappeared from the site and were replaced by very brief mission descriptions with much less information. No link was provided to the new location of the original material. The original individual mission files were combined into two PDF files (up to 1999 and from 2000) and a link to a set of HTML files for each mission and placed at a different location *http://www-paokscnasagov/kscpao/nasafact/pdf/1981-99Volume1pdf; http://www-paokscnasagov/kscpao/nasafact/pdf/Volume2 pdf;* and *http://sciencekscnasagov/shuttle/missions/missionshtml*. Most links within the HTML files do not work. This experience is indicative of the difficulties encountered when using the internet for research.

Space Science Project Web Sites: Each NASA project has a Web site of varying levels of detail and quality. Some provide extensive information about the mission and science results while others provide only basic information. Some missions have more than one Web site—one dealing with mission elements and a second dealing primarily with the science. The Web sites for the Hubble Space Telescope are particularly useful. The NASA Web site describes the mission, and the Web site sponsored by the Space Telescope Science Institute provides a great deal of detail concerning the science. Universities that co-sponsor or provide instruments to missions often have their own Web sites.

National Space Science Data Center: While not easy to navigate, the Master Catalog on the NSSDC database often provides the only available source of basic information for each mission. While not lengthy, the pages for each mission supply a basic mission description, orbital information, and a list and description of each instrument often with the names and affiliations of the Principal Investigators.

CHAPTER ONE
INTRODUCTION

CHAPTER ONE
INTRODUCTION

NASA began operating as the nation's civilian space agency in 1958 after passage of the National Aeronautics and Space Act. It succeeded the National Advisory Committee for Aeronautics (NACA). The new organization was charged with preserving the role of the United States "as a leader in aeronautical and space science and technology," expanding our knowledge of Earth's atmosphere and space, and exploring flight both within and outside the atmosphere.

The decade from 1989 to 1998 was extremely productive, as NASA added to its already considerable list of achievements. The decade was marked by assembly of the first orbiting Space Station components, launch of the first two Great Observatories, and an outstanding record of safe and fruitful missions. This volume addresses NASA's activities during the decade in the areas of launch systems, human spaceflight, and space science.

A number of groups influenced NASA's direction. Congress influenced the Agency through authorization and appropriation bills. The Executive Branch articulated the President's views on space exploration and development through the annual budget submission, other legislation, and policy directives. During the administration of President George H. W. Bush, as in the administration of President Ronald Reagan before him, the National Space Council shaped and articulated "national" space policy (as defined by the administration). Chaired by the Vice President, the Council consisted of the heads of all departments or other offices with a programmatic role or concern in federal space activities. In November 1993, President William J. Clinton established the National Science and Technology Council, a cabinet-level council serving as the principal means for the President to coordinate science, space, and technology and coordinate the diverse parts of research and development at the federal level.

In addition, a series of advisory committees, task groups, and commissions, often formed by the NASA Administrator to address specific Agency concerns, advised the Agency on the direction it deemed most advantageous and worthwhile to take and how it could solve identified problems and improve the way "it did business." These advisory committees and commissions typically consisted of individuals, both experts and non-experts in fields related to space, from diverse backgrounds such as industry, academia, the military, Congress, NASA, and other government agencies. Proceedings of these groups, as well as national policy directives, are cited in the following chapters where relevant.

Overview of the Agency

NASA is an independent federal government agency consisting of a headquarters in Washington, DC, nine Centers or installations located around the United States, and the Jet Propulsion Laboratory, a government-owned, contractor-occupied facility in Pasadena, California, operated under contract to NASA and staffed by the California Institute of Technology. NASA Headquarters consists of program and staff offices providing overall program management and administrative functions for the Agency. During the 1990s, the Agency adopted a thematic strategic enterprise approach to supplement its traditional program office structure. These strategic enterprises, led by Associate Administrators, developed strategy and policy, formulated programs, and assigned lead Centers for specific projects and activities. Although the focus and content of the enterprises changed at times, as did their names, they generally fell into the areas of aeronautics, human spaceflight, Earth science, and space science. To provide continuity when dealing with Congress, NASA retained its program office designations for its annual budget submissions to Congress. Table 1–1 shows NASA's program offices and their major functional areas as stated in the annual budget submissions broken down by appropriation.

NASA Centers operated fairly autonomously to implement Agency plans, programs, and activities as part of a program office or strategic enterprise. Each Center focused on particular types of projects, technology, and discipline areas, indicated by its designation as a Center of Excellence (see table 1–2). Installations were assigned the role of Lead Center for programs based on the Center's mission and Center of Excellence capabilities. Each Center was responsible for day-to-day program management and execution, hiring its own personnel, and awarding its own procurements.

Program and Project Development

NASA called most of its activities "programs" or "projects." The Agency defined programs as "major activities within an enterprise that have defined goals, objectives, requirements, and funding levels, and consist of one or more projects." Projects were "significant activities designated by a program and characterized as having defined goals, objectives, requirements, life-cycle costs, a beginning, and an end."[1]

NASA's programs and projects followed a sequence of events, called a life cycle, consisting of program formulation, program implementation, and several approval milestones needing to be passed. For most of the decade, the life cycle consisted of six phases (with corresponding letter designations). Formulation included Advanced Studies (Pre-Phase A), Preliminary Analysis (Phase A), and Definition (Phase B). Program implementation included Design (Phase C), Development (Phase D), and Operations (Phase E).[2]

In 1998, NASA replaced this structure with one consisting of the same two major stages—program formulation and program implementation—neither of these divided into formal phases. Program formulation included program planning, systems analysis, and technology requirements synthesis. Program implementation included program control, technical requirements management, and the design and development of technology and systems. Several reviews and evaluations took place at specific points within each stage.

Typically, funding for project formulation activities came out of research and technology funding held at the Headquarters level. Congressional funding for a specific program was received after a major review was conducted at the end of program implementation. At all stages, a prescribed set of documents, performance metrics, and evaluations were a large part of the process to ensure that requirements were achieved.[3]

NASA's Budget

NASA depends on a reasonable level of funding from Congress each year to finance its programs.[4] The federal budget process is complex and requires foresight and planning by everyone involved with the allocation of resources. This section provides an overview of the budget process. More detailed information can be found in chapter 7 of Volume VIII of the *NASA Historical Data Book, 1989–1998*.

---

[1] NASA Procedures and Guidelines (NPG) 7120.5A, "NASA Program and Project Management Processes and Requirements," Effective April 3, 1998 (canceled).
[2] NASA Handbook (NHB) 7120.5, "Management of Major System Programs and Projects Handbook," November 8, 1993 (canceled).
[3] NPG 7120.5A.
[4] The government operates on a "fiscal year" basis that runs from October 1 through September 30 of the following year. The fiscal year is called by the year in which it ends, e.g., FY 1993 runs from October 1, 1992, through September 30, 1993.

Congress funded NASA's activities each year by means of large appropriations categories. Through fiscal year (FY) 1994, four major appropriations funded the Agency. The Research and Development (R&D) appropriation funded most of NASA's programs and projects. Spaceflight, Control, and Data Communications (SFC&DC) funded operation of the Space Shuttle, some Space Station activities, and tracking and data acquisition activities. The Research and Program Management (R&PM) appropriation funded civil service salaries, regardless of the project or office in which an individual worked, as well as related expenses such as benefits, training, and travel. Construction of Facilities (C of F) funded design and construction of facilities, purchase of land, and similar activities. The Office of Inspector General appropriation funded this independent office.

In FY 1995, the appropriations categories changed to a three-appropriation structure. The new categories were Human Space Flight (HSF), Science, Aeronautics, and Technology (SAT), and Mission Support (MS). HSF funded most Space Station and Space Shuttle activities. The SAT appropriation funded most research and development programs with the exception of the Space Station and Space Shuttle. MS funded the civil service workforce, space communication services, safety and quality assurance activities, maintenance, and most activities formerly funded by the C of F appropriation. The Office of Inspector General retained its appropriation arrangement, as it had before.

NASA was required to spend its funds according to the way Congress allocated funds among the appropriation categories. Although a program office could administer activities from more than one appropriation category, the Agency could not transfer funds from one appropriation category to another without congressional notification. Table 1–3 shows the major programs within each appropriation category.

NASA's budget planning cycle lasted two years. Two years before the beginning of a fiscal year, NASA Headquarters sent programmatic and budget guidelines to each Center based on the Agency's long-range plans and budget forecasts from the Office of Management and Budget (OMB). Each Center then prepared a detailed budget, or Program Operating Plan, for the fiscal year beginning two years in the future. The Center also refined the budget for the remainder of the current fiscal year and revised the budget request for the next fiscal year that it had submitted the year before. Additionally, it provided budget figures for future years. Upon approval from each Center's comptroller and Director, this budget was forwarded to the appropriate Headquarters program or enterprise office, the NASA comptroller, and the NASA Administrator. The comptroller and Administrator finalized the budget request and submitted it to the OMB. After OMB review and further discussion with NASA, the OMB formally submitted the NASA budget request to Congress as part of the President's budget request in February of each year.

NASA prepared and submitted a draft authorization bill that went to NASA's House and Senate science committees that authorized NASA's budget. Ideally, each committee held hearings and discussed the bill with the NASA Administrator and heads of specific programs. These program heads often testified before Congress in preparation for a vote on the bill. The final bill was sent to the full House and Senate and, if necessary, a conference committee reconciled any differences between the House and Senate versions. When both houses of Congress passed the same bill, it went to the President for signature. The authorization bill limited how much could be appropriated and could set conditions on how funds were to be spent.

In some years, however, Congress did not pass an authorization bill. In those years, although Congress held authorization hearings and discussions, only an appropriations bill was passed.[5] The appropriations bill was required for NASA to actually spend funds. Without an appropriations bill at the start of a fiscal year, Congress must pass a continuing resolution allowing agencies to continue operating at a particular level of funding.

The appropriation process was similar to the authorization process, with the bills going to the proper appropriations committees for discussion, revision, and approval. However, in practice, appropriations committees usually did not review the proposed budget in as great detail as the authorization committees unless its members were especially interested in a particular program. Upon committee approval, the appropriations bills went to the full House and Senate, back to a conference committee if necessary, and finally to the President. After approval by the President, the OMB established controls on the release of the funds to the Agency.

Once NASA received control over its appropriated funds, it designated the funds for its various programs, projects, and facilities. An "account" for each item was set up allowing the Agency to commit, obligate, cost, and disburse the funds and track them as they were spent.[6] NASA scrupulously monitored all of its financial activities, first at the project and Center level and then at the Headquarters level. Its financial transactions were eventually reviewed by the congressional General Accounting Office to ensure that they were legal and followed appropriate procedures.

In FY 1995, NASA began a "full cost" accounting initiative. This initiative included all costs (both direct and indirect) associated with an activity, not just funds spent during a limited part of a program's life cycle (usually the prelaunch development phase). Before full cost was implemented, expenses associated with launch and mission operations and the cost of civil service salaries were not counted toward project costs but were instead put into a separate "launch support," or "mission operations" category. Full cost included all of these costs such as civil service salaries, the use of facilities,

---

[5] An authorization bill is not required for appropriations to be passed.
[6] "To cost" funds refers to the process of recording the total value of resources used in producing goods or rendering services.

and support services associated with the benefiting activities as part of a project's expenses, thus providing a more accurate picture of the actual cost of a project. Formulating a full cost budget allowed for full disclosure of NASA's activities and established a more defined link between funds received and funds spent. Full cost also provided the Agency with greater accountability regarding the use of its resources. For FY 1997 and FY 1998, NASA prepared dual budgets: one using full cost and one using traditional budget methods. In the next decade, NASA went completely to using full cost.

The budget tables in the following chapters show the initial amounts requested by NASA each fiscal year (two years before the start of the fiscal year for which the funds were requested) and the revised amounts (one year before the start of the fiscal year for which the funds were requested). The tables also show the programmed amount, or what the program actually had available to spend. If full cost figures are available for an activity, they are shown.

This volume addresses NASA's launch systems, human spaceflight, and space science activities. Each chapter provides a review of activities of the previous decade, an overview of the topic, budget and funding data, management structure and personnel, and a description of the systems and missions of the decade.

Table 1–1. Programs Within the R&D Appropriation

1989	1990	1991	1992	1993
Office of Space Flight	Office of Space Flight			
• Space Transportation Capability Development	• Space Station			
Office of Space Station				
• Space Station				
Office of Space Science and Applications		Office of Space Science and Applications		
• Physics and Astronomy		• Physics and Astronomy		
• Planetary Exploration		• Planetary Exploration		
• Life Sciences		• Life Sciences		
• Solid Earth Observations		• Earth Sciences		
• Environmental Observations		• Materials Processing in Space		
• Materials Processing in Space		• Communications		
• Communications		• Information Systems		
• Information Systems				
Office of Commercial Programs				
• Technology Utilization				
• Commercial Use of Space				

Table 1–1. Programs Within the R&D Appropriation (Continued)

1989	1990	1991	1992	1993
Office of Aeronautics and Space Technology • Aeronautical Research and Technology • Transatmospheric Research and Technology • Space Research and Technology		Office of Aeronautics and Space Technology • Aeronautical Research and Technology • Transatmospheric Research and Technology • Space Research and Technology • Exploration Mission Studies	Office of Aeronautics and Space Technology • Aeronautical Research and Technology • Transatmospheric Research and Technology • Space Research and Technology	
Office of Safety, Reliability, Maintainability, and Quality Assurance • Safety, Reliability, and Quality Assurance Office of Space Tracking and Data Systems • Advanced Systems	University Space Science and Technology Academic Programs Technology Academic Programs	Academic Programs		Office of Space Exploration

Table 1–2. Centers of Excellence

Center	Designated Center of Excellence	Mission Area
Ames Research Center	Information technology	Aviation operations systems and astrobiology
Dryden Flight Research Center	Atmospheric flight operations	Flight research
Goddard Space Flight Center	Scientific research	Earth science and physics and astronomy
Jet Propulsion Laboratory	Deep space systems	Planetary science and exploration
Johnson Space Center	Human operations in space	Human exploration and astro materials
Kennedy Space Center	Launch and cargo processing systems	Space launch
Langley Research Center	Structure and materials	Airframe systems and atmospheric science
Lewis Research Center	Turbomachinery	Aeropropulsion
Marshall Space Flight Center	Space propulsion	Transportation systems development and microgravity
Stennis Space Center	Propulsion testing systems	Propulsion test

Table 1–3. Program Office Functional Areas

Programs Within the R&D/Science, Aeronautics and Technology Appropriation				
1994	**1995**	**1996**	**1997**	**1998**
Office of Space Flight • Space Transportation Capability Development	Moved to Human Space Flight appropriation	Moved to Human Space Flight appropriation		
Office of Space Systems Development • Space Station	Moved to Human Space Flight appropriation			
Office of Space Science • Physics and Astronomy • Planetary Exploration			Office of Space Science (separate mission divisions were dropped)	
Office of Life & Microgravity Sciences & Applications • Life Sciences • Microgravity Science Research • Shuttle/Spacelab Payload, Mission Management and Integration	Office of Life & Microgravity Sciences & Applications • Life Sciences • Microgravity Science Research • Space Shuttle/ Spacelab Payload, Mission Management and Integration • Space Station Payload Facilities	Office of Life & Microgravity Sciences & Applications • Life Sciences • Microgravity Science Research • Space Shuttle/Spacelab Payload, Mission Management and Integration • Space Station Payload Facilities • Aerospace Medicine/ Occupational Health	Office of Life & Microgravity Sciences & Applications • Life Sciences • Microgravity Science Research • Space Shuttle/Spacelab Payload, Mission Management and Integration • Space Station Payload Facilities • Aerospace Medicine/ Occupational Health	Office of Life & Microgravity Sciences & Applications • Life Sciences • Microgravity Science Research • Space Shuttle/ Spacelab Payload, Mission Management and Integration • Aerospace Medicine/ Occupational Health • Space Product Development

Table 1–3. Program Office Functional Areas (Continued)

Programs Within the R&D/Science, Aeronautics and Technology Appropriation (Continued)				
1994	**1995**	**1996**	**1997**	**1998**
Office of Mission to Planet Earth				
Office of Advanced Concepts & Technology • Space Research and Technology • Commercial Programs • Technology Transfer • Commercial Use of Space	Office of Advanced Concepts & Technology • Advanced Concepts and Technology (combined functional areas)	Reallocated to Space Access and Technology and other programs		
		Office of Space Access and Technology • Space Access and Technology		Program office dissolved; activities moved to Aeronautics and Space Transportation Technology
	Launch Services	Included with Space Access and Technology		Moved to Human Space Flight appropriation

Table 1–3. Program Office Functional Areas (Continued)

Programs Within the R&D/Science, Aeronautics and Technology Appropriation (Continued)				
1994	1995	1996	1997	1998
Office of Aeronautics • Aeronautical Research & Technology • Transatmospheric Research & Technology	Office of Aeronautics • Aeronautical Research & Technology		Office of Aeronautical Research and Technology • Research and Technology Base • Focused Programs	Office of Aeronautics and Space Transportation Technology • Aeronautical Research and Technology • Commercial Technology/SBIR • Advanced Space Transportation Technology
Safety, Reliability, and Quality Assurance • Safety, Reliability, and Quality Assurance	Moved to Mission Support appropriation			
Space Communications • Advanced Systems	Mission Communication Services • Ground Network • Mission Control & Data Systems • Space Network Customer Service			
Academic Programs	Academic Programs • Education • Minority University Research and Education			

Table 1–3. Program Office Functional Areas (Continued)

Programs Within the Spaceflight, Control, and Data Communications Appropriation					
1989	**1990**	**1991**	**1992**	**1993**	**1994**
Office of Space Flight • Shuttle Production and Operational Capability • Space Transportation Operations Office of Space Tracking and Data Systems • Space and Ground Network, Communications and Data Systems Space Station U.S.-Russian Cooperative Program Space Shuttle Payload & Utilization Operations					Space Flight • Shuttle Production and Operational Capability • Space Transportation Operations • Launch Services

CHAPTER TWO
LAUNCH SYSTEMS

CHAPTER TWO
LAUNCH SYSTEMS

Introduction

Launch systems provide access to space, necessary for the majority of NASA's activities. During the decade from 1989–1998, NASA used two types of launch systems, one consisting of several families of expendable launch vehicles (ELV) and the second consisting of the world's only partially reusable launch system—the Space Shuttle. A significant challenge NASA faced during the decade was the development of technologies needed to design and implement a new reusable launch system that would prove less expensive than the Shuttle. Although some attempts seemed promising, none succeeded.

This chapter addresses most subjects relating to access to space and space transportation. It discusses and describes ELVs, the Space Shuttle in its launch vehicle function, and NASA's attempts to develop new launch systems. Tables relating to each launch vehicle's characteristics are included. The other functions of the Space Shuttle—as a scientific laboratory, staging area for repair missions, and a prime element of the Space Station program—are discussed in the next chapter, Human Spaceflight. This chapter also provides a brief review of launch systems in the past decade, an overview of policy relating to launch systems, a summary of the management of NASA's launch systems programs, and tables of funding data.

The Last Decade Reviewed (1979–1988)

From 1979 through 1988, NASA used families of ELVs that had seen service during the previous decade. NASA also introduced new models of ELVs and began using the fleet of Space Shuttles to launch satellites into space. NASA used three families of ELVs: the Scout, Delta, and Atlas. These ELVs were increasingly acquired from the private sector and were used to send commercial as well as scientific and other research satellites into space in compliance with national space policy. The success rate for ELV launches was high during this decade; there were only three ELV launch failures: 1984, 1986, and 1987.

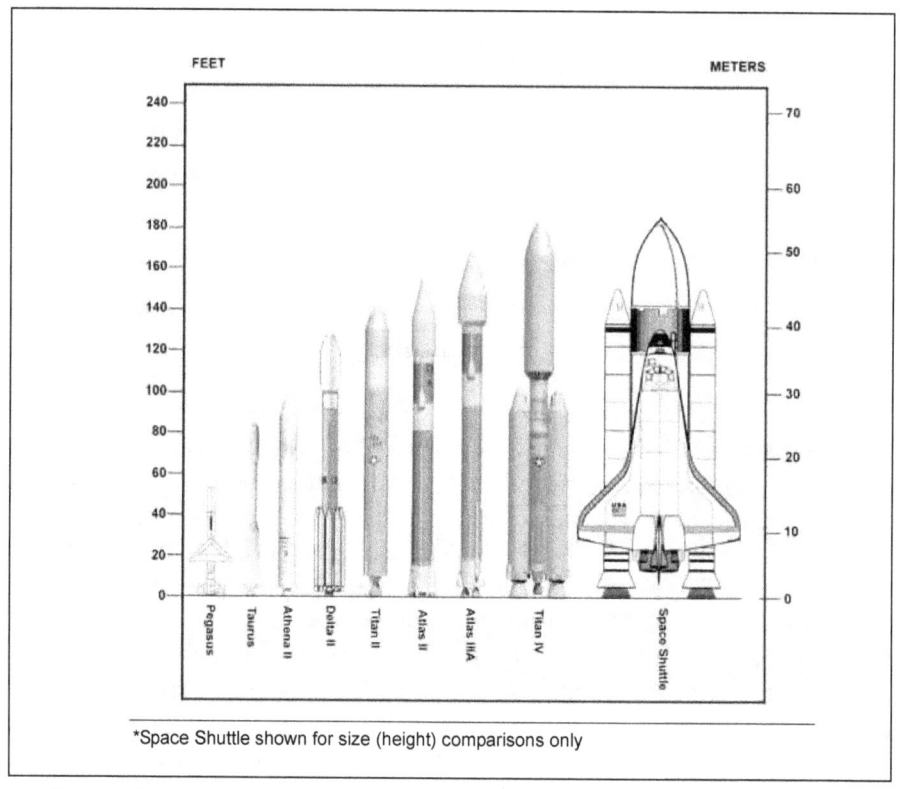

*Space Shuttle shown for size (height) comparisons only

Figure 2–1. NASA's Fleet of Launch Vehicles (1989–1998). Not Pictured: the Scout and the Conestoga. The Scout stood 75 feet (23 meters) tall, placing it between the Pegasus and Taurus. The Conestoga stood 50 feet (14 meters), making it about the same height as the standard Pegasus.

This decade marked the Space Shuttle's debut as the world's first Reusable Launch Vehicle (RLV). The Space Shuttle fleet consisted of four orbiters: the *Columbia, Challenger, Discovery,* and *Atlantis.* An earlier orbiter, the *Enterprise,* was used as a test vehicle before the Space Transportation System's first spaceflight in 1981, but it did not fly in space.

The Space Shuttle flew 26 successful missions before January 28, 1986, when the *Challenger* exploded only a few seconds into flight, taking the lives of its crew. This single tragedy defined the decade and greatly obscured the program's achievements. It would be more than two years before the Space Shuttle returned to flight in 1988.

Government policy had initially stated its intention to replace ELVs with the Shuttle as the country's prime launch vehicle. However, as early as 1984, Congress had expressed reservations about relying solely on the Shuttle. During the two years following the *Challenger* accident, NASA used ELVs exclusively to launch the Nation's satellites. When Shuttle flights resumed in 1988, NASA implemented a "mixed fleet strategy." This strategy reserved the Shuttle for those flights requiring a human presence or where only the Shuttle could handle the payloads.

Overview of NASA's Launch Systems (1989–1998)

During 1989–1998, 281 U.S. launches occurred. Of these, 215 were ELV launches, and 66 were Space Shuttle missions.[1] Twenty-seven Shuttle missions did not deploy a payload, and several other Shuttle missions were used to deploy and quickly retrieve payloads sent into space to conduct experiments. All Shuttle and most ELV launches took place from Cape Canaveral, Florida. Polar missions usually launched from Vandenberg Air Force Base, California. ELV launches took place using Athena; Atlas; Conestoga; Delta; Pegasus; Scout; Taurus; and Titan launch vehicles. The Conestoga launch took place from Wallops Flight Facility on the Eastern Shore of Virginia. Figure 2–1 shows NASA's launch vehicles.

A wide range of payload types was launched. Most were either commercial or DOD payloads and had communications or navigation purposes, although some were reconnaissance satellites. Some of the launches carried satellites for other countries, among them Japan, the Philippines, the United Kingdom, Spain, India, Korea, Canada, and various international groups of satellite owners. A smaller number of launches deployed scientific satellites; these were usually NASA missions. The success rate for all types of launches during this decade was very high. All Shuttle launches succeeded. For ELVs, the total success rate was almost 94 percent.

National space policy and legislation, either in place by 1989 or promulgated during the decade, greatly determined the direction of space launch development and activities. During the administration of President George H. W. Bush, and with Vice President Dan Quayle's special interest in space policy, NASA and other agencies and organizations undertook a large number of space transportation studies. These studies grew partly out of Congress's desire to reduce the federal budget and, in particular, NASA's budget, and partly out of the view that new launch technologies were needed.[2] These studies, and the policies and legislation reflecting them, had three primary themes: a new heavy launch system was needed to augment or replace the Shuttle; an RLV needed to be developed; and ELV launches and launch services should largely be commercial enterprises.[3]

---

[1] One joint U.S.-French launch took place from an Ariane launch vehicle.
[2] Andrew Butrica, "X-33 Fact Sheet #1, Part I: The Policy Origins of the X-33," *The X-33 History Project Home Page* (December 7, 1997), *http://www.hq.nasa.gov/office/pao/History/x-33/facts_1.htm* (accessed February 29, 2005).
[3] Advisory Committee on the Future of the U.S. Space Program, "Report of the Advisory Committee on the Future of the U.S. Space Program," December 17, 1990, *http://www.hq.nasa.gov/office/pao/History/augustine/racfup1.htm* (accessed March 14, 2005).

Executive policy statements and legislation emphasized the role of the private sector. Legislation took the government out of the business of building ELVs and supplying launch services for its primary payloads and required NASA to purchase them from commercial providers whenever possible. Policy and legislation directed the government to make national launch facilities available for private use and encouraged development of new launch systems by the private sector.[4] National policy also emphasized the importance of having a resilient and balanced launch capability so launch operations could continue even if any one system failed.

Further, restating policy set forth during President Ronald Reagan's administration,[5] Bush's policy dictated that the Shuttle would be reserved for launches requiring a human presence or the special capabilities of the Shuttle. It also stated that U.S. payloads must be launched from U.S. launch vehicles unless excepted by the President or a person designated by the President.[6]

In January 1993, William J. Clinton became President. In January 1994, the NASA Office of Space Systems Development released a study titled "Access to Space," undertaken in response to a congressional request in the NASA FY 1993 Appropriations Act. The goal of this study was to identify alternative approaches to space access that would reduce the cost of space transportation and increase safety for flight crews. The study concluded that the best option was "to develop and deploy a fully reusable single-stage-to-orbit pure-rocket launch vehicle fleet incorporating advanced technologies" and to phase out current systems "beginning in the 2008 time period."[7]

On August 5, 1994, President Clinton released a National Space Transportation Policy splitting the responsibility for space transportation between DOD and NASA. The policy gave DOD lead responsibility for improving ELVs and NASA lead responsibility for upgrading the Space Shuttle and developing and demonstrating new RLVs to replace the Space Shuttle.[8] In response, DOD initiated the Evolved Expendable Launch Vehicle program, and NASA initiated the RLV program to develop and flight-test experimental RLVs.

---

[4] A *Bill to Facilitate Commercial Access to Space, and for Other Purposes,* 100th Congress, 2nd sess., H.R. 4399, (October 14, 1988); National Space Policy Directive, NSPD-1,"National Space Policy Directives and Executive Charter," November 2, 1989, *http://www.fas.org/spp/military/docops/national/nspd1.htm* (accessed March 1, 2005); NSPD-3, "U.S. Commercial Space Policy Guidelines," February 11, 1991, *http://www.fas.org/spp/military/docops/national/nspd3.htm* (accessed March 1, 2005).

[5] The White House Office of the Press Secretary, "Presidential Directive on National Space Policy," *Aeronautics and Space Report of the President, 1998 Activities* (Washington, DC: National Aeronautics and Space Administration, 1990), p. 190.

[6] National Space Policy Directive, NSPD-2, "Commercial Space Launch Policy," September 5, 1990, *http://www.hq.nasa.gov/office/codez/new/policy/pddnspd2.html* (accessed March 1, 2005).

[7] Office of Space Systems Development, NASA Headquarters, "Access to Space Study, Summary Report," January 1994, p. i (NASA History Office file 009830).

[8] The White House, Office of Science and Technology Policy, Presidential Decision Directive (PDD), National Science and Technology Council-4 (NSTC), *National Space Transportation Policy* (August 5, 1994), *http://www.au.af.mil/au/awc/awcgate/nstc4.htm* (accessed February 28, 2005).

Clinton's policy also set guidelines for the use of foreign launch systems and components and excess ballistic missile assets for space launches. His policy also encouraged an expanded private sector role in space transportation research and development.

In September 1996, the White House released a National Space Policy stating that NASA would work with the private sector to develop flight demonstrators to make a decision about the development of a new reusable launch system. The policy also stated that NASA would acquire launch vehicles from the private sector unless the Agency's special technical abilities were needed.[9] Legislation passed in 1998 stated that the federal government would acquire space transportation services from commercial providers, except when there was a reason to use the Space Shuttle or because it was not cost effective or in the best interests of the mission. The legislation also allowed the Federal Aviation Administration (FAA) to license firms to fly vehicles back from space. Since the 1980s, private firms had been able to acquire licenses for commercial space launches; but the licenses had not provided for return from space, which had been too expensive for all but government agencies. This bill also obligated NASA's Administrator to prepare for transferring operation and management of the Space Shuttle to the private sector.[10]

Management of NASA's Launch Systems

In the decade from 1989 through 1998, NASA's launch systems included both ELVs and the Space Shuttle. NASA's launch system programs also focused on developing new ways to provide access to space by using RLVs and other advanced technologies. As in the past, the offices managing these various activities frequently shifted among organizations as NASA reorganized in an effort to more efficiently achieve its objectives. At times, management of ELVs, the Space Shuttle, and developing launch programs were all in the same organization. At other times, they were spread among different areas of the Agency.[11] For part of NASA's fourth decade, management of NASA's expendable launch systems remained with the Office of Space Flight (Code M), although it did not receive the prominence it had in past decades because providing ELV services became more of a commercial function. Management of Space Shuttle activities always remained in the Office of Space Flight.

---

[9] The White House National Science and Technology Council, "Fact Sheet–National Space Policy," PDD-NSTC-8 (September 19, 1996), *http://www.fas.org/spp/military/docops/national/nstc-8.htm* (accessed March 15, 2005).

[10] *Commercial Space Act of 1998,* 105th Congress., 1st sess., Public Law 105-303, Title II, (October 28, 1998).

[11] NASA assigned letters (called codes) as a quick way to refer to its top-level offices. The offices and codes applicable to launch systems during this decade were:

• Office of Space Flight–Code M
• Office of Space Systems Development–Code D
• Office of Advanced Concepts and Technology–Code C
• Office of Space Access and Technology–Code X
• Office of Space Science and Applications–Code E, later changed to Code S

Development programs were frequently located in other organizations. The sections that follow correspond to the major reorganizations and changes in the management structure of NASA's launch systems activities.

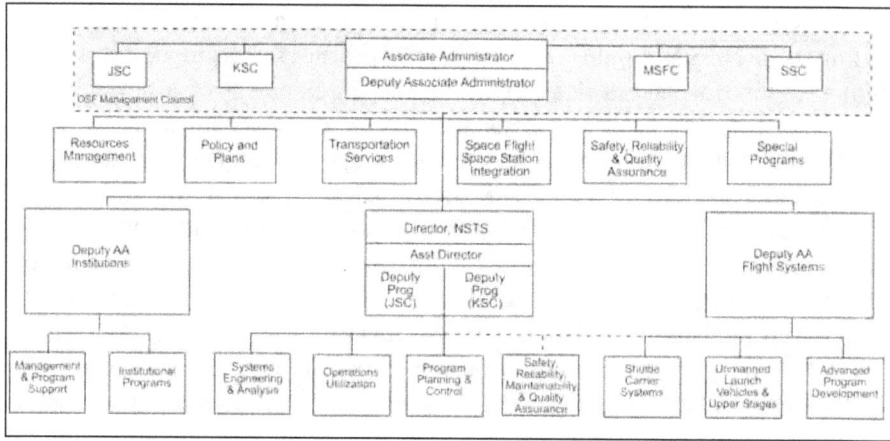

Figure 2–2. Office of Space Flight (Code M), February 1989.

Phase I: 1989–1990

The year 1989 and the first part of the 1990s saw three rapid reorganizations of the Office of Space Flight and changes in its leadership. In February 1989, the Office of Space Flight, led by Admiral Richard Truly, reorganized from its post-*Challenger* structure into an office consisting of three major divisions: 1) Institutions, headed by Richard J. Wisniewski; 2) Flight Systems, led by Joseph B. Mahon, and 3) the National Space Transportation System program (soon renamed the Space Shuttle program), headed by Arnold D. Aldrich (see Figure 2–2). Charles R. Gunn led the Unmanned Launch Vehicles and Upper Stages office in the Flight Systems Division. Aldrich left his post as head of the Shuttle program in October 1989 to become Associate Administrator of the Office of Aeronautics, Exploration and Technology and was replaced by Capt. Robert L. Crippen, initially as acting Director of the Space Shuttle program and as Director from February 1990.

Dr. William B. Lenoir, a former Space Shuttle astronaut, became Associate Administrator of the Office of Space Flight in July 1989, leaving his position as head of the Office of Space Station, a position he had held only since May 1989. In May, he had also been asked by Truly to develop a plan for consolidating the Offices of Space Flight and Space Station.[12] When President George H. W. Bush named Truly NASA Administrator, Lenoir took over leadership of the Office of Space Flight.

---

[12] "Space Station Program Leadership Selected by Truly," *NASA News* Release 98-77, May 19, 1989. (NASA History Office Folder 009610).

The February 1989 structure lasted less than a year because the office reorganized again in December and then made another small change in March 1990. The December 1989 reorganization consolidated the Office of Space Flight and Office of Space Station into a single organization consisting of four divisions that retained the name the Office of Space Flight (see Figure 2-3). Richard H. Kohrs took over the leadership of Space Station *Freedom*; Crippen, Wisniewski, and Mahon continued to head the Space Shuttle, Institutions, and Flight Systems divisions, respectively. Gunn continued as Director of Unmanned Launch Vehicles and Upper Stages. The March 1990 reorganization added a second Deputy Associate Administrator to the Office of Space Flight. In late 1990, Mahon was replaced by Michael T. Lyons as head of Flight Systems, and I. Duke Stanford became head of Institutions when Wisniewski retired from NASA. Around the same time, the heads of the divisions assumed the title of Deputy Associate Administrator of their respective organizations.

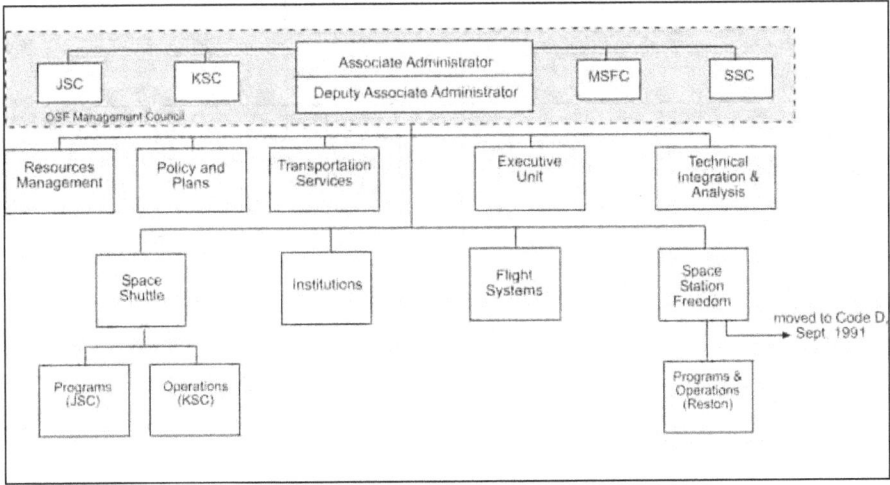

Figure 2–3. Office of Space Flight, December 1989.

Phase II: 1991–1992

The 1989 organizations remained in place until September 1991, when Administrator Truly followed the guidance of the Advisory Committee on the Future of the U.S. Space Program of December 1990, headed by Norman R. Augustine, and created a new Office of Space Systems Development (Code D).[13]

---

[13] NASA press release for September 13, 1991, that announced the formation of the new office referred to it as the Office of Space Flight Development; "New Office of Space Flight Development Announced," *NASA News* Release 91-148, September 13, 1991, *ftp://ftp.hq.nasa.gov/pub/pao/pressrel/1991/91-148.txt* (accessed March 2, 2005). Beginning with an October 3, 1991, press release, the office was referred to as the Office of Space Systems Development." "NASA Administrator Announces Key Appointments," *NASA News* Release 91-161, October 3, 1991, *ftp://ftp.hq.nasa.gov/pub/pao/pressrel/1991/91-161.txt* (accessed March 2, 2005). This name also appears on the NASA organization chart dated October 20, 1991, and in future references.

This reorganization moved several organizations from the Office of Space Flight to the new organization (see Figure 2–4). This new Space Systems Development office was responsible for Space Station *Freedom* development; large propulsion systems development, including the new National Launch System and its new space transportation main engine; other large spaceflight development; and the advanced transportation systems program planning function. Aldrich left the Office of Aeronautics, Exploration and Technology to lead the new Space Systems Development office. Dr. C. Howard Robins, Jr. was named Deputy Associate Administrator for the new office in October. The Flight Systems Division moved to the Office of Space Systems Development, with Lyons as its head. Kohrs was named head of the Space Station Freedom Division.

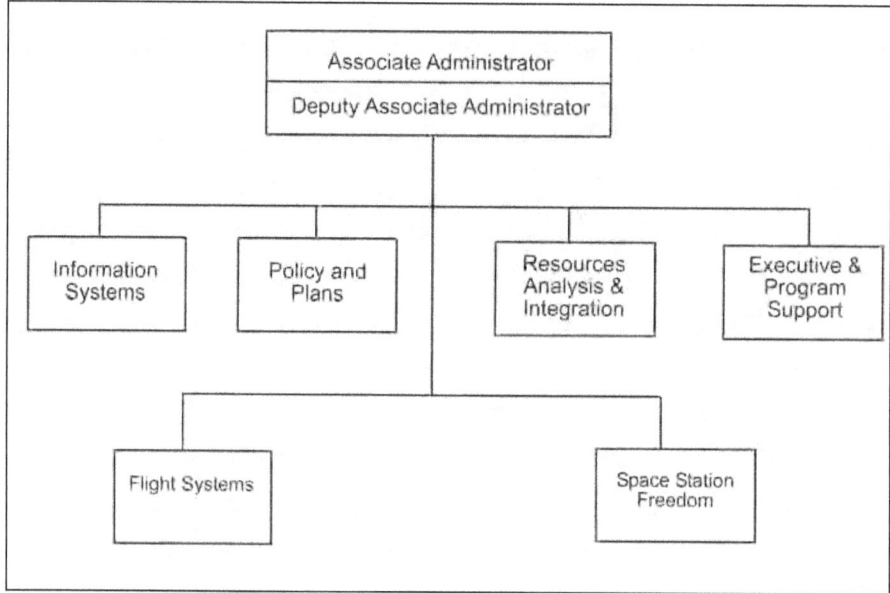

Figure 2–4. Office of Space Systems Development (Code D).

The scaled-down Office of Space Flight continued to focus on Space Shuttle operations and also retained responsibility for Space Station *Freedom/* Spacelab operations and utilization, ELV operations, and upper stages.[14] In December, Leonard S. Nicholson was named Director of the Space Shuttle program in the Office of Space Flight, replacing Crippen, who became Director of Kennedy Space Center. Lenoir remained as Associate Administrator of the Office of Space Flight until May 1992, when he resigned his leadership post and left NASA.

---

[14] "New Office of Space Flight Development Announced," *NASA New,* Release 91-148, September 13, 1991, *ftp://ftp.hq.nasa.gov/pub/pao/pressrel/1991/91-148.txt* (accessed March 1, 2005).

In April 1992, Daniel S. Goldin replaced Truly and became NASA's new Administrator. Among his first hiring decisions was the appointment of Maj. Gen. Jeremiah W. Pearson, III as Associate Administrator of the Office of Space Flight. Bryan D. O'Connor, a former NASA astronaut, was named Deputy Associate Administrator for programs within the Office of Space Flight. In June, Pearson named Thomas Utsman, who had been serving as Deputy Associate Administrator for the Office of Space Flight since June 1990, as Program Director for the Space Shuttle. In March 1993, Pearson named Brewster Shaw, Deputy Director of Space Shuttle Operations, to the position of Space Shuttle Program Manager, replacing Nicholson, who left to take the position of acting Director of Engineering at Johnson Space Center.

In the summer of 1992, management of ELVs and upper stages, still under the leadership of Gunn, moved to the Office of Space Science and Applications (OSSA) Launch Vehicles Office. This was done largely because ELVs launched space and Earth science missions, and it seemed more efficient for all aspects of these missions to be in the same organization. The Launch Vehicles Office was responsible for managing the ELV and upper stages launch services program. It maintained the NASA ELV manifest and served as the primary interface with the U.S. Air Force, foreign governments, and the ELV industry.[15]

In November 1992, NASA moved the Space Technology program, led by Gregory Reck, out of the Office of Aeronautics and Space Technology (Code R) and merged it with the Office of Commercial Programs (Code C), creating a reformulated Code C, the Office of Advanced Concepts and Technology, under Reck's leadership (see Figure 2–5). The Transportation Division within the new Code C, led by Earl VanLandingham, included several space transportation technology efforts, among them the Solid Propulsion Integrity Program (SPIP), the Advanced Launch Technology effort, and Advanced Programs.[16]

---

[15] NASA *Management Instruction* (NMI) 1102.1H, "Role and Responsibilities—Associate Administrator for Space Science and Applications," July 30, 1992; NASA Management Instruction 1102.1I, "Role and Responsibilities—Associate Administrator for Space Science and Applications," June 28, 1993; "Goldin Announces Changes in NASA Organization To Focus and Strengthen Programs and Management," *NASA News* Release 92-172, October 15, 1992, *ftp://ftp.hq.nasa.gov/pub/pao/pressrel/1992/92-172.txt* (accessed March 2, 2005).

[16] "General Statement," *National Aeronautics and Space Administration, Fiscal Year 1995 Budget Estimates*, p. AS-9.

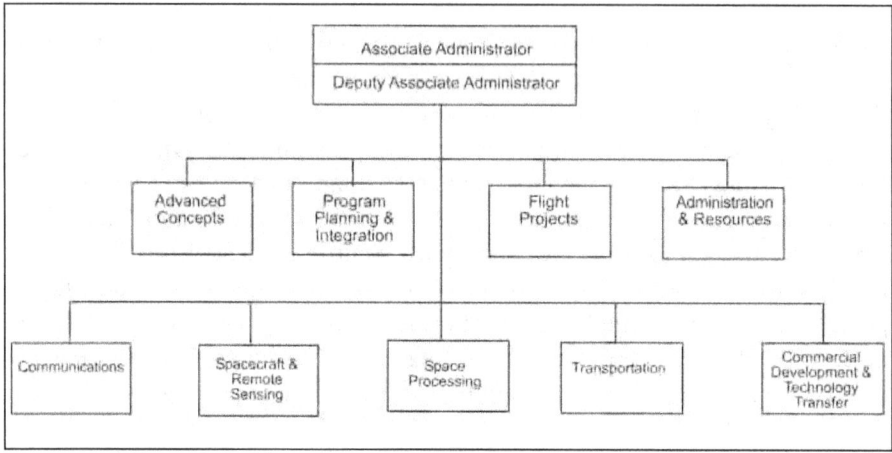

Figure 2–5. Office of Advanced Concepts and Technology (Code C).

Phase III: 1993–1996

In October 1993, Administrator Goldin announced that the Office of Space Flight would again assume responsibility for managing the Space Station Program because Space Shuttle flight activities were becoming increasingly more involved with Space Station planning.[17] This change moved the Space Station out of the Office of Space Systems Development, leaving that office without a major program to manage. O'Connor, Director of the Space Station transition since July 1993, became acting Space Station Program Director, replacing current Program Director Richard Kohrs, who retired in September 1993.[18] O'Connor remained in the position until January 1994, when Wilbur C. Trafton accepted the position.

Further changes in the Office of Space Flight took place in spring of 1994. Utsman left the position of Deputy Associate Administrator for Space Shuttle to return to Kennedy Space Center and become special assistant to the Associate Administrator in the Office of Space Flight. O'Connor, the Office of Space Flight Deputy Associate Administrator, replaced Utsman and also became the Space Shuttle Program Director, responsible for managing the Space Shuttle program. Wisniewski, who retired from NASA in 1990, returned to NASA and replaced O'Connor as Deputy Associate Administrator in the Office of Space Flight. He was responsible for resources, policy and plans, human resources, and management of the human spaceflight installations: Kennedy Space Center, Johnson Space Center, Marshall Space Flight Center, and Stennis Space Center.[19]

[17] "Goldin Announces Key Space Station Management Moves," *NASA New,* Release 93-191, October 20, 1993, *ftp://ftp.hq.nasa.gov/pub/pao/pressrel/1993/93-191.txt* (accessed March 1, 2005).

[18] See chapter 3, Human Spaceflight, for a description of space station transition to Johnson Space Center.

[19] "NASA Announces Space Flight Personnel Changes," *NASA News* Release 94-66, April 28, 1994, *ftp://ftp.hq.nasa.gov/pub/pao/pressrel/1994/94-066.txt* (accessed March 1, 2005).

A reorganization in September 1994 consolidated the Advanced Concepts and Technology Office (Code C) and Office of Space Systems Development (Code D) into a new Office of Space Access and Technology (Code X), headed by John E. Mansfield. The divisions within Code X and their heads were: Flight Integration, Jack Levine; Advanced Concepts, Ivan Bekey; Launch Vehicles, Charles Gunn; Commercial Development, Robert Norwood; Space Systems, Samuel Venneri; Space Processing, Edward Gabris; Space Transportation, Col. Gary Payton; and Management Operations, Martin Stein (see Figure 2–6).

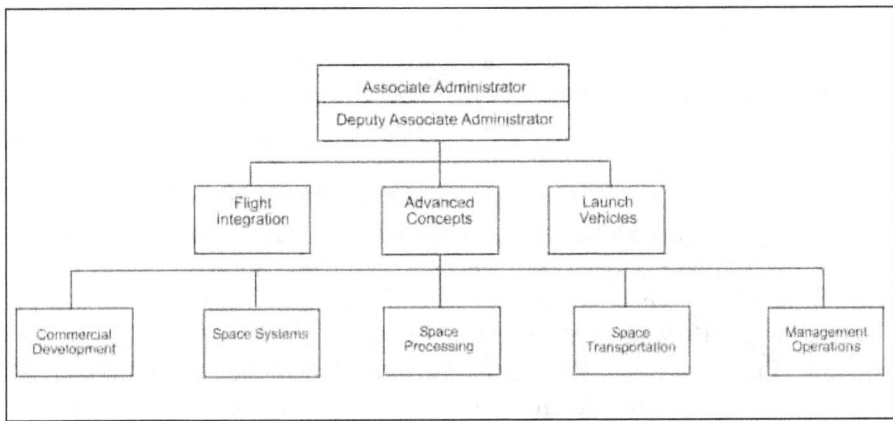

Figure 2–6. Office of Space Access and Technology (Code X), September 1994.

The Space Transportation organization in Code X managed transportation technology, advanced technology development for ELVs and the Space Shuttle, and NASA's efforts to develop an RLV. It also functioned as the single interface with DOD and other outside interests.[20] The Launch Vehicles Office consolidated NASA, the National Oceanic and Atmospheric Administration (NOAA), and international cooperative ELV mission requirements. Management and acquisition of launch services moved from the Office of Space Science and Applications to the Launch Vehicles Office as did acquisition of upper stages. Administration, procurement, and technical oversight of launch service delivery in the small and medium performance classes (Atlas E, Titan II, Pegasus, and Delta II) were handled by Goddard Space Flight Center. Launch services for the intermediate and large performance classes (Atlas I/IIAS and Titan IV/Centaur) were managed by Lewis Research Center. Kennedy Space Center had responsibility for technical oversight of vehicle assembly and testing at the launch site and for launch site spacecraft processing. Marshall Space Flight Center was responsible for managing upper stage missions.[21]

---

[20] "NASA Space Access and Technology Office Functions," *Aerospace Daily* (September 26, 1994): 480.
[21] "Office of Space Access and Technology," *National Aeronautics and Space Administration Fiscal Year 1996 Estimates*, pp. SAT 5–37.

In November 1994, Pearson resigned as Associate Administrator of the Office of Space Flight. He was replaced by NASA Chief Engineer and veteran manager Dr. Wayne Littles, who continued a review of the Shuttle work force begun by Pearson a few months earlier. At Administrator Goldin's direction, Littles was looking for any "unnecessary requirements" in the Shuttle program that could be cut and "to make sure that recent budget cuts have not affected safety."[22]

When Gunn retired in the spring of 1995, Charles J. Arcilesi took over as acting head of the Launch Vehicles Office. By summer, the Launch Vehicles Office had moved to the Office of Space Flight, and Karen Poniatowski was appointed to head the Expendable Launch Vehicles Office.

Later in the year, in October 1995, the Office of Space Flight reorganized with the goal of increasing efficiency and reducing the number of people in the organization (see Figure 2–7). In January 1996, Trafton, Director of the Space Station program, assumed additional responsibilities as the acting Associate Administrator for the Office of Space Flight, replacing Littles, who became Director of Marshall Space Flight Center. Trafton was formally named to the position in March. The position also placed Trafton in charge of the Human Exploration and Development of Space (HEDS) Enterprise, one of NASA's four Strategic Enterprises, whose mission was to "open the space frontier by exploring, using, and developing space; and to expand the human experience into the far reaches of the universe."[23] Andrew Allen became acting head of the Space Station program until Gretchen McClain took over in January 1997. In January 1996, the decision was made to transfer the ELV program from the Office of Space Access and Technology (Code X) back to the Office of Space Flight (Code M). In February 1996, O'Connor left his position of Space Shuttle Director, which he held since 1994.

---

[22] Ben Iannotta, "Littles Takes Over Space Flight Post as Pearson Quits," *Space News* (November 21–December 4, 1994): 29.

[23] Sharon M. Wong, "Strategic Management: Opening the Space Frontier," *NASA HQ Bulletin* (April 15, 1996): 5.

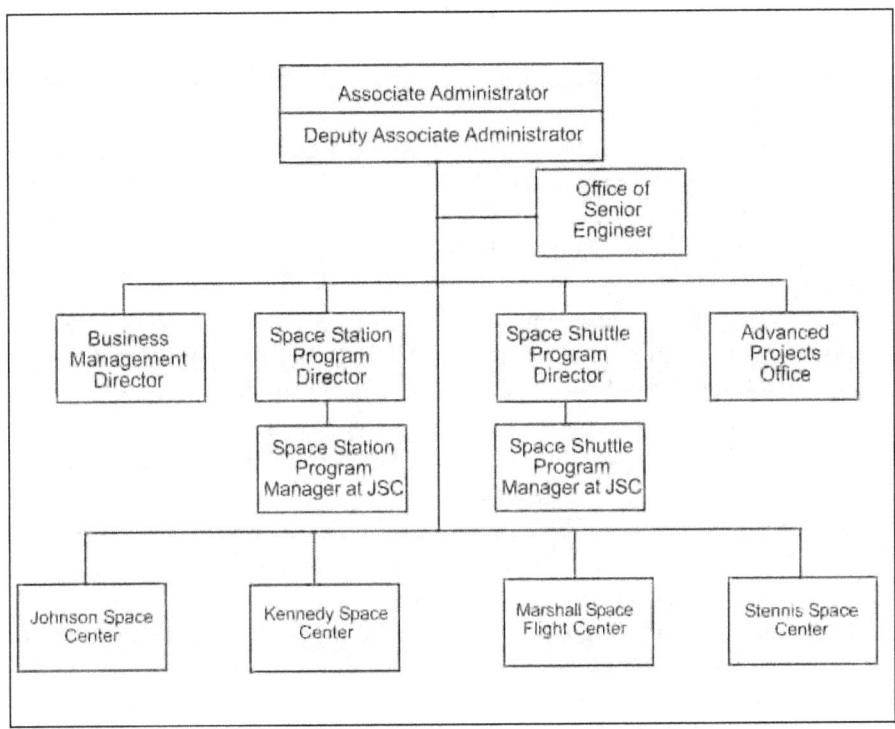

Figure 2–7. Office of Space Flight, October 1995.

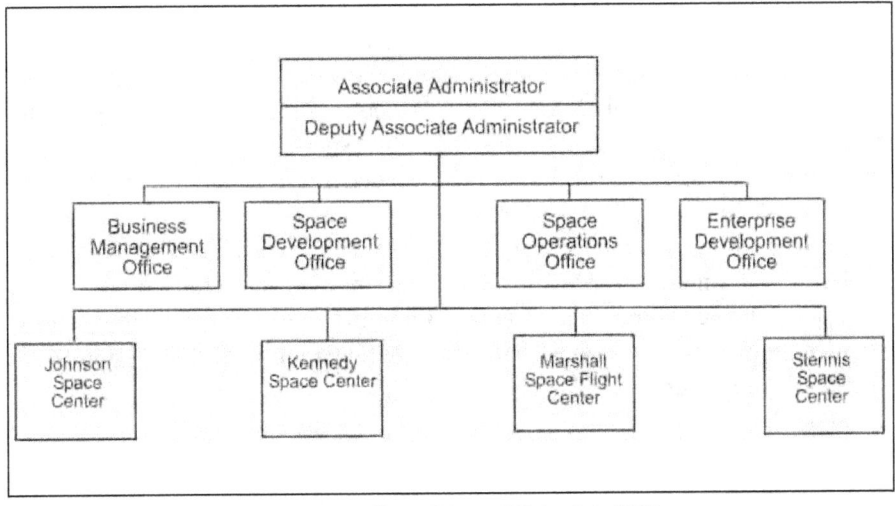

Figure 2–8. Office of Space Flight, July 1998.

Phase IV: 1996–1998

In April 1996, NASA announced plans to accelerate the downsizing of NASA Headquarters. Agency officials had previously identified more than 200 positions that could be moved from Headquarters to NASA's Field Centers; about half of the positions had already moved or were in the process of moving. In October 1996, a major Agency-wide restructuring took place that aimed to reduce NASA Headquarters staffing and transfer most technology development and commercialization activities to individual program offices and Field Centers.[24] The Office of Space Communications (Code O) merged into the Office of Space Flight, becoming another division at the same level as the Space Station program, Space Shuttle program, and Advanced Projects office.[25] The new Communications Division was headed by David W. Harris, who had previously led the Program Integration Division in the Office of Space Communications.

At the same time, the Office of Space Access and Technology (Code X), where the RLVs program was located, was disbanded. Work into space research and technology returned to Code R, now renamed the Office of Aeronautics and Space Transportation Technology. A Space Transportation Technology organization was created as well as a Space Transportation Division, both headed by Payton, who had headed the Space Transportation division in Code X. The Advanced Space Transportation office, charged with NASA's X-33 and X-34 launch vehicle technology development programs, was relocated to Code R.

Trafton resigned as Associate Administrator of the Office of Space Flight in November 1997. Joseph H. Rothenberg, Director of Goddard Space Flight Center, was appointed to the position in January 1998, becoming NASA's fourth Associate Administrator for the Office of Space Flight in little more than three years. In July 1998, the Office of Space Flight reorganized into four functional offices: 1) Operations, headed by William Readdy, which included ELVs, led by Karen Poniatowski; Space Communications, headed by Robert Spearing; and Space Operations Utilization, led by Robert L. Elsbernd; 2) Enterprise Development, led by Darrel Branscome, which included Advanced Projects, Strategic Planning, and Outreach 3) Business Management, led by Michael Reilly; and 4) Development, led by Gretchen McClain; (see Figure 2–8).

---

[24] Anne Eisele, "Restructuring Would Slash Headquarters," *Space News* (July 1–7, 1996): 4.

[25] Charles T. Force, Associate Administrator for the Office of Space Communications (Code O), had resigned from NASA in May 1996, before the announcement of the merger of Code O into Code M. "Force To Leave NASA," *NASA News* Release 96–88, May 3, 1996, *ftp://ftp.hq.nasa.gov/pub/pao/pressrel/1996/96-88.txt* (accessed March 3, 2005).

Money for NASA's Launch Systems

Budget Structure

The federal appropriation categories funding NASA space transportation and launch programs and activities changed in the mid-1990s. For the first four years of the 1989–1998 decade, they were funded by R&D and SFC&DC appropriations. With the FY 1993 budget year, NASA changed its appropriation categories to HSF, which included all Space Shuttle activities, and SAT, which included experimental or development initiatives.[26] ELVs also fell within the SAT appropriation until 1998, when it moved to HSF.

When NASA began using HSF and SAT appropriations, the names and descriptions of many of the subordinate programs and budget categories remained as they had been with the R&D and SFC&DC budget structure. Exceptions are noted below each table. If it is unclear whether a new budget category was merely a name change or whether it indicated a new or modified program, the new name is treated as a new budget category.

A large portion of NASA's budget went to fund Space Shuttle activities, and most space transportation budget categories in the annual budget relate to the Space Shuttle. Expendable launch systems received relatively little attention because NASA obtained most expendable launch services from the commercial sector. The main focus of non-Shuttle launch services in the annual budget related to the mission is to develop new and alternative reusable launch systems and reduce the cost of access to space.

In most cases, the authorization and appropriations bills funding NASA's programs addressed only major budget categories and did not provide much detail regarding where funds should be spent. Authorization bills provided more detail than appropriations bills, at least showing amounts for Space Transportation Capability and Development in the R&D appropriation and for Space Shuttle Production and Operational Capability and Space Shuttle Operations in the SFC&DC category. When the appropriation categories changed to HSF and SAT, the authorization bills typically provided amounts for Payload and Utilization Operations; Space Shuttle Safety and Performance Upgrades; Shuttle Production and Operational Capability; and Space Shuttle Operations in the HSF appropriation. In the SAT appropriation, Congress provided amounts for Advanced Concepts and Technology (Space Access and Technology). Congress only occasionally indicated that a particular amount was authorized for a specific project. Amounts for ELVs and launch services were occasionally provided separately, but many were not budgeted separately from the larger categories.

---

[26] These appropriations funded additional NASA programs.

Authorization bills provided more detail than appropriations bills, which provided almost no detail. Appropriations bills generally gave a total amount only for R&D and another for SFC&DC. After the change in appropriation categories, they gave an amount for HSF and one for SAT.

Congress based total authorized and appropriated funding on NASA's annual budget estimates provided to the President and presented to Congress. These detailed estimates formed the basis for NASA's operating plan and the amounts each program would actually spend. The House and Senate subcommittees and committees used these budget estimates for their discussions and often prepared reports dealing with the estimated amounts; but except where specific amounts were included in the authorization or appropriations bills, these reports did not legally require NASA to spend funds in a certain way except in very broad categories.

Phase I: FY 1989–FY 1992

During this period, the R&D and SFC&DC appropriations funded NASA's launch systems, as well as other NASA programs. The R&D appropriation funded Space Shuttle programs in the Space Transportation Capability Development category. These included activities such as the Tethered Satellite System; Spacelab; development and procurement of upper stages; engineering and technical base support at the human spaceflight NASA Centers (Johnson Space Center, Kennedy Space Center, Marshall Space Flight Center, and Stennis Space Center); payload operations and support equipment; studies into advanced launch systems; and other advanced programs and development activities. Space Transportation Capability Development also funded all Space Station activities (discussed in chapter 3).

The SFC&DC appropriation funded the operational activities of the Space Transportation System. The system's two major elements were Shuttle Production and Operational Capability and Space Transportation (Shuttle) Operations. Shuttle Production and Operational Capability provided for the fleet of orbiters; main engines; launch site and mission operations requirements; spares; production tooling; and related supporting activities. The appropriation also provided funds for development of an Advanced Solid Rocket Motor. Space Transportation Operations included standard operational support services for the Space Shuttle and the procurement of ELVs. This budget category funded the production of flight hardware, overhaul and repair of equipment, and labor and materials needed for flight and ground operations. The SFC&DC appropriation also was used to fund the tracking and communication systems used for all NASA flight projects.

Phase II: FY 1993–FY 1998

In FY 1993, R&D and SFC&DC budget items involving space transportation were placed into a new HSF appropriation or into the Science, Applications and Technology appropriation. The HSF appropriation included the on-orbit infrastructure (Space Station and Spacelab), transportation capability (Space Shuttle program, including operations, program support and performance, and safety upgrades), and the Russian Cooperation program (which included flight activities associated with the cooperative research flights to the Russian *Mir* Space Station). HSF appropriation activities were funded in the following major budget line items: 1) Space Station, 2) Russian Cooperation, 3) Space Shuttle, and 4) Payload Utilization and Operations. The Space Shuttle budget had two major categories: Safety and Performance Upgrades and Space Shuttle Operations. Safety and Performance Upgrades corresponded most closely with the old SFC&DC Shuttle Production and Operational Capability budget category. Payload Utilization included funding to support payloads flying on the Shuttle and Spacelab, as well as advanced technology projects and engineering technical base support for the Field Centers supporting HSF flight activities. Space Station, Russian Cooperation, and Spacelab are discussed in chapter 3. Space Shuttle and the Payload Utilization and Operations activities are discussed in this chapter.

The new SAT appropriation provided funding for NASA's research and development activities, in particular, "to extend knowledge of the Earth, its space environment, and the universe; and to invest in new technologies, particularly in aeronautics."[27] The two categories in the SAT appropriation most directly related to space transportation or launch systems were 1) Advanced Concepts and Technology (as it was called in FY 1995) or Space Access and Technology (beginning in FY 1996), and 2) Launch Services, consisting primarily of the ELV budget formerly included in the SFC&DC appropriation. Launch Services sometimes appeared in budget documents as a separate budget category under the SAT appropriation. At other times, it was shown as a subcategory in the Office of Space Science and Applications. Notes below the funding history tables that follow identify items funded from the SAT appropriation.

Funding History

For the 1989–1991 fiscal years, funding increased for launch systems and, in particular, the Space Shuttle. Payload operations and support declined slightly as did upper stages. Space Transportation Capability Development in the R&D appropriation peaked in 1991 at $763,400,000. In the SFC&DC appropriation, Space Shuttle Production and Operational Capability reached its

---

[27] "General Statement," *National Aeronautics and Space Administration, Science, Aeronautics and Technology, Fiscal Year 1995 Estimates*, p. SAT SUM-1.

high of $1,364,000,000 in 1991. Funding for Space Transportation Operations continued to rise for two more years, reaching its high of $3,085,200,000 in 1993. ELV funding, which had dropped in 1993, rose in 1994 to $300,300,000.

In 1992, the downward slide for Space Shuttle operations began as rising costs for the Space Station drained the budget. The SFC&DC Space Shuttle Production and Operational Capability authorization dropped from $1,364,000,000 in FY 1991 to $1,328,900,000 in FY 1992. The amount authorized for Space Transportation Capability Development dropped from $763,400,000 in the FY 1991 authorization to $679,800,000 in FY 1992; it rose somewhat in FY 1993 to $733,700,000 and almost to its FY 1991 level in FY 1994, reaching $7,509,300,000. The amount for Space Shuttle Operations continued to rise until FY 1994, when it dropped from $3,085,200,000 to $3,006,500,000.

In FY 1995, appropriated amounts used the new HSF appropriation categories, which covered the operational end of launch systems, and SAT for developmental areas of space transportation. It was clear that, beginning with FY 1995, the HSF budget dropped considerably. Between FY 1995 and FY 1997, the appropriated amount decreased from $5,592,900,000 to $5,362,000,000. This included a $94 million general reduction taken from Space Shuttle operations.[28] The decline reflected a concerted Clinton administration effort to reduce the deficit while dealing with greater costs for the Space Station. The FY 1996 appropriation, coming at the end of an arduous six months of discussions that included 14 continuing resolutions and two government shutdowns, allotted HSF 1 percent less than NASA's request and 1.1 percent less than its FY 1995 amount. This decrease took place even though the Space Station received 1.1 percent more than it had in FY 1995, making the cuts to the Space Shuttle program even more pronounced.[29] The amount appropriated to SAT, which handled work on new RLVs, rose in FY 1997, but dropped in FY 1998 from $711,000,000 to $696,000,000. At the same time, in FY 1998, the appropriation for HSF rose again to $5,506,500,000.

The following tables reflect the budget categories as broken down by NASA and authorized by Congress. Table 2–1 shows congressional action. Notes below the table indicate when amounts were appropriated rather than authorized. Table 2–2 shows programmed amounts. These amounts formed NASA's operating plan, i.e., what NASA budgeted for particular activities during a fiscal year. On both these tables, the reader should not assume that subordinate amounts below a major budget category equal the amount shown above in the major budget category. Some subordinate budget categories are not launch-related and are not included in these tables.

[28] "Senate Appropriators Approve $14.4 Billion for NASA," *Aerospace Daily* (July 15, 1994): 79.
[29] "Results of FY 1996 Appropriations Process," *The American Institute of Physics Bulletin of Science, Policy News,* no. 86 (May 30, 1996), *http://www.sdsc.edu/SDSCwire/v2.12/FY96results.html* (accessed March 14, 2005).

The following series of tables show the amounts NASA submitted in its annual budget estimates (see Tables 2–3 through 2–57). NASA submits a budget estimate two years before the start of each fiscal year and then a revised estimate a year later. The tables show both the original and revised estimates, separated by a forward slash. If only one amount is shown (either before or after the forward slash), NASA's budget estimate documents referenced that budget category only once—either in its original budget estimate, shown before the forward slash. or in the revised budget estimate, shown after the forward slash. If a category was mentioned in an authorization bill, that amount is shown.

Authorized and appropriated amounts come from the appropriate authorization or appropriations bill.[30] If no authorized or appropriated amount is shown for a particular category, then the bills did not address that category. Submitted and programmed amounts come from the annual NASA budget estimates. NASA appropriations were included with the Department of Veterans Affairs, Housing and Urban Development, and Independent Agencies appropriations bills for the fiscal year. If no programmed amount is shown, that year's budget did not include a programmed amount for the particular budget category. See the individual budget tables for details.

Expendable Launch Vehicles

Overview

By NASA's fourth decade, America's ELVs were obtained either from the DOD stockpile of retired rockets and modified for space launch purposes or were procured from the private sector according to criteria in NASA's FY 1991 Authorization Act and Launch Services Purchase Act (LSPA) of 1990.[31] The LSPA required NASA to purchase launch services for its primary payloads from commercial providers. This legislation quickly opened up a new market to American industry as the government no longer competed as a launch services provider. Within six months after its passage, one launch services provider, General Dynamics, had decided to fund the construction of 60 new Atlas launch vehicles although it did not yet have a single buyer for the vehicles. Other launch vehicle providers followed suit. In November 1990, NASA signed a contract with McDonnell-Douglas to provide at least three Delta IIs. In September 1991, a contract with Orbital Sciences Corporation was signed for seven Pegasus vehicles. NASA contracted with Martin Marietta in 1994 for intermediate-class launch services on Atlas vehicles, and Orbital Sciences was selected to provide ultra-lite ELV launch services the same year.

---

[30] Authorization and appropriations bills are available at *http://thomas.loc.gov.*
[31] The Launch Services Purchase Act of 1990 was Title II of the FY 1991 Authorization Act. *National Aeronautics and Space Administration Authorization Act, Fiscal Year 1991*, 101st Congress, 2nd sess., Public Law 101-611 (November 16, 1990).

In the first years following NASA's 1988 return to flight, NASA acquired ELVs noncompetitively for the scientific missions remanifested onto ELVs from the Space Shuttle. NASA acquired all subsequent ELV launch services competitively from the private sector in the small, medium, and intermediate-performance classes, which could launch payloads up to 30,000 pounds (13,600 kilograms). Larger payloads up to 39,000 pounds (17,690 kilograms) were launched aboard the Titan IV/Centaur launch vehicle, developed by Martin Marietta Corporation (later Lockheed Martin). These were acquired from the U.S. Air Force by means of a contract the Air Force had with Martin Marietta since large class launch services were not available directly from the private sector.[32]

During NASA's fourth decade, 215 launches on American ELVs and one joint U.S.-French ELV launch on a European Ariane rocket took place. Almost 94 percent of these launches succeeded. Eight families of ELVs: Athena; Atlas; Conestoga; Delta; Pegasus; Scout; Taurus; and Titan were used. They each had impressive success rates with very few failures. The large majority carried either DOD or commercial payloads. Launch vehicle performance is shown in Figure 2–9 and Table 2–58. ELV activities are summarized in the following section. Some references use the term "partial failure" to discuss specific launches. To allow inclusion in this table and in the graph that follows, each launch is classified as either a success or failure. Partial failures are explained in footnotes below the table.

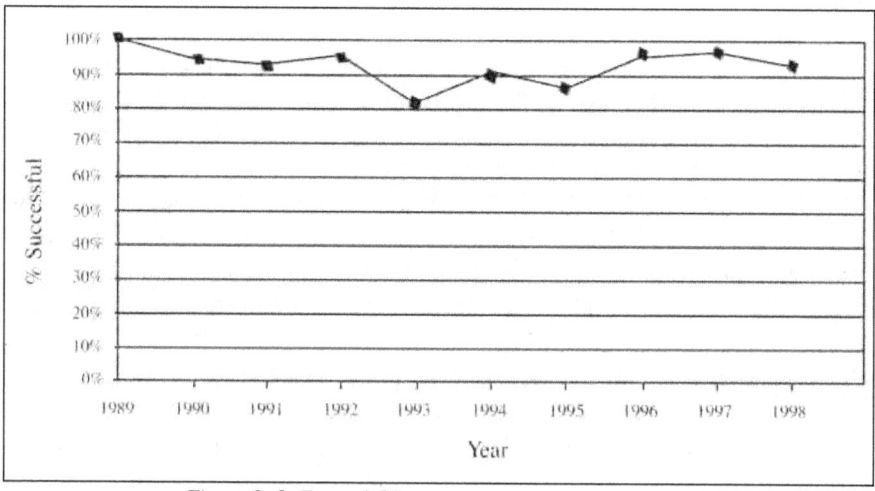

Figure 2–9. Expendable Launch Vehicle Success Rate.

---

[32] "Space Transportation Operations," *National Aeronautics and Space Administration Fiscal Year 1991 Budget Estimates*, pp. SF 2–11.

1989

In 1989, there were 13 U.S. ELV launches on known launch vehicles: 8 Deltas, 4 Titans, and 1 Atlas. All were successful. One was a NASA scientific spacecraft and one a commercial launch. The remaining launches were DOD satellites.

1990

In 1990, there were 21 U.S. ELV launches: 11 Deltas, 5 Titans, 3 Atlases, 1 Scout, and 1 Pegasus. One Titan launch failed. The launches included one joint NASA–Germany space science satellite, and two joint NASA–DOD environmental research satellites. The remaining satellites were either DOD satellites or commercial communications satellites.

1991

Twelve satellites launched on ELVs in 1991: 5 Deltas, 2 Titans, 4 Atlases, 1 Scout, and 1 Pegasus. One was a NASA launch of a meteorological satellite. The remaining satellites were either DOD or commercial spacecraft. One Atlas launch failed.

1992

Twenty-one satellites launched on American ELVs in 1992: 11 Deltas, 3 Titans, 5 Atlases, and 2 Scouts. Four payloads were science missions. The others were DOD or commercial payloads. One Atlas launch failed. In addition, a joint U.S.–French scientific spacecraft launched on a European Ariane ELV from the Kourou launch complex in French Guiana.

1993

Eighteen satellites launched on American ELVs in 1993: 7 on Deltas, 6 Atlases, 2 Titans, 2 Pegasus, and 1 Scout. One launch was a Department of Energy science payload, two were NASA earth science payloads, and the rest were DOD or commercial payloads. Both Titan launches failed.

1994

In 1994, 20 spacecraft launched on American ELVs: 3 Deltas, 5 Titans, 7 Atlases, 1 Taurus, 1 Scout, and 3 Pegasus. One was a NASA meteorological satellite, one a space science satellite, and the remainder either DOD or commercial satellites. There was one Pegasus launch failure and one Pegasus that inserted its payload into a lower-than-specified orbit.

1995

In 1995, 23 spacecraft were launched on American ELVs—3 were Deltas, 4 Titans, 12 Atlases, 2 Pegasus, 1 Athena, and 1 Conestoga. Payloads included one NASA meteorological satellite, one Canadian remote sensing satellite, two NASA science satellites, and the remainder DOD or commercial satellites. The Athena I, Conestoga, and one of the Pegasus XL launches failed.

1996

There were 26 ELV launches in 1996: 10 Deltas, 7 Atlases, 4 Titans, and 5 Pegasus launches. There were eight science payloads including one joint Italian–Dutch telescope. The rest were DOD or commercial satellites. One Pegasus XL launch failed.

1997

In 1997, 30 spacecraft were launched on ELVs: 11 Deltas, 8 Atlases, 5 Titans, 5 Pegasus, and 1 Athena. These included three Earth science payloads and two space science missions. The remainders were DOD or commercial satellites. One Delta launch failed.

1998

There were 31 ELV launches in 1998: 13 Deltas, 6 Atlases, 3 Titans, 1 Athena, 2 Taurus, and 6 Pegasus. These included five space science payloads and one meteorological payload. The rest were DOD or commercial satellites. One Titan and one Delta, the first Delta III, failed.

Expendable Launch Vehicle Characteristics

The following sections describe each family of U.S. ELVs used from 1989 through 1998. It should be noted that the figures cited in the Launch Characteristics tables are approximations and may not be accurate for all vehicles within a particular model of launch vehicle. Many factors influence detailed specifications. Each payload is different, and the payload size as well as its ultimate orbit will determine the launch vehicle configuration, including the number of stages and strap-on motors, the size of the selected fairing, and the nature of the attach fittings. Variations in payloads also determine the amount of propellant, the burn rate, thrust levels, and other parameters. Source material, although dependable, does not always state under what conditions a particular value is true. For instance, a value for thrust can indicate nominal, maximum, or average force and can exist during liftoff at sea level or in a vacuum. Different payloads and different orbits can also determine performance parameters. The maximum payload for a launch vehicle to low-Earth orbit may be different for a launch from Cape Canaveral, Florida, than for a launch from Vandenberg Air Force Base, California. There are also variations in what is considered low-Earth orbit. The *Aeronautics and Space Report of the President* and the Federal Communications Commission use a 185-kilometer (100-nautical-mile) orbit; other sources range from 144 kilometers to 196 kilometers (78 nautical miles to 106 nautical miles) or consider low-Earth orbit to be the orbit flown by the Space Shuttle.[33]

Measurements are stated in the original units used in the source material. Some measurements will appear as English units and some as metric units. The conversion to the other unit of measure follows in parentheses.

This chapter uses the following abbreviations for propellants: LH_2 = liquid hydrogen, LOX = liquid oxygen, N_2H_2 - = hydrazine, N_2O_4 = nitrogen tetroxide, RJ-1 = liquid hydrocarbon, and RP-1 = kerosene.

Athena Launch Vehicle

The Athena launch vehicle was a privately funded solid-propellant launch vehicle developed by Lockheed Martin beginning in 1993 to carry small to medium payloads into low-Earth, geostationary transfer, and interplanetary orbits. It was initially called the Lockheed Launch Vehicle (LLV) and then the Lockheed Martin Launch Vehicle (LMLV) after Lockheed merged with Martin Marietta in 1994. The core launch vehicle was called LMLV-1, later renamed Athena I. A larger version, the LMLV-2, was renamed Athena II.

Both vehicle models used a 92-inch (234-centimeter)-diameter fairing, and both used solid motors and a small liquid injection stage called the orbit adjust module as its top stage. The top stage contained the altitude control and avionics subsystems. The Athena I and Athena II both had a Castor 120 first stage, a commercial motor made by Thiokol derived from the Peacekeeper intercontinental ballistic missile first-stage motor and modified for space launch use. The Athena II's second stage was another Castor 120. The second and third stages of the Athena I were the same as the third and fourth stages of the Athena II: a Pratt & Whitney Orbus 21D motor and an orbit adjust module powered by four Primex MR-107 engines using hydrazine fuel. The orbit assist module was available with four or six propellant tanks, depending on mission requirements. Figure 2–10 shows the Athena I and Athena II configurations.

The first Athena I launch took place on August 15, 1995. This launch failed when the thrust vector control system failed. The first successful launch was on August 23, 1997, from Vandenberg Air Force Base. Its payload, the Lewis satellite, failed shortly after launch. Later launches of Athena I were planned to take place from the Kodiak Launch Complex in Alaska.

---

[33] "Glossary," NASA Life Sciences Data Archive, *http://lsda.jsc.nasa.gov/kids/L&W/glossary.htm* (accessed February 9, 2005). Also "Genesis: Search for Origins," Jet Propulsion Laboratory, *http://www.genesismission.org/glossary.html* (accessed February 9, 2005).

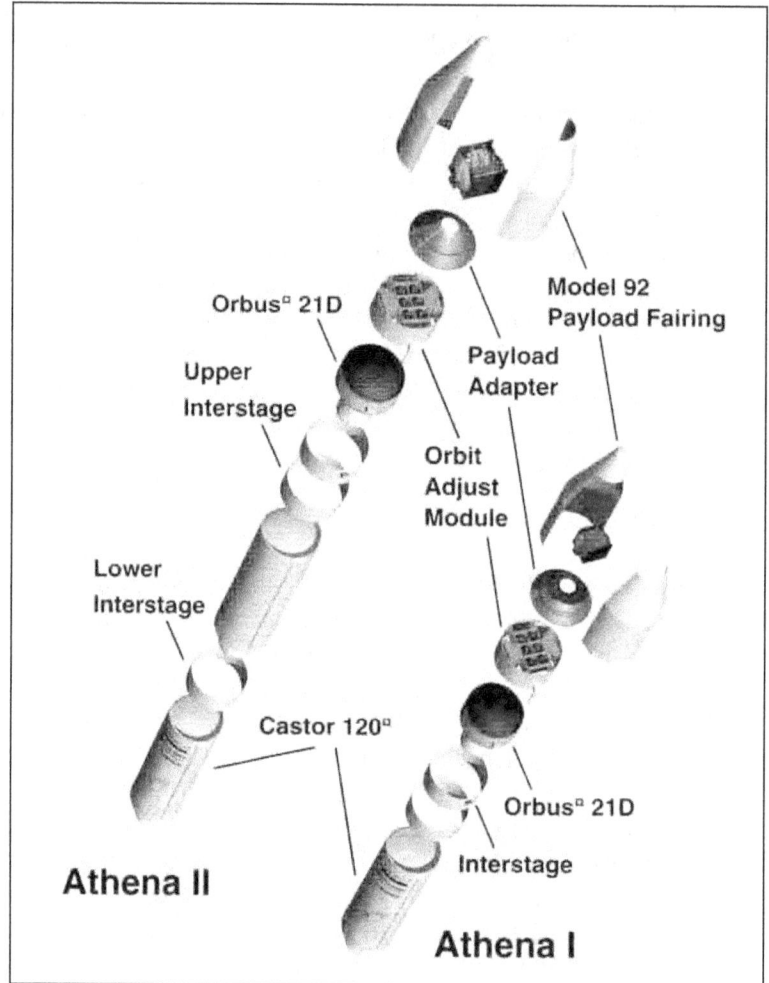

Figure 2–10. Athena I and Athena II. (Lockheed Martin)

Athena II successfully launched NASA's Lunar Prospector into orbit from Cape Canaveral Air Station on January 7, 1998. The Athena II's third stage enabled it to launch larger payloads. Table 2–59 lists Athena launches. Tables 2–60 and 2–61 list Athena I and Athena II characteristics.

The Atlas Family

The Atlas rocket was America's first intercontinental ballistic missile (ICBM). The Air Force used the missile only briefly as an ICBM, however, and made its surplus rockets available for use as space launch vehicles in the 1980s after adding an upper stage allowing the modified rockets to place various types of payloads into low-Earth orbit.

The surplus rockets were used quickly in their new role, and only a few Atlas E and Atlas G Centaur launch vehicles remained in the surplus inventory by the end of the 1980s. NASA used these remaining rockets and then started using new Atlas I, II, IIA, and IIAS launchers.

The production of Atlas rockets was government-initiated through the production of the Atlas G Centaur. The commercial sector took over launch services in June 1987. The first commercial launch took place in July 1990 with the first Atlas I rocket.[34]

The Convair Division of General Dynamics built Atlas rockets at the beginning of the program. Martin Marietta acquired Convair's launch vehicle division in 1994 and took over Atlas production until Martin Marietta merged with Lockheed in 1996 to form Lockheed Martin, the current Atlas producer. All Atlas models, except the Atlas E, used the liquid-fueled Centaur as their upper stage to provide added thrust. This upper stage, developed by General Dynamics for NASA, had been used since the 1960s. The Atlas E used solid-fueled apogee kick motors to supply extra power.

The Atlas I was the first Atlas product using a new naming convention, initiated when Lockheed began using the Atlas for commercial launches in the late 1980s. Rather than use the old letter designation for its rockets (Atlas E, F, and G, for example), Lockheed began using Roman numerals. A letter after the Roman numeral designated different variations in each family, such as Atlas IIA and Atlas IIAS, in which "S" indicated a strap-on motor.

The Atlas has been a dependable launch vehicle with only a few launch failures. Table 2–62 lists all Atlas launches between 1989 and 1998.

Atlas Characteristics

The Atlas launch vehicle system consisted of the Atlas booster (composed of a booster and a sustainer section), the Centaur upper stage, the payload fairing, and an interstage adapter located between the booster/sustainer stage and the Centaur stage. The launch vehicle was typically called a "one-and-a-half"-stage vehicle. The booster stage engines flanked the smaller sustainer engine and did not carry any propellant. The sustainer section contained propellant tanks for both the booster and sustainer burns. All engines ignited at liftoff, and the two smaller vernier engines on the Atlas E and G and Atlas I models ignited seconds later.[35] This differed from later rockets in which the stages fired sequentially.

---

[34] "Atlas," Lockheed Martin Space Systems Company, *http://www.lockheedmartin.com/wms/ findpage.do?dsp=fec&ci=14917&5c=400* (accessed July 18, 2006).
[35] The "I" in Atlas I refers to the Roman numeral "one," not the letter "I."

Atlas E

Atlas E was first used as a launcher in 1960. The last Atlas E launch took place on March 24, 1995, when it launched a military weather satellite into orbit. All Atlas E launches during this period took place from Vandenberg Air Force Base in California. The Atlas E was the only Atlas launch vehicle during this period not using a Centaur upper stage. It obtained additional boosting power from its apogee kick motor (AKM). Dimensions stated in Table 2–63 are approximate because more than one AKM model was used and fairings varied in length.

Atlas G Centaur

The Atlas G Centaur, used primarily to launch communications satellites, was an improved version of the earlier Atlas Centaur launch vehicle. It was 81 inches (2.06 meters) longer than its predecessor to allow greater fuel capacity and had increased booster thrust of 7,500 pounds (33.36 kilonewtons), leading to a total liftoff thrust of 438,877 pounds (1,950 kilonewtons).[36]

The Atlas G Centaur was first used in 1984 with an Intelsat satellite. The final Atlas G Centaur launch took place on September 25, 1989, with the launch of Fltsatcom-8. This launch marked the last NASA-managed ELV launch. From then on, NASA purchased launch services from a series of contractors. Table 2–64 shows Atlas G Centaur characteristics.

Atlas I

The Atlas I was the first of a new family of launch vehicles that could boost payloads into low-Earth orbit, geosynchronous-Earth orbit, and on interplanetary trajectories. The launch vehicle was very similar to the Atlas G Centaur, and it included two boosters, a sustainer, two vernier single-start engines, and a Centaur upper stage. An interstage adapter separated the Atlas stage from the Centaur. The vehicle had two new payload fairings, incorporated significant improvements in the guidance and control systems, and replaced analog flight control components with digital units interconnected with a digital data bus. Figure 2–11 shows an Atlas I. Table 2–65 lists Atlas I characteristics.

The first Atlas I flight took place on July 25, 1990, with the launch of the Combined Release and Radiation Effects Satellite (CRRES), a joint NASA-U.S. Air Force project. The final Atlas I launch took place on April 25, 1997, with the launch of GOES-10 into geosynchronous orbit. Although launch parameters varied slightly depending on launch date, launch time, and payload weight, Table 2–66 presents a typical launch sequence for a geosynchronous mission.

---

[36] "Atlas," GlobalSecurity.org, *http://www.globalsecurity.org/space/systems/atlas.htm* (*accessed* January 26, 2005).

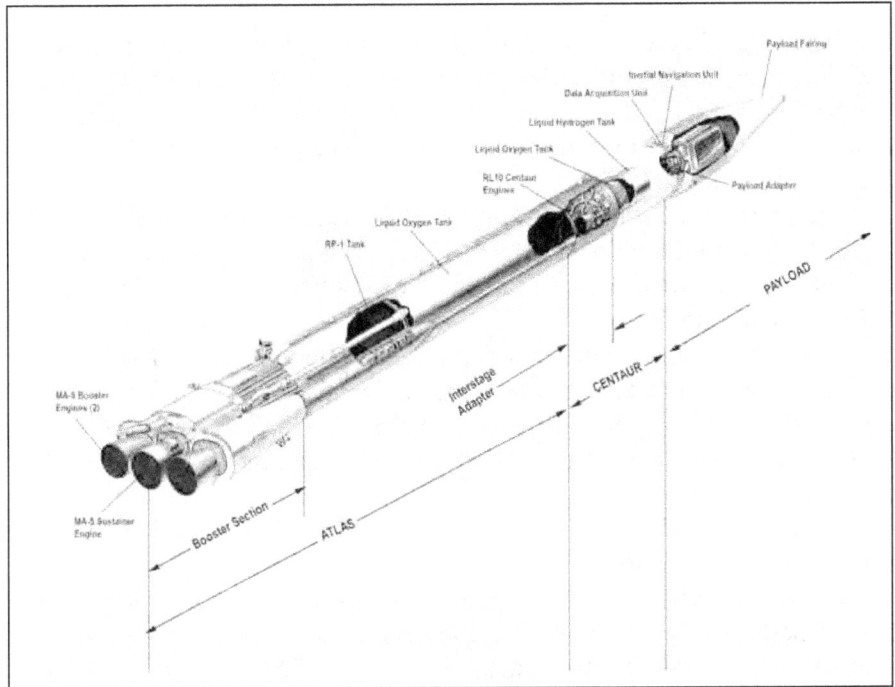

Figure 2–11. Atlas I Components.

General Dynamics produced eleven Atlas I ELVs before the program shifted to Atlas II production. Three Atlas I launches failed to propel their payloads into orbit.

Atlas II Series

The Atlas II series consisted of the Atlas II, the Atlas IIA, and the Atlas IIAS models. Development of the first of these vehicles began in June 1988. They were originally developed to launch the Air Force Defense Satellite Communications System satellites, part of the Air Force Medium Launch Vehicle II program.

The Atlas II launch vehicles were an improved version of the Atlas-Centaur rockets. They provided higher performance by using engines with greater thrust and longer fuel tanks for both the Atlas and Centaur stages. This resulted in increased payload capability. The Atlas II replaced the MA-5 propulsion system used in the Atlas I with the improved MA-5A system. The Atlas II also replaced the vernier engines of Atlas I and earlier Atlas vehicles with a hydrazine roll control system located on the Atlas II interstage that had lower-cost electronics and an improved flight computer.

The Atlas II had a longer booster than the Atlas I for greater fuel stage-one capacity and used upgraded MA-5A engines, improved structures, and a new stabilization system. It also featured a lengthened Centaur upper stage that held more fuel and thus had better upper-stage performance.[37] The Atlas II was the only Atlas to use two R-4D attitude control thrusters for attitude and orbit adjustments.

The first Atlas II flew December 7, 1991, launching Eutelsat II F3. The last Atlas II launch took place March 16, 1998, with the launch of USA 138 (UHF-8), a communications satellite for DOD that replaced the old FLTSATCOM satellites.

With Atlas II, the manufacturers changed the terminology referring to the number of stages although the configuration remained essentially the same as earlier vehicles. The vehicle was then referred to as having "two-and-a-half" stages. These stages consisted of the booster, sustainer, interstage, and Centaur upper stage.

A total of 10 Atlas II launches took place; all were successful. Table 2–67 lists Atlas II characteristics.

The Atlas IIA was the commercial version of the Atlas II. It incorporated higher performance RL10 engines and optional extendible nozzles that provided added thrust to the Centaur upper stage. The first Atlas IIA flight took place on June 8, 1992, with the launch of Intelsat-K. Through the end of 1998, 15 Atlas IIA launches took place; all were successful. Table 2–68 lists Atlas IIA characteristics.

The Atlas IIAS was similar to the earlier Atlas IIA launch vehicle except that this model used four additional strap-on Castor IVA solid rocket boosters (SRB), which provided an average thrust of 433.7 kilonewtons (97,500 pounds) each. These SRBs fired two at a time. The first pair fired at liftoff. The second pair fired during flight after the first pair had burned out, approximately 54 seconds after liftoff. Both pairs were jettisoned soon after each pair burned out. The structure of the first stage was stronger to accommodate the SRBs. Table 2–69 lists Atlas IIAS characteristics. Figure 2–12 shows the Atlas IIAS configuration for the launch of the Solar and Heliospheric Observatory (SOHO) on December 2, 1995.

The first Atlas IIAS launched Telstar 401 on December 15, 1993. Through the end of 1998, 14 Atlas IIAS launches had taken place; all were successful.

---

[37] "The Evolution of Commercial Launch Vehicles," *Fourth Quarter 2001 Quarterly Launch Report, http://ast.faa.gov/files/pdf/q42001.pdf* (accessed January 17, 2005).

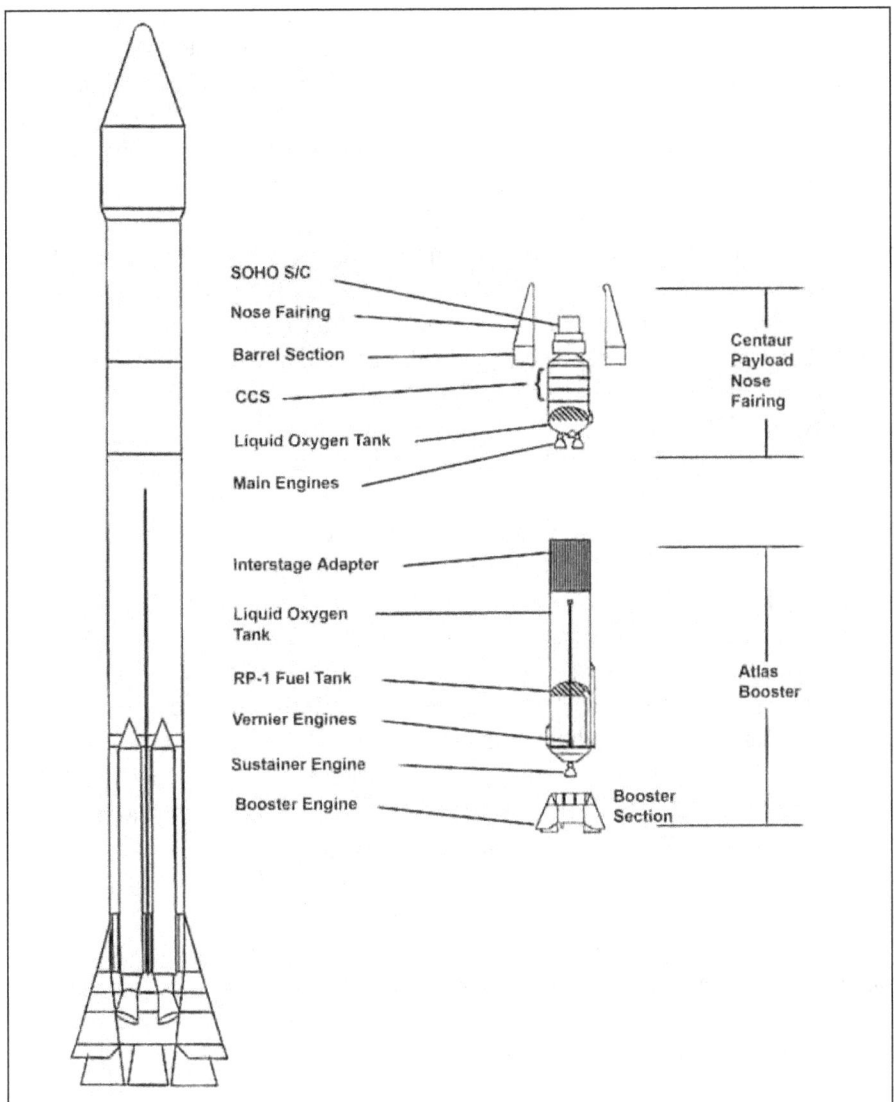

Figure 2–12. Atlas IIAS Launch Vehicle–SOHO Configuration, December 2, 1995. (NASA)

Conestoga Launch Vehicle

The Conestoga was a solid-propellant commercial launch vehicle that attempted to launch the Multiple Experiment Transporter to Earth Orbit and Return (METEOR 1) payload originally known as COMET for Commercial Experiment Transporter into low-Earth orbit in 1995. The privately funded launcher was designed to answer a need by the civilian and military community for a vehicle to launch small size orbital and suborbital payloads (500 pounds to 5,000 pounds) into low-Earth orbit. In 1982, Space Systems Inc. (SSI), managed by Mercury astronaut Donald "Deke" Slayton, successfully launched a single-stage, solid-fueled rocket as a test, ejecting its payload as planned.[38] The rocket was based on an Aerojet M56-A1, the second stage of a Minuteman ICBM. The company was the first to obtain a commercial launch license, in 1985 receiving Department of Transportation mission approval. In 1986, SSI signed the industry's first agreement to use a U.S. Government launch range, Wallops Flight Facility on the Eastern Shore of Virginia, as a commercial launch site.[39]

In November 1990, EER Systems purchased SSI, integrating it into EER's Space Systems Group. In 1991, NASA selected EER to provide Conestoga launch services for its METEOR satellite, a microgravity carrier program. METEOR was to be a recoverable payload, designed for on-orbit microgravity experiments that advanced commercial applications of materials processing and medical research.[40]

The Conestoga launch vehicle had a modular design, which provided a wide range of configurations for various purposes. It's Thiokol booster stage rockets consisted of one core Castor IVB surrounded by a combination of two to six more strap-on Castor IVA or IVB solid rocket motors. A Star 37, 48, or 63 upper stage motor sat immediately above the core booster motor. Four strap-on motors ignited at launch; they were followed by two more, and finally the central Castor. A cold gas reaction control system, located within the payload attach fitting, controlled pitch, yaw, and roll during fourth stage coast, burn, and payload separation. The hydrazine maneuvering and attitude control system sat above the Star motor upper stage and provided velocity augmentation and control of pitch, yaw, and roll. The avionics power, electronics, and guidance equipment were within the payload attach fitting. An aerodynamic payload fairing available in several sizes covered all components from the payload down to and including the Star motor. Varying the number of strap-on motors and selecting the appropriate upper stage extended the Conestoga's performance range from 500 pounds to 5,000 pounds to low-Earth orbit.[41]

METEOR was originally planned as a three-mission project with the first launch initially scheduled for September 1992. However, late delivery of the solid rocket motors delayed completion of the launch vehicle. Management difficulties with the Center for Space Transportation and Applied Research (CSTAR) at the University of Tennessee (the commercial center that had proposed the mission and which provided oversight), as well as rising costs forced NASA to reduce the project to a single mission. Agency review of the

---

[38] Deke Slayton was one of the original Mercury astronauts but was relieved of his assignment and did not fly because of a heart condition. After he was cleared to resume full flight status in 1972, he made his first spaceflight as Apollo docking module pilot of the Apollo-Soyuz Test Project mission, July 15–24, 1975.

[39] Isakowitz and Samella, *International Reference Guide to Space Launch Systems,* 2nd ed., p. 220.

[40] U.S. Department of Transportation, Federal Aviation Administration, "Special Report: U.S. Small Launch Vehicles," *Commercial Space Transportation Quarterly Launch Report, 1st Quarter 1996, http://ast.faa.gov/files/pdf/sr_96_1q.pdf* (accessed November 2, 2005).

[41] M. Daniels and B. Saaverdra, "The Conestoga Launch Vehicle—A Modular Approach to Meeting User Requirements," AIAA-94-0893, 15th American Institute of Aeronautics and Astronauts International Communications Satellite Systems Conference, February 27–March 3, 1994.

project continued into 1994 as NASA Administrator Goldin announced that NASA would refuse to continue funding the project. There also was the question of liability if the reentry module landed outside the sparsely populated Great Salt Lake desert in Utah. Congress, however, released the needed funds on condition that the contractors agreed to waive NASA's legal liability. NASA also insisted that CSTAR depart from the program. After further discussion with the three contractors providing elements of the vehicle, NASA signed a sole source, fixed-price contract with EER Systems.[42]

After further delays, launch finally took place on October 23, 1995. It was the first orbital flight from Wallops Flight Facility in 10 years.[43] After a promising liftoff, the rocket went off course when its first stage steering mechanism ran out of hydraulic fluid and became inoperable. Forty-six seconds after liftoff, 23 kilometers off Virginia's coast at an altitude of 10 kilometers, the Conestoga broke apart. The destruction resulted in the loss of the METEOR and the 14 microgravity experiments on board. EER Systems left the launch business and abandoned the project. See Table 2–70 for characteristics of the Conestoga 1620, the model that carried the METEOR.

The Delta Family

NASA has used the Delta launch vehicle since 1960 and has regularly upgraded the vehicle as the need for payload capacity grew. The vehicle has a high success rate. In the decade from 1989–1998, 82 Delta launches took place with only two failures. Table 2–71 lists all Delta launches.

The 2900 series was planned as the last Delta series. However, because the Space Shuttle was not yet ready to become an operational space launcher and NASA needed a vehicle with heavier payload capacity, the Delta 3000 series was developed in the late 1970s and early 1980s for payloads that were too heavy for Delta 2000s but did not require the Atlas-Centaur. Because the 3000 series was considered an interim vehicle for medium-weight payloads, NASA not did finance its development and production but instead bought completed vehicles for its civilian and commercial launches from McDonnell Douglas, which obtained private financing for the series. Table 2–72 lists Delta 3920 characteristics.

Delta production formally ended at the end of 1984 when its production line at Huntington Beach, California, closed. But when the *Challenger* explosion brought out the need for launch alternatives, NASA decided to resume using ELVs and reactivated the Delta production line. At the same time, President Ronald Reagan announced that the Space Shuttle would stop carrying commercial payloads.

---

[42] Andrew Butrica, "The Commercial Launch Industry, Technological Change, and Government-Industry Relations," *http://www.hq.nasa.gov/office/pao/History/x-33/butr02.htm* (accessed November 3, 2005).
[43] "Conestoga," GlobalSecurity.org, *http://www.globalsecurity.org/space/systems/conestoga.htm* (accessed November 3, 2005).

The commercial Delta era began in January 1987 when the U.S. Air Force announced its selection of McDonnell Douglas to produce seven Deltas IIs to launch its NAVSTAR Global Positioning System (GPS) satellites, originally manifested for the Space Shuttle. The initial contract expanded to 20 vehicles in 1988 when the Air Force exercised two contract options. In the interim, the remaining stock of older Deltas was modified for three missions: the Delta 4925 combined the earlier MB-3 engine with enhanced Castor IVA strap-on motors to launch the BSB-R1 and Insat 1-D satellites, and the Delta 5925 used Castors with the RS-27 engine to launch the Cosmic Background Explorer for NASA. On July 1, 1988, the Air Force officially received custody of Launch Complex 17, located at Cape Canaveral Air Force Station, Florida, from NASA and took over East Coast launch operations, ending 28 years of Delta launches managed by NASA.[44]

McDonnell Douglas built on its successful Delta 3920/PAM-D model to produce the Delta II. The first Delta II, the 6925, flew on February 14, 1989, launching the first of nine Air Force GPS satellites into orbit 20,200 kilometers (10,900 nautical miles) above Earth. NASA first contracted commercially for the Delta II in December 1990 for launch of its Geotail, Wind, and Polar science satellites, which launched in 1992, 1994, and 1996, respectively. NASA was the first U.S. government agency to procure commercial launch services.[45]

The first stage of the Delta 6925 was an 85.6-foot (26-meter)-long Extra Extended Long Tank powered by an RS-27 engine and augmented by nine Castor IVA strap-on motors. The second stage used an Aerojet AJ10-118K engine that delivered approximately 9,645 pounds (42.4 kilonewtons) of thrust. The third stage payload assist module (PAM)-D, equipped with a Thiokol Star 48B solid rocket motor, delivered approximately 15,100 pounds (67 kilonewtons) of thrust and made the vehicle suitable for geosynchronous and Earth-escape missions. Table 2–73 lists Delta II 6925 characteristics.

The versatile Delta II could be configured as a two-stage or three-stage vehicle and could launch with three or four strap-on motors as well as with the more common nine strap-ons. Both two-stage and three-stage Deltas could support 9.5-foot (2.9-meter) and 10-foot (3.05 meter)-diameter fairings. When nine strap-ons were used, six were ignited at launch and the remaining three ignited in flight. The 9.5-foot fairing was primarily designed for the three-stage Delta.[46] The 10-foot (3.05-meter) fairing was lighter than the one it replaced and was also available in a longer version for taller payloads. Typically, two-stage Deltas launched satellites to low-Earth orbit, while three-stage Delta IIs delivered payloads to geosynchronous transfer orbit or were used for deep-space missions.

---

[44] "Delta Launch Complex Transferred to Air Force," *NASA News* Release 88-99, July 15, 1988. (NASA History Office Folder 010241).

[45] "Review notes from Charles Gunn September 1, 2005.

[46] "Boeing Delta II Medium Launch Vehicle," Delta II Backgrounder, *http://www.boeing.com/ defensespace/space/delta/delta2/contour/mission_info/backgrounders/delta_2_backgrounder.htm* (accessed January 31, 2005).

Also, the Delta II could launch one or more payloads on the same launch vehicle by using a variety of payload attachments. Figure 2–13 shows the Delta II with nine strap-ons.

Several other Delta IIs were developed that eventually replaced the 6925: the 7326, 7420, 7425, the 7920, and the most powerful, the 7925. All Deltas in the 7000 series were equipped with an improved engine designated the RS-27A that boosted engine performance. Also, more power and longer Hercules (later Alliant Techsystems) graphite epoxy motors (GEMs) replaced the Thiokol Castor IVA solid rocket motors. Each GEM was 42.5 feet (13 meters) long and provided 446 kilonewtons (100,300 pounds) of thrust at liftoff (see Table 2–74 and Figure 2–14).[47] The 7925 first flew in November 1990 to launch a NAVSTAR GPS satellite. Other Delta missions launched satellites to Mars, toward asteroids and comets, and were used for Earth-observation and astronomy missions. Figure 2–15 compares the Delta 3920, Delta II 6925, and Delta II 7925.

In 1995, McDonnell Douglas began Delta III development to fulfill growing customer needs for a higher capacity commercial launch service.[48] With a payload delivery capacity to geosynchronous transfer orbit of 3,810 kilograms (8,400 pounds), the Delta III effectively doubled the performance of the Delta II. The first Delta III launch took place in 1998, but a successful launch did not occur until August 2000. Table 2–75 lists the sequence of events for a typical Delta launch to geosynchronous orbit.

Pegasus Booster

The Pegasus was the first all-new U.S. space launch vehicle since the 1970s and the only air-launched space booster vehicle attempted in the United States in approximately 30 years when the U.S. Navy attempted the unsuccessful Project Pilot. Considered the operational successor to the long-lived Scout launch vehicle in the small-payload, solid-propellant-motor category, the Pegasus was developed jointly by Orbital Sciences Corporation and Hercules Aerospace Company (later Alliant Techsystems of ATK Thiokol Propulsion Company).[49] Hercules was responsible for the design and production of the new solid rocket motors and the payload fairings. Orbital was responsible for the remaining mechanical and avionics systems, ground and flight software, the carrier aircraft interface, mission and vehicle integration, overall systems engineering, and program management. The development cost of more than $50 million was split evenly between the two partners.[50]

---

[47] Mark Cleary, "Delta II Overview," in *Delta Space Operations at the Cape, 1993–2001*, *https:// www.patrick.af.mil/heritage/DELTA%20II%20Overview.htm*, (accessed January 31, 2005).

[48] Boeing acquired the launch organization from McDonnell Douglas in 1997 and transferred production of the Delta to its facilities.

[49] Matt Bille, Pat Johnson, Robyn Kane, and Erika R. Lishock, "History and Development of U.S. Small Launch Vehicles," in *To Reach the High Frontier, A History of U.S. Launch Vehicles*, Roger D. Launius and Dennis R. Jenkins, ed. (Lexington, KY: The University Press of Kentucky, 2002), p. 214.

[50] Isakowitz et al., *International Reference Guide to Space Launch Systems*, 3rd ed., p. 279.

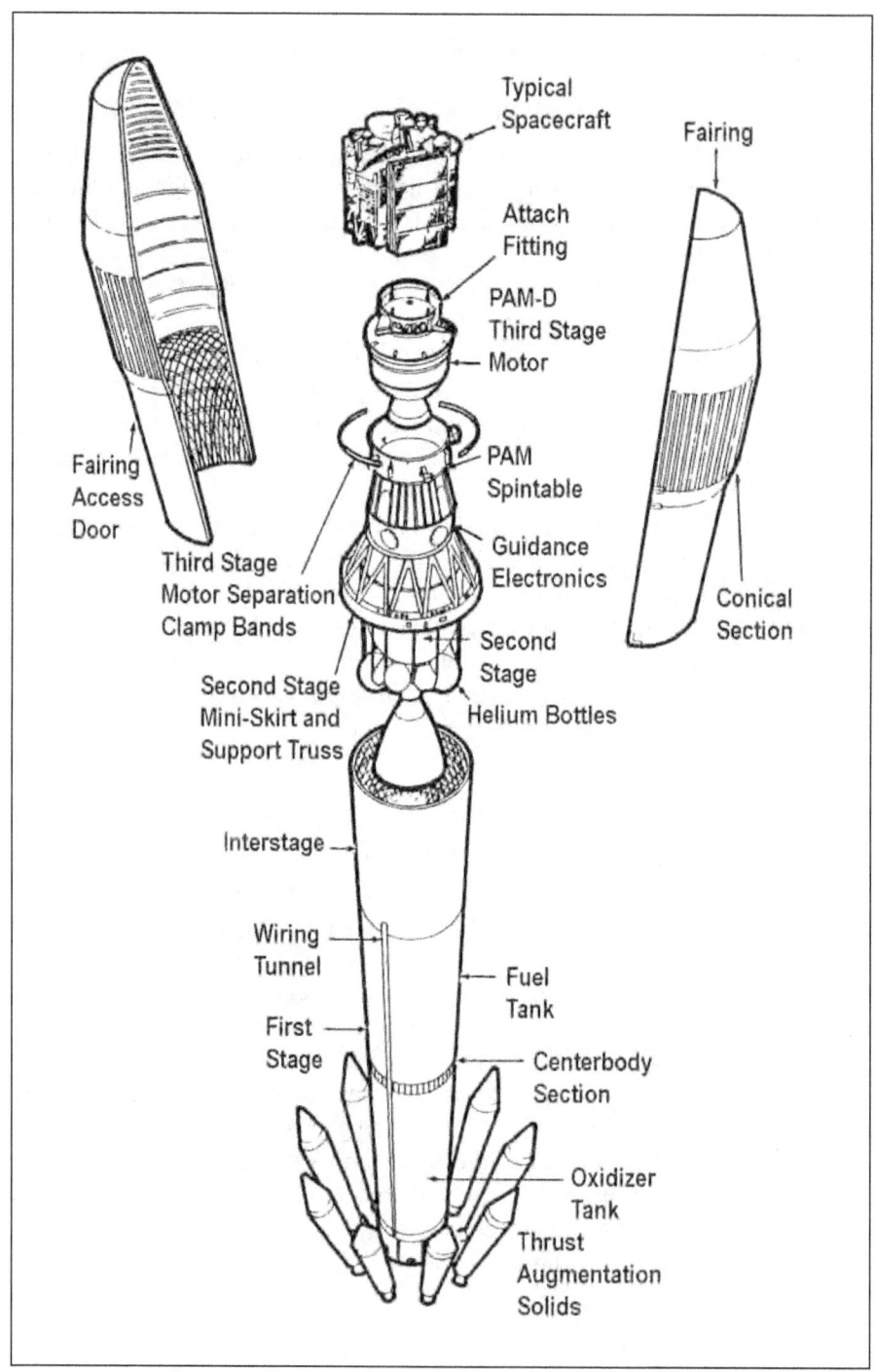

Figure 2–13. Delta II Components. (The Boeing Company)

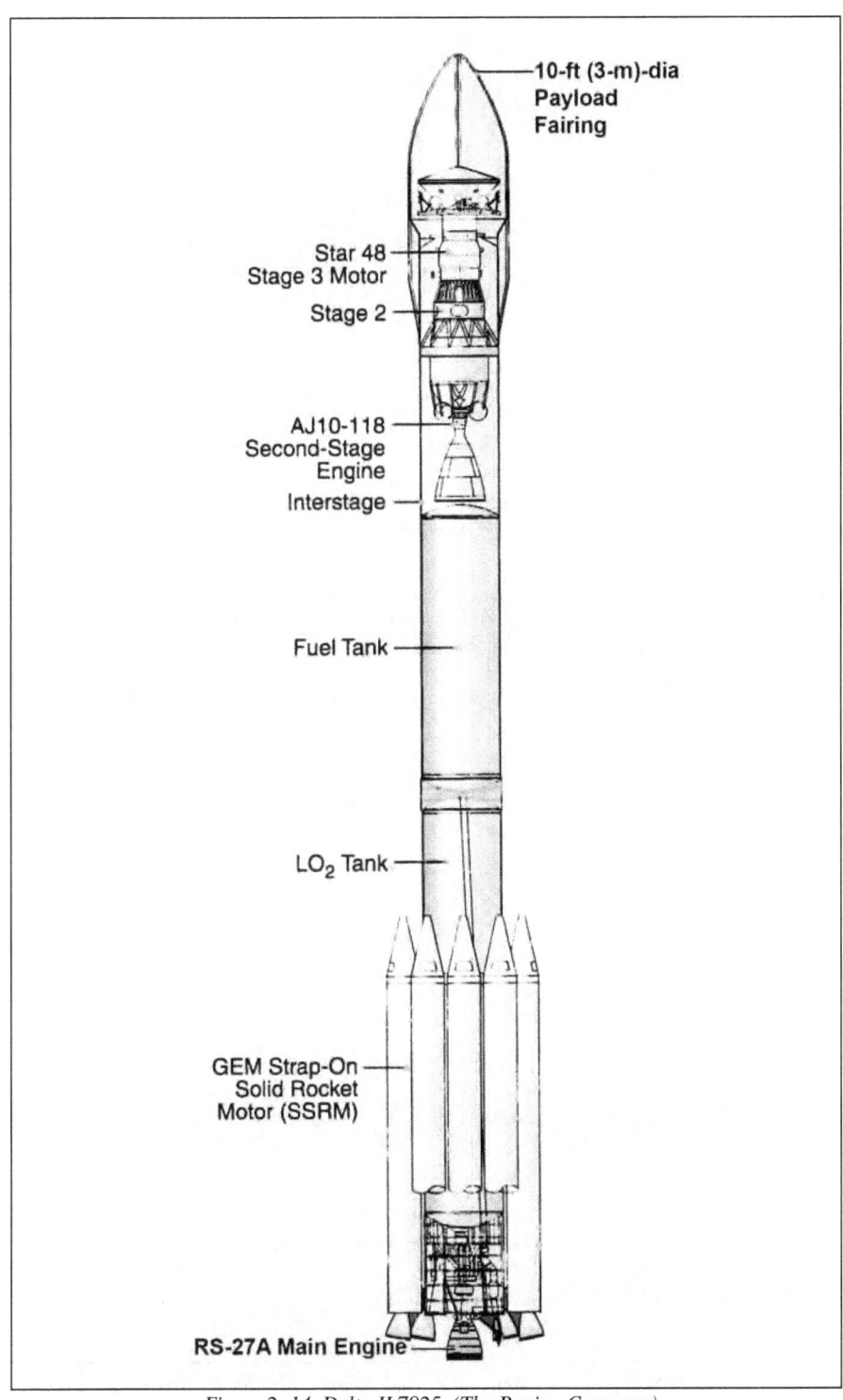

Figure 2–14. Delta II 7925. (The Boeing Company)

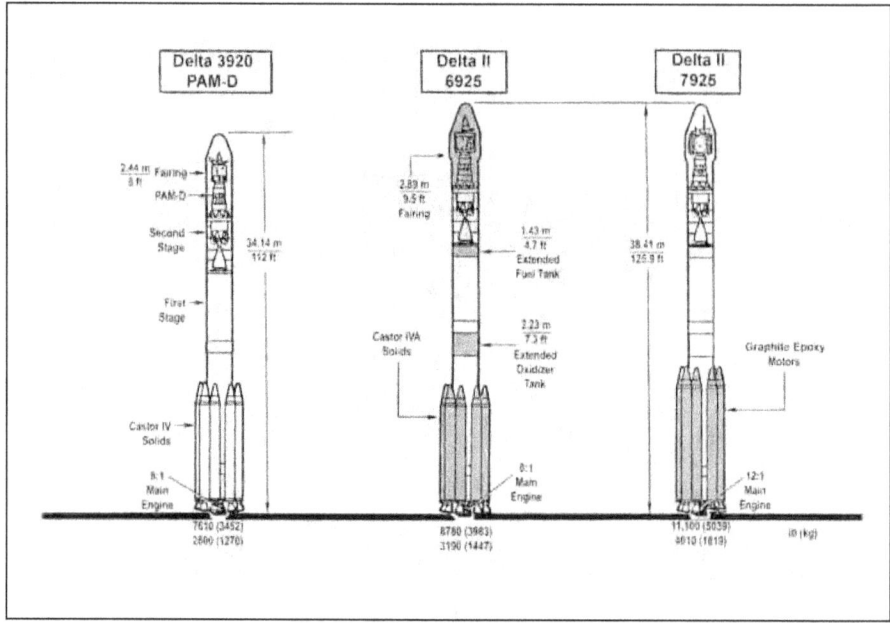

Figure 2–15. Delta 3920/PAM-D, Delta II 6925, and Delta II 7925.
(The Boeing Company)

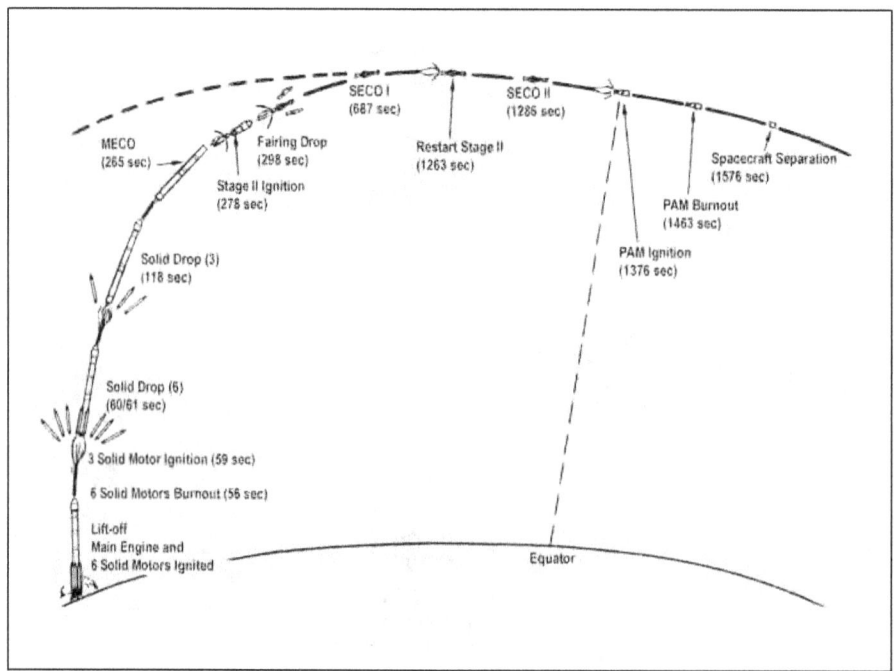

Figure 2–16. Delta II Mission and Launch Sequence Profile for a Typical
Geosynchronous Mission. (The Boeing Company)

Orbital established the Pegasus program in 1987. In 1988, the Defense Advanced Research Projects Agency awarded a contract to Orbital for one firm and five options for the rocket. In July 1990, NASA and Orbital Sciences signed an agreement in support of Orbital's commercial launch vehicle programs. This agreement allowed Orbital to enter into sub-agreements with NASA installations in which NASA would provide access to its launch support property and services on a cost-reimbursable basis.[51] In 1991, Goddard Space Flight Center selected the Pegasus to supply Small Expendable Launch Vehicle Services for its Small Explorer program, and on June 4, the Pegasus was chosen for up to 40 more launches under the Air Force Small Launch Vehicle program.[52] The Ballistic Missile Defense Organization awarded another launch contract to Orbital in July 1992. In 1994, NASA selected Orbital for its Ultralight launcher, and Spain selected the Pegasus in October 1994 to launch Minisat 01, the first West European orbital launch.[53] There have been two Pegasus models: the Standard Pegasus, which first flew on April 5, 1990, and the Pegasus XL, which was conceived in 1991 and first flew on June 27, 1994.

The first Pegasus booster rocket was launched on April 5, 1990, from Edwards Air Force Base, California, from underneath NASA's B-52 airplane in a mission originating at Dryden Flight Research Center.[54] Other launches through 1998 took place from the Canary Islands in Spain and Wallops Flight Facility, Virginia, as well as from Edwards and Vandenberg Air Force Bases, both in California, and Cape Canaveral, Florida. The B-52 launched the Pegasus until 1995, when a modified Lockheed L-1011 aircraft, the Orbital "Stargazer," replaced it. The Pegasus XL, an upgraded Pegasus that was longer, heavier, and able to boost larger payloads than the standard Pegasus, used only the L-1011 aircraft. The Pegasus's best-known achievement was its launch of the ORBCOMM communications satellites. Between 1997 and 1999, five Pegasus launches sent 32 satellites into orbit, forming the world's first private, low-Earth orbit communications network.[55] See Table 2–76 for the Pegasus flight history.

Unlike ground-launched rockets, the Pegasus was launched at an altitude of more than 40,000 feet (12,192 meters) from beneath a flying aircraft at an initial speed of Mach 0.8. This air launch offered several advantages. First, because the rocket did not require a launch pad, just a runway from which the aircraft could take off and land, it could be launched from almost anywhere around the world. Second, the booster derived a slight gain in performance (one percent to two percent) from the speed of the carrier aircraft. Third, its trajectory was flatter

[51] "NASA, Orbital Sciences Corporation Sign Agreement," *NASA News* Release 90-92, July 3, 1990, *ftp:// ftp.hq.nasa.gov/pub/pao/pressrel/1990/90-092.txt* (accessed February 2, 2005).

[52] "Milestones," Orbital, *http://www.orbital.com/About/Milestones/90_99/* (accessed February 2, 2005). Also *Aeronautics and Space Report of the President, Fiscal Year 1991 Activities* (Washington, DC: National Aeronautics and Space Administration, 1992), p. 70; Bille et al., *To Reach the High Frontier, A History of U.S. Launch Vehicles*, p. 216.

[53] Andrew Wilson, ed., *Jane's/Interavia Space Directory, 1999–2000* (Alexandria, VA: 2000), Jane's Information Group (2000), p. 237.

[54] This B-52 was the same aircraft used for the X-15 test flights in the 1960s.

[55] Bille et al., in *To Reach the High Frontier, A History of U.S. Launch Vehicles*, p. 216.

than the trajectory for ground-launched vehicles, so less power was dissipated in achieving the correct attitude for injection into orbit. Fourth, because the carrier aircraft served the same function as the first stage of a ground-launched vehicle, the rocket itself needed to carry less propellant.[56] Fifth, the fact that the launch took place above 75 percent of Earth's atmosphere reduced the energy needed to reach orbit. Finally, its air launch reduced the amount of stress the launch vehicle faced when compared with ground-launched vehicles.[57] Figure 2–17 shows a Pegasus mated to its B-52 mothership.

Figure 2–17. Pegasus Mounted Under B-52 Wing. (NASA-DFRC Photo No. EC91-348-3)

A 22-foot (6.7-meter) delta wing mounted on top of Stage 1 provided extra lift. There were three solid rocket motors in its three stages; a payload fairing; an avionics assembly; a lifting wing; an aft skirt assembly, including three movable control fins; and a payload interface system. It also could be equipped with a liquid-propellant fourth stage, the hydrazine auxiliary propulsion system (HAPS), to boost the payload into a higher orbit. The vehicle's blunt payload fairing blended into a cylindrical fuselage and ended in a flared exhaust nozzle. The wing was made of graphite composite structure, and 94 percent of the structural weight of the original model Pegasus was also graphite composite. Three control fins electromechanically actuated provided pitch, roll, and yaw control while the vehicle was still in Earth's atmosphere. When the vehicle reached the upper atmosphere, small rockets mounted in the base of each fin helped control the vehicle. Figure 2–18 shows the Pegasus vehicle.

---

[56] "Pegasus Launch Vehicle," Space & Missile Systems Center (AFMC), Department of the Air Force, *http://www.te.plk.af.mil/factsheet/pegfact.html* (accessed February 8, 2005).
[57] Matt Bille et al. in *To Reach the High Frontier, A History of U.S. Launch Vehicles*, p. 215.

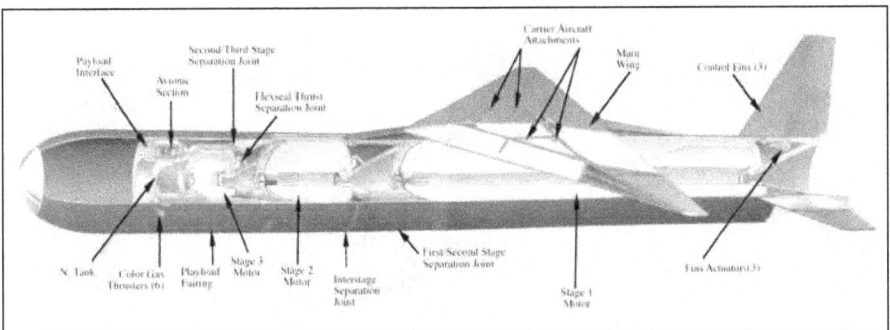

Figure 2–18. Pegasus Vehicle. (Orbital Sciences Corp.)

The standard payload fairing consisted of two graphite composite halves with a nosecap bonded to one of the halves and a separation system. The fairing separated when sequentially actuating pyrotechnic devices released the right and left halves of the fairing from a closed position and moved the halves away from either side of the payload. Pegasus could accommodate multiple payloads on the same mission. The standard fairing had a 1.17-meter (3.8-foot) diameter and was 2.13-meters (7-feet) long. If the optional HAPS was used, the fairing was 1.76-meters or 1.79-meters (5.8-feet or 5.9-feet) long. Table 2–77 lists Standard Pegasus characteristics, and Table 2–78 lists Pegasus XL characteristics.

The typical launch sequence begins with release of the Pegasus from the carrier aircraft at an altitude of approximately 11,900 meters (39,000 feet) and a speed of Mach 0.80. Approximately 5 seconds after its drop from the aircraft, when Pegasus had cleared the aircraft, Stage 1 is ignited. The vehicle quickly accelerates to supersonic speed while beginning a pull-up maneuver. Maximum dynamic pressure is experienced about 25 seconds after ignition. At approximately 20 to 25 seconds, a maneuver begins to depress the trajectory, and the vehicle's angle of attack quickly approaches zero.

Stage 1 burnout occurs at approximately 77 seconds, and Stage 2 ignition follows quickly. The payload fairing is jettisoned during Stage 2 burn as quickly as fairing dynamic pressure and payload aerodynamic heating limitations allow, about 110,000 meters (361,000 feet) and 112 seconds after drop from the aircraft. Stage 2 burnout occurs at approximately 168 seconds and is followed by a long coast, during which the payload and Stage 3 achieves orbital altitude. Stage 3 then provides the additional velocity needed to circularize the orbit. Stage 3 burnout typically occurs approximately 10 minutes after launch and 2,200 kilometers (1,200 nautical miles) downrange of the launch point.[58] Figure 2–19 shows the Pegasus XL mission profile.

---

[58] *Pegasus User's Guide*, Release 5.0, August 2000 (Orbital Sciences Corporation, 2000), p. 2-1, *http://www.orbital.com/NewsInfo/Publications/peg-user-guide.pdf* (accessed February 4, 2005).

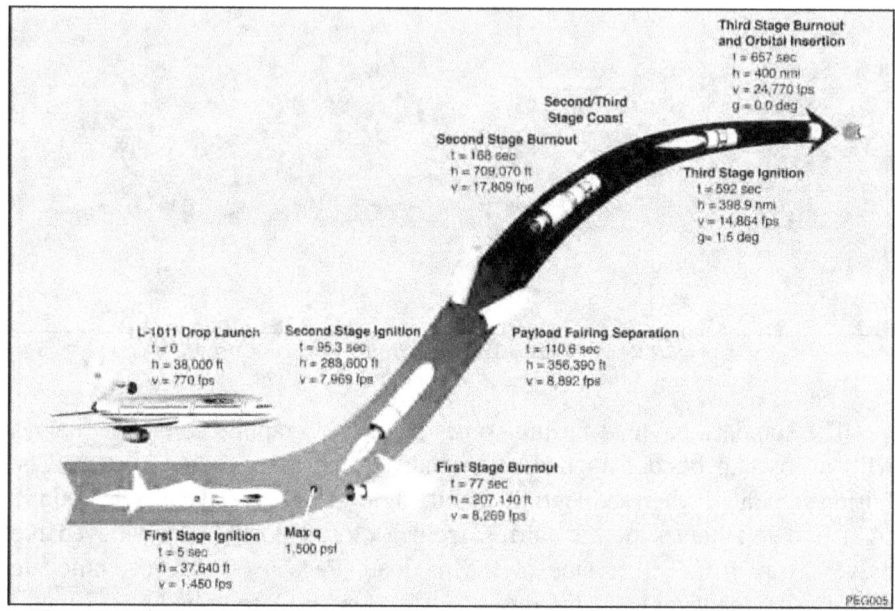

Figure 2–19. Pegasus XL Mission Profile to 741 km (400 nmi) Circular, Polar Orbit with a 227 kg (501 lb) Payload. (Orbital Sciences Corp.)

Scout Launch Vehicle

The standard Scout launch vehicle was a solid propellant, four-stage booster system.[59] It was the world's first all-solid-propellant launch vehicle and one of NASA's most reliable launch vehicles. The Scout was the smallest of the basic NASA launch vehicles. It was used for orbit, probe, and reentry Earth missions. Unlike most of NASA's larger ELVs, the Scout was assembled and the payload integrated and checked-out in the horizontal position. The Scout's first-stage motor was based on an earlier version of the Navy's Polaris missile motor. It's second-stage motor was developed from the Army's Sergeant surface-to-surface missile. The third-stage and fourth-stage motors were adapted by Langley Research Center from the Navy's Vanguard missile.[60] The Scout G1 was the last Scout model. See Table 2–79 for a list of its characteristics.

Since the first Scout launch in 1960, 118 Scout launches had taken place during almost 34 years of service. In the period 1989–1998, six missions successfully launched from Scout ELVs, all from Vandenberg Air Force Base (see Table 2–80). In addition to one NASA payload, Scout also

---

[59] Scout was an acronym for Solid Controlled Orbital Utility Test.

[60] "Scout Launch Vehicle To Retire After 34 Years of Service," *NASA News* Release 94-72, May 6, 1994, *ftp://ftp.hq.nasa.gov/pub/pao/pressrel/1994/94-072.txt* (accessed March 22, 2005). Also "Scout–Launch Vehicle," *http://www.vought.com/heritage/special/html/sscout8.html* (accessed November 4, 2005).

launched DOD payloads. The last Scout launched a military satellite on May 9, 1994. The air-launched Pegasus rocket was considered the operational replacement for the Scout.

Langley Research Center managed the Scout project from its beginning in 1958 until January 1, 1991, when management of the Scout moved to Goddard Space Flight Center. Since 1958, LTV had manufactured the NASA-developed Scout rocket under a series of government contracts that procured flight vehicles in support of NASA science missions. In December 1988, NASA and LTV signed an agreement granting the company exclusive rights to produce and market the Scout commercially. This agreement also enabled LTV to obtain access to and use of Scout launch support facilities at Wallops Flight Facility and at Vandenberg Air Force Base.

Taurus Launch Vehicle

The ground-launched Taurus, developed by Orbital Sciences Corporation, was created by adding the three stages of a Pegasus booster (without the wing and fins) atop a Peacekeeper or Castor 120 first-stage solid-propellant motor referred to as "Stage 0." An aluminum skin and stringer construction interstage extended from the forward skirt of the Castor 120 Stage 0 motor to the aft end of the Stage 1 motor. The lower part of the interstage remained with Stage 0, and the upper part of the interstage flew with the next stage. A field joint between the two sections allowed the Taurus upper stage stack to be mated to the Castor 120 Stage 0 (see Table 2–81 and Figure 2–20).

The Defense Advanced Research Programs Agency (DARPA) contracted with Orbital Sciences in 1989 to build the Taurus rapid response launch vehicle using the Pegasus as a baseline.[61] It was designed for easy transport and as a quick-reaction launch vehicle that could be launched from minimally prepared locations in just a few days.[62] The first Taurus launch took place on March 13, 1994 for a DOD mission. The commercial Taurus, developed after the successful demonstration of the military "ARPA" Taurus, used the Castor 120 first stage rather than the Peacekeeper missile, a slightly larger Orion 50S-G second stage, and a larger fairing.[63] For geosynchronous transfer orbit or deep space missions, the third stage could be replaced by a spin-stabilized Thiokol Star 37 perigee kick motor.

---

[61] Wilson, ed., *Jane's/Interavia Space Directory*, p. 240.

[62] "Taurus," *http://space.skyrocket.de/doc_lau/taurus.htm* (accessed February 9, 2005).

[63] "ARPA" Taurus was another name for the military Taurus configuration that used the Peacekeeper first stage. Isakowitz et al., *International Reference Guide to Space Launch Systems*, 3rd ed., p. 437.

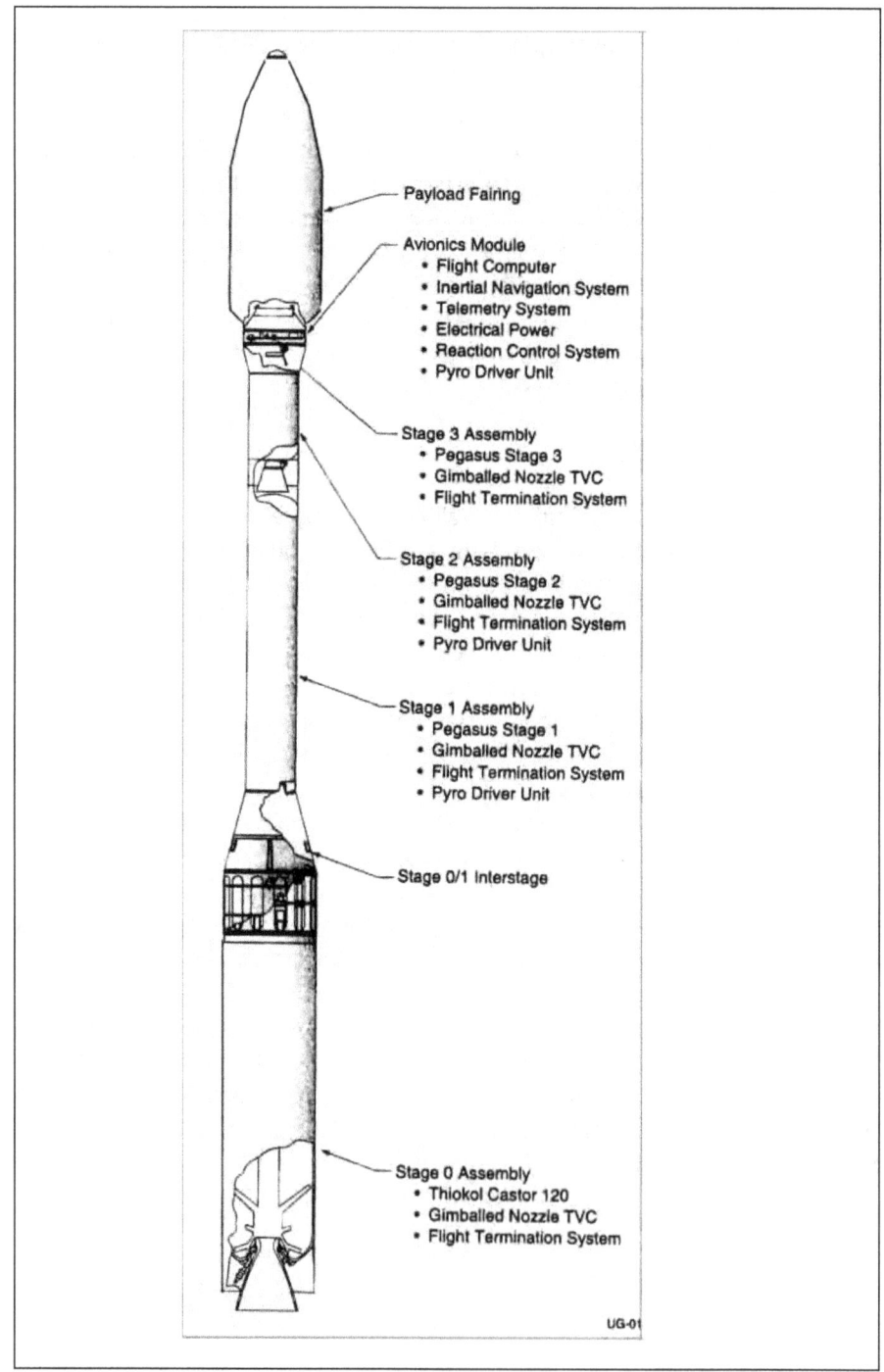

Figure 2–20. Taurus Launch Vehicle Configuration. (Orbital Sciences Corp.)

The Taurus successfully used payload fairings of 63 inches (160 centimeters) and 92 inches (234 centimeters) diameter to encapsulate the payload. Vermont Composites manufactured the 63-inch fairing, while R-Cubed Composites manufactured the 92-inch fairing. Both were bisector shells constructed of graphite/epoxy facesheets with an aluminum honeycomb core. With the addition of a structural adapter, either fairing could accommodate multiple payloads.

From 1994 through 1998, three Taurus launches took place, all from Vandenberg Air Force Base and with multiple payloads (see Table 2–82).

The Titan Family

From the earliest days of the space program, the U.S. Air Force was the primary user of the Titan, first as an intercontinental ballistic missile and later as a space launch vehicle. With its several configurations and enhanced versions, the Titan launched a wide range of military and civilian spacecraft. For a while in the mid-1980s, DOD prepared to launch its payloads exclusively from the Space Shuttle, and it seemed as if the Titan was reaching the end of its useful life. But some felt that a backup vehicle was needed, and in 1985, the Air Force placed an order with Martin Marietta for 10 launch vehicles called the complementary expendable launch vehicle (CELV) or Titan 34D-7, named for its seven-segment solid motor designed for the earlier, unsuccessful Titan IIIM. To be compatible with Shuttle payload capacity, the Titan payload fairing was increased to 5.1 meters (16.7 feet) in diameter. The 34D-7 included a Centaur upper stage and launched exclusively from Cape Canaveral. After the *Challenger* accident and the removal of DOD payloads from the Space Shuttle, the 34D-7 program grew from 10 to 41 vehicles with a mix of upper stages. The vehicles were renamed the Titan IV, and they would eventually be launched from both West and East Coast launch pads.

The Titan IV was the Nation's largest and most powerful ELV. It provided primary access to space for the heaviest and most important DOD and civil payloads. The first Titan IVA launch occurred successfully on June 14, 1989. Eventually, 22 Titan IVAs were launched, the last in August 1998. Figure 2–21 shows the first stage of the Titan IV ELV used to send NASA's Cassini spacecraft to Saturn.

Even before the first Titan IVA launch, the Air Force began looking for ways to upgrade the Titan. In October 1987, the Air Force awarded Hercules Aerospace a contract for upgraded solid rocket boosters that would have a new propellant formulation, new graphite-composite cases, and hydraulically gimbaled nozzles to replace the system used since the first Titan IIIC. The upgraded Titan motors had three segments rather than seven for greater reliability. This upgrade not only increased payload capability by 25 percent but also used fewer components, resulting in a more reliable Stage 0 booster. This model used a more efficient programmable aerospace ground equipment system to control the vehicle before launch and an improved guidance and control system, based on more accurate and lighter ring gyroscopes, manufactured by Honeywell. Mechanical and electrical interfaces to the payload were also standardized, and the design of the core vehicle could be fitted with various kits to adapt to specific payloads. Production processes were redeveloped to use a "factory-to-launch" approach. The goal was to deliver problem-free hardware requiring a minimal amount of launch site assembly and reserving the launch site for final stacking, checkout, countdown, and launch.

Development of the new motors took longer than expected, however, partly because of an explosion during the first test firing. The first flight of the new Titan IV with its new motors, now designated Titan IVB, did not occur until February 23, 1997. The new Titan IV stood 61 meters (200 feet) tall and had a lift capability of 21,680 kilograms (47,796 pounds) to low-Earth orbit and 5,760 kilograms (12,700 pounds) to geosynchronous orbit. Lockheed Martin provided overall program management, system integration, and payload integration for the program. It also built the first and second stages and the Centaur upper stage.[64]

The Air Force found a use, too, for old Titan II ICBMs. As the technology for nuclear deterrence changed, the Air Force began in July 1982 to deactivate its Titan II missiles, removing its last ICBM from its silo in Arkansas on June 23, 1987. In January 1986, the Air Force decided to begin converting some of its deactivated Titan II ICBMs for use as medium-lift space launch vehicles. From its fleet of 54 deactivated Titan IIs, the Air Force selected Martin Marietta to modify 14 for space launches from Vandenberg Air Force Base into polar orbit. Modification entailed replacing the core vehicle's warhead interface with a space payload interface and a 3-meter (9.8-foot) payload fairing and upgrading the electronics, avionics, and guidance systems using Titan III technology. An attitude control system was added for stabilization during the coast phase after second-stage shutdown and before payload separation.[65] Used for launches into polar orbit, the space launch complex at Vandenberg Air Force Base was also modified. The resulting Titan II space launch vehicle was a two-stage, liquid-fueled booster designed to provide a small-to-medium weight class capability. It could lift approximately 4,200 pounds (1,905 kilograms) into polar low-Earth circular orbit.[66] The first launch of a Titan II 23G space launch vehicle took place on September 5, 1988, from Vandenberg Air Force Base when it sent a classified payload into low-Earth orbit.

[64] Isakowitz et al., *International Reference Guide to Space Launch Systems*, 3rd ed., p. 470. Also "Titan," Lockheed Martin, *http://www.lockheedmartin.com/wms/findPage.do?dsp=fec&ci=15525&rsbci=13181&fti =0&ti=0&sc=400* (accessed December 14, 2004).

[65] Art Falconer, "Epic Proportion: The Titan Launch Vehicle," *Crosslink* (Aerospace Corporation, Winter 2002/2003): 35 (NASA History Office Folder 16680). Also Isakowitz et al., *International Reference Guide to Space Launch Systems,* 3rd ed., p. 470.

[66] "Titan II Space Launch Vehicle," Lockheed Martin (NASA History Office Folder 16680).

Figure 2–21. The first stage of the Titan IV ELV that sent NASA's Cassini Spacecraft to Saturn and its moon. Titan is Llowered into a high bay in the Vertical Integration Building at Cape Canaveral Air Station to begin stacking operations, April 14, 1997.
(NASA Photo No. KSC-97PC-640)

During 1989–1998, the Titan launched only a few civilian spacecraft. All but one were converted Titan ICBMs; the final nonmilitary Titan payload during this decade launched NASA's Cassini spacecraft to Saturn on a new Titan IVB Centaur. Table 2–83 lists all Titan launches during this period. Table 2–84 lists Titan II characteristics.

The Space Shuttle

By 1989, regular Space Shuttle flights had resumed, and 66 Shuttle flights took place in the decade from 1989–1998. Because NASA policy dictated that the Space Shuttle could be used for launches only when a human presence was required or when an ELV was not appropriate to deploy a payload, more on-board science missions took place and the Shuttle deployed fewer payloads than in the years before the *Challenger* accident. Among the Shuttle payloads were some of the most important space science projects, including the Hubble Space Telescope, the Galileo spacecraft, and the Gamma Ray Observatory.

A new orbiter, the *Endeavour*, joined the fleet of *Discovery*, *Columbia*, and *Atlantis* and began flight operations on May 7, 1992, when it blasted off on the STS-49 Intelsat VI repair mission. Table 2–85 lists all Space Shuttle missions from 1989 to 1998

In 1995, the Space Shuttle program demonstrated a new capability. In preparation for construction of the International Space Station, the crews of the Space Shuttle carried out a series of docking missions with the Russian Space Station *Mir*. U.S. astronauts lived aboard *Mir*, sometimes for several months at a time, while they acclimated themselves to living and working in space. At the end of the decade, the first Space Station mission took place when STS-88 sent materials for construction of the Station.

In November 1995, in an effort to reduce costs and increase efficiency, NASA announced its intention to pursue a non-competitive contract with the United Space Alliance (USA) that would consolidate contracts for Space Shuttle processing and operations in a single contract. USA was a joint venture between Rockwell International and Lockheed Martin Corporation. Together, these two companies held 69 percent of the dollar value of all Shuttle-related prime contracts. The consolidation virtually ensured that NASA would negotiate with the new company.[67] In April 1996, NASA signed two agreements designating USA the prime contractor for Shuttle processing work performed by Lockheed at Kennedy Space Center and Shuttle operations work performed by Rockwell at Johnson Space Center.

In September 1996, NASA entered into a contract with USA as the prime contractor for Space Shuttle and International Space Station activities to ensure that all NASA missions were successfully accomplished according to the applicable flight definition and requirements, schedule, and implementation plan. The original six-year contract ran from October 1996 through September 2002 and consisted of two phases for consolidating the existing prime contracts. During the first phase, USA assumed overall responsibility for the fleet of orbiters. During the second phase, which began in September 1997, the contracts for Kennedy Space Center base operations, the waste collection system, flight software, flight equipment, and solid rocket boosters were

---

[67] "NASA To Pursue Non-Competitive Shuttle Contract With U.S. Alliance," *NASA News* Release 95-205, November 7, 1995, *ftp://ftp.hq.nasa.gov/pub/pao/pressrel/1995/95-205.txt* (accessed April 17, 2005).

consolidated in the USA contract. As of early 2000, the remaining prime contracts—external tanks, Space Shuttle main engines, and reusable solid rocket motors—remained to be consolidated.[68]

Space Shuttle Characteristics

The Space Shuttle that NASA flew in the decade beginning in 1998 consisted of four primary elements: an orbiter spacecraft, two SRBs, an external tank to house fuel and an oxidizer, and three Space Shuttle main engines (SSMEs). Rockwell International built the orbiters. Rockwell's Rocketdyne Division built the main engines.[69]

Thiokol Corporation produced the SRB motors/ Martin Marietta Corporation built the external tank. Johnson Space Center directed the orbiter and integration contracts, while Marshall Space Flight Center managed the solid rocket booster, external tank, and Space Shuttle main engine contracts.[70] Rockwell also was the contractor for Space Shuttle operations at Johnson Space Center that included maintenance and operation of Space Shuttle facilities, flight preparation, and sustained engineering support. Lockheed Martin was responsible for Shuttle processing at Kennedy Space Center.

External Tank

The external tank held the liquid hydrogen fuel and liquid oxygen oxidizer in separate pressurized tanks and supplied them under pressure to the three main engines in the orbiter during liftoff and ascent. The main engines consumed approximately 64,000 gallons (242,266 liters) of fuel each minute. When the main engines were shut down, the external tank was jettisoned into Earth's atmosphere where it broke up and fell into a remote ocean area. The external tank was not recovered. When loaded with fuel, the external tank was the largest and heaviest element of the Space Shuttle. Built from aluminum, it also acted as the backbone for the orbiter and solid rocket boosters. The external tank was composed of three major components: the forward liquid oxygen tank, an unpressurized intertank containing most of the electrical components, and the aft liquid hydrogen tank. Characteristics of the external tank are shown in Table 2–86. Figure 2–22 shows a cutaway drawing.

---

[68] NASA Office of Inspector General, *Audit Report: Space Flight Operations Contract Phase II–Cost-Benefit Analysis*, IG-00-015, National Aeronautics and Space Administration (March 14, 2000), pp. 1–2.

[69] In December 1996, Boeing purchased the Space and Defense divisions of Rockwell International and renamed them Boeing North American. Rocketdyne had been part of Rockwell when the SSME contract was awarded. It was bought by Boeing in December 1996 when Boeing bought Rockwell. Rocketdyne became the Rocketdyne Division of Boeing North American.

[70] Detailed descriptions of all Space Shuttle components can be found in the *NSTS 1988 News Reference Manual*, September 1988, at *http://science.ksc.nasa.gov/shuttle/technology/sts-newsref/stsref-toc.html#srb-recovery* (accessed February 25, 2005) and in the *Shuttle Crew Operations Manual*, SFOC-FL0884, Rev. B, CPN-3, January 13, 2003. See also a summary in Judy Rumerman, compiler, *NASA Historical Data Book, 1979–1988, Volume V* (Washington, DC: National Aeronautics and Space Administration Special Publication 4012, 1999), pp. 33–47 and pp. 123–147. Also available at *http://history.nasa.gov/SP-4012/vol5/cover5.html*.

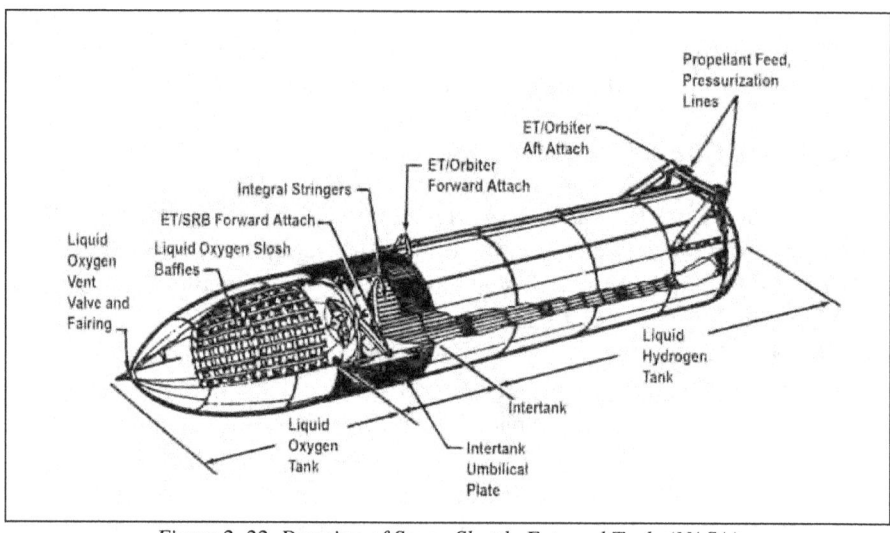

Figure 2–22. Drawing of Space Shuttle External Tank. (NASA)

Solid Rocket Booster

The solid rocket boosters were the largest solid-propellant motors ever flown and the first designed for refurbishment and reuse. The two boosters provided the main thrust to lift the Space Shuttle up off the launch pad to an altitude of about 150,000 feet (45.7 kilometers) or 24 nautical miles. The two solid rocket boosters carried the entire weight of the external tank and orbiter and transmitted the weight load through their structure to the mobile launcher platform. The solid rocket boosters were ignited after the thrust levels of the three main engines were verified. During flight, the solid rocket booster nozzles swiveled up to 6 degrees, redirecting the thrust and steering the Shuttle toward orbit. Seventy-five seconds after booster separation, SRB apogee occurred at an altitude of approximately 220,000 feet (67 kilometers) or 35 nautical miles. Impact in the Atlantic Ocean occurred approximately 122 nautical miles (226 kilometers) downrange. Table 2–87 lists solid rocket booster characteristics. Figure 2–23 shows an exploded view.

Space Shuttle Main Engine

The three Space Shuttle main engines were clustered at the tail end of the orbiter. These high-performance liquid-propellant engines were the world's first reusable rocket engines with each designed to operate for 7.5 hours over a lifespan of 55 starts. They operated with variable thrust levels in a staged combustion cycle. The engines burned liquid hydrogen as the propellant fuel and liquid oxygen as the oxidizer in a 6:1 ratio. The propellant was carried in separate tanks in the external tank and supplied to the main engines under pressure. The main engines could be throttled over a range of 65 percent to

109 percent of their rated power level in 1 percent increments. A value of 100 percent thrust corresponded to a thrust level of 375,000 pounds (1,668 kilonewtons) at sea level and 470,000 pounds (2,091.7 kilonewtons) in a vacuum. A thrust value of 104 percent (called full power) was typically used at launch, although each engine could be throttled to its maximum of 109 percent if necessary. (This power level has never been used on a Shuttle flight). All three engines received the same throttle command at the same time, normally from the orbiter general-purpose computers, although manual control of engine throttling was possible during certain contingency situations.

Firing of the three main engines began 6.6 seconds before launch. The three engines were fired at intervals of 120 milliseconds. If all three engines failed to reach at least 90 percent thrust over the next 3 seconds, a main engine cutoff command was issued automatically, followed by cutoff of all three engines. If launch proceeded normally, the engines were throttled back about 26 seconds after launch to protect the Shuttle from aerodynamic stress and excessive heating. The engines returned to full power about 60 seconds after launch and typically continued at full power for about 8.5 minutes until shortly before the Shuttle entered orbit. At about 7 minutes, 40 seconds after launch, the engines were throttled down so the vehicle and crew were not subject to forces more than 3g. The main engines operated in parallel with the solid rocket boosters during the initial ascent. After the boosters separated, the main engines continued to operate. During ascent, each engine could be gimbaled plus or minus 10.5 degrees around the yaw and pitch axes to help steer the Shuttle.[71]

The Shuttle's main engines were upgraded twice during this decade. The Block 1 SSMEs first flew on STS-70 on July 13, 1995. These engines used a new high-pressure liquid oxidizer turbopump that increased safety margins and the reliability of the Shuttle's main engines. In 1998, the Block IIA SSMEs were first used on STS-95. These upgrades increased safety and reliability and simplified manufacturing and maintenance.[72] The design had a larger nozzle throat that resulted in decreased operating pressure and temperature. To achieve the same performance as the earlier engines, the Block IIA engines typically operated at 104.5 percent thrust at launch. Figure 2–24 shows the SSME components. Table 2–88 lists SSME characteristics.

[71] David Darling, "Space Shuttle," *The Encyclopedia of Astrobiology, Astronomy, and Spaceflight, http://www.daviddarling.info/encyclopedia/S/Space_Shuttle.html* (accessed February 28, 2005).
[72] Susie Unkeless, Jack Vautin, Boeing Rocketdyne, telephone conversation, February 28, 2005. Also "STS-95 Space Shuttle Mission Chronology," *http://www-pao.ksc.nasa.gov/kscpao/chron/sts-95.htm* (accessed February 28, 2005).

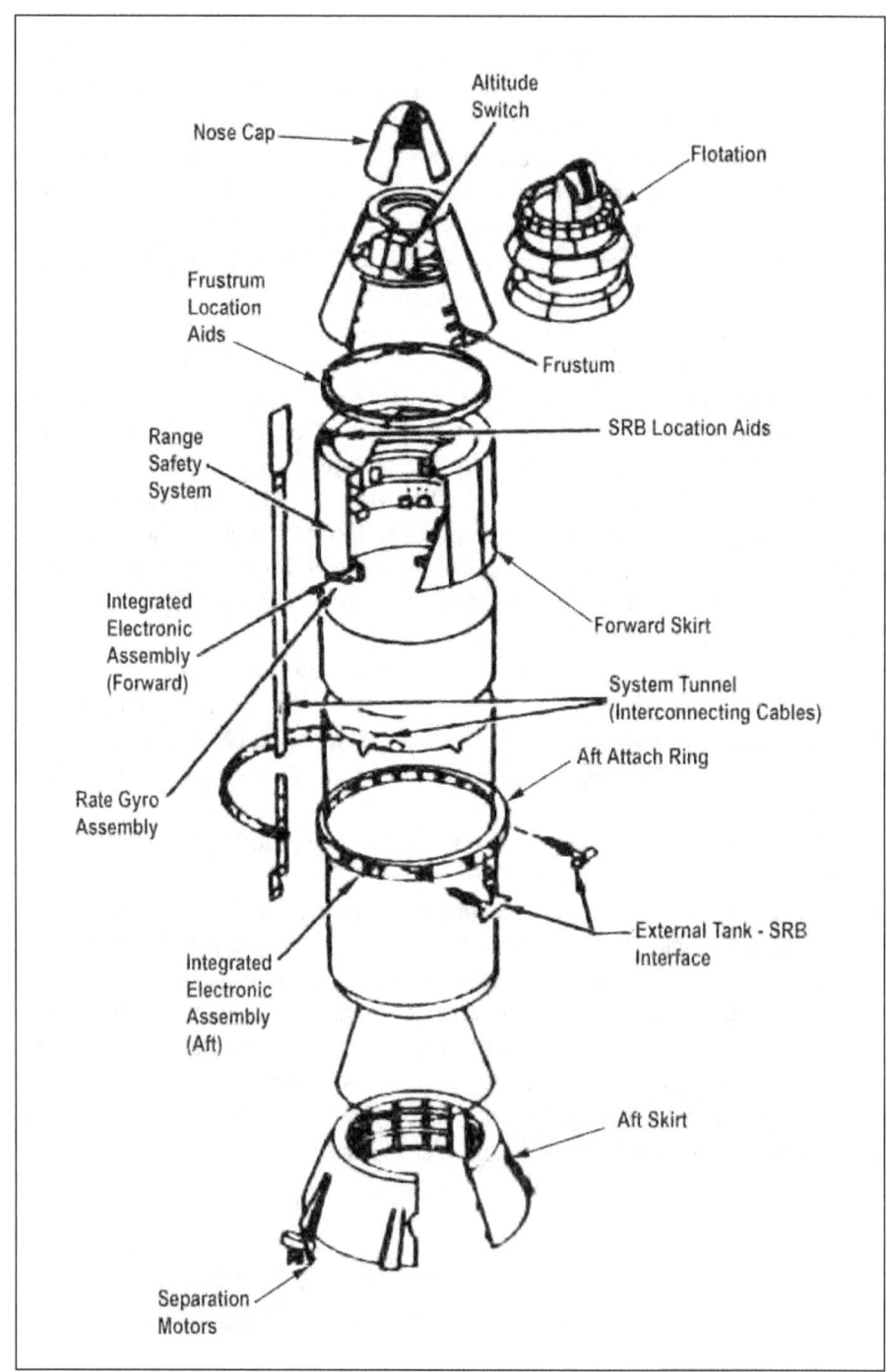

Figure 2–23. Exploded View of Space Shuttle Solid Rocket Booster.

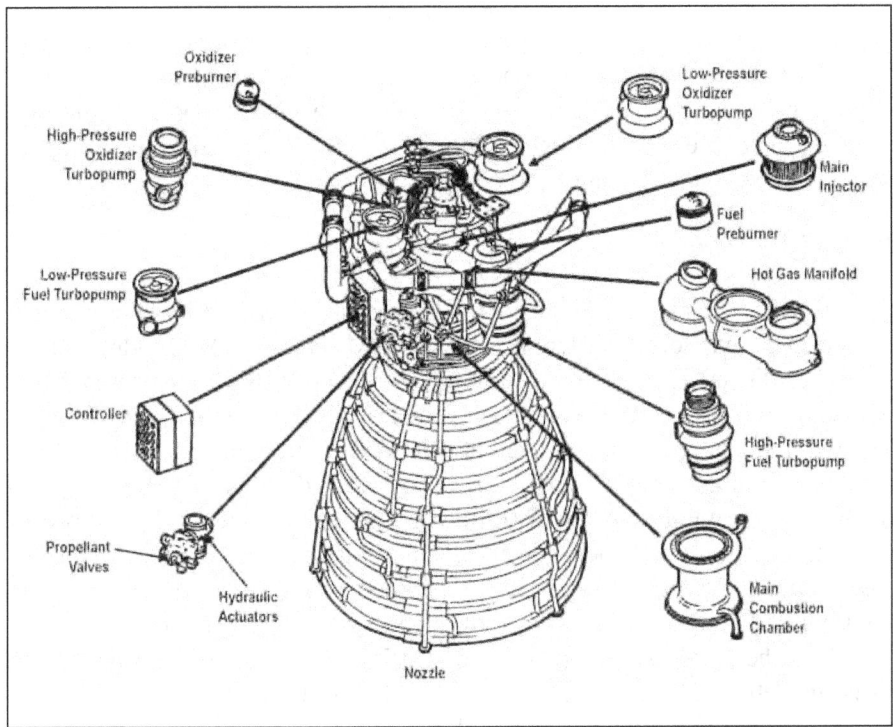

Figure 2–24. Space Shuttle Main Engine Components. (Rocketdyne)

Two orbital maneuvering system engines, mounted on either side of the upper aft orbiter fuselage, provided thrust for major orbital changes. For more precise motions in orbit, 44 small rocket engines, clustered on the Shuttle's nose and on either side of the tail, were used. Together, they were known as the reaction control system and helped Shuttle astronauts retrieve, launch, and repair satellites in orbit.

Launch and Operations

All Space Shuttle missions launched from Kennedy Space Center. The majority also landed there although, during this decade, about 39 percent landed at Edwards Air Force Base in California. See the individual Shuttle mission tables in chapter 3 for specific launch and landing information.

All satellites released from a Space Shuttle initially entered low-Earth orbit. Some remained in that orbit throughout their working lives. Many spacecraft, however, operated in geosynchronous orbit, approximately 35,790 kilometers (19,325 nautical miles or 22,300 miles) above Earth and aligned with the equator, with a speed in orbit that matched the speed of Earth's surface below. Spacecraft reached this altitude by firing an upper stage, an attached propulsion unit such as an IUS or PAM, after deployment from the Shuttle orbiter. Interplanetary explorers launched from the Space Shuttle also used an IUS. They left low-Earth orbit on trajectories that sent them out into our solar system and beyond.

Upper Stages

Upper stages were used to boost ELV and Shuttle payloads from a low-Earth orbit to geostationary transfer orbit, geosynchronous orbit, or into an interplanetary trajectory. During 1989 to 1998, NASA used three types of upper stages: the PAM, the Centaur Upper Stage, and the IUS.

Payload Assist Module

The PAM was designed to boost satellites deployed in low-Earth orbit into a higher operational orbit. Different types of PAMs were used depending on the weight of the satellite it needed to boost. A special PAM, known as PAM-D, was adapted for use with Delta launch vehicles. The PAM-DII was designed to boost Shuttle payloads into an elliptical transfer orbit after a satellite was deployed from the Shuttle's cargo bay. A specially designed PAM-S was used on the Ulysses mission for the first time in combination with an IUS to propel the spacecraft toward Jupiter. The PAM's expendable stage consisted of a spin-stabilized, solid-fueled rocket motor; a payload attach fitting to mate with the satellite; and timing, sequencing, power, and control assemblies.[73] The first launch of the PAM as the top stage of a Delta took place in 1980. The PAM made its debut flight from the Space Shuttle in 1982. Figure 2–25 shows the Ulysses spacecraft with the PAM and IUS.

Centaur Upper Stage

The Centaur was a powerful, liquid-propellant rocket—this country's first high-energy, upper-stage launch vehicle. It was developed under the direction of Lewis Research Center in the 1960s and assembled by General Dynamics. It used a liquid hydrogen-liquid oxygen propellant combination in two restartable Pratt & Whitney RL10 engines that produced more thrust for each pound of propellant burned per second than rockets using only kerosene-based hydrocarbon fuels (see Figure 2–26). The rocket was first developed to be used with the Atlas ELV, and in the decade from 1989–1998, was used on almost all Atlas launches. In the 1970s, the Centaur had been combined with the Titan III to launch larger spacecraft. Later, NASA had planned to use the Centaur to boost Shuttle payloads into higher orbits. But with the increased emphasis on safety following the *Challenger* accident, NASA determined that even with modifications, it was too dangerous to carry a liquid-propellant rocket inside a crewed spacecraft. In June 1986, the Shuttle/Centaur program was cancelled, eliminating the Centaur for use on the Shuttle.

---

[73] "Space Transportation System Payloads: Payload Assist Module," *http://science.ksc.nasa.gov/shuttle/ technology/sts-newsref/carriers.html* (accessed March 17, 2005).

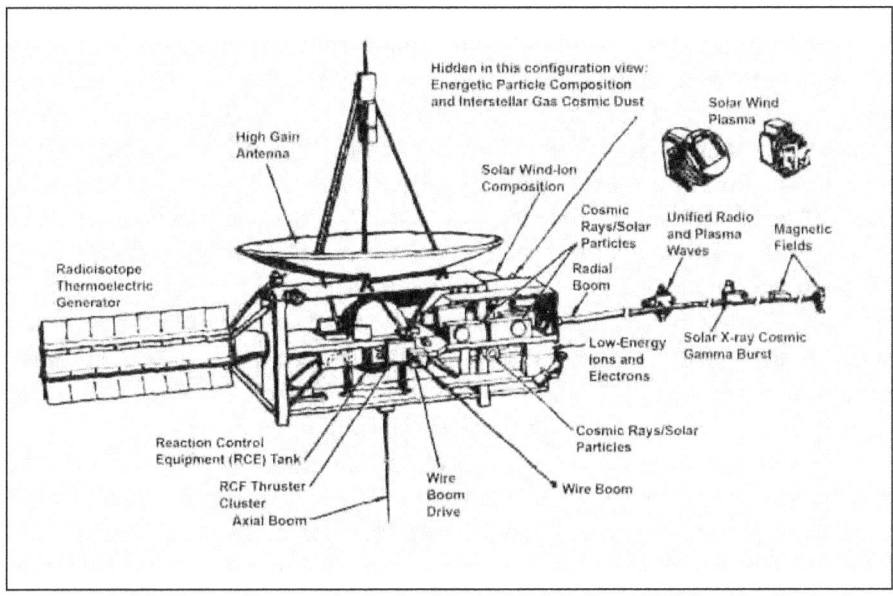

Figure 2–25. Ulysses Sits atop the Payload Assist Module-S and IUS Combination in the Vertical Processing Facility at Kennedy Space Center. (NASA/JPL-Caltech)

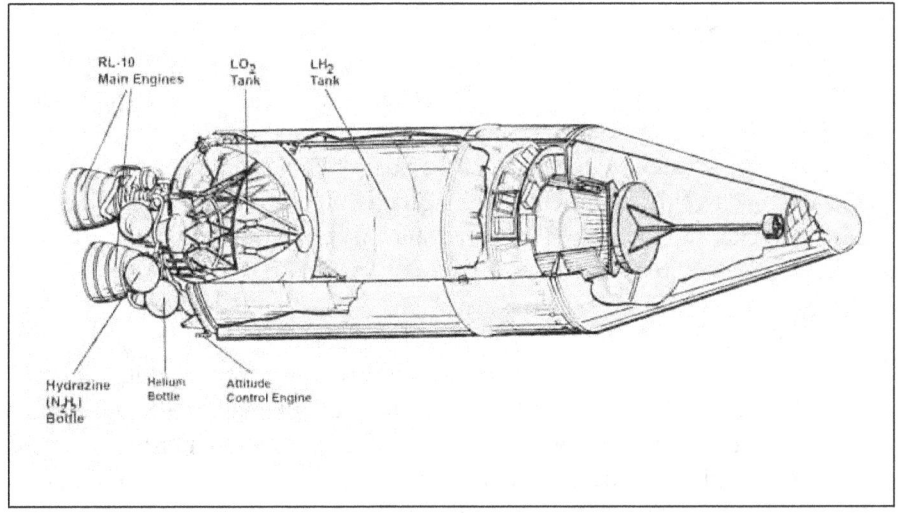

Figure 2–26. Pratt & Whitney RL10 Engine used on the Centaur Upper Stage. (Pratt & Whitney)

In the wake of the *Challenger* accident, the country's space program returned to using ELVs for all missions suitable for those launchers, and attention focused on improving the Centaur for use as an ELV upper stage. As General Dynamics began developing new Atlas launch vehicles in the late 1980s, it also improved the Centaur as its "engineers dusted off NASA studies for Centaur improvements never implemented . . . ." A pressure-fed system replaced its boost pumps, reducing complexity and cost, and its avionics system was upgraded. Designers developed a new 14-foot (4.3-meter)-diameter metal nose fairing to accommodate larger payloads. A computer-controlled pressurization system with redundant sensors was adopted, making the system more versatile and reliable.[74] This updated Centaur was first used on July 25, 1990, for the Atlas I launch of the Combined Release and Radiation Effects Satellite. See Figure 2–27 for a diagram of the Atlas-Centaur upper stage.

The Centaur G model, originally developed for launching Shuttle payloads, was updated for use on the Titan IV. Martin Marietta delivered the first Titan IV Centaur in December 1990. It had a bare metal tank, like that of the Atlas Centaur. However, the upper fuel tank was stretched to 5 meters (16.4 feet) in diameter. Most Titan IV Centaurs were used for classified DOD payloads. However, one model launched the Cassini/Huygens probe for NASA on an interplanetary trajectory in October 1997. Characteristics of the Titan Centaur are given in Table 2–89.

Inertial Upper Stage

The two-stage, solid-fueled IUS delivered a satellite to a high-stage operational orbit or to an escape trajectory for an interplanetary mission from low-Earth orbit. It extended the reach of the Space Shuttle and was also used with the Titan launch vehicle, particularly the Titan 34D and Titan IV. The IUS had two solid rocket motors, an aft skirt, an interstage, and an equipment support section where the avionics were located. It could lift 5,000 pounds (2,268 kilograms) from low-Earth to geosynchronous orbit. Figure 2–28 shows an IUS being attached to the Magellan spacecraft, which launched from STS-30 in 1989.

In a typical Titan IV-IUS launch into geosynchronous orbit, the IUS separated from the Titan's second-stage booster approximately 9 minutes after launch. Then, for the next 6 hours, 54 minutes, the IUS autonomously performed all functions to place the payload into its proper orbit. The first IUS rocket burn, which placed the payload into geosynchronous transfer orbit, occurred a little more than 1 hour into the IUS booster flight. The IUS second solid rocket motor ignited about 6.5 hours into the flight, followed by a coast phase, and then separation of the payload from the IUS after placing it into geosynchronous orbit.

---

[74] Virginia P. Dawson and Mark D. Bowles, *Taming Liquid Hydrogen: The Centaur Upper Stage Rocket, 1958-2002* (Washington, DC: National Aeronautics and Space Administration Special Publication-2004-4230, 2004), pp. 242–243.

Figure 2–27. Atlas-Centaur Upper Stage. (NASA-GRC Photo No. C-1998-02814)

In a typical Shuttle-IUS launch to geosynchronous orbit, after reaching low-Earth orbit, the Shuttle opened its payload doors and the IUS tilted outward, extending the IUS and its payload into space. After satellite and IUS checkout, the Shuttle astronauts ejected the IUS and its cargo from the orbiter. The IUS onboard computers then directed a series of maneuvers and fired the first-stage motor for approximately 140 seconds to propel the IUS and spacecraft toward the desired geosynchronous position. After a coast period of several hours, the second-stage motor ignited and burned for approximately 100 seconds, injecting the IUS into a final circularized orbit. The IUS then separated from the satellite and moved to a position where it neither collided with nor contaminated the satellite.[75] Table 2–90 lists IUS characteristics. Table 2–91 lists missions using an IUS.

Advanced Programs and Projects

Advanced Programs conducted studies and selected development efforts to support potential new programs, system improvements, and expanded capabilities for space transportation systems. The objectives were to increase reliability, cost effectiveness, and capability of spaceflight systems; continue enhancing crew safety for the Space Shuttle and Space Station; implement flight and ground systems improvements to substantially reduce the cost of spaceflight operations; and pursue technology developments to meet future human spaceflight requirements. Development efforts focused on advanced transportation, advanced operations, and satellite servicing. The two program elements were advanced operations and advanced space systems.

Tethered Satellite System

The Tethered Satellite System (TSS) program was a cooperative effort between the government of Italy and NASA. The TSS program was to enable science to be performed in the upper atmosphere and ionosphere from a satellite connected to the Space Shuttle by means of a tether up to 100-kilometers (62-miles) long. The effect of the tether passing through space also was expected to generate an electric current that could be conducted to the orbiter.

The first attempt at performing the TSS experiment took place on STS-46, launched on July 31, 1992. Due to problems with the deployment mechanism, the *Atlantis* crew could deploy the tethered satellite only 256 meters (840 feet) instead of the goal of 20 kilometers (12.4 miles).

---

[75] "Inertial Upper Stage: IUS Team," Boeing, *http://www.boeing.com/defense-space/space/ius/ius_team.htm* (accessed March 18, 2005).

Figure 2–28. The Magellan Spacecraft with its attached Inertial Upper Stage (IUS) Booster is in the Orbiter Atlantis *Payload Bay prior to closure of the doors at T-3 days to launch. Launch of Magellan and STS-30 took place on May 4, 1989. (KSC Photo No. 89PC-0469)*

A second attempt to deploy a tethered satellite from the Shuttle was made on STS-75, launched February 22, 1996. This second attempt also failed to satisfy all the mission objectives. Although the tether unreeled smoothly for almost its full length of 13 miles (21 kilometers), the 0.1-inch (0.25-centimeter)-diameter tether broke about three-fourths of a mile before reaching full length, and the Italian satellite drifted away. There was a low-power current of 3,500 volts and 500 milliamps generated by the unreeling tether in Earth's magnetic field, which satisfied the test's scientific objective.[76]

---

[76] David M. Harland, *The Story of the Space Shuttle* (Chichester, UK: Springer, Praxis Publishing, 2004), pp. 123, 137. Also Dennis R. Jenkins, *Space Shuttle: The History of the National Space Transportation System, The First 100 Missions* (Cape Canaveral, FL: Dennis R. Jenkins, 1996), p. 309.

Figure 2–29. First Test Flight of the Delta Clipper-Experimental Advanced (DC-XA), May 18, 1996. (NASA-MSFC Photo No. MSFC-9607854)

A series of less complex tethered experiments took place in 1993 and 1994 from Delta II launch vehicles. They were designed to complement to the TSS deployer when retrieval of the tether was not required. The Small Expendable Deployer System (SEDS-1) was the first of three tether experiments managed by NASA's Office of Space Systems Development Flight Demonstration Program. These experiments were more successful than the Shuttle-based attempts.

On March 29, 1993, about 63 minutes after launch, the first SEDS diagnostic payload was ejected from the Delta by springs. The tether was deployed at an altitude of 720 kilometers (447 miles) above Earth and pointing toward Earth. The tether unreeled smoothly for its full 20-kilometer (12.4-mile) length, and sensors recorded its damping motion for 14 minutes. Then the tether was cut and its 30-kilogram (66-pound) cargo floated away, ending the experiment. A second successful experiment took place from a Delta on June 26, 1993. Called the Plasma

Motor Generator, this experiment assessed the ability of a space tether to generate an electric current. The tether generated about 0.3 amp of current as it moved through Earth's magnetic field.[77] A third experiment took place on March 11, 1994, when the SEDS-2 payload unreeled to its maximum length of 19.8 kilometers (12.3 miles) in 1 hour and 48 minutes. Unlike SEDS-1, when the tether was severed, the tether on this mission remained attached to the Delta rocket, and the payload at the end of the tether transmitted for 10 hours until its battery died.[78]

Shuttle-C

Shuttle-C (cargo) was a concept for a large, uncrewed launch vehicle with a cargo canister in place of the orbiter that would make maximum use of existing Space Shuttle systems. This proposed cargo-carrying launch vehicle would be able to lift approximately 100,000 pounds to 170,000 pounds (45,300 kilograms to 77,000 kilograms) to low-Earth orbit, two to three times the capability of the Shuttle's orbiter. It could reduce by 50 percent the number of launches and length of assembly time for Space Station components. It could also carry scientific spacecraft into orbit. The vehicle would use the same type of external tank, solid rocket boosters, and main engines as the crewed Space Shuttle. Although the U.S. House Subcommittee on VA-HUD-Independent Agencies authorized $1.1 billion on a heavy-lift space cargo vehicle in FY 1991, the vehicle never moved past the study stage and was replaced by the National Launch System, another short-lived initiative.

Advanced Launch System

The Advanced Launch System (ALS) was a joint NASA-DOD program of the late 1980s that was a product of the Strategic Defense Initiative (SDI) "Star Wars" ballistic missile defense system. The program was to define concepts and develop technology for a family of uncrewed launch vehicles that would reduce the cost of putting payloads into orbit. Initially, it was projected that new heavy-lift launchers would be needed to deploy payloads of 10,000 pounds to 200,000 pounds (4,500 kilograms to 90,700 kilograms) to low-Earth orbit for the space-based elements of the SDI program. Three basic booster concepts were proposed. The least costly vehicle would use a hydrogen core and from 6 to 12 solid rocket boosters. A more costly vehicle used a liquid core and from one to six strap-on liquid rocket engines. The most expensive and most advanced ALS alternative, and the least likely because of technological uncertainty, was a winged, fully reusable booster.[79] NASA managed development of the advanced liquid cryogenic propulsion system and the advanced development program.

[77] Joel W. Powell, "Satellite Tethers Unwind," *Spaceflight*, 36 (March 1994): 97–99.

[78] "Jonathan's Space Report," no. 188 (March 14, 1994), *http://www.planet4589.org/space/jsr/back/news.188* (accessed March 21, 2005).

[79] Mark Cleary, "Future Space Operations: The Space Transportation Architecture Study and Advanced Launch System (ALS) Studies," *Military Space Operations, 1971–1992*, 45th Space Wing History Office, *https://www.patrick.af.mil/heritage/Cape/Cape4/cape4-2.htm* (accessed March 24, 2005).

However, by late 1989, the Cold War was waning, and the SDI initiative was greatly reduced in scope. In October 1989, funding cutbacks shifted emphasis to lightweight weapons, which reduced payload requirements dramatically. On December 7, 1989, the Secretary of the Air Force directed the ALS program office to terminate design efforts "as soon as possible" and suspend any new spending. The three ALS contractors, Boeing, General Dynamics, and a Martin Marietta-McDonnell Douglas team, were directed to transfer ALS technology to the existing fleet of ELVs, which all stood to benefit from technologies developed for the ALS program.[80] In January 1990, the program was downscaled to focus on propulsion technologies, particularly the Space Transportation Main Engine, although little funding was provided to pay for the project at the time.

National Launch System

One of the recommendations in the December 1990 report of the Advisory Committee on the Future of the U.S. Space Program, headed by Norman Augustine, was that the U.S. Space Program end reliance on the Shuttle. It stated that the Administration should provide funds for a "firm program for development of an evolutionary, unmanned but man-rateable, heavy-lift launch vehicle" that "should reach operational capability in time to support all but the initial phase of the Space Station deployment."[81] On January 2, 1991, Vice President Daniel Quayle directed NASA and the DOD to jointly develop a plan for a new space launch system. On April 16, 1991, the Vice President and the National Space Council directed the two organizations to "pursue the development of a new space launch system with the objective of achieving significant improvements in reliability, responsiveness, and operational efficiency."[82] This plan would meet civil and military space needs and actively consider commercial space requirements; costs would be shared equally by NASA and the Defense Department.[83]

The ensuing program, the National Launch System, sometimes called the New Launch System, replaced and combined elements of the previous Advanced Launch System and NASA's Shuttle-C programs. As stated by NASA, program goals were to: 1) develop a modular launch system with a medium-lift to heavy-lift capability, 2) facilitate evolutionary changes as requirements evolved into the 21st century, 3) use existing components from the Shuttle and ELVs to expedite initial capability and reduce development costs,

---

[80] Cleary, *Military Space Operations,*" *https://www.patrick.af.mil/heritage/Cape/Cape4/cape4-2.htm*. Also, Department of Defense and National Aeronautics and Space Administration *National Space Launch Program Report to Congress,* (March 14, 1989), p. 20.

[81] Advisory Committee on the Future of the U.S. Space Program, "Report of the Advisory Committee on the Future of the U.S. Space Program," December 17, 1990, *http://www.hq.nasa.gov/office/pao/History/augustine/racfup1.htm* (accessed March 15, 2005).

[82] *Aeronautics and Space Report of the President, Fiscal Year 1992 Activities*, p. 18.

[83] David N. Spires and Rick W. Sturdevant, "Epilogue: 'To the Very Limit of Our Ability,'" in Launius and Jenkins, *To Reach the High Frontier*, p. 488.

and 4) develop a system that, while being uncrewed initially, could be "man-rated" in the future.[84] The proposed heavy-lift space cargo vehicle would support the logistics requirements of Space Station *Freedom*. Evolution of vehicles that could support the Nation's return to the Moon and mission to Mars was also envisioned.[85] NASA's Office of Space Systems Development managed the program.

In August 1991, NASA awarded study contracts for the NLS, each valued at $500,000, to Lockheed Missiles and Space, McDonnell Douglas, and TRW Inc. The NASA FY 1992 budget request for the NLS increased the estimate from $23.9 million for the ALS and Shuttle-C to $175 million for NASA's share of the funding. It was anticipated that activities in FY 1992 would focus on beginning development of the Space Transportation Main Engine (STME) prototype, conducting definition and design studies of vehicle components and elements, and assessing requirements and design options for supporting launch facilities.[86]

Initially, the system comprised three different-sized launch vehicles with varying payload capacities to low-Earth orbit. They would be derived from a common core element consisting of the Space Shuttle external tank and a new STME. In 1992, NASA eliminated the largest rocket from the original three when a study determined that the needed modular family of vehicles should span the medium launch vehicle class up to a booster capable of supporting the Space Station's resupply missions. One of the proposed vehicles would be able to deliver 50,000 pounds (22,680 kilograms) to low-Earth orbit; the second, smaller vehicle, could deliver 20,000 pounds (9,000 kilograms).

The program continued into 1992, although funding for FY 1993 was reduced by $137 million consistent with a first launch in 2002. The remaining $28 million was earmarked to support development of the STME. However, in early 1993, the program was terminated, and no funding was included in the FY 1994 budget.

Reusable Launch Vehicles

Developing an RLV, either to supplement or replace the Space Shuttle, received a great deal of attention and significant resources during the decade beginning in 1989. The National Aerospace Plane (the X-30), a program supported strongly by President Ronald Reagan, had been initiated in 1982 as a DARPA project. Planned as a new reusable, air-breathing, single-stage-to-orbit

---

[84] "New Launch System," NASA Fact Sheet, National Aeronautics and Space Administration, Marshall Space Flight Center, August 29, 1991 (NASA History Office Folder 010274). Also, "National Launch System–NLS," FAS Space Policy Project, Military Space Programs, *http://www.fas.org/spp/military/program/launch/nls.htm* (accessed March 24, 2005).

[85] "NASA Awards Study Contracts for National Launch System," *NASA News* Release C91-gg, August 16, 1991 (NASA History Office Folder 010274).

[86] "New Launch System," *National Aeronautics and Space Administration FY 1992 Budget Estimate*, pp. RD 2-18–2-19.

hypersonic vehicle, the X-30 became a joint NASA-DARPA program in 1985. Although the project produced some important technological advances, it became too costly in a time of competing priorities, and the program was cancelled in 1994 while still in the technology development phase.

NASA Administrator Goldin joined NASA in April 1992, a time when the Shuttle and other NASA programs were under attack from Congress for their high costs. Taking advantage of the change in presidential administrations in 1993, and also to put his mark on the Agency, Goldin initiated the "Access to Space" study to identify alternative, less expensive approaches to gain access to space that would also increase safety for flight crews. Released in January 1994, the study report was followed later that year by the first executive policy specifically recommending development of an RLV. On August 5, 1994, President William Clinton issued the National Space Transportation Policy making NASA "the lead agency for technology development and demonstration of next generation reusable space transportation systems," while the DOD was given responsibility for improving ELVs.[87] The policy statement led directly to the formation of NASA's RLV Technology program.

NASA's RLV Technology program was a partnership among NASA, the U.S. Air Force, and private industry to develop a new generation of single-stage-to-orbit launch vehicles. The program consisted of the Delta Clipper-Experimental Advanced (DC-XA), X-34, X-33, and related long-term technology development efforts. RLV program managers committed themselves to developing new operations and component technologies, as well as producing an industry-Government relationship that would change the space launch industry worldwide.

DC-X

The Delta Clipper-Experimental (DC-X) program, initiated by the Ballistic Missile Defense Organization (BMDO) in 1990, supported NASA's RLV program. It successfully tested an experimental suborbital launch vehicle in a series of flight tests beginning in 1993. The early RLV efforts were conducted by the U.S. Air Force Phillips Laboratory at Kirtland Air Force Base, New Mexico, under the auspices of the BMDO Single Stage Rocket Technology program. This program's charter was to demonstrate the practicality, reliability, operability, and cost efficiency of a fully reusable rapid turnaround single-stage rocket, with the ultimate goal of aircraft-like operations of RLVs. The program focused on using existing technologies and systems to demonstrate the feasibility of building RLVs for suborbital and orbital flight that could fly into space, return to the launch site, and be serviced and ready for the next mission within three days.

---

[87] The White House, Office of Science and Technology Policy, Presidential Decision Directive, National Science and Technology Council-4, *National Space Transportation Policy*, August 5, 1994, *http://www.au.af.mil/au/awc/awcgate/nstc4.htm* (accessed March 20, 2005).

A design and risk reduction competition awarded McDonnell Douglas a $60 million contract in August 1991 to build the DC-X. The DC-X design emphasized simplified ground and flight operations and vehicle maintenance, rapid turnaround, and operational characteristics also relevant to future orbital vehicles. Table 2–92 lists its characteristics.

The flight test program took place in mid-1993. It started with low-altitude hover flights gradually increasing in altitude and duration and eventually leading to suborbital flights to approximately 18,000 feet (5,486 meters). The DC-X flew a total of eight test flights in 1993, 1994, and 1995; the 1995 flights supported NASA's RLV program. The test flight on June 27, 1994, experienced an on-board fire and successfully demonstrated the vehicle's autoland capabilities. On the July 7, 1995, flight, following a successful flight that demonstrated the vehicle's ability to turn itself around and reverse direction, the aeroshell cracked during landing, damaging the vehicle and ending the tests. At the conclusion of this test, the DC-X was officially turned over to NASA. The vehicle was returned to McDonnell Douglas for conversion into the DC-XA.[88]

The DC-XA was a modified DC-X with technology intended for use in the X-33 or X-34 RLVs being developed by NASA and industry partners. The DC-XA had a lightweight graphite-epoxy liquid hydrogen tank and an advanced graphite/aluminum honeycomb intertank built by McDonnell Douglas; an aluminum-lithium liquid oxygen tank built by Energia; and an improved reaction control system from Aerojet. These improvements reduced dry vehicle mass by 620 kilograms (1,367 pounds). NASA and the DOD operated the DC-XA under NASA's RLV program. The flight vehicle was tested at White Sands, New Mexico, during the summer of 1996. It demonstrated a short 26-hour turnaround time between its second and third flights, a record for any rocket.

The DC-XA flew until it was destroyed. During its fourth demonstration flight on July 31, 1996, a landing strut failed to extend, causing the unbalanced vehicle to tip over on the landing pad. The liquid oxygen tank exploded and there were indications of secondary explosions in the liquid hydrogen tank. The ensuing fire damaged large sections of the vehicle. An investigation board later determined that an unconnected helium pressurant line supplying hydraulic pressure to extend the landing strut caused the explosion. The program ended due to lack of funding to build a new vehicle. All flight tests are listed in Table 2–93.

---

[88] "DC-X Fact Sheet," BMDOLINK, *http://www.hq.nasa.gov/office/pao/History/x-33/dcx-facts.htm* (accessed March 22, 2005).

Figure 2–30. The X-34 Testbed Demonstrator being delivered to Dryden Flight Research Center, April 16, 1999. (NASA-DFRC Photo No. EC99-44976-31)

X-34

The X-34 program was to bridge the gap between the earlier subsonic DC-XA vehicle and the larger and higher performance X-33 demonstrator. It was structured originally as a cooperative agreement between NASA and Orbital Sciences Corporation signed in March 1995. The government team included Marshall Space Flight Center, responsible for the main propulsion system, including the Fastrac engine; Langley Research Center, responsible for wind tunnel testing and analysis; Ames Research Center, responsible for the thermal protection system; Dryden Flight Research Center; Holloman Air Force Base; White Sands Test Facility; and White Sands Missile Range, responsible for testing and flight support operations.

The proposed winged, reusable, single-stage vehicle, propelled by a kerosene/liquid oxygen engine, was expected to demonstrate key technologies. These included 1) composite primary and secondary airframe structures; 2) cryogenic insulation and propulsion system elements; 3) advanced thermal protection systems and materials; 4) low-cost avionics, including differential Global Positioning and inertial navigation systems; and 5) key operations technologies such as integrated vehicle health-monitoring and automated checkout systems. It was expected to significantly reduce mission costs for sending 1,000-pound to 2,000-pound (454-kilogram to 907-kilogram) payloads into low-Earth orbit. The vehicle would be air-dropped from beneath Orbital's L-1011 aircraft, reach speeds of Mach 8, and fly at altitudes of approximately 50 miles (80 kilometers). The vehicle would also demonstrate the ability to conduct subsonic flights through rain or fog and autonomous landings in crosswinds of up to 20 knots (23 miles per hour or 37 kilometers per hour). Characteristics of the technology demonstrator are listed in Table 2–94.

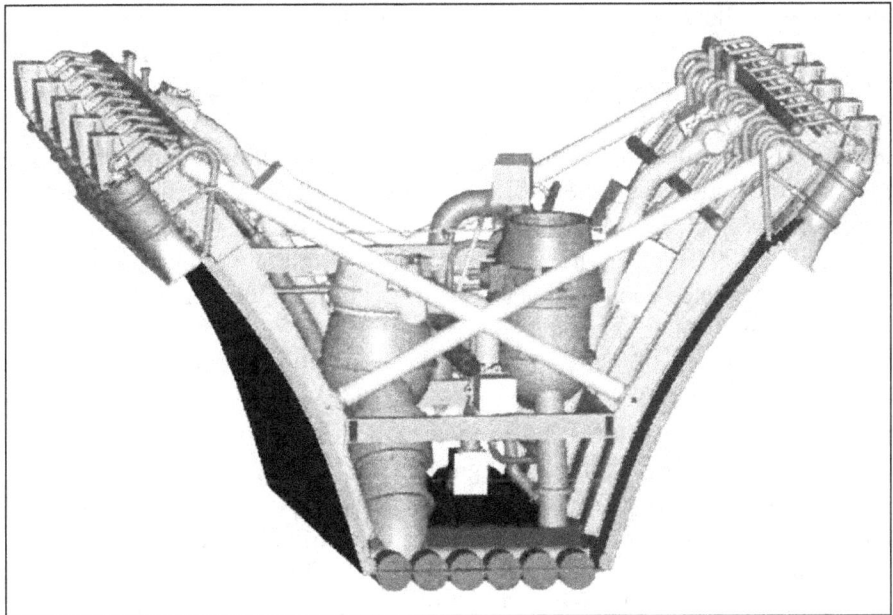

Figure 2–31. Aerospike Engine. (NASA-MSFC)

Originally, the X-34 was to progress rapidly through hardware design, flight tests planned for late 1997, and a launch expected by mid-1998. Orbital, however, withdrew from the cooperative agreement in less than a year partly because of changes in the projected profitability of the venture. NASA subsequently structured a new NASA Research Announcement in March 1996 focusing on the technology demonstration flight tests rather than on the commercial potential of the vehicle. NASA awarded the restructured fixed-price contract to Orbital in June 1996. This 30-month contract had a value of approximately $49.5 million. It included two powered test flights scheduled to begin in late 1998. NASA would spend an additional $10 million in direct support of the X-34. The contract had an option for up to 25 additional test flights after the initial contract period ended.[89]

In August 1997, a critical series of tests on the Fastrac engine were successfully completed at Marshall Space Flight Center. The Fastrac engine, only the second U.S.-made engine developed in the last 25 years, was to be the primary propulsion system for the X-34 demonstration vehicle when it began its flight tests.[90] The following May, a government-Orbital review was held final-

[89] "NASA Finalizes X-34 Contract With Orbital Sciences Corp.," NASA Marshall Space Flight Center News Releases, Release 96-161, August 30, 1996, *http://www.msfc.nasa.gov/news/news/releases/1996/96-161.html* (accessed March 23, 2005).

izing the design of the vehicle. This allowed the program to proceed with fabrication and manufacturing of systems such as structures; guidance; navigation and control; avionics; thermal protection; and main propulsion systems.[91]

To reduce program risk, NASA decided in January 1998 to modify its contract with Orbital to provide for a second flight vehicle. The modification also would allow for additional unpowered tests and more flexibility in demonstrating various technologies. The change increased the contract value by $7.7 million to purchase long lead-time hardware. NASA committed $2 million more for wind tunnel testing, additional testing and analysis, and a second leading-edge thermal protection system. An $8.5 million option called for purchase of shorter lead-time hardware, while a $1.8 million option was added for assembly.

In July 1998, the program passed a critical milestone as the first wing assembly completed qualification tests and was shipped to Orbital Sciences and mated to the X-34 test vehicle under construction. It was ultimately to fly aboard one of the two flight vehicles under construction at Orbital.

At the end of 1998, NASA exercised its option with Orbital for 25 additional test flights during a 12-month period beginning immediately after completion of the initial contract. Flights were to take place at the U.S. Army's White Sands Missile Range in New Mexico. The option was valued at more than $10 million, with government organizations performing an additional $4.7 million in work.[92]

The first of three planned X-34 technology demonstrators "rolled-out" on April 30, 1999, at Dryden Flight Research Center. The vehicle took its first test flight locked underneath the L-1011 carrier aircraft in June 1999. In August 1999, an $11 million contract for the Fastrac engine was awarded to Summa Technology. Assembly and preflight tests continued through 2000. However, in 2001, NASA decided not to add funds to the X-34 program from money dedicated to the Agency's Space Launch Initiative because the government determined that "the benefits to be derived from continuing the X-34 program did not justify the cost." This action coincided with the end of NASA's contract with Orbital Sciences Corporation. At the project's end, NASA had spent $205 million on the X-34 since its inception in 1996.

[90] "X-34 Fastrac Engine Passes Critical Tests," *NASA News* Release 97-232, August 14, 1997, *http:// www.msfc.nasa.gov/news/news/releases/1997/97-232.html* (accessed March 23, 2005).
[91] "X-34 Systems Design Freeze Completed," *NASA News* Release 97-107, May 22, 1997, *http:// www.qadas.com/qadas/nasa/nasa-hm/0872.html* (accessed March 25, 2005).
[92] "NASA Exercises X-34 Contract Option for 25 Test Flights," Marshall Space Flight Center News Releases, Release 98-251, December 18, 1998, *http://www.msfc.nasa.gov/news/news/releases/1998/98-251.html* (accessed March 23, 2005).

X-33

The X-33 program, the third RLV program, was to demonstrate a half-scale, single-stage-to-orbit vehicle that could go from launch to orbit without using multiple stages like ELVs or dropping rocket motors and fuel tanks like the Space Shuttle. Flying as fast as Mach 15, it was to decrease the per-pound cost of putting payloads into space from $10,000 to $1,000, while at the same time dramatically increasing launch vehicle safety and reliability. Ultimately, the goal of the full-size vehicle, named the "VentureStar," was to resupply the Space Station more quickly and cheaply than the Space Shuttle.

The program was a high-risk venture with unproven technologies that challenged its developers. In particular, the design required development of linear aerospike rocket engines, which had never been used in flight and had been rejected by Space Shuttle developers 25 years earlier. The program required the development of a wingless "lifting body" airframe that could keep the vehicle flying smoothly during launch and return to Earth. The program also required composite fuel tanks that could withstand the pressures of a space launch while filled with pressurized liquid hydrogen at a temperature of -423°F (-253°C).[93]

NASA initiated this NASA-industry partnership through a Cooperative Agreement Notice for Phase I concept definition and design of a technology demonstrator vehicle, the X-33, issued in January 1995. In March, NASA signed cooperative agreements with three companies—Lockheed Advanced Development Company (the Skunk Works), McDonnell Douglas Aerospace, and Rockwell International Corporation—to design the vehicle. The agreement called for NASA to work with each of these companies over the next 15 months on vehicle concept definition and design. The government would provide approximately $7 million to each of the companies, and each company was expected to match the investment.

Each company produced a design concept: all the vehicles would take off vertically, but only the McDonnell Douglas concept would land vertically. The others landed horizontally like an airplane.

At the beginning of April 1996, NASA issued another Cooperative Agreement Notice for Phase II of the project: the design, fabrication, and flight test of the X-33 demonstrator. It was planned that Phase II of the project would culminate in flight demonstration testing of the X-33 to begin in early 1999. NASA and industry would share costs during this phase. This was the first time a cooperative agreement rather than a conventional contract was used for a program of this size.

After a selection process of only a few months (due to an innovative paperless procurement process), on July 2, 1996, amid much fanfare, Vice President Albert Gore announced that NASA had selected Lockheed Martin

---

[93] "NASA's Billion-Dollar Shuttle Replacement May Never Fly," *CNN.com/Space* (September 25, 2000), *http://archives.cnn.com/2000/TECH/space/09/25/troudledspaceship.ap/index.html* (accessed March 24, 2005).

to build the X-33 test vehicle. According to the terms of the agreement, by March 1999, Lockheed Martin would design, build, and conduct the first test flight of the remotely piloted demonstration vehicle and would conduct at least 15 flights by December 1999. Major components would include a more robust metal heat shield in place of the Space Shuttle's tiles and an updated aerospike engine. The X-33 design was based on a lifting body shape that would be launched vertically like a rocket and land horizontally like an airplane. NASA had budgeted $941 million for the effort. Lockheed Martin initially invested $220 million of its own funds in the design. Figure 2–33 shows an artist's concept of the X-33 and VentureStar. Table 2–95 lists X-33 characteristics.

In 1997, the project successfully passed two important milestones. The Critical Design Review (CDR), held in October, ended 51 subsystem and component CDRs that had been held earlier that year. It allowed the program to proceed with fabrication of the remaining components, completion of subsystems, and assembly of the subscale prototype launch vehicle. Earlier in the year, the project had needed to resolve issues regarding aerodynamic stability and control and vehicle weight by modifying the design of the vehicle's canted and vertical fins. The project also planned to reduce weight by using composite materials and densified propellants.[94] In November, NASA completed the environmental impact statement process, which allowed all 15 test flights to proceed from the launch site at Haystack Butte on the eastern part of Edwards Air Force Base, California, and land at Michael Army Air Field, Dugway Proving Ground, Utah, and Malmstrom Air Force Base near Great Falls, Montana.[95]

The next major milestone was completion of flight-testing of the thermal protection system (TPS) materials. The tests took place in June 1998 at Dryden Flight Research Center on its F-15B Aerodynamic Flight Facility aircraft. The plane reached an altitude of 36,000 feet (10,973 kilometers) and a top speed of Mach 1.4 during the tests. The materials in the TPS included metallic Inconel tiles, soft Advanced Flexible Reusable surface insulation tiles, and sealing materials.

---

[94] "X-33 Program Successfully Completes Critical Design Review," *NASA News* Release 97-250, October 31, 1997, *ftp://ftp.hq.nasa.gov/pub/pao/pressrel/1997/97-250.txt* (accessed March 15, 2005).
[95] "NASA Completes X-33 Environmental Impact Statement Process," *NASA News* Release 97-254, November 5, 1997, *ftp://ftp.hq.nasa.gov/pub/pao/pressrel/1997/97-254.txt* (accessed March 15, 2005).

Figure 2–32. Artist's concept showing the relative size of the X-33 (left) and the proposed operational VentureStar. The VentureStar was to be twice the size of the half-scale X-33. (NASA-DFRC Photo No. ED97-43938-1)

Meanwhile, NASA's Office of Inspector General was investigating the program. The first inquiry examined whether NASA's use of a cooperative agreement on the X-33 program was appropriate for the program and "whether the agreement effectively defined roles, responsibilities, and rights of the government and industry partners." A secondary objective was to determine whether NASA implemented and managed the program consistent with congressional guidance. An audit determined that obligated funds for Lockheed Martin had not been recorded in a timely manner, a potential violation of federal law. Consequently, the Inspector General concluded that reports and financial statements "did not accurately reflect the financial status" of the program.[96] The next year, another audit from the Office of Inspector General examined whether the government had adequately addressed the cost of the project and its cost risk and cost estimate. The audit concluded that a better risk analysis "would have alerted NASA decision-makers to the probability of cost overruns" that "put NASA's investment . . . at risk."[97]

---

[96] Office of Inspector General, National Aeronautics and Space Administration, "Audit Report: X-33 Funding Issues," IG-99-001, November 3, 1998, pp. 1–3, *http://www.hq.nasa.gov/office/oig/hq/ig-99-001es.htm* (accessed March 19, 2005).

[97] Office of Inspector General, National Aeronautics and Space Administration, "Audit Report: X-33 Cost Estimating Processes," IG-99-052, September 24, 1999, pp. i–ii, *http://www.hq.nasa.gov/office/oig/hq/ig-99-052es.htm* (accessed March 15, 2005).

In 1999, the X-33 program experienced a setback when the composite materials used for its liquid hydrogen fuel tank failed during testing. An investigation into the cause of the failure determined that the composite technology was not "mature enough" for such a use. Lockheed Martin proposed replacing the composite tanks with aluminum tanks, which NASA agreed to if Lockheed Martin could obtain Space Launch Initiative funding. However, it was determined that the benefit did not justify the cost, and NASA cancelled the program in 2001 before proceeding to the next phase.[98] NASA investment in the X-33 program totaled $912 million, staying within its 1996 budget projection for the program. Lockheed Martin originally committed to invest $220 million in the X-33 and, during the life of the program, increased that amount to $357 million. In response to the cancellation, Lockheed Martin chose not to continue developing the VentureStar. A criticism of both the X-34 and X-33 programs was that NASA had not developed risk management plans until well after the programs had begun. Table 2–96 lists the chronology for NASA's RLV development.

[98] Leonard David, "NASA Shuts Down X-33, X-34 Programs," *Space.com*, *http://www.space.com/ missionlaunches/missions/x33_cancel_010301.html* (accessed March 22, 2005).

Table 2–1. Authorized/Appropriated Budget (FY 1989–FY 1998) (in thousands of dollars)

	1989	1990	1991	1992	1993	1994
R&D[a]	4,191,700	5,366,050	5,600,000	6,413,800	7,089,300	7,509,300
Space Transportation Capability Development[b]	606,600	651,500	763,400	679,800[c]	733,700[d]	751,600[e]
Upper Stages	156,200	—	—	—	—	—
Orbital Maneuvering Vehicle	—	—	45,400,000	—	—	—
Shuttle-C	—	—	40,000,000[f]	—	—	—
SFC&DC[g]	4,364,200	4,614,600	6,319,132	5,157,075	5,086,000	4,878,400
Space Shuttle Production and Operational Capability	1,335,500	1,340,300	1,364,000[h]	1,328,900	1,315,800	1,069,200
Advanced Solid Rocket Motor	51,000	35,000	—	375,000	315,000	150,000[i]
Safety Enhancements	—	75,000	—	—	—	—
Space Transportation (Shuttle) Operations	2,365,400	2,544,900	2,831,400[j]	2,970,600	3,085,200	3,006,500
Expendable Launch Vehicles (Launch Services)	—	169,500	229,200[k]	291,000	207,500	300,300

	1995	1996	1997[l]	1998[m]
Human Space Flight[n]	5,573,900[o]	5,456,600	5,362,900	5,506,500
Payload and Utilization Operations[p]	346,200[q]	315,000	271,800	247,400
Space Shuttle Safety and Performance Upgrades	3,309,000[r]	837,000	636,000	483,400
Space Shuttle Production and Operational Capability				
Space Shuttle Operations	—	2,341,800	2,514,900	2,494,400
Launch Services	313,700	—	—	—
Science, Aeronautics and Technology[s]	5,901,200	5,928,900	5,762,100	5,690,000
Advanced Concepts and Technology/Space Access and Technology	623,000[t]	639,800	711,000	696,600[u]
Advanced Space Transportation	—	193,000	324,700	—

Table 2–1. Authorized/Appropriated Budget (FY 1989–FY 1998) (in thousands of dollars) (Continued)

	1995	1996	1997[l]	1998[m]
X-33 Advanced Technology Demonstration Vehicle	—	—	—	333,500
Follow-on to X-33 Focused Technology Demonstration	—	—	—	150,000
Experimental Vehicle Procurement	—	—	—	150,000

a Total R&D amounts were stated in the appropriations bills, not in the authorization bills. R&D amounts shown did not equal the amounts shown in subcategories. Amounts for subordinate categories were from authorization bills unless otherwise noted.

b Amounts authorized for Space Transportation Capability Development included the Spacelab category, addressed in chapter 3, Human Spaceflight.

c Includes $40 million authorized for propulsion technology development and $10 million authorized for launch vehicle design studies, including single-stage-to-orbit vehicles.

d Specified $30 million for development of the Space Transportation Main Engine.

e Included $21 million to develop improvements in existing ELVs (including development of a single-engine version of the Centaur upper stage rocket) and $21.4 million to support development of advanced launch technologies, including single-stage-to-orbit technologies and components.

f Required in FY 1991 authorizations bill. Does not appear in programmed amounts in NASA's budget.

g Amounts for SFC&DC were stated in appropriations bill, not in authorization bills. Amounts for subordinate categories were from authorization bills unless otherwise noted.

h Of such funds, $45 million for FY 1991 was to be used for the Space Shuttle main engine, solid/rocket booster/solid rocket motor, external tank, orbiter, and the Assured Shuttle Availability program.

i For termination of program.

j Of these funds, $4 million was to be made available for the provision of launch services for eligible satellites in accordance with Section 6 of the Commercial Space Launch Act Amendments of 1988, Public Law 100-657.

k Launch Services (ELVs) transferred to Office of Space Science and Applications under the R&D appropriation.

l In the FY 1997 and FY 1998 authorization, Congress included NASA's authorization in a bill titled the Omnibus Civilian Science Authorization Act. H.R. 3322 was passed by the House. It was referred to Senate committee but was not acted upon by the Senate.

m Authorization bill H.R. 1275 passed by House. Referred to Senate committee but was not acted upon by Senate.

n Beginning with the FY 1995 estimate, and computation of FY 1993 programmed amounts, many R&D and SFC&DC amounts that involved human spaceflight moved to a single, new appropriation called Human Space Flight.

o Appropriated amount per Conference Committee.

p Formerly Space Transportation Capability Development.

q Included $40,000,000 to develop improvements in existing ELVs (including the development of a single-engine version of the Centaur upper stage rocket) and $46,000,000 to support development of advanced launch technologies, including single-stage-to-orbit technologies and components as well as other Space Transportation Capability Development/Payload and Utilization Operations budget categories.

r Amount included total Space Shuttle category: Production and Operational Capability and Operations.

s SAT appropriation used for some launch systems items.

t Included $40 million for single-stage-to-orbit technology development program, $13.6 million for University Space Engineering Research Centers, and $12.5 million for Small Spacecraft Technology Initiative.

u Called Advanced Space Transportation Technology in H.R. 1275.

Table 2–2. Programmed Budget (FY 1989–FY 1998) (in thousands of dollars)

	1989	1990	1991	1992	1993[a]	1994	1995	1996	1997	1998
R&D/SAT					Human Space Flight[b]					
Space transportation capability	674,000	558,142	602,467	739,711	442,300	405,600	320,100[d]	323,000	265,300	205,400
Development/payload and utilization operations[c]										
Upper stages	131,600	79,700	82,467	41,200	47,700[e]	6,900	—	—	—	—
Engineering and technical base	160,600	181,600	208,500	210,800	214,200	180,400	165,600	169,700	144,600	102,900
Payload operations & support equipment	60,700	65,461	101,200	130,100	95,200	85,100	44,000[f]	40,600	45,900	46,700
Tethered satellite system	26,400	27,300	21,900	16,400	4,000	7,400	7,400	1,800	—	—
Orbital maneuvering vehicle	73,000	75,681	—	—	—	—	—	—	—	—
Advanced programs	52,700	34,700	35,200	34,700	—	—	—	—	—	—
Advanced projects					16,100	7,200	12,200	24,200	34,700	46,700
Advanced space transportation[g]					114,600[h]	109,100[i]	162,100	234,000	—	417,100
Advanced concepts and technology					464,900	502,400	605,400	—	—	—
Advanced transportation technology[j]	81,400		23,900[k]	28,000	10,000[l]	20,000	—[m]	—	—	—
Reusable launch vehicle–systems engineering and analysis						3,500	3,800	—	—	—
Reusable launch vehicle–technology program						28,400	75,500	—	—	—
Reusable launch vehicle–initial flight demo program (FDP)						2,300	49,200	—	—	—

Table 2–2. Programmed Budget (FY 1989–FY 1998) (in thousands of dollars) (Continued)

	1989	1990	1991	1992	1993[a]	1994	1995	1996	1997	1998
X-33 advanced technology demonstrator	—	—	—	—	—	—	—	157,500	262,000	319,300
X-34 technology demonstration program	—	—	—	—	—	—	—	30,000	20,500	26,700
Transportation technology support	—	—	—	—	—	54,900	33,600	29,500[n]	34,400	62,100
SFC&DC/HSF										
Shuttle production and operational capability/safety and performance upgrades[o]	1,121,600	1,194,949	1,313,945	1,296,400	1,131,000	1,009,700	710,000	658,400	496,000	568,400
Orbiter	159,000	148,300	186,300	158,800	—	—	—	—	—	—
Orbiter improvements	—	—	—	—	235,000	204,300	194,800	271,400	159,900	232,500
Systems integration	34,500	15,000	10,700	7,200	—	—	—	—	—	—
Extended duration orbiter	20,000	23,700	25,000	10,700	—	—	—	—	—	—
Structural spares	20,300	22,900	66,000	57,600	—	—	—	—	—	—
Orbiter spares	48,000	28,200	26,800	13,800	—	—	—	—	—	—
Flight operations upgrades	—	—	—	—	121,100	109,900	54,300	73,400	66,000	40,300
Launch site equipment	104,100	105,700	101,200	93,100	80,100[p]	81,700	50,200	24,200	58,600	115,400
Mission operations and support capability[q]	153,500	177,349	136,045	148,100	—	—	—	—	—	—
Space Shuttle main engine upgrades	—	—	—	—	320,300	355,500	318,900	234,100	196,000	170,100
Solid rocket booster[r]	121,000	72,500	50,400	34,900	—	—	—	—	—	—
Solid rocket booster improvements	—	—	—	—	1,400	23,500	39,100	7,200	800	1,200
External tank	7,000	2,700	—	—	—	—	—	—	—	—

Table 2–2. Programmed Budget (FY 1989–FY 1998) (in thousands of dollars) (Continued)

	1989	1990	1991	1992	1993[a]	1994	1995	1996	1997	1998
Super lightweight tank	—	—	—	—	—	50,800	41,200	30,700	6,000	700
Construction of facilities[s]	—	—	—	—	178,100	34,300	12,300	17,400	—	8,200
Advanced solid rocket motor	51,000	160,400	309,100	315,000	195,000	149,700[t]	—	—	—	—
Assured Shuttle availability[u]	—	—	—	104,600	—	—	—	—	—	—
Space transportation operations/ Space Shuttle operations[v]	2,612,700	2,632,400	2,752,400	3,029,300	2,857,200	2,549,000	2,444,300	2,485,400	—	2,344,400
Mission support[w]	230,900	252,600	275,000	295,900	361,000[x]	316,000	287,700	358,900	46,200	814,700
Integration	285,000	303,200	317,900	315,400	200,000	199,000	169,500	142,500	—	—
Support	182,500	194,900	194,300	196,600	—	—	—	—	—	—
Orbiter	314,100	397,800	414,500	430,700	477,000	387,900	358,700	378,500	—	507,900
Space Shuttle main engine	403,200	438,200	402,400	322,100	239,900	189,200	163,300	185,000	208,300	173,400
Solid rocket booster	704,100	458,600	577,400	542,000	172,000	158,200	163,000	153,300	151,200	152,200
Redesigned solid rocket motor	—	—	—	—	409,400	396,400	370,700[y]	395,700	412,800	360,200
External tank	295,000	344,600	378,100	354,100	300,200	252,200	305,000	327,500	352,500	336,000
Launch and landing operations	534,600	541,000	595,200	642,900	697,100[z]	650,100	621,400	544,000	801,400	—[aa]
Launch operations	481,600	484,000	539,200	578,600	—	—	—	—	—	—
Payload and launch support	53,000	57,000	56,000	64,300	—	—	—	—	—	—
Expendable launch vehicles and services	66,500	139,700	229,200[ab]	155,800	180,801	84,600	255,600	245,300	240,600[ac] 84,700[ad]	212,900[ae] 27,600[af] 39,400[ag]
Small class	13,900	11,900	14,100	32,600	25,272[ah]	10,400	—	—	—	—
Medium class	45,000	75,400	97,300	58,100	61,451[ai]	43,000	—	—	—	—

Table 2–2. Programmed Budget (FY 1989–FY 1998) (in thousands of dollars) (Continued)

	1989	1990	1991	1992	1993[a]	1994	1995	1996	1997	1998
Intermediate class	6,300	49,600	108,100	45,000	41,100[aj]	43,000	—	—	—	—
Large class	1,300	2,800	9,700	20,100	5,278[ak]	—	—	—	—	—
Launch services mission support	—	—	—	—	—	37,100	—	—	—	—

a Beginning with the FY 1995 estimate, and computation of FY 1993 programmed amounts, all R&D and SFC&DC amounts that involved human spaceflight moved to a single, new appropriation called Human Space Flight.

b HSF appropriation except where noted otherwise.

c Budget for Space Transportation Capability Development includes Spacelab category. This is addressed with budget information in chapter 3, Human Spaceflight.

d Became Payload and Utilization Operations under HSF appropriations category.

e SAT appropriation, Office of Space Science.

f Renamed Payload Processing and Support.

g SAT appropriation within budget category Advanced Concepts and Technology. Items include Advanced Transportation Technology, Technology Assessment and Development, Advanced Technology Maturation, In-Space Transportation, and Single Engine Centaur.

h SAT appropriation.

i Part of Office of Space Access and Technology.

j Called Advanced Launch Systems in FY 1989 and FY 1990. Called Advanced Launch Technology in FY 1995 budget estimate. Advanced Transportation Technology includes: advanced launch system—civil needs, advanced launched system—propulsion, Shuttle-C studies (funding provided within Advanced Programs under Advanced Transportation line item), and heavy-lift vehicle studies.

k Called New Launch System beginning in FY 1993.

l SAT appropriation.

m Advanced Space Transportation Technology was also supported by $15 million in FY 1996 and $12 million in FY 1997, funded within the Engineering and Technical Base program of the Office of Space Flight.

n Renamed Advanced Space Transportation Program.

o Changed to Safety and Performance Upgrades in FY 1993.

p Included Launch Site Equipment upgrades (HSF appropriation) from FY 1993.

q Called Mission Support Capability beginning with FY 1992 estimate.

r Included safety upgrades and improvements to solid rocket booster and redesigned solid rocket motor.

s Construction of Facilities funding for Space Shuttle projects was provided to refurbish, modify, replace, and restore facilities at Office of Space Flight Centers to improve performance, address environmental concerns of the older facilities, and to ensure their readiness to support Space Shuttle Operations.

t Termination funding.

u Name changed to Safety and Obsolescence Upgrade beginning with FY 1994 budget estimate.

v Name changed to "Space Shuttle Operations" beginning in FY 1993.

w   Name of category changed to "mission operations" in FY 1992. Description of function was unchanged.

x   Called Mission and Crew Operations.

y   Name changed to Reusable Solid Rocket Motor. Description of activity remained the same.

z   Not broken down into smaller budget categories. Includes payload and launch support.

aa   Combined with Mission Support.

ab   Expendable Launch Vehicles and Services were officially transferred to the Office of Space Science and Applications (OSSA). Actual (appropriated) costs were charged to OSSA in FY 1991.

ac   Space Science ELVs and launch support.

ad   Earth Science ELVs and launch support.

ae   Moved to HSF appropriation.

af   Space Science launch support.

ag   Earth Science launch support.

ah   SAT appropriation.

ai   SAT appropriation.

aj   SAT appropriation.

ak   SAT appropriation.

Table 2–3. Space Transportation Capability Development/Payload and Utilization Operations Funding History (in thousands of dollars)

Year (Fiscal)	Submission	Authorization	Programmed
1989	631,300/681,000	606,600	674,000
1990	639,000/562,381	651,500	558,142
1991	773,400/602,500	763,400	602,467
1992	879,800/731,456	679,800[a]	739,711
1993	863,700/649,216	733,700[b]	442,300
1994	649,200/412,600	751,600[c]	405,600
1995[d]	356,200/320,100	819,300	320,100
1996	315,600/315,000	315,000	323,000
1997	271,800/275,300	271,800	—[e]
1998	227,400/205,400	247,400	205,400

a Included $40,000,000 authorized for propulsion technology development and $10,000,000 authorized for launch vehicle design studies, including single-stage-to-orbit vehicles.

b Specified $30,000,000 for development of the Space Transportation Main Engine for use with the Advanced/New Launch System.

c Included $21,000,000 to develop improvements in existing ELVs (including development of a single-engine version of the Centaur upper stage rocket) and $21,400,000 to support development of advanced launch technologies, including single-stage-to-orbit technologies and components.

d Payload and Utilization Operations budget category (beginning FY 1995) included same subcategories as Space Transportation Capability Development (Spacelab, Tethered Satellite System, Payload Processing and Support, Advanced Projects, and Engineering and Technical Base).

e No programmed amount shown.

Table 2–4. Upper Stages Funding History
(in thousands of dollars)

Year (Fiscal)	Submission	Programmed
1989	146,200/138,800	131,600
1990	88,600/84,600	79,700
1991	91,300/82,200	82,467
1992	108,500/62,256	41,200
1993[a]	56,500/47,700	47,700[b]
1994	51,100/43,600	6,900
1995	31,800/15,200	—
1996	18,300/—[c]	—

[a] Moved to Expendable Launch Vehicle category in FY 1993, which had been relocated to OSSA in FY 1993; was a SAT appropriation.
[b] SAT appropriation, Office of Space Science.
[c] Funding for mission-unique launch services was now included under the budget request for the benefiting program. Funding support for management oversight of the entire Launch Services program rested with the Launch Vehicles Office (LVO), which was now part of the newly formed Office of Space Access and Technology. The LVO aggregated NASA, NOAA, and international cooperative ELV mission requirements. The administration, procurement, and technical oversight of launch services in the small and medium performance classes were managed by Goddard Space Flight Center (Pegasus XL, Med-lite, and Delta II). Intermediate launch services (Atlas I/IIAS) were managed by Lewis Research Center. Upper stages were managed by Marshall Space Flight Center. Kennedy Space Center was delegated responsibility for technical oversight of vehicle assembly and testing at the launch site by Goddard and Lewis and was responsible for spacecraft processing at the launch site.

Table 2–5. Engineering and Technical Base Funding History
(in thousands of dollars)

Year (Fiscal)	Submission	Programmed
1989	158,900/155,400	160,600
1990	189,800/181,600	181,600
1991	218,500/208,500	208,500
1992	235,200/215,800	210,800
1993	224,200/214,155	214,200
1994	203,400/180,400	180,400
1995	176,400/165,600	165,600
1996	171,700/171,700	169,700
1997	151,500/148,600	144,600
1998	102,900/102,900	102,900

Table 2–6. Payload Operations and Support Equipment[a] Funding History (in thousands of dollars)

Year (Fiscal)	Submission	Programmed
1989	67,300/64,700	60,700
1990	81,100/66,700	65,461
1991	122,500/101,500	101,200
1992	144,500/119,100	130,100
1993	153,600/92,100	95,200
1994	95,400/92,100	85,100
1995	62,600/36,300[b]	44,000
1996	30,300/40,600	40,600
1997	42,700/41,700	45,900
1998	51,600/43,900	46,700

[a] Name of category changed to Payload Processing and Support in FY 1995.
[b] Reduction reflected closing of four of the 10 payload processing facilities before the end of the year.

Table 2–7. Advanced Programs Funding History (in thousands of dollars)

Year (Fiscal)	Submission	Programmed
1989	45,000/52,700	52,700
1990	48,700/33,600	34,700
1991	53,200/35,200	35,200
1992	53,800/39,300	34,700
1993	57,700/32,897	—
1994	60,700[a]/—	—

[a] Most elements moved to Advanced Space Technology.

Table 2–8. Advanced Projects Funding History (in thousands of dollars)

Year (Fiscal)	Submission	Programmed
1993	—[a]	16,100
1994	7,200/—	7,200
1995	15,200/12,200	12,200
1996	12,200/12,200	24,200
1997	15,200/34,700	34,700
1998	58,700/46,700	46,700

[a] Budget category not established at time of budget submission.

Table 2–9. Tethered Satellite System Funding History
(in thousands of dollars)

Year (Fiscal)	Submission	Programmed
1989	23,800/26,400	26,400
1990	19,900/24,000	27,300
1991	17,900/21,900	21.900
1992	12,600/16,400	16,400
1993	3,400/3,400	4,000
1994	—7,400[a]	7,400
1995	9,700/7,400	7,400
1996	3,800/3,800	1,800

[a] The Tether mission was flown on STS-46 in August 1992. No further Tether missions were manifested when the initial budget estimate was prepared. In 1993, it was determined that a reflight could be readily accomplished and several improvements to enhance the probability of success were recommended. The reflight was manifested for early 1996.

Table 2–10. Orbital Maneuvering Vehicle Funding History
(in thousands of dollars)

Year (Fiscal)	Submission	Programmed
1989	96,500/73,000	73,000
1990	107,000/76,281	75,681
1991	85,400/—[a]	—

[a] A decision was made to terminate the Orbital Maneuvering Vehicle program in June 1990. Consistent with congressional direction, no FY 1991 funding was provided.

Table 2–11. Advanced Concepts and Technology/Space Access and Technology Funding History (in thousands of dollars)[a]

Year (Fiscal)	Submission	Authorization	Programmed
1993	—[b]	—	464,900
1994	495,300/495,300	—	562,400
1995	608,400/642,400	623,000[c]	605,400
1996	705,600/641,300	639,800	—
1997	725,000/—	711,000	—
1998	—[d]	696,600[e]	—

[a] Funding categories dealing with advanced transportation varied frequently. Included Advanced Space Transportation, Launch Vehicle Support, and other budget categories not relevant to Launch Systems.
[b] Budget category not established at time of budget submission.
[c] $40,000,000 for single-stage-to-orbit technology development program, $13,600,000 for University Space Engineering Research Centers, and $12,500,000 for Small Spacecraft Technology Initiative.
[d] Budget category not shown in budget submission or programmed amount.
[e] Called Advanced Space Transportation Technology in H.R. 1275, "Civilian Space Authorization Act, Fiscal Years 1998 and 1999."

Table 2–12. Advanced Space Transportation Funding History (in thousands of dollars)[a]

Year (Fiscal)	Submission	Authorization	Programmed
1993	—	—	114,600
1994	—[b]/121,900	—	109,100
1995	103,100/162,100	—	162,100
1996	193,000/188,500	193,000	234,000
1997	324,700/336,700	324,700	336,700
1998	396,600/417.100	—	417,100

[a] Categories varied depending on year and active projects. At times they included Advanced Launch Technology, Reusable Launch Vehicles, Transportation Technology Support, New Launch System, Single-Stage Centaur, and other categories.
[b] Budget category not established at time of budget submission.

Table 2–13. Advanced Transportation Technology/New Launch System
Funding History (in thousands of dollars)[a]

Year (Fiscal)	Submission	Authorization	Programmed
1989	13,000/81,400[b]	6,500	81,400
1990	5,000/(10,500)[c]	—	—[d]
1991[e]	53,900/23,900[f]	40,000[g]	23,900[h]
1992	175,000[i]/38,000[j]	—	28,000
1993[k]	125,000[l]/10,042[m]	—	10,000[n]
1994	—	—	20,000

[a] Called Advanced Launch System in congressional documents.
[b] Renamed Advanced Launch Systems. This was a joint NASA–DOD program with the objective of defining a new heavy-lift capability based on advanced technology that would reduce the cost of placing payloads in space. NASA had responsibility for the civil requirements not addressed by the joint ALS baseline design. The original FY 1989 budget estimate of $13 million, reflected only the civil requirements. The revised estimate of $81.4 million reflected both a reduced civil requirement of $6.5 million and the propulsion element estimated at $74.9 million.
[c] Funding was deleted in FY 1990 legislation. Total funding for Advanced Launch Systems, including NASA-managed elements, was included in the DOD budget request. NASA's Advanced Launch Systems propulsion advanced development effort was provided through reimbursable funding rather than appropriation transfers.
[d] No programmed amount shown.
[e] Included Advanced Launch Systems, Shuttle-C, Heavy Lift Vehicle Studies.
[f] This revised estimate was consistent with congressional direction. It was accommodated primarily through deferral of the Assured Crew Return Vehicle Phase B study and other program realignments.
[g] Amount specified for Shuttle-C. Other uses for funds not listed.
[h] Called New Launch System beginning in FY 1993.
[i] Increase reflected plans to proceed with the initial stages of a new launch system. Program planning for FY 1992 was not completed when the estimate was prepared. It was thought that the focus of FY 1992 activities would be initiating development of the Space Transportation Main Engine prototype, conducting definition and design studies of vehicle components and elements, and assessing requirements and design options for supporting launch facilities. Technologies and operational approaches that could reduce per-flight costs and increase system robustness would also be pursued.
[j] The budget reduction supported a change to a 2002 first launch schedule. An equal amount was budgeted by DOD.
[k] Called New Launch System beginning in FY 1993.
[l] Funding level (along with an equal amount from DOD) allowed completion of the preliminary design effort for the New Launch System.
[m] The New Launch System (formerly Advanced Transportation Technology) was to be a joint program with the DOD to develop a new family of launch vehicles that would improve national launch capability with reductions in operating costs and improvements in launch system reliability, responsiveness, and mission performance. Initial efforts focused on developing the Space Transportation Main Engine (STME) since this was the common element of all configuration. The reduction in the FY 1993 budget estimate terminated the effort on the NLS while retaining options to develop the STME and/or examine alternative engine technologies.
[n] SAT appropriation.

Table 2–14. Reusable Launch Vehicle–Systems Engineering and Analysis Funding History (in thousands of dollars)

Year (Fiscal)	Submission	Programmed
1994	—[a]	3,500
1995	—/4,600	3,800
1996	4,700/500	—[b]

[a] Budget category not established at time of budget submission.
[b] No programmed amount shown.

Table 2–15. Reusable Launch Vehicle–Technology Program Funding History (in thousands of dollars)

Year (Fiscal)	Submission	Programmed
1994	—[a]	28,400
1995	—/76,900	75,500
1996	59,300/49,500	—[b]

[a] Budget category not established at time of budget submission.
[b] No programmed amount shown.

Table 2–16. Reusable Launch Vehicle–Initial Flight Demonstration Program (FDP) Funding History (in thousands of dollars)

Year (Fiscal)	Submission	Programmed
1994	—[a]	2,300
1995	—/47,000	49,200
1996	60,000/109,000	—[b]
1997	266,100/—[c]	—

[a] Budget category not established at time of budget submission.
[b] No programmed amount shown.
[c] No revised budget submission for this category.

Table 2–17. X-33 Advanced Technology Demonstrator Funding History (in thousands of dollars)[a]

Year (Fiscal)	Submission	Authorization
1996	—[b]	157,500
1997	—/246,800	262,000
1998	333,500/318,300	319,300

[a] Part of Reusable Launch Vehicle Program.
[b] Budget category not established at time of budget submission.

Table 2–18. X-34 Technology Demonstration Program Funding
History (in thousands of dollars)[a]

Year (Fiscal)	Submission	Programmed
1996	—[b]	30,000
1997	—/36,700	20,500
1998	43,100/26,700	26,700

[a] Part of Reusable Launch Vehicle Program.
[b] Budget category not established at time of budget submission.

Table 2–19. Transportation Technology Support Funding History
(in thousands of dollars)

Year (Fiscal)	Submission	Programmed
1994	—[a]	54,900
1995	—/33,600	33,600
1996	34,000/29,500	29,500[b]
1997	16,600/53,200	34,400
1998	43,100/26,100	62,100

[a] Budget category not established at time of budget submission.
[b] Renamed Advanced Space Transportation Program.

Table 2–20. Space Shuttle Production and Operational Capability/
Safety and Performance Upgrades Funding History
(in thousands of dollars)[a]

Year (Fiscal)	Submission	Authorization	Programmed
1989	1,400,500/1,128,200	1,335,500	1,121,600
1990	1,305,300/1,119,500	1,340,300	1,194,949
1991	1,302,000/1,327,800	1,364,000[b]	1,313,945
1992	1,288,900/1,276,400	1,328,900	1,296,400
1993	1,021,800/1,053,016	1,315,800	1,131,000
1994	1,189,600/978,700	1,069,200	1,009,700
1995	903,900/739,800	3,309,000[c]	710,000
1996	837,000/663,400	837,000	658,400
1997	636,000/636,000	636,000	496,000
1998	483,400/553,400	483,400	568,400

[a] Included orbiter improvements, propulsion upgrades (SSME upgrades, SRB improvements, super lightweight tank), flight operations and launch site equipment upgrades, advanced solid rocket motor, and construction of facility budget categories are displayed separately below.
[b] $45 million to be used only for 1) SSME, 2) SRB/SRM, 3) ET, 4), orbiter, and 5) assured Shuttle availability.
[c] Amount for total Space Shuttle budget category, including both Production and Operational Capability and Operations.

Table 2–21. Orbiter (Orbiter Operational Capability) Funding History (in thousands of dollars)

Year (Fiscal)	Submission	Programmed
1989	181,000/155,800	159,000
1990	157,500/125,900	148,300
1991	113,400/144,900	186,300
1992	143,300/162,100	158,800
1993	196,900/179,516	—[a]

[a] No programmed amount shown.

Table 2–22. Systems Integration (Orbiter Operational Capability) Funding History (in thousands of dollars)

Year (Fiscal)	Submission	Programmed
1989	17,000/30,500	34,500
1990	9,000/15,400	15,000
1991	11,100/11,400	10,700
1992	19,900/9,100	7,200
1993	26,900/13,400	—[a]
1994	14,400/—[b]	—

[a] No programmed amount shown.
[b] No revised budget submission in this category.

Table 2–23. Orbiter Improvements Funding History (in thousands of dollars)

Year (Fiscal)	Submission	Programmed
1993	—[a]	235,000
1994	215,500/127,100	204,300
1995	191,800/194,800	194,800
1996	227,900/258,700	271,400
1997	169,900/169,900	159,900
1998	137,300/232,500	232,500

[a] Budget category not established at time of budget submission.

Table 2–24. Extended Duration Orbiter (Orbiter Operational Capability) Funding History (in thousands of dollars)

Year (Fiscal)	Submission	Programmed
1989	10,000/20,000	20,000
1990	157,500[a]/125,900	23,700
1991	15,000/25,000	25,000
1992	18,500/10,500	10,700
1993	21,700/22,600	—[b]
1994	67,700[c]/—[d]	—

[a] Funding increase reflected expectations of development of a 28-day extended duration orbiter.
[b] Budget category not established at time of budget submission.
[c] Funding was to cover the payback costs to the prime contractor for use of the cryogenic pallet kit to extend on-orbit stay time capability from the baseline 7 to 10 days to 14 to 16 days. It also was to initiate the required modifications on *Endeavour* and *Atlantis*.
[d] No revised budget request submitted in this category.

Table 2–25. Structural Spares (Orbiter Operational Capability) Funding History (in thousands of dollars)

Year (Fiscal)	Submission	Programmed
1989	57,300/20,300[a]	20,300
1990	15,200/25,200	22,900
1991	52,400/71,000	66,000
1992	78,300/50,600	57,600
1993	51,000/35,000	—[b]
1994	35,000/—[c]	—

[a] Reduction in funding reflected slower start of structural spares program than expected.
[b] Programmed amount not shown.
[c] No revised budget request submitted.

Table 2–26. Orbiter Spares (Orbiter Operational Capability) Funding History (in thousands of dollars)

Year (Fiscal)	Submission	Programmed
1989	54,000/55,200	48,000
1990	30,300/27,100	28,200
1991	21,700/23,300	26,800
1992	13,800/13,800	13,800
1993	9,000/9,000	—[a]
1994	—[b]	—

[a] No programmed amount shown.
[b] Activity was concluded in FY 1993.

Table 2–27. Flight Operations Upgrades
Funding History (in thousands of dollars)

Year (Fiscal)	Submission	Programmed
1993	—[a]	121,000
1994	—/107,700	109,900
1995	110,900/63,900	54,300
1996	89,000/69,400	73,400
1997	69,500/89,000	66,000
1998	51,500/70,600	40,300

[a] Budget category not established at time of budget submission.

Table 2–28. Launch Site Equipment (Launch and Mission Support)
Funding History (in thousands of dollars)

Year (Fiscal)	Submission	Programmed
1989	142,000/109,400	104,100
1990	98,500/89,400	105,700
1991	114,900/110,400	101,200
1992	79,400/85,100	93,100
1993	86,000/69,000	80,100[a]
1994	81,700/68,500	81,700
1995	76,100/40,600	50,200
1996	43,800/21,100	24,200
1997	45,50026,000	58,600
1998	40,800/67,500	115,400

[a] Launch site equipment upgrades.

Table 2–29. Mission Support Capability (Launch and Mission Support)
Funding History (in thousands of dollars)

Year (Fiscal)	Submission	Programmed[a]
1989	93,500/42,100[b]	—[c]
1990	75,600/—[d]	—

[a] Combined into Mission Operations and Support Capability.
[b] Mission support decreased as program reserves were deleted to comply with the FY 1989 appropriations general reduction.
[c] No programmed amount shown.
[d] Combined with Mission Operations Capability into Mission Operations and Support Capability.

Table 2–30. Mission Operations Capability (Launch and Mission Support) Funding History (in thousands of dollars)

Year (Fiscal)	Submission	Programmed
1989	108,200/112,700	—[a]
1990	166,900/—[b]	—

[a] Combined into Mission Operations and Support Capability.
[b] Combined with Mission Support Capability into Mission Operations and Support Capability.

Table 2–31. Mission Operations and Support Capability (Launch and Mission Support) Funding History (in thousands of dollars)[a]

Year (Fiscal)	Submission	Programmed
1989	—[b]	153,500
1990	242,500/169,900	177,349
1991	150,600/142,600	136,045
1992	190,700/176,600	148,100
1993	124,700/109,100	—[c]
1994	105,400/—[d]	—

[a] Combined Mission Support Capability and Mission Operations Capability.
[b] Budget category not established at time of budget submission.
[c] No programmed amount shown.
[d] No revised request submitted.

Table 2–32. Space Shuttle Main Engine Upgrades Funding History (in thousands of dollars)

Year (Fiscal)	Submission	Programmed
1993	—[a]	320,300[b]
1994	—/287,900[c]	355,500
1995	380,500/354,200	318,900
1996	357,200/251,300	234,100
1997	309,500/324,500	196,000
1998	231,200/170,700	170,100

[a] Budget category not established at time of budget submission.
[b] Unclear where this programmed amount originated. Some activities, e.g., the alternate turbopump and the large throat main combustion chamber, were specifically named in the SFC&DC SSME budget line item. However, "other upgrades" were not specified, so it is not clear where the costs for these previously resided since the amount was larger than the entire amount budgeted for SSME under the SFC&DC appropriation.
[c] New budget category under HSF appropriation.

Table 2–33. Solid Rocket Booster (Propulsion Systems)
Funding History (in thousands of dollars)

Year (Fiscal)	Submission	Programmed
1989	105,000/123,700	121,000
1990	106,700/75,300	72,500
1991	82,900/50,400	50,400
1992	48,600/38,200	34,900[a]
1993	43,100/30,200	—[b]

[a] Included SRB safety upgrades.
[b] No programmed amount shown.

Table 2–34. Solid Rocket Booster Improvements
Funding History (in thousands of dollars)

Year (Fiscal)	Submission	Programmed
1993	43,100/30,200	1,400
1994	52,500[a]/23,200	23,500
1995	51,600/34,400	39,100
1996	69,000/1,400	7,200
1997	2,100/800	800
1998	6,600/3,500	1,200

[a] Included improvements to redesigned solid rocket motor.

Table 2–35. External Tank (Propulsion Systems)
Funding History (in thousands of dollars)

Year (Fiscal)	Submission	Programmed
1989	7,000/7,000	7,000
1990	2,700/2,700[a]	2,700

[a] Closeout of production funding for external tank tooling and equipment to support manufacturing rate capability requirements took place in FY 1990.

Table 2–36. Super Lightweight Tank
Funding History (in thousands of dollars)

Year (Fiscal)	Submission	Programmed
1994	—[a]/49,500	50,800
1995	80,700/39,600	41,200
1996	32,700/44,100	30,700
1997	31,200/17,500	6,000
1998	9,200/1,800	700

[a] Budget category not established at initial time of budget submission.

Table 2–37. Construction of Facilities
Funding History (in thousands of dollars)

Year (Fiscal)	Submission	Programmed
1993	—[a]	178,100
1994	—/33,500	34,300
1995	12,300/12,300	12,300
1996	17,400/17,400	17,400
1997	8,300/8,300	8,300
1998	6,800/6,800	8,200

[a] Budget category not established at time of budget submission.

Table 2–38. Advanced Solid Rocket Motor (Propulsion Systems)
Funding History (in thousands of dollars)

Year (Fiscal)	Submission	Authorization	Programmed
1989	88,000/51,000	51,000	51,000
1990	121,300/125,400	35,000	160,400
1991	309,100/309,100	—[a]	309,100
1992	200,000/315,000[b]	375,000	315,000
1993	—[c]/195,000	315,000	195,000
1994	280,000/179,700	150,000[d]	149,700[e]

[a] Not stated in authorization bill.
[b] Funding for the Advanced Solid Rocket Motor was increased $115 million based on congressional direction. However, the program was terminated.
[c] Due to the tight budget environment, the Advanced Solid Rocket Motor was not included in the initial FY 1993 budget request. Congress reinstated funding in the FY 1993 appropriation at a lower funding level than for the previous year.
[d] For termination of program.
[e] Reflected program termination.

Table 2–39. Assured Shuttle Availability Funding History
(in thousands of dollars)

Year (Fiscal)	Submission	Programmed
1992	122,300/104,600	104,600
1993	138,900/89,500[a]	—
1994	140,200/—	—

[a] Name changed to Safety and Obsolescence Upgrades. Included items that moved to other Safety and Performance Upgrades categories in next fiscal year, e.g., alternative turbopump, large throat main combustion chamber, hardware interface module replace, cable plant upgrades, and multifunction electronic display system.

Table 2–40. Space Transportation (Space Shuttle) Operations Funding History (in thousands of dollars)

Year (Fiscal)	Submission	Authorization	Programmed
1989	2,405,400/2,390,700	2,365,400	2,612,700
1990	2,732,200/2,636,036	2,544,900	2,632,400
1991	3,118,600/3,019,200	2,831,400[a]	2,752,400
1992	3,023,600/2,943,400	2,970,600	3,029,300
1993	3,115,200/3,015,953	3,085,200	2,857,200
1994	3,006,500/2,570,600	3,006,500	2,549,000
1995	2,420,100/2,415,297	3,309,000[b]	2,444,300
1996	2,394,800/2,485,400	2.341,800	2,485,400
1997	2,514,900/1,514,900	2,514,900	2,464,900
1998	2,369,400/2,494,400	2,494,400	2,344,400

[a] In accordance with the Commercial Space Launch Act Amendments of 1988, less than or equal to $4 million was made available for the provision of launch services for eligible satellites. *Commercial Space Launch Act Amendments of 1988*, 100th Congress., 1st sess., Public Law 100-657 (November 15, 1988).
[b] Amount was for total Space Shuttle costs, including both Shuttle Operations and Production and Operational Capability.

Table 2–41. Mission Support (Flight Operations) Funding History (in thousands of dollars)

Year (Fiscal)	Submission	Programmed
1989	215,400/230,500	230,900
1990	247,500/253,700	252,600
1991[a]	280,500/276,500	275,000
1992	318,800/260,400	295,900
1993	338,400/329,117	361,000[b]
1994	330,900/322,800	316,000
1995	298,400/253,700	287,700
1996	284,600/358,900	358,900
1997	289,700/289,700	46,200[c]
1998	289,100/94,500	814,700[d]

[a] Name of category changed to Mission Operations.
[b] Called Mission and Crew Operations.
[c] Reflected transfer of flight operations to consolidated United Space Alliance contract from Boeing and Lockheed Martin contracts.
[d] Renamed Mission and Launch Operations. Included costs for Launch and Landing Operations.

Table 2–42. Integration (Flight Operations) Funding History
(in thousands of dollars)

Year (Fiscal)	Submission	Programmed
1989	264,100/268,800	285,000
1990	300,300/314,100	303,200
1991	335,600/319,900	317,900
1992	342,300/315,400	315,400
1993	163,000/146,000	200,000
1994	151,700/211,200	199,000
1995	190,500/168,400	169,500
1996	152,200/142,500	142,500
1997	141,200/141,200	124,700
1998	126,200/107,000	—[a]

[a] Combined with Orbiter budget category.

Table 2–43. Support (Flight Operations) Funding History
(in thousands of dollars)

Year (Fiscal)	Submission	Programmed
1989	180,600/186,400	182,500
1990	224,800/191,000	194,900
1991	199,500/205,100	194,300
1992	251,400/214,500	196,600
1993	239,000/277,000	—[a]
1994	285,200—[b]	—

[a] No programmed amount shown.
[b] No revised budget request submitted.

Table 2–44. Orbiter (Flight Hardware) Funding History
(in thousands of dollars)

Year (Fiscal)	Submission	Programmed
1989	339,400/301,300	314,100
1990	351,800/370,200	397,800
1991	397,800/442,900	414,500
1992	441,700/390,400	430,700
1993	522,700/540,853	477,000
1994	508,900/364,100	387,900
1995	292,800/359,800	358,700
1996	352,700/378,500	378,500
1997	375,400/375,400	367,900
1998	376,700/356,100	507,900[a]

[a] Included both orbiter and integration budget categories.

Table 2–45. Space Shuttle Main Engine (Propulsion Systems) Funding History (in thousands of dollars)

Year (Fiscal)	Submission	Programmed
1989	511,800/400,500	403,200
1990	496,600/438,200	438,200
1991	430,900/388,300	402,400
1992	374,100/362,200	322,100
1993	314,600/263,200	239,900
1994	245,400/191,800[a]	189,200
1995	144,400/149,200	163,300
1996	145,600/185,000	185,000
1997	172,300/182,300	208,300
1998	184,900/204,600	173,400

[a] Change to HSF appropriation from SFC&DC took place with budget estimate for FY 1995 and revised FY 1994 budget request. Old SFC&DC SSME budget category included both production of SSME and upgrades and safety. New budget category under HSF appropriation was only for shuttle operations and did not include upgrades and safety, which was budgeted separately.

Table 2–46. Solid Rocket Booster (Flight Hardware) Funding History (in thousands of dollars)

Year (Fiscal)	Submission	Programmed
1989	382,500/516,800	704,100
1990	537,000/487,500	458,600
1991	691,300/572,900	577,400
1992	592,400/541,300	542,000
1993	556,700/559,100	172,000
1994	515,700/156,400[a]	158,200
1995	144,900/162,200	163,000
1996	164,200/153,300	153,300
1997	174,800/150,400	151,200
1998	157,700/135,500	152,200

[a] Reduction reflected creation of new budget category: Redesigned Solid Rocket Motor.

Table 2–47. Redesigned Solid Rocket Motor (Flight Hardware)
Funding History (in thousands of dollars)

Year (Fiscal)	Submission	Programmed
1993	—	409,400
1994	—[a]/368,900	396,400
1995	373,100/365,997	370,000
1996	355,400/395,700	395,700
1997	402,900/427,000	412,800
1998	434,600/380,400	360,200

[a] No initial FY 1994 budget request for Redesigned Solid Rocket Motor

Table 2–48. External Tank (Flight Hardware) Funding History
(in thousands of dollars)

Year (Fiscal)	Submission	Programmed
1989	313,300/294,600	295,000
1990	347,700/347,500	344,600
1991	378,100/377,500	378,100
1992	382,900/365,400	354,100
1993	375,900/298,200	300,200
1994	340,000/305,300	252,200
1995	379,600/329,600	305,000
1996	328,000/327,500	327,500
1997	348,700/339,000	352,400
1998	359,700/341,300	336,000

Table 2–49. Launch and Landing Operations Funding History
(in thousands of dollars)

Year (Fiscal)	Submission	Programmed
1989	514,600/506,800	534,600
1990	492,500/471,500	541,000
1991	606,600/596,200	595,200
1992	694,400/628,300	642,900
1993	639,900/690,800	697,100[a]
1994	696,400/650,100	650,100
1995	596,400/626,400	621,400
1996	612,100/544,000	544,000
1997	609,900/609,900	801,400
1998	605,300/720,200	—[b]

[a] Included launch operations and payload and launch support.
[b] Combined with Mission Support.

Table 2–50. Launch Operations (Launch and Landing Operations) Funding History (in thousands of dollars)

Year (Fiscal)	Submission	Programmed
1989	456,600/452,800	481,600
1990	492,500/471,500	484,000
1991	546,400/537,500	539,200
1992	629,300/567,500	578,600
1993	581,100/632,000	—
1994	637,500/—[a]	—

[a] No revised estimate submitted for this budget category.

Table 2–51. Payload and Launch Support (Launch and Landing Operations) Funding History (in thousands of dollars)

Year (Fiscal)	Submission	Programmed
1989	58,000/54,000	53,000
1990	61,100/58,700	57,000
1991	60,200/57,700	56,000
1992	64,800/60,800	64,300
1993	58,800/58,800	—
1994	58,900/—[a]	—

[a] No revised estimate submitted for this category.

Table 2–52. Expendable Launch Vehicles and Services[a]
Funding History (in thousands of dollars)

Year (Fiscal)	Submission	Authorization	Programmed
1989	195,500/85,500	—[b]	66,500
1990	169,500/141,836	169,500	139,700
1991	229,200/229,200	229,200	229,200[c]
1992	341,900/195,300	291,000	155,800
1993	217,500/180,801[d]	207,500	180,801
1994	300,300/313,500	300,300	84,600
1995	340,900/95,800	313,700	255,600
1996	74,200/254,300	—[e]	245,300
1997	253,500/240,600	—	240,600: Space Science 84,700: Earth Science
1998	236,300/215,900: Space Science 34,800: Earth Science	—	212,900[f] 27,600: Space Science 39,400: Earth Science

[a] Included funds for upcoming missions in all classes of ELVs.
[b] Not stated in authorization bill.
[c] Expendable Launch Vehicles and Services was officially transferred to the OSSA. Actual (appropriated) costs were charged to OSSA (SAT) in FY 1991.
[d] Included amounts budgeted for upper stages.
[e] Not stated in authorization bill.
[f] Moved to HSF appropriation from SAT.

Table 2–53. Small Class (Expendable Launch Vehicles and Services)
Funding History (in thousands of dollars)

Year (Fiscal)	Submission	Programmed
1989	—[a]	13,900
1990	26,300/12,100	11,900
1991	15,000/14,800	14,100
1992	33,700/33,100	32,600
1993	27,900/25,272	25,272
1994	26,200/16,800[b]	10,400
1995	31,400[c]/4,000	—
1996	10,800/—[d]	—

[a] Budget category not established at time of budget submission.
[b] SAT appropriation from revised budget estimate.
[c] SAT appropriation.
[d] Budget category no longer appeared in budget.

Table 2–54. Medium Class (Expendable Launch Vehicles and Services) Funding History (in thousands of dollars)

Year (Fiscal)	Submission	Programmed
1989	—a	45,000
1990	86,200/76,036	75,400
1991	102,90098,700	97,300
1992	81,500/61,100	58,100
1993	67,300/61,451	61,451b
1994	77,500/93,500c	43,000
1995	116,200d/35,600	—
1996	31,000/—e	—

a Budget category not established at time of budget submission.
b SAT appropriation.
c SAT appropriation from revised budget estimate.
d SAT appropriation.
e No revised estimate submitted for this budget category.

Table 2–55. Intermediate Class (Expendable Launch Vehicles and Services) Funding History (in thousands of dollars)

Year (Fiscal)	Submission	Programmed
1989	—a	6,300
1990	54,900/50,400	49,600
1991	101,100/106,000	108,100
1992	156,500/85,000b	45,000c
1993	54,800/41,000	41,100
1994	63,200/63,200d	43,000
1995	70,200e/26,000	—

a Budget category not established at time of budget submission.
b Funding was decreased partially because the launch of TDRS-7, originally scheduled to take place from an ELV, was assigned to the Shuttle.
c SAT appropriation.
d SAT appropriation from revised budget estimate.
e SAT appropriation.

Table 2–56. Large Class (Expendable Launch Vehicles and Services) Funding History (in thousands of dollars)

Year (Fiscal)	Submission	Programmed
1989	—[a]	1,300
1990	2,100/3,300	2,800
1991	10,200/9,700	9,700
1992	70,200/16,100[b]	20,100
1993	11,000/5,278	5,278[c]
1994	82,300[d]/86,400[e]	—
1995	91,300[f]/—	—

[a] Budget category not established at time of budget submission.
[b] Funding reduction due to deferral of the Cassini launch to October 1997, thereby reducing the funding requirements for the Titan IV/Centaur vehicle.
[c] SAT appropriation.
[d] Increase in large-class ELV funding requests was for Titan IV vehicle needed to support the Cassini mission, scheduled for an October 1997 launch. These funds also supported the required Centaur upper stage, with both vehicle elements purchased as a package from the U.S. Air Force.
[e] SAT appropriation from revised budget estimate.
[f] SAT appropriation.

Table 2–57. Launch Services Mission Support Funding History (in thousands of dollars)

Year (Fiscal)	Submission	Programmed
1994	—[a]	37,100
1995	—/37,000	—[b]
1996	37,600/—	—

[a] Budget category not established at time of budget submission.
[b] No programmed amount shown.

Table 2–58. Expendable Launch Vehicle Success Rate by Year and Launch Vehicle

Year	Athena	Atlas-G Centaur/E	Atlas I/IIA/ IIAS	Conestoga	Delta	Pegasus	Scout	Taurus	Titan	Total
1989		1/1			8/8				4/4	13/13
1990		2/2	1/1		11/11	1/1	1/1		4/5	20/21
1991		2/2	1/2		5/5	1/1	1/1		2/2	12/13
1992			4/5		11/11		2/2		3/3	20/21
1993		1/1	4/5		7/7	2/2	1/1		1/2	15/18
1994		2/2	5/5		3/3	1/3[a]	1/1	1/1	5/5	18/20
1995	0/1	1/1	11/11	0/1	3/3[b]	1/2			4/4	20/23
1996			7/7		10/10	4/5			4/4	25/26
1997	1/1		8/8		10/11	5/5			5/5	29/30
1998	1/1		6/6		12/13	6/6		2/2	2/3	29/31
Totals	2/3	9/9	47/50	0/1	79/82	21/25	6/6	3/3	34/37	202/215

[a] The Pegasus launch on May 19, 1994 did not reach its intended orbit and was classified in most sources as a "partial failure." It is counted as a failure in this table.

[b] The Delta launch on August 5, 1995 placed the Koreasat spacecraft in a lower than expected orbit. It still allowed the mission to achieve most of its objectives, although it shortened the satellite's useful life. It is counted as a success in this table.

Table 2–59. Athena Launches (1989–1998)

Launch Date	Mission	Vehicle Type	Comment
August 15, 1995	Gemstar 1	Athena I	Failed
August 23, 1997	Lewis	Athena I	Launch successful but spacecraft failed
January 7, 1998	Lunar Prospector	Athena II	Successful lunar mission

Table 2–60. Athena I Characteristics[a]

	Stage 1	Stage 2	Orbit Assist Module	Payload Fairing Envelope	Total
Length	10.7 m (35.2 ft)	3.0 m (10.0 ft)	1.0 m (3.3 ft)	6.1 m (20 ft)	18.9 m (61.9 ft)
Diameter	2.3 m (7.7 ft)	2.3 m (7.7 ft)	2.3 m (7.7 ft)	7.7 ft (2.3 m) outer; 6.75 ft (2.1 m) inner	
Inert mass	4,375 kg (9,650 lb)	1,030 kg (2,280 lb)	360 kg (790 lb)		
Gross mass	53,100 kg (117,100 lb)	10,810 kg (23,840 lb)	596 kg or 715 kg (1,310 lb or 1,570 lb)	535 kg (1,180 lb)	66,300 kg (146,100 lb)
Propulsion	Castor 120 Motor	Orbus 21D motor	Four Primex MR-107		
Propellant	HTPB	HTPB	Hydrazine		
Propellant mass	48,700 kg (107,400 lb)	8,780 kg (21,560 lb)	236 kg or 354 kg (520 lb or 780 lb)		
Avg. thrust	Sea level: 1,450 kN (325,900 lb) Vac.: (1,604 kN (360,500 lb)	187 kN (42,400 lb)	Initially 890 N (200 lb), decreases with time		1,450 kN (325,900 lb) at liftoff
Nominal burn time	83.4 sec	150 sec	1,500 sec (depends on mission)		
Max. payload	545 kg–820 kg (1,200 lb–1,805 lb) to low-Earth orbit depending on launch inclination				
Contractor	Thiokol	Pratt & Whitney	Primex Technologies	Lockheed Martin	Lockheed Martin

[a] Steven Isakowitz, Joseph P. Hopkins, Jr., and Joshua B. Hopkins, *International Reference Guide to Space Launch Systems*, 3rd ed., (Reston, Virginia: American Institute of Aeronautics and Astronautics, 1999), pp. 40–47.

Table 2–61. Athena II Characteristics[a]

	Stage 1	Stage 2	Stage 3	Orbit Assist Module	Payload Fairing Envelope	Total
Length	10.7 m (35.2 ft)	10.7 m (35.2 ft)	3.0 m (10.0 ft)	1.0 m (3.3 ft)	6.1 m (20 ft)	28.2 m (93.2 ft)
Diameter	2.3 m (7.7 ft)	2.3 m (7.7 ft)	2.3 m (7.7 ft)	2.3 m (7.7 ft)	7.7 ft (2.3 m) outer; 6.75 ft (2.1 m) inner	
Inert mass	4,375 kg (9,650 lb)	4,375 kg (9,650 lb)	1,030 kg (2,280 lb)	360 kg (790 lb)		
Gross mass	53,100 kg (117,100 lb)	53,100 kg (117,100 lb)	10,810 kg (23,840 lb)	596 kg or 715 kg (1,310 lb or 1,570 lb)	535 kg (1,180 lb)	120,700 kg (266,100 lb)
Propulsion	Castor 120 Motor	Castor 120 Motor	Orbus 21D motor	Four Primex MR-107		
Propellant	HTPB	HTPB	HTPB	Hydrazine		
Propellant mass	48,700 kg (107,400 lb)	48,700 kg (107,400 lb)	8,780 kg (21,560 lb)	236 kg or 354 kg (520 lb or 780 lb)		
Avg. thrust	Sea level: 1,450 kN (325,900 lb) Vac.: 1,604 kN (360,500 lb)	Sea level: 1,450 kN (325,900 lb) Vac.: 1,604 kN (360,500 lb)	187 kN (42,400 lb)	Initially 890 N (200 lb), decreases with time		1,450 kN (325,900 lb) at liftoff
Nominal burn time	83.4 sec	83.4 sec	150 sec	1,500 sec (depends on mission)		
Max. payload	1,575 kg to 2,065 kg (3,470 lb to 4,520 lb) to low-Earth orbit depending on orbital inclination					
Contractor	Thiokol	Thiokol	Pratt & Whitney	Primex Technologies	Lockheed Martin	Lockheed Martin

[a] Isakowitz, et al., *International Reference Guide to Space Launch Systems*, 3rd ed., pp. 40–47.

Table 2–62. Atlas Launches (1989–1998)

Vehicle Number[a]	Mission	Launch Date (Based on GMT)	Vehicle Type[b]	Comments
AC-68	Fltsatcom F-8	September 25, 1989	Atlas G Centaur	Launched by NASA/industry team for Navy. Last in NASA inventory of Atlas G Centaur rockets.
S/N 28	USA 56, 57, 58	April 11, 1990	Atlas E	DOD meteorological satellite. Classified mission.
AC-69	Combined Release and Radiation Effects Satellite (CRRES)	July 25, 1990	Atlas I	NASA–DOD mission. First Atlas I launch.
S/N 61	USA 68 (DMSP-10)	December 1, 1990	Atlas E	Defense Meteorological Satellite Program (DMSP) satellite.
AC-70	Yuri 3H (BS 3H)	April 18, 1991	Atlas I	Failed when one of the two Centaur engines did not start.
S/N 50	NOAA-12	May 14, 1991	Atlas E	NOAA weather satellite.
S/N 53	USA 73 (DMSP-11)	November 28, 1991	Atlas E	DOD weather satellite.
AC-102	Eutelsat II F3	December 7, 1991	Atlas II	European communications satellite.
AC-101	USA 78 (DSCS III-06)	February 11, 1992	Atlas II	Defense Satellite Communications Systems (DSCS) III satellite.
AC-72	Galaxy 5	March 14, 1992	Atlas I	Commercial communications satellite.

Table 2–62. Atlas Launches (1989–1998) (Continued)

Vehicle Number[a]	Mission	Launch Date (Based on GMT)	Vehicle Type[b]	Comments
AC-105	Intelsat-K	June 10, 1992	Atlas IIA	International communications satellite. First Atlas IIA launch.
AC-103	USA 82	July 2, 1992	Atlas II	DSCS III satellite.
AC-71	Galaxy 1R	August 22, 1992	Atlas I	Failed when one of the two Centaur engines did not start.
AC-74	UFO 1 (UHF-1)	March 25, 1993	Atlas I	Military communications satellite. Failed to reach operational orbit.
AC-104	USA 93 (DSCS III F8)	July 19, 1993	Atlas II	DSCS III satellite.
S/N 34	NOAA-13	August 9, 1993	Atlas E	NOAA weather satellite.
AC-75	UHF-2	September 3, 1993	Atlas I	U.S. Navy communications satellite.
AC-106	DSCS III	November 28, 1993	Atlas II	Military communications satellite.
AC-108	Telstar 401	December 16, 1993	Atlas IIAS	Communications satellite. First Atlas IIAS launch.
AC-73	GOES-8	April 13, 1994	Atlas I	NOAA weather satellite.
AC-76	UFO 3 (UHF-3)	June 24, 1994	Atlas I	U.S. Navy communications satellite.
AC-107	DBS-2	August 3, 1994	Atlas IIA	Communications satellite.
S/N 20	DMSP F-12	August 29, 1994	Atlas E	DMSP satellite.
AC-111	Intelsat 703	October 6, 1994	Atlas IIAS	International communications satellite.

Table 2–62. Atlas Launches (1989–1998) (Continued)

Vehicle Number[a]	Mission	Launch Date (Based on GMT)	Vehicle Type[b]	Comments
AC-110	Orion 1	November 29, 1994	Atlas IIA	German communications satellite.
S/N 11	NOAA-14	December 30, 1994	Atlas E	NOAA weather satellite.
AC-113	Intelsat 704	January 10, 1995	Atlas IIAS	International communications satellite.
AC-112	UFO-4 (USA 108)	January 29, 1995	Atlas II	U.S. Navy communications satellite.
AC-115	Intelsat 705	March 22, 1995	Atlas IIAS	International communications satellite.
S/N 45	USA 109 (DMSP-F13)	March 24, 1995	Atlas E	DMSP/F13 satellite. Last Atlas E launch.
AC-114	AMSC-1 (MSAT)	April 7, 1995	Atlas IIA	Provide mobile telephone communication.
AC-77	GOES-9	May 23, 1995	Atlas I	NOAA geostationary weather satellite.
AC-116	UHF 6 (USA 111)	May 31, 1995	Atlas II	Navy communications satellite.
AC-118	USA 113 (DSCSIII B5)	July 31, 1995	Atlas IIA	DSCS III satellite.
AC-117	JCSat 3	August 29, 1995	Atlas IIAS	Japanese communications satellite.
AC-119	UFO-6 (USA 114)	October 22, 1995	Atlas II	Military communications satellite.

Table 2–62. Atlas Launches (1989–1998) (Continued)

Vehicle Number[a]	Mission	Launch Date (Based on GMT)	Vehicle Type[b]	Comments
AC-121	Solar and Heliospheric Observatory	December 2, 1995	Atlas IIAS	NASA-European Space Agency space science mission.
AC-120	Galaxy 3-R	December 15, 1995	Atlas IIA	Commercial communications satellite.
AC-126	Palapa C-1	February 1, 1996	Atlas IIAS	Indonesian communications satellite.
AC-122	Inmarsat 3 F1	April 3, 1996	Atlas IIA	International communications satellite.
AC-78	Beppo-SAX	April 30, 1996	Atlas I	Italian Dutch telescope.
AC-125	UFO 7 (USA 127)	July 25, 1996	Atlas II	Military communications satellite.
AC-123	GE-1	September 8, 1996	Atlas IIA	Commercial communications satellite.
AC-124	Hot Bird 2	November 21, 1996	Atlas IIA	European communications satellite.
AC-129	Inmarsat 3 F3	December 18, 1996	Atlas IIA	Communications satellite.
AC-127	JCSat 4	February 17, 1997	Atlas IIAS	Japanese communications satellite.
AC-128	Tempo 2	March 8, 1997	Atlas IIA	Commercial communications satellite.
AC-79	GOES-10	April 25, 1997	Atlas I	NOAA geostationary weather satellite. Last Atlas I launch.
AC-133	Superbird C	July 28, 1997	Atlas IIAS	Japanese communications satellite.

Table 2–62. Atlas Launches (1989–1998) (Continued)

Vehicle Number[a]	Mission	Launch Date (Based on GMT)	Vehicle Type[b]	Comments
AC-146	GE-3	September 4, 1997	Atlas IIAS	Communications satellite.
AC-135	EchoStar 3	October 5, 1997	Atlas IIAS	Communications satellite.
AC-131	USA 133 (Lacrosse 3)/USA 135 (Defense Satellite Communications System 3 and Falcon Gold)	October 25, 1997	Atlas IIA	Military satellite.
AC-149	Galaxy 8i	December 8, 1997	Atlas IIAS	Communications satellite.
AC-109	USA 137 (Capricorn)	January 29, 1998	Atlas IIA	Military satellite.
AC-151	Intelsat 806	February 28, 1998	Atlas IIAS	International communications satellite.
AC-132	USA 138, UHF F8	March 16, 1998	Atlas II	Military communications satellite. Last Atlas II launch.
AC-153	Intelsat 805	June 18, 1998	Atlas IIAS	International communications satellite.
AC-134	Hotbird 5	October 9, 1998	Atlas IIA	Communications satellite.
AC-130	UHF F9	October 20, 1998	Atlas IIA	Military communications satellite.

[a] Atlas Centaur vehicle numbers from Jean-Jacques Serra and Gunter Krebs, "Atlas Centaur, Atlas Centaur Launches," *The Satellite Encyclopedia, http://www.tbs-satellite.com/tse/online/lanc_atlas_centaur.html* (accessed January 24, 2005).

[b] "Atlas Launches," International Launch Services Launch Archives, *http://www.ilslaunch.com/launches* (accessed January 25, 2005).

Table 2–63. Atlas E Characteristics[a]

	1-1/2 Stages (Booster and Sustainer)	Star Apogee Kick Motor (AKM)	Fairing	Total
Length	21.3 m (69.8 ft)	0.94 m (3.1 ft)	6.9 m (22.5 ft)	Up to 28.1 m (92.1 ft)
Diameter	3.05 m (10 ft)	0.94 m (3.1 ft)	2.1 m (7 ft)	
Gross mass	121,000 kg (266,759 lb)	47.7 kg (105 lb) (weight of motor)	735 kg (1,620 lb) assembly case after depletion of fuel	121,000 kg (266,759 lb)
Propulsion	MA-3 system consisting of two LR 89-NA-5 boosters, one LR 105-NA-5 sustainer, and two LR 101-NA-7 vernier engines (VE)	TE-M-364-15 motor		
Propellant	LOX-RP-1-1	Solid		
Propellant mass	112,900 kg (248,902 lb)	666 kg (1,468 lb)		
Liftoff thrust	Booster: 1,470 kN (330,000 lb) Sustainer: 267 kN (60,000 lb) Each vernier engine: 3.0 kN (670 lb)	42.4 kN (9,532 lb)[b]		1,743 kN (391,842 lb)
Burn time (average)	Booster: 120 sec, Sustainer: 309 sec	45 sec		
Max. payload	2,090 kg (4,608 lb) to 195-km (105-nmi) orbit from polar launch with dual TE-364-4 engines; 1,500 kg (3,307 lb) to 195-km orbit from polar launch with single TE 374-4 engine			
Contractors	Rocketdyne	Thiokol		General Dynamics
Remarks	Atlas E in this decade was used primarily to launch meteorological satellites into polar or geosynchronous orbit			

[a] Steven J. Isakowitz and Jeff Samella, *International Reference Guide to Space Launch Systems*, 2nd ed. (Washington, DC: American Institute of Aeronautics and Astronautics, 1991), pp. 206–211.

[b] "NOAA-D," Friends and Partners in Space, (downloaded to Friends and Partners from NASA Spacelink), *http://www.friends-partners.org/oldfriends/jgreen/noaa.html* (accessed January 25, 2005).

Table 2–64. Atlas G Centaur Characteristics[a]

	Atlas G Booster and Sustainer	Centaur Stage	Total
Length	22.2 m (72.8 ft)	9.15 m (30 ft)	38.0 m (125 ft) (includes fairing)
Diameter	3.05 m (10 ft)	3.05 m (10 ft)	3.05 m (10 ft)
Gross mass	145,700 kg (321,200 lb)	15,600 kg (34,300 lb)	166,140 kg (366,276 lb)[b] at liftoff (includes fairing)
Propulsion	MA-5 system consisting of two LR-89-NA-7 boosters, one LR-105-NA-7 sustainer, and two vernier engines	Two RL10A-3-3A[c] multiple-start engines and 12 small hydrogen peroxide thrusters	
Propellant	Oxidizer: LOX Fuel: RP-1	Oxidizer: LOX Fuel: LH2	
Propellant mass	138,300 kg (305,000 lb)	13,900 kg (30,600 lb)	
Liftoff thrust	Booster: 1,680 kN (377,500 lb) Sustainer: 269 kN (60,600 lb) Each vernier engine: 3 kN (670 lb)	146.8 kN (33,000 lb) vacuum	1,950 kN (438,877 lb)
Nominal burn time	Booster: 174 sec, Sustainer: 266 sec	402 sec	
Max. payload	6,100 kg (13,448 lb) to 185 km (100 nmi) orbit; 2,360 kg (5,203 lb) to geosynchronous transfer trajectory[d]		
Contractors	Rocketdyne	Pratt & Whitney	General Dynamics
Remarks	The lower booster and sustainer stage was integrated electronically with the Centaur upper stage		

[a] Isakowitz and Samella, *International Reference Guide to Space Launch Systems*, 2nd ed., pp. 206–210. The Atlas G was almost identical to the Atlas I. Isakowitz does not list Atlas G specifications, and Atlas I specifications are used in this table unless a different reference specific to Atlas G is noted.

[b] Federal Aviation Administration, "The Evolution of Commercial Launch Vehicles," *Fourth Quarter 2001 Launch Report*, *http://ast.faa.gov/files/pdf/q42001.pdf* (accessed January 25, 2005). Also "Atlas G," *Encyclopedia Astronautica, http://www.astronautix.com/lvs/atlasg.htm* (accessed January 25, 2005).

[c] According to Pratt & Whitney records, the RL10-3-3A engine was used on the launch of Fltsatcom F-8 on September 25, 1989. A number of earlier Atlas G launches used the RL10-3-3 engine.

[d] *Aeronautics and Space Report of the President, 1988 Activities*, (Washington, DC: National Aeronautics and Space Administration, 1990), p. 184.

Table 2–65. Atlas I Characteristics[a]

	Atlas I First Stage	Centaur Stage	Total
Length	22.2 m (72.8 ft)	9.15 m (30 ft)	Up to 43.9 m (144 ft) with large fairing
Diameter	3.05 m (10 ft)	3.05 m (10 ft)	
Gross mass	145,700 kg (321,200 lb)	15,600 kg (34,000 lb)	164,300 kg (362,200 lb)
Propulsion	MA-5 propulsion system consisting of two LR-89-NA-7 boosters, one LR-105-NA-7 sustainer, and two vernier single-start engines	Two RL10A-3-3A multiple start engines and 12 small hydrogen peroxide thrusters	
Propellant	Oxidizer: LOX Fuel: RP-1	Oxidizer: LOX Fuel: LH2	
Propellant mass	138,300 kg (305,000 lb)	13,900 kg (30,600 lb)	
Liftoff thrust	Booster: 1,680 kN (377,500 lb) Sustainer: 269 kN (60,600 lb) Each vernier engine: 3 kN (670 lb)	146.8 kN (33,000 lb) vacuum	1,950 kN (438,877 lb)[b]
Nominal burn time	Booster: 174 sec, Sustainer: 266 sec	402 sec	
Max. payload	6,580 kg (14,500 lb) to low-Earth orbit; 2,610 kg (5,754 lb) to geosynchronous transfer orbit; 4,300 kg (9,480 lb) to sun synchronous orbit[c]		
Contractors	Rocketdyne	Pratt & Whitney	General Dynamics
Remarks	An aluminum interstage adapter with a length of 3.96 m (13 ft), diameter of 3.05 m (10 ft), and mass of 477 kg (1,052 lb) supported the Centaur until separation took place.		

[a] Isakowitz and Samella, *International Reference Guide to Space Launch Systems*, 2nd ed., pp. 206–210.
[b] Liftoff thrust refers only to thrust produced by the Atlas stage. Thrust produced by the Centaur upper stage is produced approximately 4 minutes, 40 seconds after liftoff.
[c] *Aeronautics and Space Report of the President, Fiscal Year 1992 Activities* (Washington, DC: National Aeronautics and Space Administration, 1993), p. 94.

Table 2–66. Typical Atlas Launch Events Sequence for a Geosynchronous Mission[a]

Event	Time After Liftoff	Altitude Miles (Km)	Downrange Miles (Km)	Speed (mph/km per hr)
Liftoff	T-0			
Atlas booster engine cutoff	2 min 35 sec	37 (60)	54 (87)	6,527 (10,504)
Jettison Atlas booster engine	2 min 38 sec	38 (61)	59 (95)	6,590 (10,606)
Jettison Centaur insulation panel	3 min 0 sec	50 (80)	70 (113)	6,967 (11,212)
Jettison nose fairing	3 min 36 sec	67 (108)	154 (248)	7,746 (12,466)
Atlas sustainer/ vernier engines cutoff	4 min 27 sec	85 (137)	258 (415)	9,326 (15,009)
Atlas/Centaur separation	4 min 29 sec	86 (138)	266 (428)	9,330 (15,015)
First Centaur main engine start	4 min 40 sec	89 (143)	286 (460)	9,306 (14,977)
Centaur main engine cutoff	9 min 53 sec	94 (151)	1,298 (2,088)	17,953 (28,893)
Second Centaur main engine start	24 min 53 sec	212 (341)	5,366 (8,636)	17,487 (28,143)
Second Centaur main engine cutoff	26 min 29 sec	241 (388)	5,836 (9,392)	22,535 (36,267)
Centaur/payload separation	28 min 44 sec	334 (538)	6,566 (10,567)	22,262 (35,827)

[a] "CRRES Press Kit," Press Kit, July 1990, (from NASA Spacelink), *http://www.flyaria.com/document/html/mission/crres/cr.htm* (accessed July 18, 2006).

Table 2–67. Atlas II Characteristics[a]

	Atlas II Stage	Centaur II Upper Stage	Total
Length	24.9 m (81.7 ft)	9.15 m (30 ft)	47.5 m (156 ft) with large fairing
Diameter	3.05 m (10 ft)	3.05 m (10 ft)	
Gross mass	165,700 kg (365,300 lb)	15,600 kg (34,300 lb)	187,600 kg (413,500 lb)
Propulsion	MA-5A system with one two-chamber RS-27 booster engine and one RS-56SA sustainer engine	Two 10A-3-3A cryogenic multiple start engines	
Propellant	Oxidizer: LOX Fuel: RP-1	Oxidizer: LOX Fuel: LH2	
Propellant mass	155,900 kg (345,500 lb)	13,900 kg (30,000 lb)	
Avg. thrust	Booster: 1,840 kN (414,000 lb) Sustainer: 269 kN (60,500 lb)	146.8 kN (33,000 lb) (vacuum)	2,110 kN (474,500 lb)
Nominal burn time	Booster: 172 sec, Sustainer: 283 sec	402 sec	
Max. payload	6,580 kg (14,500 lb) to low-Earth orbit from Cape Canaveral; 5,510 kg (12,150 lb) to low-Earth orbit from Vandenberg AFB; 2,810 kg (6,200 lb) to geosynchronous transfer orbit		
Contractors	Rocketdyne	Pratt & Whitney	General Dynamics/ Lockheed
Remarks	The Atlas was integrated with the Centaur vehicle by an interstage adapter weighing 482 kg (1,067 lb) and measuring 3.05 m (10 ft) in diameter and 4 m (13 ft) long		

[a] Isakowitz and Samella, *International Reference Guide to Space Launch Systems*, 2nd ed., pp. 206–210.

Table 2–68. Atlas IIA Characteristics[a]

	Atlas IIA Stage	Centaur IIA Upper Stage	Payload Fairing	Total
Length	24.9 m (81.7 ft) + 4-m (13-ft) interstage	10 m (33 ft)	Large: 12.0 m (39.4 ft); extended: 12.9 m (42.4 ft)	47.4 m with large payload fairing and interstage
Diameter	3.05 m (10 ft)	3.05 m (10 ft)	4.2 m (13.7 ft)	
Inert mass	9,800 kg (21,605 lb) + 545-kg (1,202-lb) interstage	2,200 kg (4,850 lb)		
Gross mass	166,700 kg (367,510 lb) (includes interstage)	18,980 kg (41,844 lb)	Large: 2,085 kg (4,600 lb); extended: 2,255 kg (4,970 lb)	187,500 kg (413,366 lb) with large payload fairing
Propulsion	MA-5A system with one two-chamber RS-27 booster engine and one RS-56SA sustainer engine	Two RL10A-4 cryogenic multiple start engines		
Propellant	Oxidizer: LOX Fuel: RP-1	Oxidizer: LOX Fuel: LH2		
Propellant mass	156,400 kg (344,800 lb)	16,780 kg (37,000 lb)		
Avg. thrust	Booster: 1,854 kN (416,000 lb) sea level Sustainer: 266 kN (59,800 lb) sea level	185.2 kN (41,635 lb)		2,140 kN (481,200 lb) at liftoff
Nominal burn time	Booster: 165 sec, Sustainer: 274 sec	370 sec		
Max. payload	6,192 kg (13,651 lb)–7,316 kg (16,129 lb) to low-Earth orbit with large fairing depending on launch inclination; 3,066 kg (6,760 lb) to geosynchronous transfer orbit with large fairing			
Contractors	Rocketdyne	Pratt & Whitney		Lockheed Martin
Remarks	The operational Atlas IIA had uprated RL10 engines with optional nozzle extensions for the Centaur stage.			

[a] Isakowitz, et al., *International Reference Guide to Space Launch Systems*, 3rd ed., pp. 54, 68–73. Also, International Launch Services, *Atlas Launch System Mission Planner's Guide, Rev. 7* (December 1998), pp. 1–6, A9, *http://www.ilslaunch.com/missionplanner/pdf/amps_r7.pdf* (accessed January 27, 2005).

Table 2–69. Atlas IIAS Characteristics[a]

	Atlas IIAS Stage	Centaur IIAS Upper Stage	Payload Fairing	Total
Length	24.9 m (81.7 ft) + 4-m (13-ft) interstage SRB: 13.6 m (44.6 ft)	10 m (33 ft)	Large: 12.0 m (39.4 ft) Extended: 12.9 m (42.4 ft)	47.4 m with large payload fairing
Diameter	3.05 m (10 ft) SRBs: 102-cm (40 in)	3.05 m (10 ft)	4.2 m (13.7 ft)	
Inert mass	9,800 kg (21,605 lb) + 545-kg (1,202-lb) interstage	2,200 kg (4,850 lb)		
Gross mass	166,700 kg (367,510 lb) (includes interstage) SRBs: 11,567 kg (25,500 lb) (each fueled)	18,980 kg (41,850 lb)	Large: 2,085 kg (4,600 lb) Extended: 2,255 kg (4,970 lb)	237,200 kg (522,900 lb)
Propulsion	MA-5A system with one two-chamber RS-27 booster engine, and one RS-56SA sustainer engine augmented with four Castor IVA SRBs	Two 10A-4 cryogenic multiple-start engines		
Propellant	Oxidizer: LOX, Fuel: RP-1	Oxidizer: LOX, Fuel: LH2		
Propellant mass	156,400 kg (344,800 lb)	16,780 kg (37,800 lb)		
Thrust	Booster: 1,854 kN (416,000 lb) sea level Sustainer: 266 kN (59,800 lb) sea level SRBs: 433.7 kN (97,500 lb) each	185.2 kN (41,635 lb)		3,000 kN (676,200 lb)

Table 2–69. Atlas IIAS Characteristics[a] (Continued)

	Atlas IIAS Stage	Centaur IIAS Upper Stage	Payload Fairing	Total
Nominal burn time	Booster: 163 sec Sustainer: 289 sec	370 sec		
Max. payload	6,192 kg (15,900 lb) to 7,360 kg (19,000 lb) to low-Earth orbit depending on launch inclination; 3,719 kg (8,200 lb) to geosynchronous transfer orbit			
Contractors	Rocketdyne Thiokol: SRBs	Pratt & Whitney		Lockheed Martin

[a] Isakowitz, et al., *International Reference Guide to Space Launch Systems*, 3rd ed., pp. 54, 68–73. Also, International Launch Services, *Atlas Launch System Mission Planner's Guide, Rev. 7* (December 1998), pp. 1–6, A–9.

Table 2–70. Conestoga 1620 Characteristics[a]

	Booster Solid Rocket Motor Stage	Upper Stage	Payload Fairing	Total
Length	30 ft (9.12 m)	6.8 ft (2.07 m)	16 ft (4.88 m)	50 ft (15.24 m)
Diameter	3.3 ft (1.0 m)	4.1 ft (1.25 m)	72 in (1.83 m)	
Gross mass	Each: 25,100 lb (11,400 kg)	4,765 lb (2,161 kg)	Varies	192,700 lb (87.407 kg)
Propulsion	Two Castor IVA and four Castor IVB strap-on motors plus one Castor IVB core strap-on motor	Star 48V motor		
Propellant	Hydroxyl-terminated polybutadiene (HTPB)[b]	HTPB		
Propellant mass	Each: 22,300 lb (10,100 kg)	4,430 lb (2,010 kg)		
Thrust	Each: 111,000 lb (493,700 kN)	15,355 lb (68,300 N)		355,600 lb (1,581 kN)
Max. payload	5,000 lb to low-Earth orbit			
Contractors	Thiokol	Thiokol		EER Systems

a Isakowitz and Samella, International Reference Guide to Space Launch Systems, 2nd ed., pp. 221–224.
b This definition of HTPB as hydroxyl-terminated polybutadiene comes from the NASA Kennedy Space Center acronym list at www.ksc.nasa.gov/facts/acronyms.html. Other reliable acronym lists, including the NASA Scientific and Technical Information acronym list (http://www.sti.nasa.gov/acronym/h.html) define HTPB as hydroxy-terminated polybutadiene.

Table 2–71. Delta Launches (1989–1998)[a]

Delta Mission No.	Mission	Launch Date (GMT)	Vehicle Type	Comments
184	NAVSTAR II-1 GPS	February 14, 1989	Delta II/6925	Global Positioning System (GPS). First Delta II launch.
183	SDI Delta Star	March 24, 1989	Delta/3920-8	Last Delta 3920 launch.
185	NAVSTAR II-2 GPS	June 10, 1989	Delta II/6925	Second Block II NAVSTAR GPS satellite.
186	NAVSTAR II-3 GPS	August 18, 1989	Delta II/6925	Third Block II NAVSTAR GPS satellite.
187	BSB-R1	August 27, 1989	Delta/4925-8	Launched for British Satellite Broadcasting. First commercial licensed NASA U.S. space launch.
188	NAVSTAR II-4 GPS	October 21, 1989	Delta II/6925	Fourth Block II NAVSTAR GPS satellite.
189	COBE	November 18, 1989	Delta/5920	Cosmic Background Explorer. Last NASA-owned Delta.
190	NAVSTAR II-5 GPS	December 11, 1989	Delta II/6925	Fifth Block II NAVSTAR GPS satellite.
191	NAVSTAR II-6 GPS	January 24, 1990	Delta II/6925	Sixth Block II NAVSTAR GPS satellite.
192	SDI-LACE/RME (LOSAT)	February 14, 1990	Delta II/6920-8	Part of Strategic Defense Initiative program testing.
193	NAVSTAR II-7 GPS	March 26, 1990	Delta II/6925	Seventh Block II NAVSTAR GPS satellite.
194	Palapa B-2R	April 13, 1990	Delta II/6925-8	Indonesian communications satellite.
195	ROSAT	June 1, 1990	Delta II/6920-10	Röentgen Satellite. Joint German, U.S., and British space science mission.
196	INSAT-1D	June 12, 1990	Delta/4925-8	Indian communications and weather satellite. Last Delta I launch.
197	NAVSTAR II-8 GPS	August 2, 1990	Delta II/6925	Eighth Block II NAVSTAR GPS satellite.
198	BSB-R2 (Thor 1)	August 18, 1990	Delta II/6925	Launched for British Satellite Broadcasting.

Table 2–71. Delta Launches (1989–1998)[a] (Continued)

Delta Mission No.	Mission	Launch Date (GMT)	Vehicle Type	Comments
199	NAVSTAR II-9 GPS	October 1, 1990	Delta II/6925	Ninth Block II NAVSTAR GPS satellite.
200	INMARSAT-2 (F1)	October 30, 1990	Delta II/6925	International Maritime Satellite Organization.
201	NAVSTAR II-10 GPS	November 26, 1990	Delta II/7925	First Delta 7925. Tenth Block II NAVSTAR GPS satellite.
202	NATO IV-A	January 8, 1991	Delta II/7925	Military communications satellite.
203	INMARSAT-2 (F2)	March 8, 1991	Delta II/6925	International Maritime Satellite Organization.
204	ASC-2	April 13, 1991	Delta II/7925	Communications satellite.
205	Aurora II	May 29, 1991	Delta II/7925	Communications satellite.
206	NAVSTAR II-11 GPS and LOSAT-X	July 4, 1991	Delta II/7925	Eleventh Block II NAVSTAR GPS satellite and DOD mission.
207	NAVSTAR II-12 GPS	February 23, 1992	Delta II/7925	Twelfth Block II NAVSTAR GPS satellite.
208	NAVSTAR II-13 GPS	April 10, 1992	Delta II/7925	Thirteenth Block II NAVSTAR GPS satellite.
209	Palapa B4	May 14, 1992	Delta II/7925-8	Indonesian communications satellite.
210	EUVE	June 7, 1992	Delta II/6920-10	Extreme Ultraviolet Explorer.
211	NAVSTAR II-14 GPS	July 7, 1992	Delta II/7925	Fourteenth Block II NAVSTAR GPS satellite.
212	1) Geotail, 2) DUVE	July 24, 1992	Delta II/6925	1) Joint NASA-Japanese Institute of Space and Astronomical Science mission; 2) DUVE (Diffuse Ultraviolet Experiment) was attached to the 2nd stage.
213	SATCOM C-4	August 31, 1992	Delta II/7925	Comsat.

Table 2–71. Delta Launches (1989–1998)[a] (Continued)

Delta Mission No.	Mission	Launch Date (GMT)	Vehicle Type	Comments
214	NAVSTAR II-15 GPS	September 9, 1992	Delta II/7925	Fifteenth Block II NAVSTAR GPS satellite.
215	DFS 3 Kopernikus	October 12, 1992	Delta II/7925	Communications satellite launched by McDonnell Douglas for Germany.
216	NAVSTAR II-16 GPS	November 22, 1992	Delta II/7925	Sixteenth Block II NAVSTAR GPS satellite.
217	NAVSTAR II-17 GPS	December 18, 1992	Delta II/7925	Seventeenth Block II NAVSTAR GPS satellite.
218	NAVSTAR II-18 GPS	February 3, 1993	Delta II/7925	Eighteenth Block II NAVSTAR GPS satellite.
219	NAVSTAR II-19 GPS and SEDS-1	March 30, 1993	Delta II/7925	Nineteenth Block II NAVSTAR GPS satellite and Small Expendable Deployer System tether experiment.
220	NAVSTAR II-20 GPS	May 13, 1993	Delta II/7925	Twentieth Block II NAVSTAR GPS satellite.
221	NAVSTAR II-21 GPS and PMG	June 26, 1993	Delta II/7925	Twenty-first Block II NAVSTAR GPS satellite and Plasma Motor Generator was tethered to the 2nd stage.
222	NAVSTAR II-22 GPS	August 30, 1993	Delta II/7925	Twenty-second Block II NAVSTAR GPS satellite.
223	NAVSTAR II-23 GPS	October 26, 1993	Delta II/7925	Twenty-third Block II NAVSTAR GPS satellite.
224	NATO IVB	December 8, 1993	Delta II 7925	Military communications satellite. Launched commercially by McDonnell Douglas.
225	Galaxy I-R	February 19, 1994	Delta II/7925-8	Communications satellite launched commercially by McDonnell Douglas.

Table 2–71. Delta Launches (1989–1998)[a] (Continued)

Delta Mission No.	Mission	Launch Date (GMT)	Vehicle Type	Comments
226	NAVSTAR II-24 GPS and SEDS-2	March 10, 1994	Delta II/7925	Twenty-fourth Block II NAVSTAR GPS satellite and SED-2 tether experiment.
227	Wind	November 1, 1994	Delta II/7925-10	International Solar Terrestrial Physics/ Global Geospace Science program.
228	Koreasat-1	August 5, 1995	Delta II/7925	Partial failure; booster failed to separate.[b]
229	RADARSAT and SURFSAT	November 4, 1995	Delta II/7920-10	Canadian remote sensing mission and Student Undergraduate Research Fellowship Satellite.
230	RXTE	December 30, 1995	Delta II/7920-10	Rossi X-ray Timing Explorer.
231	Koreasat-2	January 14, 1996	Delta II/7925	Korean communications satellite.
232	NEAR	February 17, 1996	Delta II/7925-8	Near Earth Asteroid Rendezvous.
233	Polar	February 24, 1996	Delta II/7925-10	Space physics satellite.
234	NAVSTAR II-25 GPS	March 28, 1996	Delta II/7925	Twenty-fifth Block II NAVSTAR GPS satellite.
235	Middlecourse Space Experiment (MSX)	April 24, 1996	Delta II/7920-10	USA 118.
236	Galaxy IX	May 24, 1996	Delta II/7925-8	Commercial communications satellite.
237	NAVSTAR II-26 GPS	July 16, 1996	Delta II/7925	Twenty-sixth Block II NAVSTAR GPS satellite.
238	NAVSTAR II-27 GPS	September 12, 1996	Delta II/7925	Twenty-seventh Block II NAVSTAR GPS satellite.
239	Mars Global Surveyor	November 7, 1996	Delta II/7925	Remote sensing mission of Mars.
240	Mars Pathfinder	December 4, 1996	Delta II/7925	Planetary spacecraft with rover.

Table 2–71. Delta Launches (1989–1998)[a] (Continued)

Delta Mission No.	Mission	Launch Date (GMT)	Vehicle Type	Comments
241	GPS BIIR-01 (NAVSTAR 2R-1)	January 17, 1997	Delta II/7925	Failed due to split in the casing of one of the solid rocket motors.
242	MS-1 Iridium® (5 satellites)	May 5, 1997	Delta II/7920-10C	Communications satellites.
243	Thor II	May 20, 1997	Delta II/7925	Norwegian communications satellite.
244	MS-2 Iridium® (5 satellites)	July 9, 1997	Delta II/7920-10C	Communications satellites.
245	NAVSTAR GPS-IIR2	July 23, 1997	Delta II/7925	Block IIR NAVSTAR GPS satellite.
246	MS-3 Iridium® (5 satellites)	August 21, 1997	Delta II/7920-10C	Communications satellites.
247	Advanced Composition Explorer (ACE)	August 25, 1997	Delta II/7920-8	Space science mission.
248	MS-4 Iridium® (5 satellites)	September 27, 1997	Delta II/7920-10C	Communications satellites.
249	NAVSTAR II-28 GPS	November 6, 1997	Delta II/7925	Twenty-eighth block II NAVSTAR GPS satellite.
250	MS-5 Iridium® (5 satellites)	November 9, 1997	Delta II/7920-10C	Communications satellites.
251	MS-6 Iridium® (5 satellites)	December 20, 1997	Delta II/7920-10C	Communications satellites.
252	Skynet 4D	January 10, 1998	Delta II/7925	British military communications satellite.
253	Globalstar-1 (4 satellites Space Systems/Loral)	February 14, 1998	Delta II/7420	Communications satellites.

Table 2–71. Delta Launches (1989–1998)[a] (Continued)

Delta Mission No.	Mission	Launch Date (GMT)	Vehicle Type	Comments
254	MS-7 Iridium® (5 satellites)	February 18, 1998	Delta II/7920-10C	Communications satellites.
255	MS-8 Iridium® (5 satellites)	March 30, 1998	Delta II/7920-10C	Communications satellites.
256	Globalstar-2 (4 satellites Space Systems/Loral)	April 24, 1998	Delta II/7420-10C	British military communications satellites.
257	MS-9 Iridium® (5 satellites)	May 17, 1998	Delta II/7920-10C	Communications satellites.
258	Thor III	June 10, 1998	Delta II/7925	European communications satellite.
259	Galaxy X	August 27, 1998	Delta III/8930	Failed. Exploded 80 seconds after liftoff. First Delta III launch.
260	MS-10 Iridium® (5 Satellites)	September 8, 1998	Delta II/7920-10C	Communications satellites.
261	Deep Space 1 and Sedsat	October 24, 1998	Delta II/7326[c]	New Millennium Program and Students for the Exploration and Development of Space Satellite secondary payload.
262	MS-11 Iridium® (5 satellites)	November 6, 1998	Delta II/7920-10C	Communications satellites.
263	BONUM-1	November 22, 1998	Delta II/7925	Russian television satellite.
264	Mars Climate Orbiter	December 11, 1998	Delta II/7425	Interplanetary spacecraft.

a "Delta Launch Record," *http://www.boeing.com/defense-space/space/delta/record.htm* (accessed January 31, 2005).
b Koreasat-1 was able to achieve orbit. The Delta booster, however, placed the satellite in a lower-than-specified orbit, thus shortening its useful life.
c New variant of Delta II that used three solid Alliant GEM-40 strap-ons rather than nine.

Table 2–72. Delta 3920/PAM-D Characteristics

	Strap-ons (each)[a]	Stage 1	Stage 2	Stage 3 (Payload Assist Module)	Total
Length	9.07 m (30 ft)	22.8 m (75 ft) (includes second stage)	6 m (19.6 ft)	2 m (6.6 ft)	35.5 m (116 ft) including fairing
Diameter	1.02 m (3.3 ft)	2.4 m (8 ft)	2.4 m (8 ft)	1.25 m (4.1 ft)	
Gross mass	10,530 kg (23,215 lb)	85,076 kg (187,560 lb)	6,930 kg (15,331 lb)	1,122 kg (2,474 lb)	
Propulsion	Nine Thiokol Castor IV TX 526-2 strap-on motors	Rocketdyne RS-27 assembly consisting of one RS27 A/B main engine and two LR101-NA-11 vernier engines	Aerojet AJ10-118K engine	Thiokol Star 48 motor	
Propellant	HTPB	Oxidizer: LOX Fuel: RP-1	Aerozine-50 and N_2O_4	HTPB	
Propellant mass	9,373 kg (20,664 lb)	79,380 kg (175,000 lb)	6,004 kg (13,236 lb)	1,909 kg (4,200 lb)	
Avg. Thrust	428 kN (96,218 lb)	1,030 kN (231,553 lb)	44 kN (9,815 lb)	66.6 kN (14,972 lb)	
Nominal burn time	57 sec	224 sec	431 sec	44 sec	
Max. payload		3,045 kg (6,713 lb) to low-Earth orbit; 1,275 kg (2,800 lb) to geosynchronous transfer orbit; 2,135 kg (4,700 lb) to circular sun-synchronous orbit (polar launch)[b]			
Contractors	Thiokol	Rocketdyne	Aerojet	Thiokol	McDonnell Douglas

[a] Jean-Jacques Serra, "Castor," *The Satellite Encyclopedia*, http://www.tbssatellite.com/tse/online/lanc_castor.html (accessed April 7, 2005).
[b] *Aeronautics and Space Report of the President, 1989–1990 Activities* (Washington, DC: National Aeronautics and Space Administration, 1991), p. 160.

Table 2–73. Delta II 6925 Characteristics[a]

	Strap-ons (each)	Stage 1	Stage 2	Stage 3 (Payload Assist Module)	Total
Length	11.2 m (36.3 ft)	26.1 m (85.6 ft)	6 m (19.6 ft)	2 m (6.7 ft)	Up to 38.1 m (125 ft) including fairing
Diameter	1.0 m (3.3 ft)	2.44 m (8 ft)	2.44 m (8 ft)	1.25 m (4.1 ft)	
Gross mass	Ground lit: 11,700 kg (25,800 lb) Air lit: 11,900 kg (26,100 lb)	101,700 kg (224,210 lb)	6,997 kg (15,400 lb)	2,141 kg (4,721 lb)	220,000 kg (480,000 lb)
Propulsion	Nine Castor IVA solid rocket motors	Rocketdyne RS-27 assembly consisting of one RS2701A/B main engine and two LR101-NA-11 vernier engines	Aerojet AJ10-118K engine	Thiokol Star 48B motor	
Propellant	HTPB	Oxidizer: LOX Fuel: RP-1	Aerozine-50 and N2O4	HTPB	
Propellant mass	10,100 kg (22,300 lb)	96,100 kg (211,900 lb)	6,076 kg (14,400 lb)	2,009 kg (4,430 lb)	
Avg. thrust	427.1 kN (97,700 lb) at sea level 478.3 kN (108,700 lb) vac.	911 kN (204,800 lb) (sea level)	42.4 kN (9,645 lb)	66.4 kN (15,100 lb)	2,620 kN at liftoff (595,000 lb)
Nominal burn time	56.2 sec	265 sec	440 sec	54.8 sec	
Max. payload		5,039 kg (11,100 lb) to low-Earth orbit; 1,819 kg (4,000 lb) to geosynchronous transfer orbit, 3,175 kg (7,000 lb) to sun synchronous orbit[b]			
Contractors	Thiokol	Rocketdyne	Aerojet	Thiokol	McDonnell Douglas

[a] Isakowitz and Samella, *International Reference Guide to Space Launch Systems*, 2nd ed., pp. 234–237.
[b] *Aeronautics and Space Report of the President, Fiscal Year 1992 Activities*, p. 94.

Table 2–74. Delta 7925 Characteristics[a]

	Strap-on Solid Rocket Motors	Stage 1	Stage 2	Stage 3 (Payload Assist Module)[b]	Total
Length	13.0 m (42.5 ft)	26.1 (85.6 ft)	6 m (19.6 ft)	2 m (6.7 ft)	38.2 m–38.9 m (125.2 ft–126.5 ft) depending on fairing
Diameter	1.0 m (3.3 ft)	2.4 m (8 ft)	2.4 m (8 ft)	1.25 m (4.1 ft)	
Gross mass	13,080 kg (28,840 lb) each	101,800 kg (224,400 lb)	6,954 kg (15,331 lb)	2,217 kg (4,887 lb)	231,870 kg (511,190 lb)
Propulsion	Nine Hercules GEM 40 solid rocket motors; some configurations used three or four motors	Rocketdyne RS-27 assembly consisting of one RS27A/B main engine and two LR101-NA-11 vernier engines	Aerojet AJ10-118K engine	Thiokol Star 48B motor	
Propellant	HTPB	Oxidizer: LOX Fuel: RP-1	Aerozine-50 and N2O4	HTPB	
Propellant mass	11,765 kg (25,940 lb)	96,100 kg (211,900 lb)	6,004 kg (13,236 lb)	2,009 kg (4,430 lb)	
Avg. thrust	Sea level: 446 kN (100,300 lb); air-lit: 516.2 kN (116,100 lb) each	890 kN (200,000 lb) (sea level)	44 kN (9,815 lb)	66.4 kN (14,927 lb)	3,110 kN (699,250 lb) at liftoff
Nominal burn time	63.3 sec	261 sec	431 sec	87.1 sec	
Max. payload	3,895 kg (8,590 lb) to 5,140 kg (11,330 lb) to low-Earth orbit depending on launch inclination; 3,220 kg (7,100 lb) to Sun-synchronous orbit; 1,870 kg (4,120 lb) to geosynchronous transfer orbit				
Contractors	Alliant Techsystems	Rocketdyne	Aerojet	Thiokol	McDonnell Douglas

[a] Isakowitz et al., International Reference Guide to Space Launch Systems, 3rd ed., pp. 112, 115–118.
[b] No PAM upper stage was used for low-Earth orbit missions.

Table 2–75. Representative Delta II Mission Profile Events

Event	Mission Elapsed Time (sec)
Main engine and six solid motors ignited, liftoff	0.0
Mach 1	32.4
Maximum dynamic pressure	49.7
Solid motor burnout (6 of 9)	56
Solid motor ignition (3 of 9)	59
Jettison 6 solid motors	60/61
Jettison 3 solid motors	118
Stage 1 main engine cutoff (MECO)	265
Stage 1-2 separation	271.4
Stage 2 ignition	278
Payload fairing jettison	298
Stage 2 engine first cutoff 1 (SECO 1)	687
Stage 2 restart ignition	1263
Second cutoff–Stage 2 (SECO 2)	1286
Stage 2-3 separation	1300
Stage 3/PAM ignition	1376
Stage 3/PAM burnout	1463
Spacecraft separation	1576

Table 2–76. Pegasus Launches (1989–1998)

Launch Date	Vehicle Model	Customer(s)	Payload	Type of Mission
April 5, 1990	Standard	NASA, DOD	PegSat, USA 55 (SECS)	Flight test instrumentation and atmospheric research. Navy experimental satellite.
July 7, 1991	Standard with HAPS	DOD	MicroSat 1, 2, 3, 4, 5, 6, and 7	Tactical communications network. Achieved mission objectives at lower orbit than planned.[a]
February 9, 1993	Standard	1) INPE Brazil	1) SCD-1	1) Data communications.
		2) Orbital Sciences Corp.	2) OXP-1	2) Experimental communications satellite.
April 25, 1993	Standard	1) Department of Energy-sponsored	1) ALEXIS	1) Array of Low Energy X-ray Imaging Sensors. Satellite was damaged at launch, delaying communication with ground by six weeks.
		2) Orbital Sciences Corp.	2) OXP-2[b]	2) Experimental communications satellite.
May 19, 1994	Standard with HAPS	DOD	STEP-2	Technology validation. Satellite placed in lower than expected orbit.
June 27, 1994	XL	DOD	STEP-1	Technology validation. Mission failed.
August 3, 1994	Standard	DOD	APEX	Advanced Photovoltaic and Electronic Experiments. Space physics technology validation.
April 4, 1995	Standard (Hybrid)	1) ORBCOMM	1) FM1 & FM2	1) Communications.
		2) NASA	2) MicroLab 1	2) Atmospheric research.
June 22, 1995	XL	DOD	STEP-3	Technology validation. Mission failed.

Table 2–76. Pegasus Launches (1989–1998) (Continued)

Launch Date	Vehicle Model	Customer(s)	Payload	Type of Mission
March 8, 1996	XL	DOD	REX-2	Radiation experiment. Technology validation.
May 16, 1996	Standard (Hybrid)	U.S. Air Force	MSTI-3	Miniature Sensor Technology Integration. Technology validation.
July 2, 1996	XL	NASA	TOMS-EP	Total Ozone Mapping Spectrometer Earth Probe. Atmospheric research.
August 21, 1996	XL	NASA	FAST	Fast Auroral Snapshot Explorer. Space physics research.
November 4, 1996	XL	NASA	SAC-B HETE-1	Space physics research. Spacecraft did not separate from third stage. Mission failed.[c]
April 21, 1997	XL	INTA Spain	MINISAT 01	Space physics research. Spain's first satellite, also release of funeral ashes.[d]
August 1, 1997	XL	Orbital Sciences Corp./ NASA	OrbView-2 (SeaStar)	Ocean color imaging, Sea-viewing Wide Field-of-view Sensor (SeaWiFS) Project.
August 29, 1997	XL	DOD	FORTE	Technology validation.
October 22, 1997	XL	DOD	STEP-4	Technology validation.
December 23, 1997	XL with HAPS	ORBCOMM-1	ORBCOMM 5-12	Eight low-Earth orbit communications satellites.
February 25, 1998	XL	1) NASA, 2) Teledesic	1) SNOE, 2) T1	1) Student Nitric Oxide Explorer. 2) Commercial communications satellite.
April 1, 1998	XL	NASA	TRACE	Transition Region and Coronal Explorer. Solar physics.

Table 2–76. Pegasus Launches (1989–1998) (Continued)

Launch Date	Vehicle Model	Customer(s)	Payload	Type of Mission
August 2, 1998	XL with HAPS	ORBCOMM-2	ORBCOMM 13-20	Eight low-Earth orbit communications satellites.
September 23, 1998	XL with HAPS	ORBCOMM-3	ORBCOMM 21-27	Eight low-Earth orbit communications satellites.
October 22, 1998	Standard (Hybrid)	INPE Brazil	SCD-2	Data communications.
December 5, 1998	XL	NASA	SWAS	Submillimeter Wave Astronomy Satellite. Space physics.

[a] The 356-km by 455-km (192-nmi by 246-nmi) orbit fell short of the planned 720-km (389-nmi) circular orbit because a problem at first stage separation caused a guidance error. Orbital Sciences listed the mission as a "success" and stated that the inclination was on target, allowing mission objectives to be met (*Orbital Sciences Corporation Spacecraft History*, table 2–2, pp. 2–8). The satellites' customer, the Defense Advanced Research Projects Agency (DARPA), said that the rocket's guidance system compensated for the low orbit and repositioned the satellites to an elliptical orbit that ensured DARPA's objectives were met. *Aviation Week & Space Technology*, July 22–July 24, 1991 (NASA History Office Folder 010788). However, other references call it a "failure," or "partial failure." (Mark Wade, Astronautix.com, *http://www.astronautix.com/lvs.pegasus.htm* (accessed February 8, 2005)). Also "Launching on Pegasus," Small Satellites home page, *http://centaur.sstl.co.uk/SSHP/launcher/launch_pegasus.html* (accessed February 8, 2005) and "Pegasus," *The Satellite Encyclopedia*, *http://tbs-satellite.com/tse/online/lanc_pegasus.html* (accessed February 8, 2005).

[b] This payload is listed on Jonathan's Space Report, *http://planet4589.org/space/log/launchlog.txt* (accessed February 23, 2005) but does not appear on the Pegasus Mission History list produced by Orbital Sciences: *http://www.orbital.com/SpaceLaunch/Pegasus/pegasus_history.htm* (accessed February 3, 2005). It appears that the payload did not separate from the Pegasus third stage.

[c] SAC-B was unable to deploy its solar arrays because the spacecraft did not separate from the Pegasus third stage due to a battery failure in the Pegasus third stage. HETE remained sealed in the interior of the dual payload support structure. SAC-B solar arrays were deployed via ground commands but were unable to generate enough power to keep the satellite's batteries charged. Both died due to power failure within days of launch. "SAC-B/HETE Spacecraft No Longer Operational," *NASA News* Release 96-231, November 7, 1996, *ftp://ftp.hq.nasa.gov/pub/pao/pressrel/1996/96-231.txt* (accessed February 10, 2005); "Partial Launch Vehicle and Spacecraft Re-enter Earth's Atmosphere," Goddard Space Flight Center Top Story, April 4, 2002, updated April 7, 2002, *http://www.gsfc.nasa.gov/topstory/20020401hetereenter.html* (accessed February 10, 2005); "SAC-B" Gunter's Space Page *http://skyrocket.de/space/doc_sdat/sac-b.htm* (accessed February 23, 2005) and "HETE 1, 2," Gunter's Space Page *http://space/skyrocket.de/doc_sdat/hete.htm* (accessed April 14, 2006).

[d] Among the 24 capsules of funeral ashes taken aloft and put into orbit from this Pegasus were those of the 1960s icon, Timothy Leary, and Gene Roddenberry, creator of "Star Trek." Marlise Simons, "A Final Turn-On Lifts Timothy Leary Off," *New York Times*, April 22, 1997, A1.

Table 2–77. Standard Pegasus Characteristics[a]

	Stage 1	Stage 2	Stage 3	Total
Length	8.9 m (29 ft)	2.3 m (7.5 ft)	1.3 m (4.3 ft)	15.2 m[b] (50 ft)
Diameter	1.28 m (4.2 ft) without 6.7-m (22-ft) wingspan	1.28 m (4.2 ft)	0.97 m (3.2 ft)	
Liftoff mass	13,417 kg (29,579 lb)	3,367 kg (7,423 lb)	897 kg (1,978 lb)	
Propulsion	Orion 50S motor	Orion 50 motor	Orion 38 motor	
Propellant	HTPB	HTPB	HTPB	
Propellant mass	12,160 kg (26,808 lb)	3,024 kg (6,667 lb)	771 kg (1,700 lb)	
Nominal burn time[c]	72.4 sec	73.3 sec	68.4	
Thrust (max. vac.)	580.46 kN (130,493 lb)	138.64 kN (31,168 lb)	35.81 kN (8,050 lb)	
Max. payload[d]	380 kg into 185-km orbit; 280 kg into 185-km polar orbit from Vandenberg Air Force Base; 210 kg into sun-synchronous orbit from Vandenberg Air Force Base			
Contractor	Hercules	Hercules	Hercules	Orbital Sciences

[a] All vehicle characteristics are from *NASA SELVS Pegasus Launch System Payload User's Guide*, Release 2.00 (Orbital Sciences Corporation, June 1994), pp. 2–7, unless otherwise indicated.

[b] "Pegasus Launch Vehicle," Space & Missile Systems Center, Department of the Air Force, *http://www.te.plk.af.mil/factsheet/pegfact.html* (accessed February 8, 2005). Included aft skirt assembly, interstage, and fairing.

[c] At 21°C (70°F).

[d] *Aeronautics and Space Report of the President, Fiscal Year 1994 Activities* (Washington, DC: National Aeronautics and Space Administration, 1995), p. 91.

Table 2–78. Pegasus XL Characteristics[a]

	Stage 1	Stage 2	Stage 3	Total
Length	10.3 m (34 ft)	3.1 m (10.2 ft)	1.3 m (4.3 ft)	16.9 m (55.4 ft) including interstage and fairing
Diameter	1.28 m (4.2 ft) without 6.7-m (22-ft) wingspan	1.28 m (4.2 ft)	1 m (3.3 ft)	
Liftoff mass	16,383 kg (36,118 lb)	43,411 kg (95,705 lb)	896 kg (1,975 lb)	23,130 kg (26,742 lb)
Propulsion	Orion 50S XL motor	Orion 50 XL motor	Orion 38 motor	
Propellant	HTPB	HTPB	HTPB	
Propellant mass	15,014 kg (33,100 lb)	3,925 kg (8,653 lb)	770 kg (1,698 lb)	
Nominal burn time[b]	68.6 sec	69.4 sec	68.5 sec	
Thrust (max. vac.)	726 kN (163,211 lb)	196 kN (44,063 lb)	36 kN (8,093 lb)	
Payload capacity[c]	460 kg (1,014 lb) into 185-km orbit; 350 kg (772 lb) into 185-km polar orbit from Vandenberg Air Force Base; 335 kg (739 lb) into sun-synchronous orbit from Vandenberg Air Force Base			
Contractor	Alliant Techsystems	Alliant Techsystems	Alliant Techsystems	Orbital Sciences
Remarks	All XL launches have taken place from the L-1011 "Stargazer" aircraft			

[a] *Pegasus User's Guide*, Release 5.0, August 2000 (Orbital Sciences Corporation, 2000), pp. 2–4, *http://www.orbital.com/NewsInfo/Publications/peg-user-guide.pdf* (accessed February 4, 2005).
[b] At 21°C (70°F).
[c] *Aeronautics and Space Report of the President, Fiscal Year 1999 Activities* (Washington, DC: National Aeronautics and Space Administration, 2000), p. 97.

Table 2–79. Scout G1 Characteristics[a]

	Stage 1	Stage 2	Stage 3	Stage 4	Total
Length	9.94 m (32.6 ft)	6.56 m (21.5 ft)	3.28 m (10.8 ft)	1.97 m (6.5 ft)	23 m (75 ft) including transition and payload sections
Diameter	1.01 m (3.3 ft) max.	0.79 m (2.6 ft)	0.75 m (2.5 ft)	0.5 m (1.7 m)	
Launch mass	14,255 kg (31,361 lb)	4,424 kg (9,753 lb)	1,395 kg (3,075 lb)	302 kg (665.8 lb)	
Propulsion	Algol IIIA motor	Castor IIA motor	Antares IIIA motor	Altair IIIA motor	
Propellant	Solid	Solid	Solid	Solid	
Propellant mass	12,684 kg (27,965 lb)	3,762 kg (8,294 lb)	1,286 kg (2,835 lb)	275 kg (606.3 lb)	
Avg. thrust	467.1 kN (105,112 lb)	284.3 kN (63,971 lb)	83.1 kN (18,698 lb)	25.4 kN (5713 lb)	
Nominal burn time	56 sec	35 sec	44 sec	29 sec	
Payload capacity	175 kg (386 lb) to a 185-km (100-nmi) orbit				
Prime Contractor	Vought Corp. (LTV Corp.)				

[a] "Scout–Launch Vehicle," Vought Corp., *http://www.vought.com/heritage/special/html/sscout1.html* (accessed February 9, 2005). "Scout Launch Vehicle Program," Langley Research Center Fact Sheet, last updated November 24, 2004, *http://www.nasa.gov/centers/langley/news/factsheets/Scout.html* (accessed February 9, 2005).

Table 2–80. Scout Launches (1989–1998)

Mission No.	Launch Date	Vehicle Type	Customer(s)	Payload	Comment
212C	May 9, 1990	Scout G1	Reimbursable DOD	MACSAT (Multiple Access Comsat)	Two NAVY/DARPA communications satellites
216C	June 29, 1991	Scout G	DOD	REX	Air Force Radiation Experiment
215C	July 3, 1992	Scout G1	NASA	SAMPEX	Solar, Anomalous and Magnetospheric Particle Explorer, first Small Explorer mission
210C	November 21, 1992	Scout G1	Ballistic Missile Defense Organization and U.S. Air Force	MSTI I (Miniature Sensor Technology Integration)	Atmospheric studies
217C	June 25, 1993	Scout G1	U.S. Air Force	RADCAL	Radar Calibration Satellite
218	May 9, 1994	Scout G1	Ballistic Missile Defense Organization and U.S. Air Force	MSTI II	Tracking and Earth observation studies. Last Scout launch

Table 2–81. Taurus 2210 Characteristics[a]

	Stage 0[b]	Stage 1	Stage 2	Stage 3	Total
Length	12.8 m (41.9 ft)	8.6 m (28.3 ft)	3.1 m (10.1 ft)	1.3 m (4.4 ft)	27.9 m (91.4 ft) including interstage and fairing
Diameter	2.4 m (7.8 ft)	1.3 m (4.2 ft)	1.3 m (4.2 ft)	1.0 m (3.2 ft)	2.4 m (7.8 ft)
Liftoff mass	53,424 kg (117,800 lb)	13,242 kg (29,200 lb)	3,379 kg (7,450 lb)	875 kg (1,930 lb)	73,000 kg (161,000 lb)[c]
Propulsion	Castor 120 motor	Orion 50S-G motor	Orion 50 motor	Orion 38 motor[d]	
Propellant	HTPB	HTPB	HTPB	HTPB	
Propellant mass	49,024 kg (108,100 lb)	12,154 kg (26,800 lb)	3,027 kg (6,674 lb)	771 kg (1,700 lb)	
Thrust (avg. vac.)	1,615 kN (363,087 lb)	471 kN (106,000 lb)	115 kN (25,910 lb)	13.8 kN (7,155 lb)	
Nominal burn time	82.5 sec	72.4 sec	75.1 sec	68.5 sec	
Payload capacity	1,400 kg (3,086 lb) into 185-km orbit (100-nmi); 1,080 kg (2,381 lb) into 185-km (100-nmi) polar orbit from Vandenberg Air Force Base; 255 kg (562 lb) into geosynchronous transfer orbit; 1,020 kg (2,249 lb) into sun-synchronous orbit from Vandenberg Air Force Base[e]				
Contractor	Thiokol	Alliant Techsystems	Alliant Techsystems	Alliant Techsystems	Orbital Sciences Corp.

a   Taurus User's Guide, Release 3.0 (Orbital Sciences Corporation, September 1999), pp. 2–5, http://www.orbital.com/NewsInfo/Publications/taurus-user-guide.pdf (accessed February 9, 2005). Liftoff masses and total length were not available from the Taurus User's Guide and were obtained from Isakowitz et al., International Reference Guide to Space Launch Systems, 3rd ed., pp. 440–441.

b   The first stage was known as "Stage 0."

c   Included interstage and fairing.

d   This stage could be replaced by a spin-stabilized upper stage using Thiokol's Star 37FM perigee kick motor for insertion into geosynchronous transfer orbit.

e   Aeronautics and Space Report of the President, Fiscal Year 1999 Activities, p. 97.

Table 2–82. Taurus Launches (1989–1998)

Launch Vehicle	Launch Date	Mission	Comments
Taurus ARPA	March 13, 1994	STEP-0 (USA-101), Darpasat (USA-102)	DOD mission
Taurus 2210	February 10, 1998	1) Celestis 2 2) Geosat Follow-on (GFO) 3) ORBCOMM FM-3, FM-4	1) funeral ashes disposal 2) military Earth science 3) communications satellite
Taurus ARPA	October 3, 1998	STEX, ATEX (USA-141)	DOD mission

Table 2–83. Titan Launches (1989–1998)

Titan Launch Vehicle	Launch Date (GMT)	Mission	Comments
34D	May 10, 1989	USA 37	DOD satellite.
IV	June 14, 1989	USA 39	Defense Support Program satellite. IUS booster.
34D	September 4, 1989	USA 43, 44	Defense Satellite Communications System payload.
II	September 6, 1989	USA 45	DOD satellite.
III	January 1, 1990	Skynet 4A/JCSat 2	U.K. defense communications satellite/Japanese communications satellite. First commercial Titan III launch.
III	March 14, 1990	Intelsat 6 F-3	International communications satellite. Second stage reached correct orbit but failed to deploy payload. Satellite separated itself from kick stage and was rescued and reboosted by astronauts on STS-49 mission in May 1992.[a]
IVA	June 8, 1990	USA 59, 60, 61, 62	DOD satellite.
III	June 23, 1990	Intelsat 6 F-4	International communications satellite.
IVA	November 13, 1990	USA 65	DOD satellite. IUS booster.
IVA	March 8, 1991	USA 69	DOD satellite.
IVA	November 8, 1991	USA 72, 74, 76, 77	DOD satellite.
II	April 25, 1992	USA 81	DOD satellite.
III	September 25, 1992	Mars Observer	NASA space science mission launched by refurbished Titan ICBM. Transfer orbit kick stage.
IVA	November 28, 1992	USA 86	DOD satellite.
IVA	August 2, 1993	USA	DOD satellite. Failed. Explosion destroyed vehicle.

Table 2–83. Titan Launches (1989–1998) (Continued)

Titan Launch Vehicle	Launch Date (GMT)	Mission	Comments
II	October 5, 1993	Landsat-6	Earth science mission launched by refurbished Titan ICBM. Failed to achieve orbit due to a ruptured hydrazine manifold that stopped fuel from reaching the satellite's stabilizing engines, preventing its ability to attain a stable orbit.
II	January 25, 1994	Clementine	DOD satellite.
IVA	February 7, 1994	Milstar	Military communications satellite. First Titan IV with Centaur upper stage.
IVA	May 3, 1994	DSP	Defense Support Program satellite. Centaur upper stage.
IVA	August 27, 1994	USA 105	DOD satellite. Centaur upper stage.
IVA	December 22, 1994	USA 107	Defense Support Program satellite. IUS booster.
IVA	May 14, 1995	USA 110	DOD satellite.
IVA	July 10, 1995	USA 112	DOD satellite. Centaur upper stage.
IVA	November 6, 1995	USA 115	Military communications satellite. Centaur upper stage.
IVA	December 5, 1995	USA 116	Military reconnaissance.
IVA	April 24, 1996	USA 118	DOD satellite. Centaur upper stage.
IVA	May 12, 1996	USA 119, 120, 121, 122, 123, 124	DOD satellite.
IVA	July 3, 1996	USA 125	Military reconnaissance.
IVA	December 20, 1996	USA 129	DOD satellite.
IVB	February 23, 1997	USA 130	First Titan IVB launch. DOD satellite. IUS booster.
II	April 4, 1997	USA 131, DMSP F14	DOD satellite.
IVB	October 15, 1997	Cassini/Huygens	NASA space science mission. Centaur upper stage.

Table 2–83. Titan Launches (1989–1998) (Continued)

Titan Launch Vehicle	Launch Date (GMT)	Mission	Comments
IVA	October 24, 1997	USA 133	DOD satellite. Centaur upper stage.
IVA	November 8, 1997	USA 136	DOD satellite. Centaur upper stage.
IVB	May 8, 1998	USA 139	Military reconnaissance. Centaur upper stage.
II	May 13, 1998	NOAA-15	NOAA meteorological satellite.
IVA	August 12, 1998	USA	DOD satellite. Centaur upper stage. Last Titan IVA launch. Failed.

ᵃ Isakowitz et al., *International Reference Guide to Space Launch Systems*, 3rd ed., p. 453.

Table 2–84. Titan II Characteristics[a]

	Stage 1	**Stage 2**
Length	70 ft (21.3 m)	24 ft (7.3 m)
Diameter	10 ft (3.0 m)	10 ft (3.0 m)
Launch mass	269,000 lb (122,016 kg)	65,000 lb (29,484 kg)
Propulsion	Two LR87-AJ-5	One LR 91-AJ-5
Propellant	Aerozine 50, N2O4	Aerozine 50, N2O4
Propellant mass	260,000 lb (117,934 kg)	59,000 lb (27,215 kg)
Thrust (vac.)	474,000 lb (2,100 kN)	100,000 lb (450 kN)
Nominal burn time	147 sec	182 sec
Payload capacity	4,200 lb (1,905 kg) to polar low-Earth orbit	
Contractor	Aerojet Techsystems (engines)	
	Lockheed Martin (vehicle refurbishment)	

[a] "Titan II Space Launch Vehicle," Fact Sheet, United States Air Force, *http://www.losangeles.af.mil/ SMC/PA/Fact_Sheets/ttn2_fs.htm* (accessed February 14, 2005) and Isakowitz et al., *International Reference Guide to Space Launch Systems,* 3rd ed., pp. 457–458.

Table 2–85. Space Shuttle Flights (1989–1998)

Mission	Date	Orbiter	Payload	Comment
STS-29	March 13–March 18, 1989	*Discovery*	Tracking and Data Relay Satellite (TDRS)-4	NASA communications satellite.
STS-30	May 4–May 8, 1989	*Atlantis*	Magellan	First launch of interplanetary spacecraft. Attached to IUS booster.
STS-28	August 8–August 13, 1989	*Columbia*	DOD payload	
STS-34	October 18–October 23, 1989	*Atlantis*	Galileo	Attached to IUS booster, deployed on trajectory toward Jupiter. Space science mission.
STS-33	November 23–November 26, 1989	*Discovery*	DOD payload	
STS-32	January 9–January 20, 1990	*Columbia*	DOD communications satellite Syncom IV-5	Also retrieved the Long Duration Exposure Facility.
STS-36	February 28–March 4, 1990	*Atlantis*	DOD payload	
STS-31	April 24–April 29, 1990	*Discovery*	Hubble Space Telescope	First "Great Observatory." Space science mission.
STS-41	October 6–October 10, 1990	*Discovery*	European Space Agency-sponsored Ulysses	Attached to IUS and Payload Assist Module S (PAM-S) boosters.
STS-38	November 15–20, November 1990	*Atlantis*	DOD payload	
STS-35	December 2–December 11, 1990	*Columbia*	No deployed payload	Astro-1 Spacelab mission.
STS-37	April 5–April 11, 1991	*Atlantis*	Gamma Ray Observatory	Second "Great Observatory." Space science mission.

Table 2–85. Space Shuttle Flights (1989–1998) (Continued)

Mission	Date	Orbiter	Payload	Comment
STS-39	April 28–May 6, 1991	*Discovery*	Deployed and retrieved Strategic Defense Initiative Organization's Infrared Background Signature Survey experiment, mounted on the Shuttle Pallet Satellite (SPAS)-II platform	First unclassified DOD-dedicated Space Shuttle mission.
STS-40	June 5–June 14, 1991	*Columbia*	No deployed payload	Life sciences mission.
STS-43	August 2–August 11, 1991	*Atlantis*	TDRS-5	NASA communications satellite.
STS-48	September 12–September 18, 1991	*Discovery*	Upper Atmosphere Research Satellite	Earth science mission.
STS-44	November 25–December 1, 1991	*Atlantis*	Defense Support Program Satellite	
STS-42	January 22–January 30, 1992	*Discovery*	No deployed payload	International Microgravity Laboratory (IML)-1.
STS-45	March 24–April 2, 1992	*Atlantis*	No deployed payload	Atmospheric Laboratory for Applications and Science (ATLAS)-1.
STS-49	May 2–May 16, 1992	*Endeavour*	Captured and redeployed Intelsat VI satellite after repair	First flight of *Endeavour*.
STS-50	June 25–July 9, 1992	*Columbia*	No deployed payload	U.S. Microgravity Laboratory (USML)-1.
STS-46	July 31–August 8, 1992	*Atlantis*	European Space Agency European Retrievable Carrier (EURECA)	Also deployed tethered Italian satellite, which did not deploy as planned.

Table 2–85. Space Shuttle Flights (1989–1998) (Continued)

Mission	Date	Orbiter	Payload	Comment
STS-47	September 12–September 20, 1992	*Endeavour*	No deployed payload	Spacelab-J (First Japanese Spacelab).
STS-52	October 22–November 1, 1992	*Columbia*	Laser Geodynamic Satellite II	Joint U.S.-Italy mission. Also U.S. Microgravity Payload (USMP)-1.
STS-53	December 2–December 9, 1992	*Discovery*	DOD payload	Last classified payload.
STS-54	January 13–January 19, 1993	*Endeavour*	TDRS-6	NASA communications satellite.
STS-56	April 8–April 17, 1993	*Discovery*	Deployed and retrieved Shuttle Pointed Autonomous Research Tool for Astronomy (SPARTAN)-201	Also ATLAS-2 science mission.
STS-55	April 26–May 6, 1993	*Columbia*	No deployed payload	German Spacelab D-2.
STS-57	June 21– July 1, 1993	*Endeavour*	Retrieved EURECA	Also commercial SPACEHAB laboratory.
STS-51	September 12–September 22, 1993	*Discovery*	1) Advanced Communications Technology Satellite (ACTS), 2) Orbiting and Retrievable Far and Extreme Ultraviolet Spectrograph (ORFEUS)-SPAS deployed and retrieved	
STS-58	October–18 November 1, 1993	*Columbia*	No deployed payload	Spacelab life sciences mission.
STS-61	December 2–December 13, 1993	*Endeavour*	Hubble Space Telescope retrieved and redeployed	First Hubble servicing mission.

Table 2–85. Space Shuttle Flights (1989–1998) (Continued)

Mission	Date	Orbiter	Payload	Comment
STS-60	February 3–February 11, 1994	*Discovery*	Deployed two payloads from Get Away Special (GAS) canisters	SPACEHAB mission. Wake Shield Facility-1 not deployed as planned.
STS-62	March 9–March 19, 1994	*Columbia*	No deployed payload	1) USMP-2, 2) Office of Aeronautics and Space Technology (OAST)-2 experiments.
STS-59	April 9–April 20, 1994	*Endeavour*	No deployed payload	Space Radar Laboratory (SRL)-1.
STS-65	July 9–July 23, 1994	*Columbia*	No deployed payload	Last *Columbia* mission before scheduled modification and refurbishment. Carried IML-2.
STS-64	September 9–September 20, 1994	*Discovery*	Deployed and retrieved SPARTAN-201	Also LIDAR In-Space Technology Experiment.
STS-68	September 30–October 11, 1994	*Endeavour*	No deployed payload	SRL-2.
STS-66	November 3–November 14, 1994	*Atlantis*	Deployed and retrieved German Cryogenic Infrared Spectrometers and Telescopes for the Atmosphere (CRISTA)-SPAS	Also ATLAS-3 science mission.
STS-63	February 3–February 11, 1995	*Discovery*	Deployed and retrieved SPARTAN-204	Performed approach and fly-around of *Mir.* Also SPACEHAB mission.
STS-67	March 2–March 18, 1995	*Endeavour*	No deployed payload	Astro-2 mission.
STS-71	June 27–July 6, 1995	*Atlantis*	No deployed payload	100th U.S. human spaceflight. Docked with *Mir.*
STS-70	July 13–July 22, 1995	*Discovery*	TDRS-7	NASA communications satellite. Last TDRS deployed.

Table 2–85. Space Shuttle Flights (1989–1998) (Continued)

Mission	Date	Orbiter	Payload	Comment
STS-69	September 7–September 18, 1995	Endeavour	Deployed and retrieved SPARTAN 201 and Wake Shield Facility-2	First dual deployment and retrieval.
STS-73	October 20–November 5, 1995	Columbia	No deployed payload	USML-2.
STS-74	November 12– November 20, 1995	Atlantis	No deployed payload	Docked with Mir.
STS-72	January 11–January 20, 1996	Endeavour	Deployed and retrieved SPARTAN OAST flyer	Also captured and returned Japanese satellite.
STS-75	February 22–March 7, 1996	Columbia	Deployed tethered satellite (3-day duration before tether broke)	USMP-3.
STS-76	March 22–March 30, 1996	Atlantis	No deployed payload	Docked with Mir.
STS-77	May 19–May 29, 1996	Endeavour	Deployed and retrieved SPARTAN-207/Inflatable Antenna Experiment	Commercial SPACEHAB mission.
STS-78	June 20–July 7, 1996	Columbia	No deployed payload	Life and Microgravity Spacelab.
STS-79	September 19–September 26, 1996	Atlantis	No deployed payload	Docked with Mir.
STS-80	November 19–December 7, 1996	Columbia	Deployed and retrieved ORFEUS-SPAS and Wake Shield Facility-3	
STS-81	January 12–January 22, 1997	Atlantis	No deployed payload	Docked with Mir.
STS-82	February 11–February 21, 1997	Discovery	Retrieved and redeployed Hubble Space Telescope	Second Hubble servicing mission.
STS-83	April 4–April 8, 1997	Columbia	No deployed payload	Microgravity Science Laboratory-1 (MSL-1) postponed.

Table 2–85. Space Shuttle Flights (1989–1998) (Continued)

Mission	Date	Orbiter	Payload	Comment
STS-84	May 15–May 24, 1997	*Atlantis*	No deployed payload	Docked with *Mir.*
STS-94	July 1–July 17, 1997	*Columbia*	No deployed payload	Reflight of MSL–1.
STS-85	August 7–August 19, 1997	*Discovery*	Deployed and retrieved German CRISTA-SPAS-2	
STS-86	September 25–October 6, 1997	*Atlantis*	No deployed payload	Docked with *Mir.*
STS-87	November 19–December 5, 1997	*Columbia*	Deployed and retrieved SPARTAN-201	Also USMP-4 Spacelab.
STS-89	January 22–January 31, 1998	*Endeavour*	No deployed payload	Docked with *Mir.*
STS-90	April 17–May 3, 1998	*Columbia*	No deployed payload	Final scheduled flight of Spacelab. Neurolab mission.
STS-91	June 2–June 12, 1998	*Discovery*	No deployed payload	Docked with *Mir.*
STS-95	October 29–November 7, 1998	*Discovery*	Deployed and retrieved SPARTAN-201	Also SPACEHAB module. Carried Hubble Orbiting Systems Test (HOST) platform. John Glenn flight.
STS-88	December 4–December 15, 1998	*Endeavour*	Satelite de Aplicaciones Cientifico (SAC)-A for Argentina	First Space Station mission.

Table 2–86. External Tank Characteristics[a]

Component	Characteristics
Propellants	LOX/LH2
Length	153.8 ft (46.9 m)
Diameter	27.6 ft (8.4 m)
Gross liftoff weight	1,655,600 lb (760,947 kg)
Inert weight of lightweight tank	66,000 lb (29,937 kg)
Inert weight of super lightweight tank[b]	58,500 lb (26,535 kg)
Liquid oxygen max. weight	1,361,936 lb (617,764 kg)
Liquid oxygen tank weight (empty)	12,000 lb (5,443 kg)
Liquid oxygen tank volume	19,563 cu ft (553,963 liters)
Liquid oxygen tank length	49.3 ft (15 m)
Liquid oxygen tank diameter	27.6 ft (8.4 m)
Liquid hydrogen max. weight	227,641 lb (103,256 kg)
Liquid hydrogen tank diameter	27.6 ft (8.4 m)
Liquid hydrogen tank length	96.7 ft (29.5 m)
Liquid hydrogen tank volume	53,518 cu ft (1,515,461 liters)
Liquid hydrogen tank weight (empty)	29,000 lb (13,154 kg)
Intertank length	22.5 ft (6.9 m)
Intertank diameter	27.6 ft (8.4 m)
Intertank weight	12,100 lb (5,488 kg)
Prime contractor	Martin Marietta/Lockheed Martin since 1994

[a] "External Tank," *NSTS 1988 News Reference Manual*, September 1988, *http://science.ksc.nasa.gov/ shuttle/technology/sts-newsref/et.html* (accessed February 25, 2005).

[b] The super lightweight external tank was first used on STS-91 in June 1998. "Super Lightweight External Tank," Space Shuttle Technology Summary, NASA Marshall Space Flight Center, FS-2003-06-70-MSFC, *http://www.nasa.gov/centers/marshall/pdf/100423main_shuttle_external_tank.pdf* (accessed February 25, 2005).

Table 2–87. Solid Rocket Booster Characteristics[a]

Component	Characteristics
Length	149.16 ft (45.5 m)
Diameter	12.17 ft (3.7 m)
Propellant weight (each solid rocket motor)	1,100,000 lb (500,000 kg)
Inert weight (each SRB)	192,000 lb (89,090 kg)
Thrust at launch	3,300,000 lb (14,679 kN)
Propellant mixture	Ammonium perchlorate oxidizer, aluminum fuel, iron oxide, polymer, epoxy

[a] "Solid Rocket Boosters," *NSTS 1988 News Reference Manual,* September 1988, *http://science.ksc.nasa.gov/shuttle/technology/sts-newsref/srb.html* (accessed February 25, 2005).

Table 2–88. Space Shuttle Main Engine Characteristics[a]

Component	Characteristics
Length	14 ft (4.3 m) at nozzle exit
Diameter	7.5 ft (2.3 m) at nozzle exit
Approx. weight (each)	7,000 lb (3,175 kg)
Number of engines	Three on each orbiter
Range of thrust level	65%–109% of rated power level
Thrust (100%)	Each engine: 375,000 lb (1,668 kN) at sea level, 470,000 lbs (2,091 kN) in vacuum
Thrust (109%)	417,300 lb (1,856 kN) at sea level, 513,250 lb (2,283 kN) in vacuum[a]
Operating life	7.5 hours and 55 starts
Propellant	Fuel: LH_2, Oxidizer: LOX, in a 6:1 ratio
Nominal burn time	522 secb[b]
Prime contractor	Boeing Rocketdyne

[a] "Main Propulsion System," *NSTS 1988 News Reference Manual,* September 1988, *http://science.ksc.nasa.gov/shuttle/technology/sts-newsref/sts-mps.html* (accessed February 25, 2005).
[b] Boeing Rocketdyne, the engine manufacturer, lists the maximum thrust of each engine in vacuum at 512,950 lb (2,282 kN).
[b] Isakowitz et al., *International Reference Guide to Space Launch Systems,* 3rd ed., p. 407.

Table 2–89. Titan Centaur Upper Stage Characteristics[a]

Component	Characteristics
Length	29.45 ft (9 m)
Diameter	14.2 ft (4.3 m)
Thrust	33,000 lb (15,000 kg)
Propellants	Cryogenic–Liquid Oxygen and Liquid Hydrogen Stage
Propellant weight	46,000 lb (20,865 kg)
Propulsion	Two Pratt & Whitney restartable RL10 engines
Contractor	Lockheed Martin Space Systems

[a] "Titan," Lockheed Martin, *http://www.lockheedmartin.com/wms/findPage.do?dsp=fec&ci=15525& rsbci=0&fti=0&ti=0&sc=400* (accessed March 17, 2005).

Table 2–90. Inertial Upper Stage Characteristics

Component	Characteristics
Length	17 ft (5.18 m)
Diameter	9.25 ft (2.8 m)
Weight	32,500 lb (14,742 kg)
Propulsion	Two solid-fueled United Technologies motors
Propellant weight	First stage: 21,400 lb (9,797 kg) Second stage: 6,000 lb (2,722 kg)
Thrust	First stage: 42,000 lb (188,496 N) Second stage: 18,000 lb (80,784 N)
Contractor	Boeing

Table 2–91. Inertial Upper Stage Launches

Date	Vehicle	Payload
March 13, 1989	STS-29	Tracking and Data Relay Satellite-4
May 4, 1989	STS-30	Magellan
June 14, 1989	Titan IV	Defense Support Program satellite
September 4, 1989	Titan 34D	Defense Satellite Communications System satellite
October 18, 1989	STS-34	Galileo
November 23, 1989	STS-33	DOD payload
October 6, 1990	STS-41	Ulysses
November 13, 1990	Titan IV	Defense Support Program satellite
August 2, 1991	STS-43	Tracking and Data Relay Satellite-5
November 24, 1991	STS-44	Defense Support Program satellite
January 13, 1993	STS-54	Tracking and Data Relay Satellite-6
December 22, 1994	Titan IV	Defense Support Program satellite
July 13, 1995	STS-70	Tracking and Data Relay Satellite-7
February 24, 1997	Titan IV	Defense Support Program satellite

Table 2–92. DC-X Characteristics[a]

Component	Characteristics
Width	13-1/3 ft (4 m) at base, conical shape
Height	40 ft (12.2 m)
Weight (empty)	20,000 lb (9,072 kg)
Weight (with propellants)	41,600 lb (18,870 kg)
Propellants	LOX and LH2
Propulsion	Four RL10A5 rocket engines
Thrust	13,500 lb each (60,000 N)
Reaction controls	Four 440 lb (1,957 N)-thrust gaseous oxygen, gaseous hydrogen thrusters
Contractor	McDonnell Douglas

[a] "DC-X Fact Sheet," BMDOLink, Delta Clipper-Experimental Fact Sheet, Office of External Affairs, April 1993, *http://www.hq.nasa.gov/office/pao/History/x-33/dcx-facts.htm* (accessed March 22, 2005).

Table 2–93. DC-X and DC-XA Flight Tests[a]

Flight	Launch Date	Duration (sec)	Altitude (m/ft)	Description
DC-X Test Flights				
1	August 18, 1993	59	46/151	Verified flight control systems and vertical landing capabilities
2	September 11, 1993	65.8	92/302	Ascent and landing mode control and ground effects survey
3	September 30, 1993	72.2	370/1,214	180-degree roll; aerostability data
4	June 20, 1994	135.9	870/2,854	Full propellant load; radar altimeter in control loop
5	June 27, 1994	77.9	790/2,592	In-flight abort after gaseous hydrogen explosion; vehicle demonstrated autoland capabilities
6	May 16, 1995	123.6	1,330/4,364	Continued expansion of flight envelope; constant angle of attack
7	June 12, 1995	132	1,740/5,709	First use of reaction control system thrusters; angle of attack from 0 to 70 degrees
8	July 12, 1995	124	2,500/8,202	Final flight of DC-X; demonstrated turnaround maneuver; aeroshell cracked during 14 ft/sec landing
DC-XA Test Flights				
1	May 18, 1996	62	244/801	First flight of DC-XA; aeroshell caught fire during slow landing
2	June 7, 1996	63.6	590/1,936	Maximum structural stresses with 50 percent full LOX tank
3	June 8, 1996	142	3,14010,302	26-hour rapid turnaround demonstration; new altitude and duration record
4	July 31, 1996	140	1,250/4,101	Landing strut 2 failed to extend; vehicle tipped over and LOX tank exploded; vehicle destroyed

a "The Delta Clipper Experimental: Flight Testing Archive," *http://www.hq.nasa.gov/office/pao/History/x-33/dcxfile.htm* (Web site created by Kirk Sorensen) (accessed March 22, 2005).

Table 2–94. X-34 Characteristics[a]

Component	Characteristics
Length	58.3 ft (17.8 m)
Wingspan	27.7 ft (8.4 m)
Weight unfueled	18,000 lb (8,165 kg)
Main propulsion	One NASA (Marshall Space Flight Center)-designed Fastrac engine
Propellant	LOX/RP-1
Propellant weight	30,000 lb (13,600 kg)
Thrust	60,000 lb (27,216 kg)
Nominal burn time	154 sec (without throttling)
Maximum speed	Mach 8
Maximum altitude	Approximately 50 miles (80 km)
Prime contractor	Orbital Sciences Corporation

[a] "X-34 Demonstrating Reusable Launch Vehicle Technologies," Historical Fact Sheet, NASA Marshall Space Flight Center, *http://www.nasa.gov/centers/marshall/news/background/facts/x-34.html* (accessed March 22, 2005).

Table 2–95. X-33 Characteristics[a]

Component	Characteristics
Length	69 ft (21 m)
Width	77 ft (23.5 m)
Takeoff weight	285,000 lb (129,274 kg)
Propellant	LH2/LOX
Fuel weight	210,000 lb (95,254 kg)
Main propulsion	Two J-2S linear aerospike engines
Take-off thrust	410,000 lb (185,973 kg)
Maximum speed	Mach 13+
Contractors	Lockheed Martin (prime) Rocketdyne (engines) Rohr (thermal protection systems) Allied Signal (subsystems) Sverdrup (ground support equipment)

[a] "X-33 Advanced Technology Demonstrator," Historical Fact Sheet, Marshall Space Flight Center, *http://www.nasa.gov/centers/marshall/news/background/facts/x33.html* (accessed March 22, 2005).

Table 2–96. Reusable Launch Vehicle Chronology[a]

Date	Event
1990	Ballistic Missile Defense Organization initiated DC-X program.
August 1991	McDonnell Douglas won a $60 million contract to build the DC-X.
August 18, 1993	Flight tests of DC-X were begun.
January 1994	NASA's Access to Space study released. The study recommended that development of an advanced technology, single-stage-to-orbit, fully reusable rocket launch vehicle become a NASA goal.
February 1994	NASA released a series of NASA Research Announcements to industry for RLV component technology. This program laid the groundwork for technologies to be demonstrated during the X-33 flight program.
May 31, 1994	NASA identified $1 million for the DC-XA test program in addition to $990,000 dollars transferred to the DC-X program earlier in 1994. Enabled acceptance of DC-X vehicle from the Air Force.
June 20, 1994	First flight of DC-X under second phase of program took place.
June 27, 1994	DC-X test demonstrated the vehicle's autoland capabilities.
July 1994	Eighteen cooperative agreements were signed with industry in the areas of structures, thermal protection, and advanced propulsion.
August 5, 1994	President William J. Clinton issued National Space Transportation Policy (NSTC-4) for the RLV Technology program. It called for NASA to formulate an implementation plan by October 5, 1994 for Administration review.
October 31, 1994	NASA's FY 1995 Operating Plan established a new Space Access and Technology Program and funded the RLV program at $93.5 million.
November 7, 1994	The Administration approved the NASA Implementation Plan for the President's National Space Transportation Policy. The plan accelerated the X-33 schedule and called for NASA to select an X-33 technology demonstrator by July 1996.
January 12, 1995	NASA issued two Cooperative Agreement Notices requesting proposals for the development of technology demonstrators for an RLV program.
March 1995	NASA and Orbital Sciences Corporation signed a cooperative agreement for the X-34.

Table 2–96. Reusable Launch Vehicle Chronology[a] (Continued)

Date	Event
March 29, 1995	NASA signed three cooperative Phase I agreements to design the X-33, the next generation space booster. Agreements were signed with Lockheed Advanced Development Company (Skunk Works), McDonnell Douglas Aerospace, and Rockwell International Corporation. NASA provided approximately $7 million to each industry partner, with each investing a matching sum.
May 16, 1995	DC-X test flights were begun in support of NASA's RLV program.
July 7, 1995	Last test flight of DC-X took place. The aeroshell cracked during landing. The vehicle was turned over to NASA and sent to McDonnell Douglas for modifications to the DC-XA.
December 15, 1995	NASA issued a draft Cooperative Agreement Notice for the design, fabrication, and flight test of the X-33 advanced technology demonstrator.
March 1996	The President's FY 1997 budget highlighted that the RLV was a science and technology investment. The RLV was cited as a way to significantly cut the cost of reaching space.
March 1996	NASA issued a new NASA Research Announcement for the X-34.
April 1, 1996	NASA issued a Cooperative Agreement Notice for demonstration of single-stage-to-orbit (SSTO) technologies through the design, fabrication, and flight test of an X-33 advanced technology demonstrator.
May 8, 1996	The DC-XA completed a series of ground tests at the U.S. Army White Sands Missile Range in preparation for flight tests.
May 18, 1996	The DC-XA began a new set of test flights.
June 1996	NASA awarded a contract valued at approximately $50 million to Orbital Sciences Corporation for the X-34.
June 14, 1996	A full-scale segment of a graphite-composite wing designed for an RLV was successfully "tested to failure" at Langley Research Center. This was the first structural test of a full-scale component designed and fabricated to validate the use of graphite-composite primary structures for RLVs. The purpose of the test was to determine the maximum load the wing-box could carry as well as to understand how it would fail.
July 2, 1996	Vice President Al Gore announced at the Jet Propulsion Laboratory in Pasadena, California, that Lockheed Martin had been selected to build the X-33 test vehicle, called VentureStar. Lockheed Martin won the competition for the X-33 Phase II contract over contenders McDonnell Douglas and Rockwell International.

Table 2–96. Reusable Launch Vehicle Chronology[a] (Continued)

Date	Event
July 31, 1996	A landing strut on the DC-XA failed to extend. The vehicle tipped over and exploded due to an open pressurant line. The vehicle was destroyed.
October 1, 1996	NASA filed Notice of Intent 96-118 with the Federal Register of its intention to prepare an environmental impact statement (EIS) and to conduct scoping meetings for the development and testing of the X-33 vehicle. The EIS addressed environmental issues associated with fabrication, assembly, testing, and preparation of the flight operations and landing sites associated with the X-33 flight vehicle.
November 13, 1996	Gary Payton, NASA's Director of Space Transportation, and T. K. Mattingly, Vice President for Lockheed Martin's RLV Program, held an informal meeting to discuss the program status and answer questions on the X-33, which was undergoing its Preliminary Design Review that week in California to formalize the engineering baseline of the X-33 vehicle before moving on to the detailed design phase.
November 1, 1996	Langley Research Center conducted thermal-mechanical tests toward the development of a durable, lightweight, cryogenic insulation system for possible use on future RLVs.
December 18, 1996	A three-day Preliminary Design Review (PDR) was completed for the X-33 operations segment and ground systems segment. Individual PDRs already had been conducted on the aerospike engine, the hydrogen tank, the structure, and most subsystems.
January 21, 1997	Langley Research Center issued a press release about X-33 wind tunnel testing during Phase I in the 22-Inch Mach 20 Helium Tunnel at Langley.
January 23, 1997	NASA held a public meeting in Idaho Falls, Idaho, to gather public comment on its plan to conduct flight tests of the X-33. The meeting was part of NASA's EIS process in support of the X-33 program. The formal process had begun on October 7, 1996, after NASA published a Notice of Intent 96-118 in the Federal Register. The Idaho Falls meeting was the 12th NASA public meeting to discuss the potential environmental impact of the X-33 test flights. Earlier meetings were held in towns neighboring proposed takeoff and landing sites in Southern California, Utah, Washington, and Montana.
February 20, 1997	A 7.75 percent scale model of the X-33 completed two weeks of wind tunnel tests in the 5.1-meter transonic wind tunnel at the Air Force's Arnold Engineering Development Center at Arnold Air Force Base, Tullahoma, Tennessee, according to Space Log, March 10 to March 16, 1997.
March 1997	An aluminum and stainless steel model of the X-33 was tested in Langley's Low-Turbulence Pressure Tunnel.

Table 2–96. Reusable Launch Vehicle Chronology[a] (Continued)

Date	Event
March 6, 1997	NASA announced that surveying was underway at Edwards Air Force Base, California, in preparation for the construction of the X-33 launch site. Sverdrup Corporation, the X-33 team's launch facility contractor, was undertaking the surveying of the launch site at Haystack Butte. Construction of the launch pad and facilities was expected to be completed by September 9, 1998. Launch facility activation, which included verification of the launch pad fueling system, was scheduled to be completed by October 1, 1998.
April 10, 1997	NASA announced that an aluminum and stainless steel scale model of the X-33–about 38 cm (15 in) long by 38 cm (15 in) wide–was undergoing extensive wind tunnel testing at Langley's 16-Foot Transonic Tunnel through mid-April 1997.
April 16, 1997	Continuing wind tunnel testing was carried out at Marshall Space Flight Center to correct an X-33 control deficiency at low supersonic speeds (Mach 1 to Mach 2). Adding canards appeared to be the only viable solution to date.
April 30, 1997	Marshall Space Flight Center announced that, it had conducted hot-fire tests of components for the X-33 linear aerospike engine in its Propulsion Laboratory's East Test Area. The test apparatus consisted of three hydrogen-cooled thrust cells constructed to represent a section of the X-33 engine, which was to have two banks of 10 side-by-side thrusters. Test results were to be reviewed with Rocketdyne, which built the test thrust cells and was to build the X-33 aerospike engine.
Mid-April–May 1997	Wind tunnel testing of a scale model X-33 in the Langley's Research Center's Unitary Wind Tunnel at supersonic speeds ranging from Mach 1.5 to Mach 4.5 continued from mid-April to early May. Wind tunnel testing also continued through May at Marshall Space Flight Center.
May 1997	A "tiger team" was working full-time on reducing the dry weight (without fuel) of the X-33 by 5,000 lb (2,268 kg) to 6,000 lb (2,722 kg). The team sorted through more than 400 recommendations of ways to reduce the weight.
May 21, 1997	The "tiger team" working on the X-33 weight problem gave a presentation. Weight reduction recommendations were ranked according to minor, medium, or major cost and schedule impacts. The team indicated that weight could be reduced by about 8,000 lb (3,629 kg) to 11,000 lb (4,990 kg), but the X-33 project costs and schedule would be affected.

Table 2–96. Reusable Launch Vehicle Chronology[a] (Continued)

Date	Event
June 1997	Additional wind tunnel testing of X-33 models took place in Langley's Hypersonic Facilities Complex. Also, X-33 wind tunnel testing started in Langley's 14-by-22-Foot Subsonic Tunnel in mid-June.
June 24, 1997	Aerospace Daily reported that "typical development problems" had led to postponement of the first X-33 test flight from March 1999 to July 1999, and slippage of the Critical Design Review (CDR) from September to an unspecified time in the fall. A critical problem behind the postponement was fabrication of the liquid-hydrogen fuel tank. In addition, Aerospace Daily reported that the Lockheed Martin Skunk Works had consolidated X-33 project management at Palmdale, California, and Jerry Rising had been named Vice President for X-33 and RLVs. Rising replaced T.K. Mattingly, who transferred to Lockheed Martin's aeronautical division at corporate headquarters in Bethesda, Maryland.
June 27, 1997	NASA released the draft EIS.
July 1997	In mid-July, wind tunnel testing of X-33 models in Langley's 14-by-22-Foot Subsonic Tunnel was concluded.
July 3, 1997	Aerospace Daily reported on X-33 progress, based on an interview with Lockheed Martin X-33 Vice President Jerry Rising. The Skunk Works was considering use of a colder, denser cryogenic propellant and had dropped plans to add canards for vehicle stability in the low transonic range (Mach 1 and Mach 2) in favor of changes in the tail structure. Weight growth was under attack by a special "tiger team."
August 1997	A critical series of tests on the X-34 Fastrac engine was successfully completed at Marshall Space Flight Center.
August 26, 1997	The Linear Aerospike SR-71 Experiment was mounted on a NASA SR-71 aircraft at Dryden Flight Research Center, Edwards, California, in preparation for the experiment's first flight, then scheduled for September.
August 26, 1997	Aerospace Daily reported that a gas generator adapted for the X-33 aerospike engine from a J-2 Saturn rocket engine had undergone 14 hot-fire tests at Marshall Space Flight Center.
August 28, 1997	Langley Research Center conducted load tests of a full-scale segment of a composite intertank structure for the X-33 program.

Table 2–96. Reusable Launch Vehicle Chronology[a] (Continued)

Date	Event
September 11, 1997	Aerospace Daily reported on X-33 progress. Five of eight 100-lb (45.4 kg) liquid hydrogen tank panels had been fabricated by Alliant Techsystems in a Utah plant, and tests of the composite seams were proceeding without any surprises. The liquid oxygen tank had been welded together. Removing the turbo alternator removed a "big hunk" of vehicle weight. Cooling the liquid oxygen and hydrogen propellants to temperatures lower than normal cut overall vehicle weight further and allowed the X-33 to carry additional fuel.
September 18, 1997	A two-day CDR of the X-33 thermal protection system by Rohr at its Chula Vista, California, facility ended
September 24, 1997	The two-day CDR of the X-33 aerospike engine (known also as the XRS-2200 engine) ended. The CDR took place at Rocketdyne's DeSoto campus in Chatsworth, California, where the X-33 engines were being designed.
September 26, 1997	NASA released the Final EIS for the X-33 and named the preferred flight testing launch and landing sites.
October 31, 1997	NASA announced that the X-33 had completed the five-day vehicle CDR successful, a major event in X-33 evolution. With completion of the CDR, NASA gave the Lockheed Martin Skunk Works approval to proceed with the fabrication of all remaining components and the assembly of the flight vehicle. The package of CDR technical information contained roughly 2,750 charts in 11 volumes.
October 31, 1997	The first successful flight of the Linear Aerospike SR-71 Experiment (LASRE) at Dryden Flight Research Center took place.
November 4, 1997	NASA completed its Record of Decision on the X-33 EIS and announced an intention to proceed with the preferred X-33 flight test program as described in the Final EIS issued October 3, 1997.
November 14, 1997	Groundbreaking ceremony took place at the future X-33 launch site on Edwards Air Force Base.
January 1998	NASA decided to modify its contract with Orbital Sciences Corporation to provide for a second X-34 flight vehicle. The modification also allowed for additional unpowered tests and more flexibility in demonstrating various technologies.
January 1, 1998	A faulty control system in the X-33 construction hangar set off water canons intended to fight fires. A crew of about a dozen worked on New Year's Eve to dry out the X-33 construction area. No permanent damage resulted, and work continued as usual.

Table 2–96. Reusable Launch Vehicle Chronology[a] (Continued)

Date	Event
January 14, 1998	Construction of the X-33 launch site at Haystack Butte progressed. Sverdrup completed rough grading of the launch site. The new road to the launch site was drivable but, like the site, was still at subgrade level.
January 21, 1998	Sverdrup completed rough grading of the X-33 launch site.
February 11, 1998	The first major X-33 component, the liquid oxygen tank, was delivered to the Palmdale, California, hangar where construction of the vehicle was taking place. An Airbus A300-600ST made the delivery.
February 12, 1998	The ground cold flow test of the LASRE was performed. This test included one normal cold flow and one emergency systems cold flow. The emergency systems cold flow tested the effects of control system power loss during flight. The liquid oxygen tank pressurized normally during the first (normal cold flow) test, validating the repair made to the vent system. The emergency test appeared to have been successful. A data review was scheduled for February 18, 1998.
February 25, 1998	A routine X-33 quarterly review took place at Marshall Space Flight Center. Presentations surveyed current progress.
February 25, 1998	Launch site construction continued to progress as all Edwards Air Force Base infrastructure (roads, power, water, and communications) was extended to the site.
March 4, 1998	A NASA SR-71 completed its first cold flow flight as part of the LASRE at Dryden Flight Research Center, Edwards, California.
March 11–March 12, 1998	The NASA Independent Annual Review of the X-33 program took place. X-33 technical and cost performance was surveyed. A final report detailing findings and conclusions was to be briefed to the NASA Program Management Council on April 15, 1998. The review indicated that Lockheed Martin's Skunk Works had addressed many of the concerns that arose during the September 1997 Independent Annual Review. NASA's Gary Payton and Gene Austin were pleased with the review results.
March 20, 1998	During a project review held at the Rocketdyne facility in Canoga Park, California, Rocketdyne made known certain schedule hazards that had developed with two of their suppliers, Weldmac and CFI. It was reported that, in the worst case, aerospike engine deliveries might slip three to five months. Rocketdyne was looking into their suppliers' difficulties to mitigate risk to the program schedule.
April 8, 1998	With the exception of some fastener shortages, the center thrust structure of the X-33 vehicle was now complete.

Table 2–96. Reusable Launch Vehicle Chronology[a] (Continued)

Date	Event
April 19, 1998	The liquid oxygen tank was moved into the main assembly fixture. The move took less than an hour and was completed two days ahead of schedule.
May 18, 1998	NASA's F-15B Aerodynamic Flight Facility fighter aircraft, based at Dryden Flight Research Center, flight-tested thermal protection materials intended for use on the X-33 to determine the durability of the materials, specifically measuring the shear and shock loads to which the materials were exposed. The materials tested included metallic Inconel tiles, soft Advanced Flexible Reusable Surface Isolation tiles, and sealing materials.
June 8, 1998	Aerospace Daily reported that "Lockheed Martin was carrying a 'three-month hazard' on the linear aerospike engine it will need to power the X-33 testbed next summer, but Rocketdyne had developed workarounds and fixes to get the engine back on track," cited Jerry Rising, Lockheed Martin Program Manager.
June 8, 1998	Aerospace Daily reported that leakage into the structure of the subscale aerospike mounted on NASA's SR-71 Blackbird had delayed the first hot-fire test of the engine "a few weeks."
June 8, 1998	Aerospace Daily reported that X-33 Program Manager Jerry Rising and X-34 Program Manager Bob Lindberg threatened to not allow their X vehicles to fly unless Congress passed indemnification legislation protecting them against third-party liability in case of an accident during flight testing.
June 10, 1998	NASA announced that pictures of the X-33 vehicle and launch site, taken every 15 minutes from three digital cameras, would be posted on an Internet site. The images from two cameras would show the vehicle's primary assembly structure, the side-by-side tooling structures for the X-33's upper thermal protection system, and the vehicle's upper internal support structure, while the third camera would focus on the vehicle's launch pad. The vehicle images would not be current, delayed one day.
June 30, 1998	NASA announced completion of the F-15B flight testing of thermal protection materials for the X-33 at Dryden Flight Research Center, Edwards, California. The six flights tested the durability of the materials at hypersonic velocities. The F-15B reached an altitude of 36,000 ft (10,973 M) and a top speed of Mach 1.4. The material samples tested included metallic Inconel tiles, soft Advanced Flexible Reusable Surface Insulation tiles, and sealing materials.

Table 2–96. Reusable Launch Vehicle Chronology[a] (Continued)

Date	Event
July 1998	The X-34 program passed a critical milestone as the first wing assembly completed qualification tests and was shipped to Orbital Sciences Corporation and mated to the X-34 test article under construction.
July 6, 1998	Aerospace Daily, in an article titled "Wagons Ho!" reported that the Lockheed Martin Skunk Works had abandoned flying the X-33 back to its launch pad at Edwards Air Force Base in favor of trucking the experimental aircraft overland, "because the Shuttle program won't give up one of its two Boeing 747s for ferry flights."
July 22, 1998	Difficulties with fabricating the X-33 liquid hydrogen tanks continued. As a result, delivery dates for the two tanks slipped from July 31 and September 2 to mid-October and mid-November, respectively. The impact of these delays on vehicle assembly was still being assessed.
July 29, 1998	Aerojet recommended to NASA and Lockheed Martin that they use a thruster configuration that included a nozzle made of columbium to correct for the thermal problems that had caused nozzles to burn through in earlier tests. Using columbium nozzle parts would not increase the X-33's net weight; however, preparing the parts would require a long lead time. To minimize schedule impact, Aerojet proposed delivering the thrusters without nozzles to allow continuation of vehicle assembly and supplying the columbium nozzles at a later date.
August 5, 1998	The X-33 System Architecture Review (SAR) and Optimized Design Review (ODR) were held in Palmdale with representatives from each Skunk Works partner, NASA, and the "Gray Beards" attending. The "Gray Beards" panel of experts was composed mainly of NASA senior personnel led by Del Freeman of Langley Research Center.
August 26, 1998	AlliedSignal delivered the X-33 nose landing gear strut. It was to be modified into the X-33 configuration for a test fit. This same test fit already had been accomplished for the main trunion pivots and the drag link attachments without any problems.

Table 2–96. Reusable Launch Vehicle Chronology[a] (Continued)

Date	Event
September 2, 1998	Spence M. (Sam) Armstrong, recently named NASA Associate Administrator for the Office of Aeronautics and Space Transportation Technology (Code R), revealed a reorganization during a staff briefing that would dilute the responsibilities of Gary Payton, who, as Deputy Associate Administrator for Space Transportation Technology, currently headed the X-33, X-34, and advanced space transportation programs, moving him more into the aeronautics half of the Office. Payton would occupy a lower position, Division Director, under the proposed reorganization, which was scheduled to take place on October 1. Payton had championed single-stage-to-orbit vehicles for many years. The change seriously jeopardized the status of the program within the NASA hierarchy.
September 11, 1998	Aerospace Daily reported that the pending reorganization of NASA's Office of Aeronautics and Space Transportation Technology "raised the hackles" of Rep. Dana Rohrabacher of California, a long-time champion of single-stage-to-orbit technology and chairman of the NASA authorization subcommittee. Rohrabacher expressed his concerns in a letter to NASA Administrator Daniel Goldin.
September 23, 1998	A nine-panel thermal protection system array was test-fitted on the bottom of the X-33 during the previous week by a joint team of B.F. Goodrich and Skunk Works technicians. The metallic panels were equipped with the new secondary seal designs. One panel also was removed from the center of the array to prove that any panel could be replaced.
September 27, 1998	Continuing difficulties with fabrication of the two liquid hydrogen tanks were experienced. A cure cycle was lost during the first doubler installation process on tank #2. The tank was removed early from the cure cycle after blowing a bag at the end of a ramp-up point. The combination of the out time and this cure cycle resulted in an unacceptable strength impact to the bond joints. The doublers were removed over the weekend (September 26–27) and could be replaced with existing materials. Loss of the cure cycle delayed fabrication of tank #2 by 30 days. Construction of the vehicle structure and electronics continued.

Table 2–96. Reusable Launch Vehicle Chronology[a] (Continued)

Date	Event
October 2, 1998	The X-33 engine testing program began. At 12:13 a.m. Central Time, the first successful aerospike engine-related test took place at Stennis Space Center. The test intended to calibrate the liquid hydrogen and liquid oxygen fuel turbopumps, check facility settings, and verify valve timing to prime the gas generator. The test lasted 2.81 seconds, and no flaws or anomalies were detected. The tested powerpack hardware consisted of the main power-generating and pumping components of the aerospike engine, including the liquid oxygen and liquid hydrogen turbopumps, a gas generator for the turbopump drive, vehicle connect lines, and interconnecting flight ducts. These powerpack tests were critical to the development of the linear aerospike engine because they allowed various performance levels to be tested in parallel with the design and construction of the engine. Full-scale engine tests were scheduled to take place at Stennis Space Center in late 1998.
October 7, 1998	B.F. Goodrich completed the last major testing of the metallic panels for the X-33 thermal protection system at Marshall Space Flight Center.
October 14, 1998	NASA announced the reorganization of NASA's Code R—the Office of Aeronautics and Space Transportation Technology—under Associate Administrator Spence M. Armstrong, to the Office of Aero-Space Technology. In the NASA press release, it was reported that Armstrong stated that "Goldin wanted me to personally be an advocate for the Reusable Launch Vehicle programs to effect a cheaper means of access to space." The press release did not mention Gary Payton's changed role within Code R or on the X-33 program.
October 14, 1998	Boeing presented its estimate to complete engine delivery. Boeing's plan transferred $36 million from the VentureStar RLV to the half-scale X-33 by eliminating the fabrication, assembly, and testing of the RLV power pack. By adding a second engine test stand in Phase III (the program is presently in Phase II), Boeing developed a schedule that would support a first flight of the VentureStar within six months of the Skunk Works schedule. In addition, Boeing declined additional investment in the project. The $36 million transferred from the VentureStar to the X-33 was the same amount as the additional X-33 costs caused by Boeing's delay in delivering the aerospike engine.

Table 2–96. Reusable Launch Vehicle Chronology[a] (Continued)

Date	Event
October 21, 1998	The first two upper thermal protection system panels arrived at the hangar from B.F. Goodrich's Riverside plant. They were to be test fitted on the forward-most position of the liquid oxygen tank. Repair patches for liquid hydrogen tank #1 had been completed and shipped, while work continued on the second tank.
October 23, 1998	NASA announced that it and Lockheed Martin would hold a media teleconference on Tuesday, October 27, with program officials Gary Payton, NASA Deputy Associate Administrator for Space Transportation Technology, NASA Headquarters; Gene Austin, NASA X-33 Program Manager; Jerry Rising, Lockheed Martin Skunk Works Vice President for the X-33 and VentureStar; and Cleon Lacefield, Lockheed Martin Skunk Works X-33 Program Manager. A similar teleconference took place the previous October to update the media on the status of the program following the CDR. This teleconference was expected to announce a six-month delay in the X-33 flight tests.
October 27, 1998	In a joint NASA and Lockheed Martin media teleconference, Jerry Rising announced that the first flight of the X-33 would be delayed six months until December 1999 because of late delivery of the aerospike engine in September 1999. This delay in engine delivery would cost an additional $36 million. Lockheed Martin expected Rocketdyne to absorb the additional cost. Lockheed Martin had cut all overtime on the program and planned to cut project personnel to reduce escalating costs.
October 28, 1998	The two leeward #1 composite panels were delivered to the hangar for a fit check on the vehicle. The two leeward #2 panels were to be shipped on November 1 for a fit check. Once the panel fit checks were made, all composite panels would be shipped back to B.F. Goodrich's Riverside plant for completion. Also, a successful cure cycle on liquid hydrogen tank #2 was accomplished October 24–25. As a result, all lobe skins were bonded on both tanks.
November 6, 1998	NASA released the Hawthorne Report, named after the Boston firm, Hawthorne, Krauss, and Associates, LLC.[b] The firm conducted a study titled "Analysis of Potential Alternatives to Reduce NASA's Cost of Human Access to Space." NASA intended to use the Hawthorne Report and the Space Transportation Architecture Study currently under way as guides for planning future space launchers. The Hawthorne report strongly supported the economics of commercial RLVs over continued use of the Space Shuttle. Hawthorne also urged NASA to exercise caution in setting up loan guarantees to support development of commercial RLVs.

Table 2–96. Reusable Launch Vehicle Chronology[a] (Continued)

Date	Event
November 11, 1998	The X-33's electronics achieved an important milestone when Sanders shipped two Vehicle Health Monitoring computers to the Skunk Works. Also, the Skunk Works identified a potential winner of the contract to transport the X-33 over land.
November 13, 1998	A test of the aerospike engine power pack took place at 100 percent power over a period of 30 seconds during the week ending November 13. A 250-second test was planned for the following week.
November 18, 1998	Work began on the ballast bulkhead assembly. Faced with a continually slipping schedule at the Sunnyvale plant, subcontractor Alliant and the Lockheed Martin Skunk Works formulated a plan to speed up work. Shift schedules were changed to double manpower.
November 20, 1998	The announcement was made that NASA and Lockheed Martin had terminated the LASRE. The LASRE sought to obtain data on the aerospike engine intended for use on the X-33 and VentureStar by mounting half of a scale-model aerospike engine on the back of an SR-71 aircraft and studying the effects of gas flow. The modified SR-71 carried out seven LASRE test flights. Those flights, however, tested only cold flow gas conditions; all hot flow experiments were now cancelled. Two flights collected aerodynamic data on the combination of the aerospike engine with the SR-71 aircraft. In two other flights, gaseous helium and liquid nitrogen were cycled through the test rig to test its plumbing and, in three more flights, liquid oxygen flowed through the system. The two hot-fire test flights planned to validate computer models of aerospike performance in flight were now cancelled. The LASRE had been repeatedly delayed by hardware and other problems. Cancellation of the LASRE allowed any remaining funds to be used by the Skunk Works to cover X-33 cost overruns.
November 24, 1998	In its December 2, 1998, issue, Aerospace Daily reported that on November 24, 1998, Boeing's Rocketdyne Division completed the first four tests of its XRS-2200 linear aerospike engine at Stennis Space Center. In these tests, the engine's turbomachinery and gas generator were run at full power and then throttled back to 57 percent power.
December 1998	Construction of the X-33 Flight Operations Center was completed a little more than 12 months after groundbreaking. The center was located on the eastern portion of Edwards Air Force Base.

Table 2–96. Reusable Launch Vehicle Chronology[a] (Continued)

Date	Event
December 2, 1998	Aerospace Daily reported that NASA's Office of Inspector General, in an audit titled "X-33 Funding Issues" (IG-99-001), found that Marshall Space Flight Center allowed $56 million in year-end obligations for the X-33 to go unrecorded in FYs 1996 and 1997, thereby giving Congress an inaccurate picture of the program's status at the end of those two years. The Inspector General reported that Marshall contract officers had established an arrangement with Lockheed Martin to delay billing for completed X-33 work until the following fiscal year. In FY 1996, that amounted to $22 million, and in FY 1997, to $34 million. The Inspector General maintained that obligations "should be recorded not later than NASA's acceptance of the completed milestone work" and recommended that NASA adjust its financial records to reveal the X-33 program's financial status "fully and accurately," and that NASA review the funding and payment practices used on the X-33 program to ensure that they met the requirements of the Antideficiency Act and internal controls.
December 4, 1998	Senior NASA staff, Boeing representatives, and X-33 project personnel from Boeing, Rocketdyne, and Lockheed Martin attended a meeting at Lockheed Martin's corporate headquarters in Bethesda, Maryland, to discuss development of the X-33 aerospike engine. At the meeting, Boeing proposed to downsize the ground portion of the propulsion demonstration program to use the resulting savings to fund X-33 engine cost overruns. Three teams were formed to evaluate Boeing's proposal and to assess opportunities that NASA Centers might have to mitigate the impact on technology development. The results of these independent team assessments were to be reviewed in mid-January.
December 16, 1998	Construction of the X-33 continued. Both liquid hydrogen tanks completed cures. Two gaseous oxygen tanks and two methane tanks belonging to the auxiliary propellant system were installed on the liquid oxygen tank. The thrust structure was nearly complete. Some clearance issues had emerged during installation of the nose gear support structure. Power pack assembly No. 2 was completed and sent to Stennis Space Center for testing, while power pack assembly No. 1 was still having problems. Construction of the X-33 launch site continued. The four vehicle hold-down posts were installed onto the rotating launch mount. The diesel generator for the site's electrical supply was run for the first time. The Vehicle Positioning System was unpacked and set up for testing. Sanders completed delivery of the Operations Control Center hardware. The X-33 launch site was now complete.

Table 2–96. Reusable Launch Vehicle Chronology[a] (Continued)

Date	Event
December 18, 1998	NASA exercised an option with Orbital Sciences for 25 additional test flights during a 12-month period beginning immediately after completion of the initial contract. The option was valued at more than $10 million, with government organizations performing an additional $4.7 million in work.
March 2001	The X-33 and X-34 programs were cancelled.

[a] Material in this table relating to the X-33 for the years 1996–1998 is drawn largely from Andrew Butrica, "Key X-33 Events," *http://www.hq.nasa.gov/office/pao/History/x-33/1998.htm.* (accessed March 15, 2005).

[b] Hawthorne, Krauss & Associates, LLC, "Analysis of Potential Alternatives to Reduce NASA's Cost of Human Access to Space," September 30, 1998, ftp://ftp.hq.nasa.gov/pub/pao/reports/1998/Hawrep.pdf (accessed May 18, 2005).

CHAPTER THREE

HUMAN SPACEFLIGHT

CHAPTER THREE
HUMAN SPACEFLIGHT

Introduction

NASA's human spaceflight undertakings seek to bring the frontier of space fully within the sphere of human activity, bringing people and machines together to overcome challenges of distance, time, and environment.[1] This chapter discusses NASA's human spaceflight activities during the decade from 1989 through 1998, focusing on Space Shuttle missions and the Space Station. It reviews the prior decade's activities; presents an overview of events during the 1989–1998 decade; summarizes the management and budget for human spaceflight at NASA; provides detailed information about each Space Shuttle mission; describes Space Shuttle payload accommodations; and discusses development of the Space Station.[2]

Most material in this chapter is based on primary NASA documents and Web-based NASA materials. These include pre- and post-launch mission operation reports, press kits and press releases, key personnel announcements, and various reports and plans issued by the Agency. Where applications activities are Shuttle-based, the Space Shuttle mission archives and mission chronologies have been consulted. The NASA projects have provided plentiful amounts of data. Most have comprehensive Web sites, and many also publish information booklets and fact sheets. Partner agencies, such as the European Space Agency (ESA), also publish printed and online material about their joint activities with NASA, as do the academic and private-sector institutions and organizations that are the homes of researchers and investigators. Most budget

---

[1] NASA Policy Directive (NPD) 1000.1, *NASA Strategic Plan* (Washington, DC: National Aeronautics and Space Administration, 1998), p. 26.
[2] Details of Spacelab missions are included in chapter 2, Earth Science and Applications, of Volume 8 of the *NASA Historical Data Book, 1989–1998.*

material comes from the annual budget estimates generated by the NASA Office of the Chief Financial Officer and from federal budget legislation. Other government agencies and organizations including the General Accounting Office, the Congressional Research Service, and the National Oceanic and Atmospheric Administration also issue reports and documents used as reference material. Measurements are presented in the unit used in the original reference (metric or English); conversions are in parentheses.

The Last Decade Reviewed (1979–1988)

The decade from 1979 through 1988 saw the inauguration of Space Shuttle flights in 1981, opening a new era of human spaceflight that had been on hold since the end of the Apollo era. Twenty-seven Shuttle flights took place during the decade; twenty-six were successful. The one unsuccessful flight, mission STS-51-L, set the tone for the remaining years of the decade, as the crew of the *Challenger* lost their lives in a catastrophic accident. Immediately after the accident, NASA began a far-reaching examination of the tragedy, using the findings of the independent Rogers Commission, appointed by President Ronald Reagan, and the NASA STS-51-L Data and Design Analysis Task Force to implement a set of recommendations that improved both the technical and management aspects of the human spaceflight program and increased the emphasis on safety. Two successful Shuttle missions at the end of the decade marked NASA's return to flight, as they demonstrated NASA's resilience and its determination to learn from the worst accident it ever experienced.

The 26 successful Shuttle flights deployed a variety of payloads from the government and commercial sectors and performed an array of scientific and engineering experiments. Four on-board Spacelab missions studied everything from plant life and monkey nutrition to x-ray emissions from clusters of galaxies.

Space Station development also began during the decade. In 1984, President Ronald Reagan directed NASA to develop and build a permanently manned Space Station and have it in place within a decade. NASA joined with partners in Europe (ESA), Canada, and Japan to begin developing Space Station *Freedom*. At the end of the decade, NASA and its partners had completed the Definition and Preliminary Design Phase and begun the Design and Development Phase.

Overview of Human Spaceflight (1989–1998)

During 1989–1998, NASA's human spaceflight activities focused on the Space Shuttle both as a launch vehicle and as a venue for a wide range of experiments. NASA also focused on developing the largest free-flying facility ever, the Space Station. NASA launched 66 Space Shuttle missions during the decade. These missions launched satellites into space, conducted on-board experiments, and performed rendezvous and docking exercises as part of the Shuttle-*Mir* program, the first phase of the International Space Station program. Spacelab activities on board the Shuttle and *Mir* began in 1983 with STS-9 and concluded in 1998. This series of international missions paved the way for research aboard the Space Station. Over its 17-year flight history, 22 Spacelab missions hosted payloads in practically every research discipline in which NASA engaged except those associated exclusively with planetary exploration. Between 1989 and 1998, 18 Spacelab missions flew.

This chapter summarizes each Shuttle mission and describes the on-board payloads and experiments on each mission. Descriptions of launched spacecraft, as well as descriptions of payloads launched and retrieved by the Shuttle, can be found in the chapters relating to space science, applications, and communications. Table 3–51 provides a summary list of all Space Shuttle missions with their major payloads.

Space Station development was undoubtedly the most ambitious NASA human spaceflight activity during the decade. NASA initiated the International Space Station program, but the orbiting laboratory was designed as an international undertaking, with participation by ESA, Japan, Canada, Italy (both as ESA member and NASA contractor), Russia from 1993, and Brazil (to a limited extent). The project, initially named Space Station *Freedom* by President Ronald Reagan in 1988, was the object of regular debate over its cost and scientific merit and, at the start of the decade, had already undergone redesigns in an effort to reduce its cost.

In 1993, President William J. Clinton, concerned by the cost and determined to reduce the federal deficit, ordered NASA to redesign the Station to make it simpler, smaller, and cheaper. The chosen redesign, with fewer capabilities, was first called Space Station *Alpha*. It became the International Space Station (ISS) with development spread over three phases. The ISS had Russia as a full-fledged partner contributing the first element, the service module, and a number of other essential components. The program also streamlined construction and management in the United States by assigning Johnson Space Center to be the host Center and eliminating *Freedom*'s complex work package structure with its independent contractors.[3] Instead, it consolidated all work under a single prime contractor, Boeing, with responsibility for the entire project.

The first phase of ISS development, lasting through 1998, consisted primarily of the Shuttle-*Mir* program in which U.S. astronauts spent months at a time aboard the orbiting Russian *Mir* space station. The purpose of these missions was to accustom American astronauts to living in space for long periods, provide additional experience with spacecraft rendezvous and docking, and develop good working relations between U.S. and Russian

---

[3] NASA used the term "host Center" to describe the role of Johnson Space Center in the Space Station program and the term "lead Center" to describe its role in the Space Shuttle program.

crew members. The second and third phases, which began at the end of 1998, comprised ISS assembly. Phase II consisted of initial on-orbit construction. It began with launch of the first ISS elements in 1998, providing initial living quarters and life support systems, and ending with launch of a three-person crew that marked the beginning of permanent ISS habitation. Phase III, the "assembly complete" phase, was to consist of remaining assembly, including the addition of laboratory modules, attaching a robotic arm, and crews of up to seven members.[4]

The program experienced continuous problems and delays due both to financial problems with its Russian partner and to an overly ambitious schedule and significant cost overruns by the U.S. prime contractor. Russian contributions were intended to be "enhancing" rather than "enabling," but it was clear that the country's contributions were needed for assembly to proceed. Russia lacked the funds to pay their prime contractor, causing years of schedule delays for both individual elements and project completion.

Although occurring years later than originally planned, on-orbit assembly began before the end of the decade, with successful deployment of the first two elements in 1998. The Russian Zarya (paid for with U.S. funds) was launched in November from Russia; the first U.S. module, Unity, was successfully delivered by Shuttle and joined with Zarya early in December 1998.

Management of Human Spaceflight Programs

The management and organizational structure of both the Space Shuttle and Space Station programs changed frequently as the technical nature of the programs evolved and as NASA sought to make the programs and their management more efficient. Some of these changes merely consolidated existing organizations or gave them new names that better reflected their functions; others eliminated divisions or offices; other changes established new divisions or offices. The sections that follow describe many of these changes.

NASA used a letter designation (called a "code") as an easy way to refer to its top-level organizations, or "offices." When an office was first formed, there was usually a connection between the assigned letter and the office's function (for instance, Code M was Manned Spaceflight), but in general, any connection became less likely over time as offices were created and eliminated, and many new letter designations were chosen merely because the letter was available. In the area of human spaceflight, the following letter designations were used during the decade from 1989 to 1998 and are mentioned in this chapter:

- Office of Space Flight–Code M
- Office of Space Station–Code S

---

[4] "Assembling a World-Class Orbiting Laboratory, Phases Two and Three," *http://spaceflight.nasa.gov/station/reference/fel/phases2_3.html* (accessed November 21, 2005). Also "ISS Program Phases," Boeing, *http://www.boeing.com/defense-space/space/spacestation/overview/program_phases.html* (accessed November 21, 2005).

- Office of Space Systems Development–Code D
- Office of Life and Microgravity Sciences and Applications–Code U

For general information about NASA's organizational structure, see chapter 1 of this volume and the chapter titled Facilities and Installations in Volume VIII of the *NASA Historical Data Book* for a description of each NASA Center.

Management of the Space Shuttle Program

The Headquarters Office of Space Flight (Code M) managed the Space Shuttle program. Chapter 2 describes the various reorganizations as well as personnel assignments and transfers within that organization. Briefly, in 1989, the Office of Space Flight organized into three major divisions: Institutions, Flight Systems, and the National Space Transportation System (NSTS) program. NSTS was soon renamed the Space Shuttle program. In December 1989, the Office of Space Flight added management of the Space Station program to its other responsibilities, moving it from an independent organization.[5] In 1991, several organizations related to development of new space transportation systems, including Space Station development, moved from the Office of Space Flight to a new organization, the Office of Space Systems Development (Code D). The operational aspects of the Shuttle program remained in Code M as did Spacelab and Space Station *Freedom* operations and utilization. In October 1993, the Office of Space Flight again assumed responsibility for the entire Space Station program.

The Office of Space Flight reorganized in October 1995 into four major offices: the Business Management Office, the Space Station Program Office, the Space Shuttle Program Director, and the Advanced Projects Office. A major Agency restructuring in October 1996 merged the Office of Space Communications into the Office of Space Flight. In July 1998, the final reorganization of the decade took place as the Office of Space Flight organized into four functional offices: Operations, Enterprise Development, Business Management, and Development.

Although overall management of the Space Shuttle program resided at NASA Headquarters, several NASA Centers had particular responsibilities relating to the program. Johnson Space Center in Houston, Texas, was designated the Space Shuttle Program Lead Center and managed development and operation of the Space Shuttle. Johnson Space Center was responsible for flight crew operations; mission operations; extravehicular activity; mission support; program safety and mission assurance; and design and development

---

[5] NASA Management Instruction 1102.5E, "Roles and Responsibilities–Associate Administrator for Space Flight," Effective December 29, 1989 (NASA History Office Folder 14829).

of the orbiter and crew-related government-furnished equipment. The Johnson Customer and Flight Integration office managed integration of the customer's payload into the Shuttle.

The Space Shuttle program manager at Johnson had full responsibility and authority to operate and conduct the program. Among the elements within this person's area of authority and responsibility were: overall program requirements and performance; total program control, including budget, schedule, and program content; approval of critical hardware waivers and deviations; budget authorization adjustments that exceeded a predetermined level; informing the Johnson Space Center director of program content and status; and integration of payloads with the orbiter.

Representatives of the Space Shuttle program elements, projects, and directorates supporting program activities were also part of the management team. They were located at various NASA Centers.

Kennedy Space Center in Cape Canaveral, Florida, was the launch site and primary landing site for the Shuttle. The Center was responsible for design, development, and operation of the launch and landing site facilities and support equipment; ground turnaround testing and maintenance of the orbiter; payload processing and installation into the orbiter; retrieval and disassembly of the solid rocket boosters; and conduct of all prelaunch and launch countdown activities required for each Space Shuttle mission. The launch integration manager at Kennedy was responsible for final vehicle preparation and return of the orbiter for processing for its next flight; managing the Certification of Flight Readiness process; presenting and scheduling of the Flight Readiness Review; the final launch decision process including final authority to commit to launch; and chairing the Mission Management Team before launch.

Marshall Space Flight Center in Huntsville, Alabama, through its Space Shuttle Projects Office, managed design, development, and integration of the solid rocket boosters, external tanks, and the Space Shuttle main engines.

Goddard Space Flight Center in Greenbelt, Maryland, managed the worldwide NASA communications network, including the Tracking Data and Relay Satellite System used to maintain communications with the Shuttle. In addition, Goddard oversaw the Get Away Special (GAS) program and several other small payload carrier programs. Stennis Space Center in Mississippi was responsible for testing the Shuttle's main engines.

Figure 3–1 shows the Space Shuttle program organization. Figure 3–2 is an expanded diagram showing the project elements assigned to Johnson Space Center, Marshall Space Flight Center, and Kennedy Space Center that together support the manager of the Space Shuttle program at Johnson Space Center in carrying out the program's responsibilities.

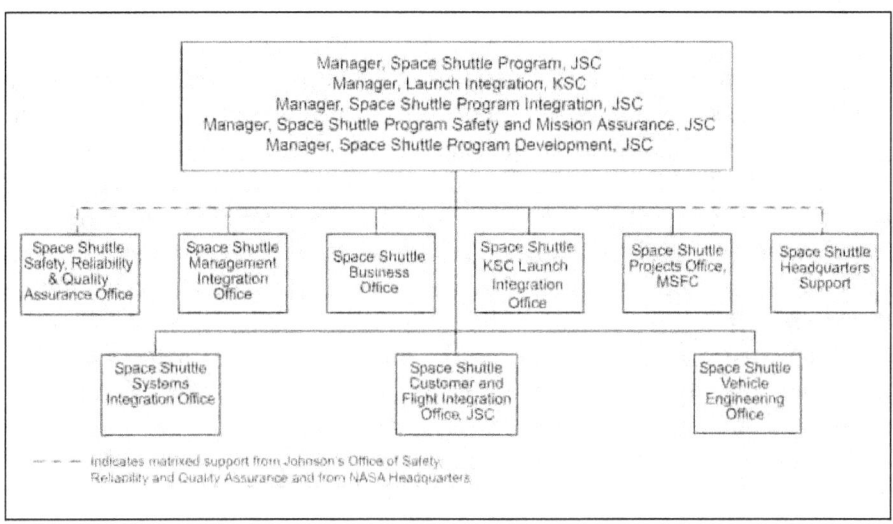

Figure 3–1. Space Shuttle Program Organization, December 1997.[6]

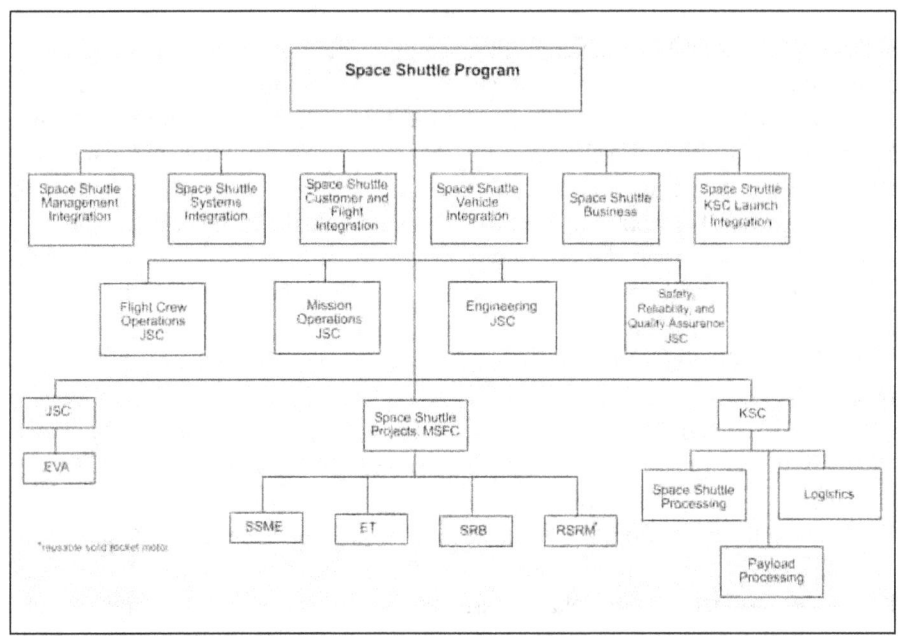

Figure 3–2. Space Shuttle Program Elements and Projects, December 1997.[7]

---

[6] Derived from "Space Shuttle Program Description and Requirements Baseline; Program Definition and Requirements," NSTS 07700 Volume I, Rev. G, December 17, 1997, pp. 3–17, *http://pbma.nasa.gov/docs/ public/pbma/bestpractices/bp_jsc_44.pdf* (accessed June 28,2005).

[7] Derived from "Space Shuttle Program Description and Requirements Baseline; Program Definition and Requirements," NSTS 07700 Volume I, Rev. G, December 17, 1997, pp. 3–18, *http://pbma.nasa.gov/docs/ public/pbma/bestpractices/bp_jsc_44.pdf* (accessed June 28, 2005).

Spacelab Management

Spacelab missions on the Space Shuttle and on *Mir* were the precursor to ISS activities. All major preparatory events leading to a Spacelab mission generally fell under the responsibility of four NASA departments: Headquarters, the Mission Management Office, the Mission Science Office, and payload element developers. NASA Headquarters was generally responsible for establishing mission objectives, sending out Announcements of Opportunity, and reviewing the experiment proposals. In the middle of the decade, the Life Sciences Flight program and the Space Shuttle/Spacelab Mission Management and Integration program in the Office of Life and Microgravity Sciences and Applications selected, defined, developed, and conducted in-space medical and biological research. These organizations also performed the mission planning, integration, and execution of all NASA-Spacelab, NASA-*Mir* Research Program, and attached Space Shuttle payloads.

The Spacelab management team was responsible for overseeing all aspects of hardware integration and coordination of all mission-related support activities. The mission manager served as the interface between the payload element developers' management and the Space Shuttle Program Office to maximize the mission objectives consistent with science requirements and Spacelab and orbiter system constraints. The Mission Science Office was responsible for organizing and coordinating all activities associated with payload specialist selection and experiment development. The payload element developers, reporting to the mission manager, were responsible for the design, fabrication, test, and formal turnover of experiment hardware, software, and experiment operating procedures.

Other offices that provided oversight management functions for a Spacelab mission included the Johnson Space Center Space Shuttle Program Office, the Marshall Space Flight Center Spacelab Management Office, Kennedy Space Center Launch Site Support Management, and Goddard Space Flight Center Communications and Data Support.

Marshall Space Flight Center was NASA's lead Spacelab Center and provided project management oversight for Spacelab hardware. The Center developed selected Spacelab hardware and provided technical and programmatic monitoring of the international Spacelab development effort. Marshall was also responsible for managing many Spacelab missions, including developing mission plans; integrating payloads; training payload crews; and controlling payload operations. The Payload Operations Control Center, which controlled Spacelab, was located at Marshall.

Management of the Space Station Program

The Space Station program underwent numerous organizational and management changes between 1989 and 1998 as it changed from Space Station *Freedom* to the ISS, brought Russia on board as a full-fledged partner, moved its center of operations from Reston, Virginia, in the Washington, DC, area to Johnson Space Center in Texas, and scaled down the size and complexity of the program. The following sections address the program structure and management. Note that the phases used below represent changes in management or management structure and are used to organize the discussion. They do not correspond with the NASA's three formal phases of Space Station development discussed later in this chapter.

Phase I: 1989–1992

Beginning in 1984, the Office of Space Station (Code S), an independent program office, managed the Space Station Freedom program (see Figure 3–3). Management was spread among three levels.

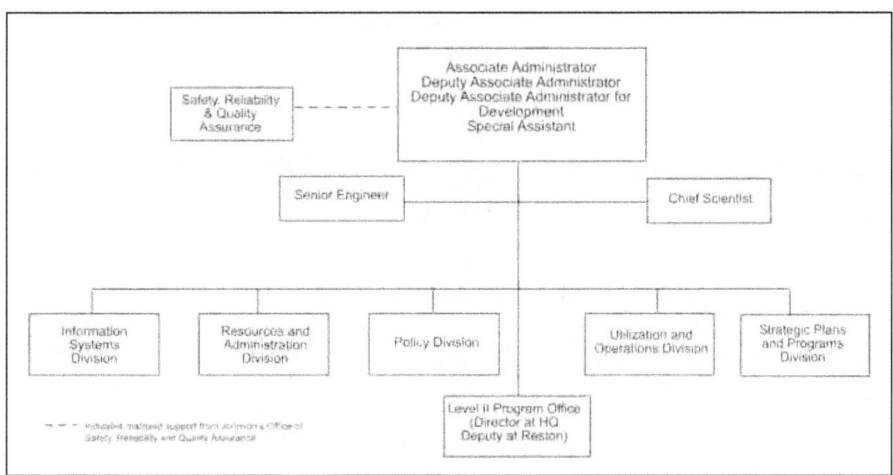

Figure 3–3. Headquarters Office of Space Station (Code S), December 1988.[8]

Level I comprised the Office of the Associate Administrator for the Office of Space Station (Code S) at NASA Headquarters. The Associate Administrator was responsible for overall program management and strategic planning. Level I was responsible for defining and controlling program requirements, schedule, milestones, and resources. The Level I divisions were Information Systems, Resources and Administration, Policy, Utilization and Operations, and Strategic Plans and Programs.

---

[8] "Roles and Responsibilities–Associate Administrator for Space Station," NASA Management Instruction 1102.12A, Ch. 1, Attachment A, December 16, 1988.

Level II consisted of the Space Station Program Office in Reston, Virginia. It was responsible for development of the Space Station, the operational capability of flight and ground systems, and the control of internal and external interfaces. The director of the Space Station program headed this office and was responsible for day-to-day management. Four offices—Safety/Product Assurance, Program Support, Program Integration, and Program Requirements and Assessment—and five groups—Program Control, Program Information Systems Services, Program Utilization and Operations, Program Systems Engineering and Integration, and International Programs—comprised Level II. NASA's accounting and procurement offices provided additional support.

Level III comprised the four work package Centers at the NASA Field Centers and their contractors. They were responsible for design, development, testing, and evaluation; operation of hardware and software systems; and element, evolution, and engineering support. A Space Station Project Office was located at each work package Center. The project manager of each Level III office reported to the director of the Space Station program at Level II. Figure 3–4 shows the three-tiered structure as it existed in April 1989.

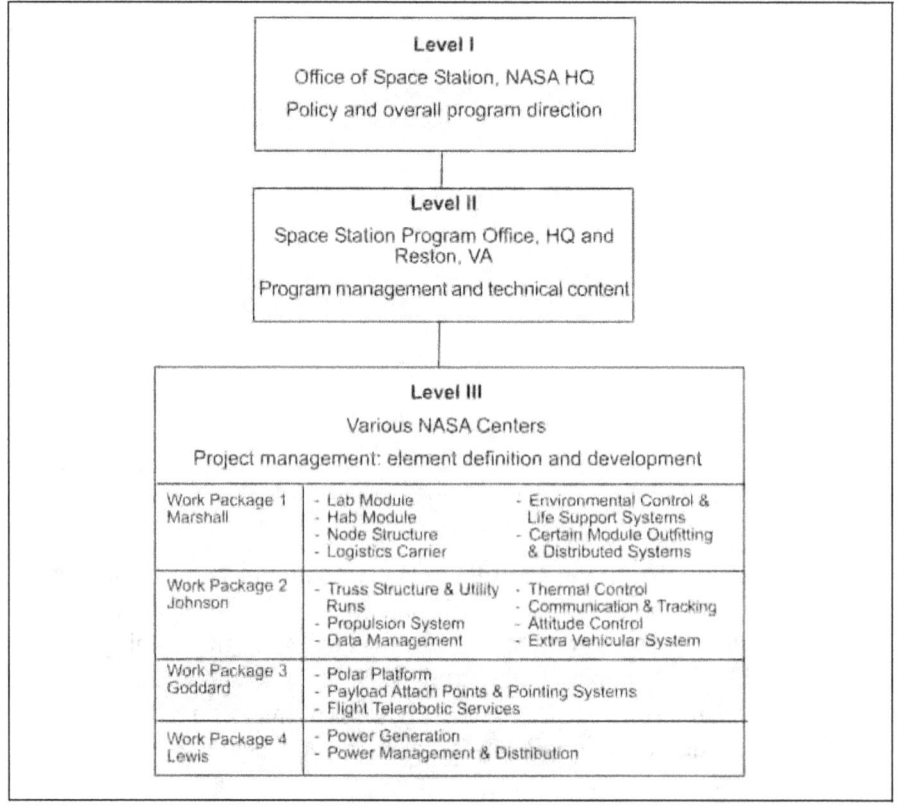

Figure 3–4. Tiered Space Station Organizational Structure, April 1989.[9]

9   *Space Station Freedom Media Handbook* (Washington, DC: Technical & Administrative Services Corporation, 1989), p. 10.

Level III Field Centers

Marshall Space Flight Center

Marshall Space Flight Center in Huntsville, Alabama, was the Work Package 1 Center. Work Package 1 included the design and manufacture of the astronauts' living quarters (Habitation Module); the U.S. Laboratory Module and logistics elements for resupply and storage; node structures connecting the modules; the Environmental Control and Life Support System; and the Internal Thermal Control and Audio/Video Systems in the pressurized modules.

Marshall also provided technical direction for the design and development of the engine elements of the propulsion system and was responsible for operations capability development associated with the Station's payload operations and planning. Boeing Aerospace was the Work Package 1 prime contractor. Figure 3–5 shows the Marshall Space Station organization.

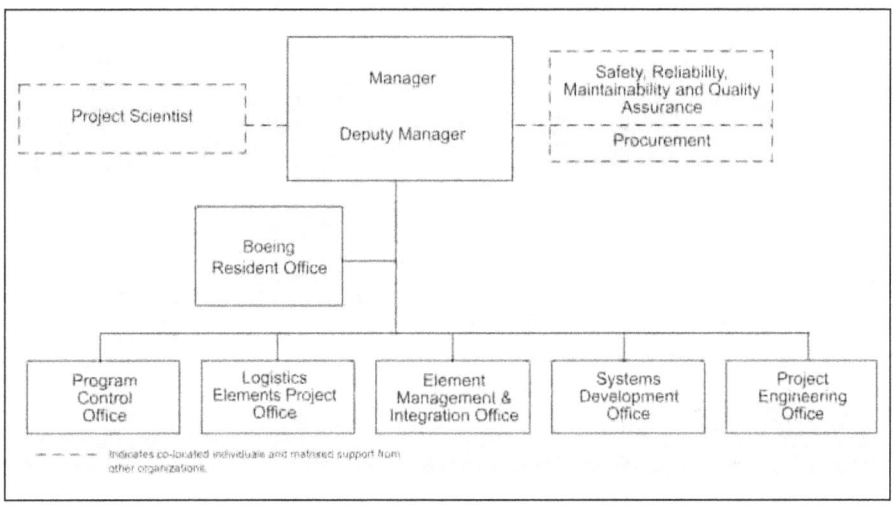

Figure 3–5. Marshall Space Flight Center Space Station Organization, April 1989.[10]

Johnson Space Center

Johnson Space Center, near Houston, Texas, was responsible for managing the design, development, test, and engineering of Work Package 2 flight elements and systems. These included the integrated truss assembly, propulsion assembly, mobile transporter system, outfitting of the resource node structures provided by Work Package 1, extravehicular system, and the external thermal control system. The extravehicular activity system included the extravehicular mobility unit (the spacesuit), associated life support, and other support equipment. Johnson was also responsible for the attachment systems for docking the Space Shuttle with the Space Station as well as the attachment

---

[10] *Space Station Freedom Media Handbook,* 1989, p. 33.

systems needed for logistics supply modules; the guidance, navigation, and control system; the communications and tracking system; the data management system; and the airlocks. The Center's prime contractor was McDonnell Douglas Astronautics.

Johnson provided technical direction for the design and development of all human space subsystems. These included crew quarters restraints and mobility aids; health care; operational and personal equipment; portable emergency provisions; workstations; galley and food management; personal hygiene; lighting; wardroom; stowage; and housekeeping/trash management. It was also responsible for providing a portion of the Canadian Space Agency's Mobile Servicing System training for the Space Station crew and Johnson ground support personnel. Figure 3–6 shows Johnson's Space Station organization.

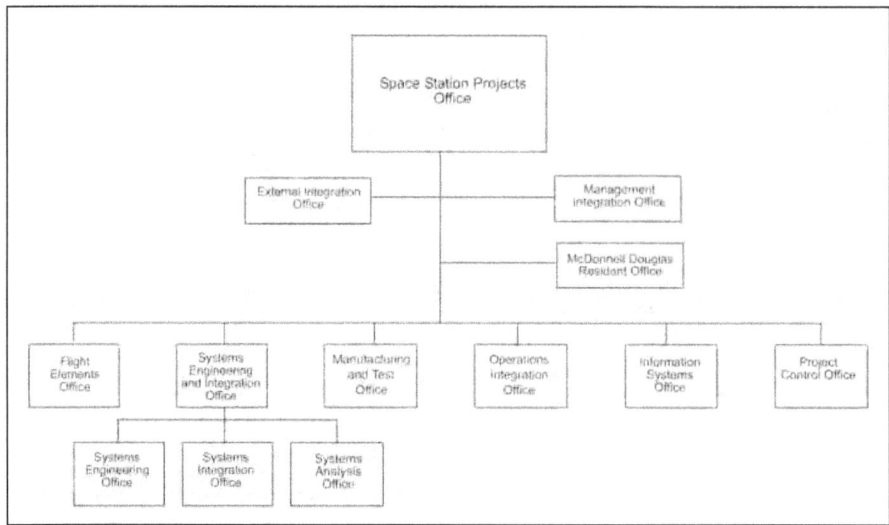

Figure 3–6. Johnson Space Center Space Station Organization, April 1989.[11]

Goddard Space Flight Center
Goddard Space Center in Greenbelt, Maryland, was responsible for managing the design, development, test, and engineering of Work Package 3 flight elements and systems. Goddard and its prime contractor, the Astro-Space Division of General Electric Company, were to manufacture the servicing facility, the flight telerobotic servicer, the accommodations for attached payloads, and the U.S. uncrewed free-flyer platforms. However, as part of the 1991 reorganization, Work Package 3 and Goddard's participation in the Space Station Freedom program were terminated. Figure 3–7 shows Goddard's Space Station organization as of 1989.

---

[11] *Space Station Freedom Media Handbook*, 1989, p. 50.

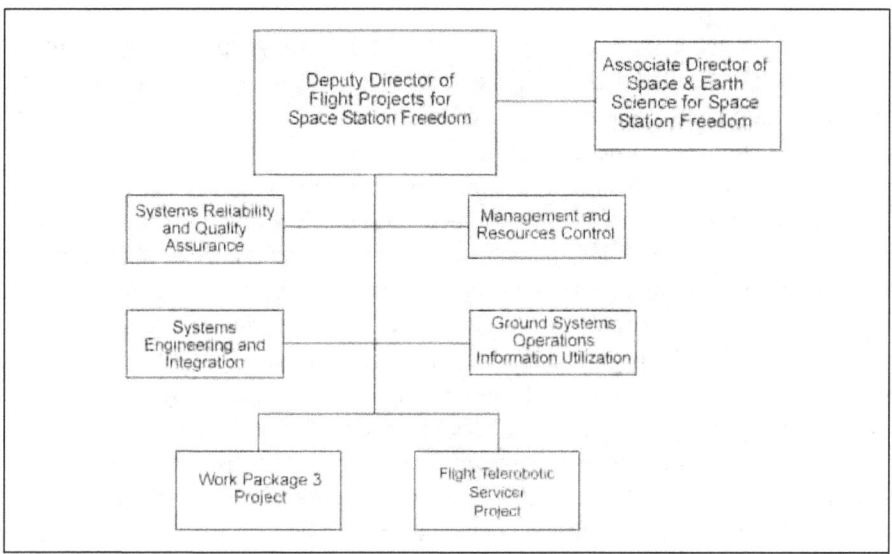

Figure 3–7. Goddard Space Flight Center Space Station Organization, April 1989.[12]

Lewis Research Center

Lewis Research Center in Cleveland, Ohio, was responsible for the Work Package 4 portions of the Space Station, consisting of the design and development of the entire electric power system; photovoltaic power generation subsystem; energy storage subsystem; solar power module; and primary power distribution. The Power Systems Facility at Lewis provided the capability to develop, test, and evaluate prototype power systems hardware for the program. Figure 3–8 shows Lewis's Space Station organization.

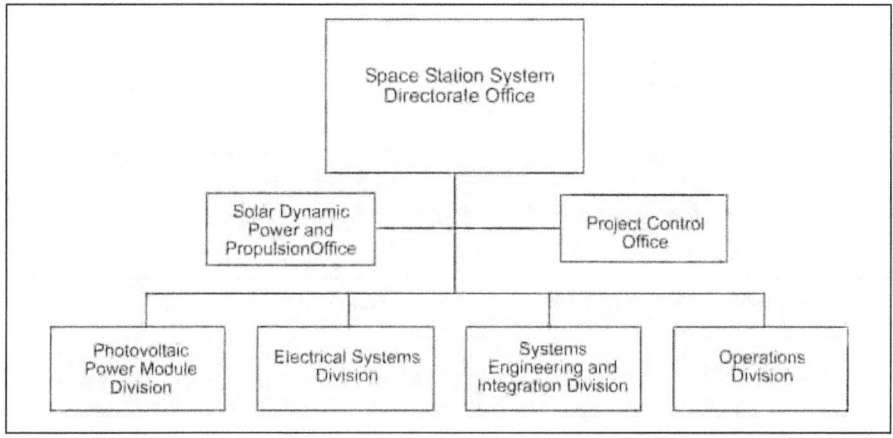

Figure 3–8. Lewis Research Center Space Station Organization.[13]

---

[12] *Space Station Freedom Media Handbook*, 1989, p. 54.
[13] *Space Station Freedom Media Handbook*, 1989, p. 70.

Kennedy Space Center

Although not a work package Center, the Kennedy Space Station Project Office was devoted to Space Station systems engineering and integration, ground support equipment management, operations and customer support, project control, and logistics systems. Because NASA was the Agency responsible for integrating both international and U.S. elements and systems with the Shuttle, Kennedy was the focal point for prelaunch and launch activities. The Center was responsible for launch sites; launch site common ground support equipment; facilities to support prelaunch and postlanding processing; payload processing and logistics; management and operations of integrated logistics systems; and the Space Station Processing Facility. Technicians from the Space Station partners would provide technical and hands-on support for the integration of the international elements at Kennedy. The Kennedy Space Station *Freedom* test teams would provide launch site final acceptance testing to verify major interfaces, provide confidence tests of critical systems, and verify end-to-end operations between the flight elements and ground control Centers. Figure 3–9 shows Space Station project organization at Kennedy.[14]

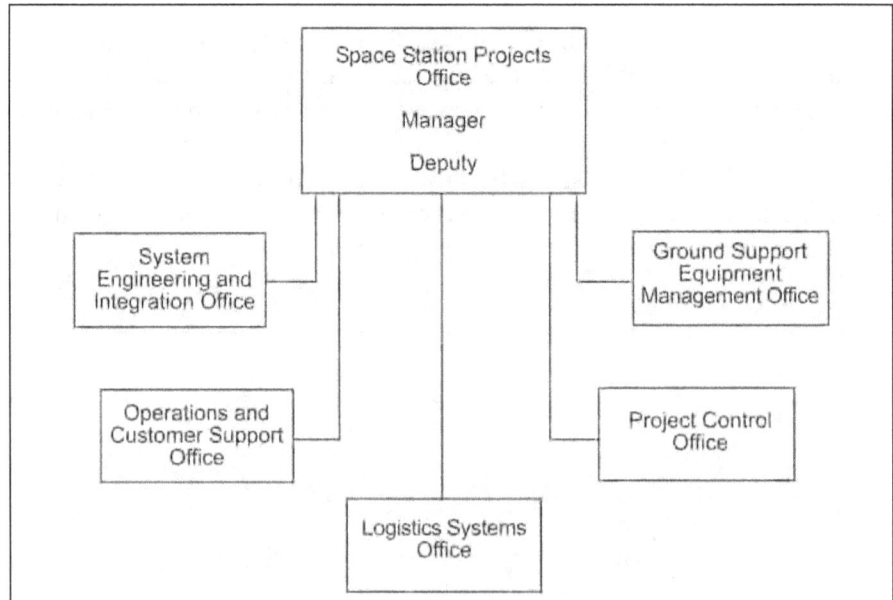

Figure 3–9. Kennedy Space Center Space Station Organization, April 1989.[15]

[14] *Space Station Freedom Media Handbook* (Washington, DC: Technical & Administrative Services Corporation, 1992), pp. 9, 24–71.
[15] *Space Station Freedom Media Handbook*, 1989, p. 74.

Space Station Management

James B. Odom was Associate Administrator for the Office of Space Station at NASA Headquarters from March 1988 until his retirement on April 30, 1989. When Odom retired, Thomas L. Moser served briefly as acting Associate Administrator until he left NASA in mid-May 1989. On May 18, NASA acting Administrator Richard Truly named Dr. William B. Lenoir Associate Administrator for the Office of Space Station effective 1 June. On July 13, 1989, he was also appointed acting Associate Administrator for the Office of Space Flight, filling the position held by Truly, before he became NASA Administrator. Truly asked Lenoir to develop a plan to consolidate the Office of Space Flight and the Office of Space Station. Henry Hartsfield was assigned temporary duty to direct the Space Flight/Space Station Integration Office, replacing Robert Parker.[16]

In Reston, Virginia, E. Ray Tanner became Director of the Space Station Freedom Program Office on January 3, 1989.[17] On May 18, 1989, Truly named Richard Kohrs Director of Space Station Freedom at NASA Headquarters, and Tanner Deputy Director of Space Station Freedom Program and Operations in Reston, Virginia. James Sisson was named as acting Deputy Director for the Space Station Freedom Program Office, moving from his position as Deputy Program Manager for the Space Station Freedom Program Office.[18]

At the beginning of October 1989, Sisson, moved from acting Deputy Director, Program and Operations of the Space Station Freedom Program Office, to the position of Deputy Manager, Space Station Freedom Program and Operations. Robert W. Moorehead became the new Deputy Director, Program and Operations, of the Space Station Freedom Program Office.[19]

In November 1989, Truly announced the consolidation of the Space Station Program and Space Shuttle Program into a combined organization named the Office of Space Flight (Code M), effective December 1989.[20] (See Figure 2–3 in chapter 2 of this volume.) The combined organization made sense, NASA explained, because the Space Station would be launched using the Shuttle and assembled in-orbit. Having both activities under a single Associate Administrator would improve communications and decision making in several key areas, including the assembly sequence and Space Shuttle-Space Station interfaces. In addition, the astronauts who would assemble and operate the Station would more directly influence its design.[21]

---

[16] The integration office had been established in 1987 to facilitate integration of the Space Station into the Space Transportation System.

[17] "Tanner Named Director Space Station Freedom Program," *NASA News* Release 88–175, December 29, 1988 (NASA History Office Folder 009610).

[18] "Space Station Program Leadership Selected by Truly," *NASA News* Release 89–77, May 19, 1989 (NASA History Office Folder 009610). Tanner retired from NASA on July 15, 1989.

[19] "Moorehead Named Space Station Freedom Program Deputy," *NASA News* Release 89–155, October 2, 1989 (NASA History Office Folder 009610).

[20] NASA Management Instruction 1102.5E, "Roles and Responsibilities–Associate Administrator for Space Flight," Effective December 29, 1989 (NASA History Office Folder 14829).

[21] Office of Space Flight, National Aeronautics and Space Administration, "Space Station Level II Management and Integration Status," June 1990, pp. 4–6 (NASA History Office Folder 009524).

Richard Kohrs headed the Space Station Freedom Program Office, one of the two major divisions in the Office of Space Flight. Richard A. Thorson, at Johnson Space Center, became Deputy Program Manager of Space Station Freedom Systems Integration Office. Level II Integration Offices were established at Marshall Space Flight Center and Johnson Space Center. James Sisson became Manager of the Element Integration Office at Marshall. Jesse F. Goree, Jr., became acting Manager of the Systems Integration Office at Johnson. Figure 3–10 shows the management structure of the Space Station Freedom Program Office as of May 1990. It shows the same tiered structure that had been in place since the program's inception. Level I provided overall program direction and policy. Level II provided day-to-day management and overall system engineering and integration. The Level III Field Center project offices directed the design and development of the hardware and software, which was performed by the contractor teams below them.

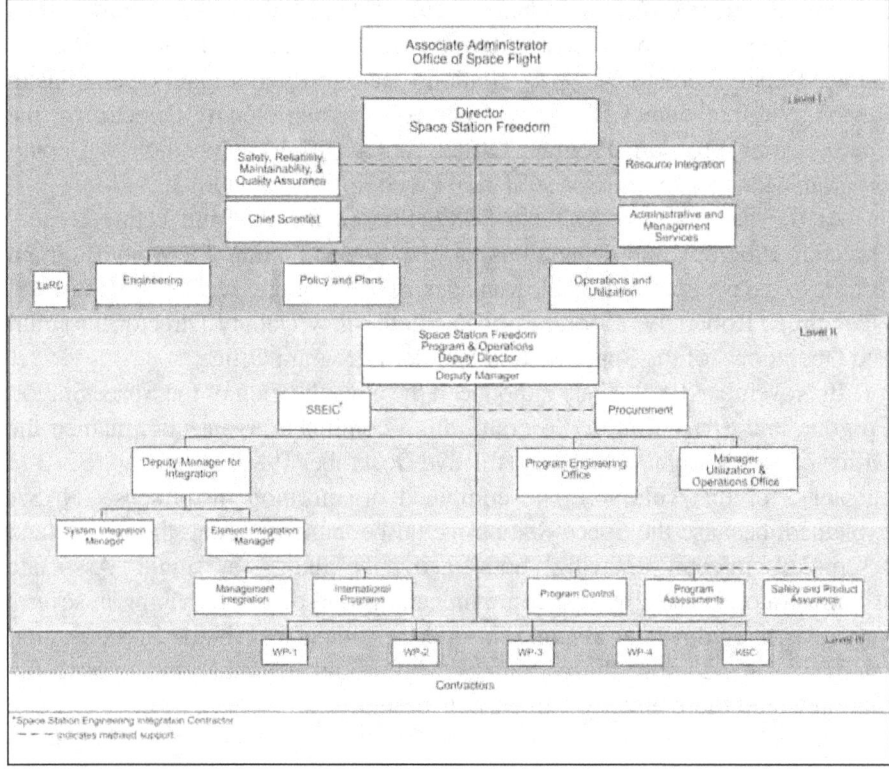

Figure 3–10. Space Station Freedom Program Office, May 1990.[22]

In December 1990, the Advisory Committee on the Future of the U.S. Space Program, led by Norman Augustine, issued a report that, among its recommendations, advised separating operations from development and

---

[22] "Space Station Level II Management and Integration Status," Office of Space Flight, National Aeronautics and Space Administration, June 1990, (NASA History Office Folder 009524).

grouping the Space Station program with other development programs headed by a NASA Associate Administrator for spaceflight development. It also recommended locating "a strong and independent project office reporting to headquarters" near the NASA Center that had the most work for the project.[23] As a result, on September 13, 1991, Truly announced plans to create a new office to be named the Office of Space Systems Development (Code D). The new organization would have responsibility for Space Station *Freedom* development as well as other development programs (see Figure 3–11). The Office of Space Flight would retain responsibility for Space Station *Freedom*-Spacelab operations, the Space Shuttle program, and other areas of spaceflight operations.[24] On October 3, Truly named Arnold D. Aldrich Associate Administrator for the new organization, and Dr. C. Howard Robins, Jr., as Deputy. Richard Kohrs was named Deputy Associate Administrator for Space Station *Freedom*.

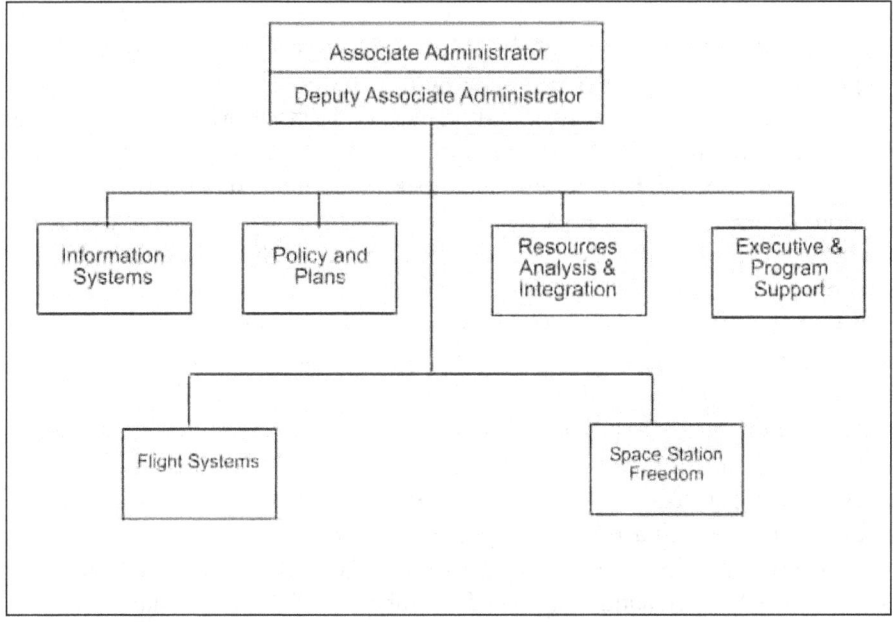

Figure 3–11. Office of Space Systems Development (Code D).

[23] Advisory Committee on the Future of the U.S. Space Program, "Report of the Advisory Committee on the Future of the U.S. Space Program, December 1990," *http://www.hq.nasa.gov/pao/History/augustine/racfup1.htm* (accessed March 15, 2005).
[24] "New Office of Space Flight Development Announced," *NASA News* Release 91–148, September 13, 1991 (NASA History Office Folder 009610).

Phase II: 1992–1994

On March 30, 1992, William Lenoir announced that he would be leaving his position as head of the Office of Space Flight and retiring from NASA in May. On April 28, the new NASA Administrator, Daniel S. Goldin, announced the appointment of Major General Jeremiah W. Pearson, III as Associate Administrator of the Office of Space Flight.

During 1992, Space Station *Freedom* continued to use a three-tiered management structure, although the program had moved from the Office of Space Flight (Code M) to the newly established Office of Space Systems Development (Code D). Responsibilities of the three levels remained essentially unchanged after the move, with the exception that there were now only three Work Package centers—Marshall Space Flight Center, Johnson Space Center, and Lewis Research Center.[25] Level I consisted of the Associate Administrator for the Office of Space Systems Development (Code D) at NASA Headquarters; Level II, the Deputy Director, Program and Operations in Reston, Virginia; and Level III, the NASA Field Centers' Space Station Freedom Project Offices (see Figure 3–12). The managers of these Level III project offices reported to the Deputy Director of the Space Station program on Level II.

Outside the Work Package structure, the Johnson Mission Operations Directorate was responsible for training of Space Station *Freedom* crew and ground controllers and for around-the-clock operational support of the Space Station. Kennedy Space Center was responsible for processing of payloads for flights to *Freedom* on the Shuttle. This included the required assembly, servicing, integration and testing of payload hardware and software, and the requisite operations associated with a Shuttle launch. Contractors were responsible for design; development; testing; evaluation; operation of hardware and software systems; and element, evolution, and engineering support. A number of international partners were also providing various Station elements. Figure 3–13 maps the three Work Packages and the contributions of the international partners with the various Space Station *Freedom* elements as published the 1992 Space Station Freedom Strategic Plan.[26]

At this time, the Level II program office headed by Richard Kohrs was located at NASA Headquarters. On December 1, 1992, NASA announced its intention to consolidate some management functions and move the Headquarters-based program office, led by Kohrs, to Reston, Virginia. NASA also announced that the Agency would create a contractor-led joint vehicle integration team based at Johnson Space Center and staffed by the three Space

[25] *Space Station Freedom Media Handbook*, 1992, p. 8. Goddard Space Flight Center and its prime contractor, GE Astro-Space, originally were to manufacture the servicing facility, the flight telerobotic servicer, accommodations for attached payloads, and the U.S. uncrewed free-flyer platforms. However, in 1991, these elements were either terminated or transferred to other NASA organizations, and this Work Package was dissolved.

[26] *NASA Space Station Freedom Strategic Plan 1992* (undated), p. 19 (NASA History Office Folder 16941).

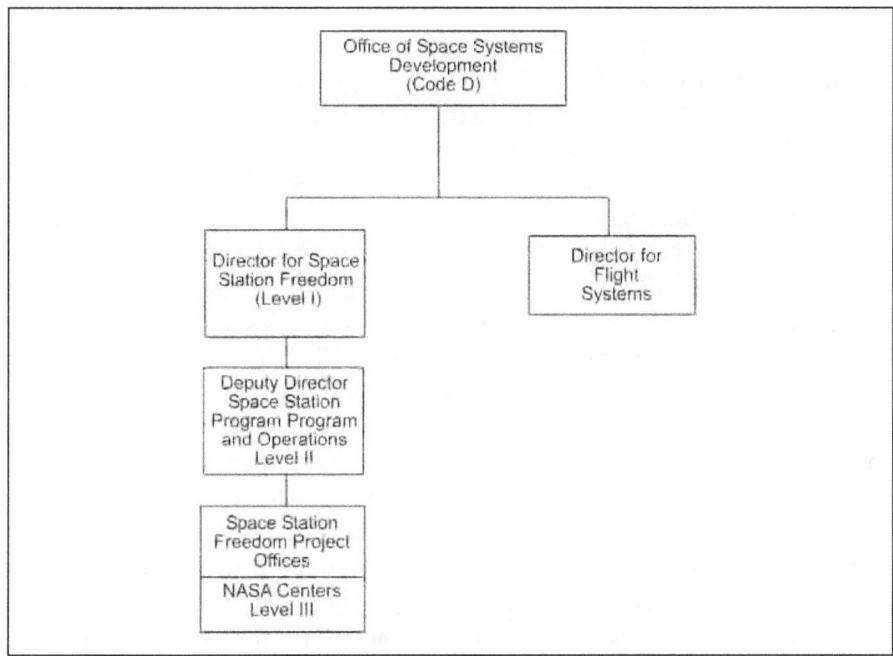

Figure 3–12. Space Station Freedom *Three-Level Program Management, 1992.*

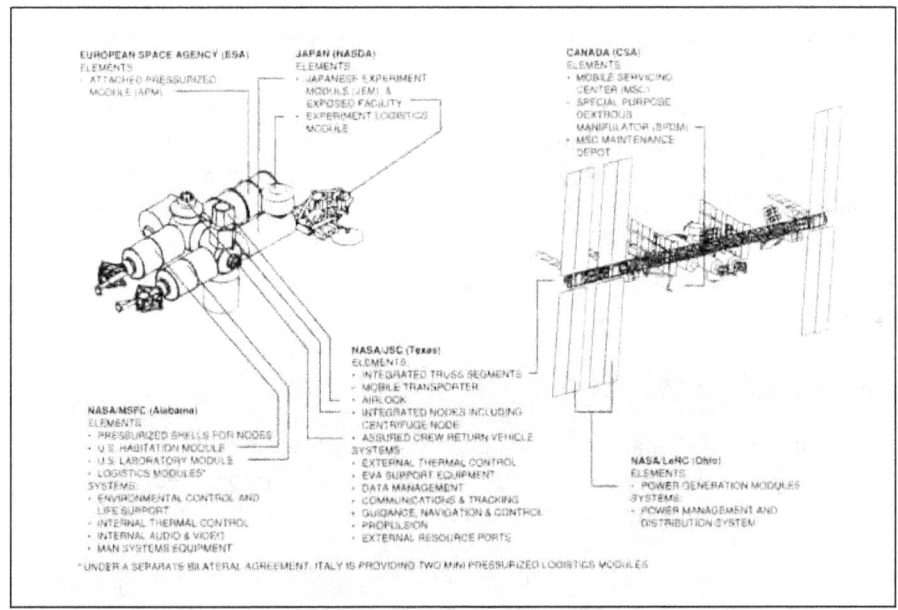

Figure 3–13. Space Station Freedom *Work Package and International Partner Development Responsibilities.*[27]

---

[27] *Space Station Freedom Media Handbook*, 1992, p. 22.

Station prime contractors—Boeing, McDonnell Douglas, and Rocketdyne—and Grumman's Space and Electronics Group, which had an engineering and integration contract with the office in Reston, Virginia. The team would ensure the "successful building and deployment" of the Space Station.[28]

In early 1993, President William J. Clinton called for NASA to redesign the Space Station, reducing the complexity of both the Station itself and its management structure to reduce cost and produce greater returns on NASA's investment.[29] In March, Administrator Goldin announced a number of changes relating to the redesigned Station, given the name Space Station *Alpha*. These changes affected the Station's management as well as its workforce level and location. In July, Goldin announced that Bryan O'Connor would head the Space Station transition to the redesigned Station. Goldin also stated that the number of civil servants needed for the Space Station would be reduced from approximately 2,300 to 1,000. At the same time, NASA announced that it would recruit 300 positions to staff the new Space Station Program Office at a host Center "yet to be determined." In August, Goldin announced the selection of Johnson Space Center in Houston, Texas, as host Center and Boeing as the prime contractor. The new program office had all implementation responsibilities: the design, development, and the physical and analytical integration of the Space Station as the program evolved into operations. The new organization structure would have about 1,000 civil servants, consisting of about 300 civil servants at the program office at Johnson and the other 700 positions spread among all involved NASA Centers, including Johnson.[30] Approximately 800 Space Station contractors working near the Reston office either lost their jobs or were invited to relocate.

In September 1993, Space Station Director Kohrs retired from NASA. In October, Goldin announced that the Space Station program would move from the Office of Space Systems Development (Code D) back to the Office of Space Flight (Code M). Jeremiah W. Pearson III managed the integration of the two programs. William Shepherd became Space Station Program Manager at Johnson Space Center. O'Connor, Director of the Space Station transition, was named acting Space Station Program Director.

Phase III: 1994–1998

A number of management changes took place in January 1994 as Space Station *Alpha* transitioned into the ISS. Wilbur C. Trafton became Deputy Associate Administrator for the Space Station.[31] Trafton was assisted by

---

[28] "Management Changes Made to Space Station Program," *NASA News* Release 92–214, December 1, 1992 (NASA History Office Folder 009610).
[29] "Organizational Changes to Enhance Programs, Relations," *NASA News* Release 93–044, March 11, 1993. "Special Announcement," March 11, 1993 (NASA History Office Folder 009610).
[30] "Space Station Host Center and Prime Contractor Announced," *NASA News* Release 93–148, August 17, 1993 (NASA History Office Folder 009610).
[31] Trafton was the sixth person to run the Space Station program since it began in 1984. His predecessors were John Hodge, Philip Culbertson, Andrew Stofan, James Odom, and most recently, Richard Kohrs.

Randy Brinkley at Johnson Space Center in Houston, Texas. Brinkley was appointed Space Station Program Manager responsible for managing all United States-Russian activities and working with Russia to implement United States-Russian activities for Phase I and Phase II of the Space Station program.[32] William Shepherd was named Deputy Program Manager at Johnson. Pamela McInerney served as acting head of the Space Station Headquarters Office during much of 1994. After McInerney left the position, it remained vacant until Joyce Carpenter took the position in early 1995. She was replaced in the fall of 1995 by Gretchen McClain.

In October 1995, the Office of Space Flight reorganized with the goal of simplifying its structure and increasing its organizational efficiency. Figure 3–14 shows its structure in October 1995.

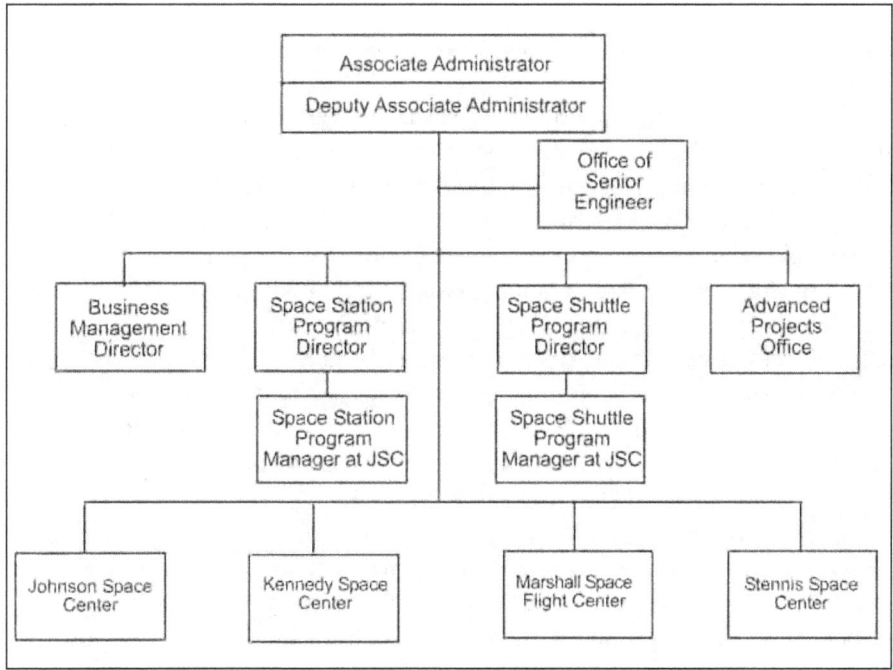

Figure 3–14. Office of Space Flight (Code M), October 1995.

In January 1996, Trafton assumed additional responsibilities, first becoming acting Associate Administrator for the Office of Space Flight at NASA Headquarters and then Associate Administrator in March. Andrew Allen became acting Space Station Program Director until Gretchen McClain took over in January 1997. Other changes at the same time included naming William Shepherd Deputy Manager for Space Station International Affairs to lead the

---

[32] Phase I and Phase II of the Space Station program consisted of the Shuttle-*Mir* flights and the first set of Space Station assembly flights. See the detailed description of Space Station development later in this chapter.

integration of all Russian issues for the Station. Douglas Cooke, Manager of the Vehicle Office, was named acting Deputy Manager for the program. Denny Kross took Cooke's place as acting Manager of the Vehicle Office, and Lauri Hansen was named acting Deputy Manager of the Vehicle Office.[33] In June, Program Manager Randy Brinkley established three new positions to help prepare for launch and flight operations: Kevin Chilton became Deputy for operation, Douglas Cooke was named Deputy for technical development, and Dan Tam became acting Deputy for business management.[34]

In November 1997, Trafton resigned from NASA, leaving his position as Associate Administrator of the Office of Space Flight. In January 1998, Administrator Goldin named Joseph Rothenberg, Director of Goddard Space Flight Center, to the job. Rothenberg became NASA's fourth human spaceflight Associate Administrator in little more than three years, closing out this decade.

Life Sciences and Microgravity Sciences Management

Life sciences and microgravity sciences were closely tied to human spaceflight. The 1992 Space Life Sciences Strategic Plan stated that NASA's life sciences program "significantly contributed to NASA's manned and unmanned exploration of space" during the last 30 years. The plan also stated that the life sciences program maintained a "close working relationship with the Office of Space Flight on operational issues dealing with crew health . . .[and] with the newly formed Office of Space Systems Development in conducting the research and development to support the operation and utilization of Space Station *Freedom* . . ."[35] Similarly, later in the decade, when NASA adopted the enterprise approach to organize its themes, the Microgravity Sciences and Applications Division and the Microgravity Research program supported the Human Exploration and Development of Space Enterprise.

Until early 1993, the Life Sciences Division was located within NASA's Office of Space Science and Applications. The Life Sciences Division focused on activities that dealt with understanding how living systems responded to the space environment; the search for the origin, evolution, and distribution of life in the universe; the development of the scientific and technological foundations for expanding the human presence beyond Earth orbit and into the solar system; and providing operational medical support to all space missions involving humans. Results from the division's research helped maintain astronaut health and productivity, understand the response of

---

[33] "Space Station Office Makes Managerial Changes," *Space News Roundup,* Johnson Space Center (January 12, 1996): p. 1 (NASA History Office Folder 009610).

[34] "Brinkley Establishes Key Management Positions," *Space News Roundup,* Johnson Space Center (June 24, 1996): p. 4 (NASA History Office Folder 009610).

[35] *Space Life Sciences Strategic Plan, 1992,* National Aeronautics and Space Administration, Space Life Sciences.

biological mechanisms to weightlessness, and design controlled ecological life support systems.[36] In 1989, Arnauld Nicogossian led the division, having become its head in 1983.

The Microgravity Science and Applications Division was also a division within the Office of Space Science and Applications. It appeared as a budget line within the larger Materials Processing in Space budget category. The division aimed to foster the development of near-Earth space as a natural resource by exploiting microgravity and other unique attributes that might be attained in an orbiting spacecraft. From 1989 to 1991, the division was led by acting Director Robert Schmitz. In 1991, Robert Rhome was appointed Division Director.

In March 1993, in a series of organizational changes, Administrator Goldin announced a new Office of Life and Microgravity Sciences and Applications (OLMSA), designated Code U. The new office was to "assure the right emphasis in the [Space Station] redesign effort . . . The redesigned Space Station must provide for significant long duration space research in materials and life sciences during this decade." It was also responsible for instituting NASA policies and procedures for the protection of human research subjects.[37]

Harry Holloway was appointed the first OLMSA Associate Administrator, and Nicogossian was appointed OLMSA Deputy Associate Administrator for spaceflight activities, moving from his position as Chief Medical Officer in the Office of Space Flight.[38] Rhome continued leading the Microgravity Sciences and Applications Division, which moved from the Office of Space Sciences and Applications to OLMSA. The division's focus was to increase understanding of the effects of gravity on biological, chemical, and physical systems using both spaceflight and ground-based experiments. Joan Vernikos became head of the Life and Biomedical Sciences and Applications Division in the Microgravity Division, and Edmond Reeves was appointed initially as acting head of the Flight Systems Division and then as Division Director in 1994. In mid-1993, the Occupational Health and Aerospace Medicine Division, led by Marshall S. Levine, was added to OLMSA. Earl Ferguson took over leadership of the Occupational Health and Aerospace Medicine Division on an acting basis in the spring of 1994 when Levine became Director of the Occupational Health Office.

In May 1996, Holloway left his position, and Nicogossian became initially the new acting OLMSA Associate Administrator and then Associate Administrator in June 1997. Beth McCormick became Deputy Associate Administrator. In early 1997, James Collier was appointed the new head of the Aerospace Medicine Division, which had split off from the Division of

---

[36] "Estimates Life Sciences Program Budget Summary," Research and Development Fiscal Year 1989 p. RD 4–2.

[37] NASA Management Instruction 7100.8B "Protection of Human Research Subjects", August 8, 1995.

[38] "Organizational Changes to Enhance Programs, Relations," NASA News Release 93–044, March 11, 1993. Also "Special Announcement," March 11, 1993 (NASA History Office Folder 009610).

Occupational Health and Aerospace Medicine Division. The Space Development and Commercial Research Division, led by Edward Gabris, was also established. In 1998, Richard Williams moved to lead the Aerospace Medicine Division; Mark Uhran became acting head of the Flight Systems Office; and Raymond Whitten replaced Gabris as head of the Space Development and Commercial Research Division.

Money for Human Spaceflight

This section discusses funding for Spacelab, Space Station, and life and microgravity sciences. The budget tables that follow show budget requests and programmed amounts for NASA's human spaceflight programs (other than the Space Shuttle, which is addressed in chapter 2, Launch Systems). Since NASA typically submits an original and revised budget request before Congress acts on a budget, both amounts are indicated and separated by a forward slash. Where no amount appears, there was no submission. Programmed amounts are determined after the end of a fiscal year and reflect the amounts actually available to be spent. Occasionally, a budget category is established during a fiscal year. When this happens, there will be a programmed amount shown but no budget request for that budget category. Funds for these activities often were transferred from another project's budget through a "reprogramming" of funds during the year. All amounts come from the annual budget requests prepared by the NASA Office of the Chief Financial Officer.

Spacelab

Spacelab funds were included in the Space Transportation Capability Development budget category in the Research and Development and Human Spaceflight appropriations. Some Spacelab funding also came from the Shuttle/ Spacelab Payload Mission Management and Integration budget category and from various life and microgravity sciences budget categories. Spacelab development funding supported space-based and ground support equipment and hardware to expand Spacelab capabilities and ensure its continued operational availability. Spacelab operations support funding included mission planning and integration and flight and ground operations. It also funded smaller payloads including the Get Away Specials and Hitchhiker payloads.

The level of funding for Spacelab generally corresponded with the number and complexity of scheduled Spacelab missions and whether funds would be received from other Spacelab participants, such as Japan and Germany. The final Spacelab mission occurred in April 1998, and funds for Spacelab began to fall in FY 1997 and were reduced to a very low level in FY 1998.

Space Station

Funding for the Space Station mirrored the contentiousness of the entire program. President Ronald Reagan first proposed the program, and although it generally received support from later presidents, members of Congress continued to question the validity of the program and the advisability of spending large sums of money on a program whose scientific and political benefits were doubtful. They also questioned NASA's dependence on the contributions and cooperation of international partners who had their own financial and political problems, particularly Russia. Thus, in almost every session of Congress, motions were introduced to cancel the program; although all of these motions were defeated, on one occasion, the margin to continue the program was only a single vote. Congress kept an extremely close eye and tight rein on the program, generally providing only one year of funding at a time and requiring NASA to annually justify new funding requests.

The Space Station took a large portion of NASA's Research and Development and HSF appropriated funds, reaching a high of 42 percent in FY 1998 (see Table 3–1). This spending caused some resentment among the science community, as some considered the research benefits of the orbiting laboratory limited. From FY 1989 to FY 1998, funds appropriated annually for the Space Station grew from a low of $900 million to a high of almost 2.4 billion, a factor of more than two and one-half. This growth occurred even though redesigns reduced the size and complexity of the Station. The funding need brought Russia into the partnership to assume some of the expenses. The Space Station also received some funding from the Science, Aeronautics and Technology appropriation (not reflected in this table), which effectively increased the percent of the total NASA budget dedicated to the program.

In 1993, President William J. Clinton ordered NASA to redesign the Space Station to reduce program costs while still providing significant research capabilities. The chosen redesign came with a total budget cap of $17.4 billion and a fixed annual budget of $2.1 billion, although these limits were not set in law. The President's annual cap was below the annual ceiling of $2.8 billion identified by NASA Station designers in their three proposed design options. To accommodate the lower ceiling but stay within the $17.4 billion total, NASA regularly slipped the delivery dates for both individual elements and a completed Station.

Beginning in 1995 when NASA's appropriation categories were restructured and the Research and Development appropriation category was eliminated, Space Station-related activities were funded from the HSF and the Science, Aeronautics and Technology appropriations. Activities funded in the HSF appropriation included the development and operation of the Space Station and the flight support component of the Russian cooperation program of joint flights to *Mir*. Space Station-related funding from the Science

Aeronautics and Technology appropriation provided for development, operation, and science research associated with the scientific, technology, and commercial payloads being built for Space Station use or in conjunction with the Mir program. The largest amount came from the Office of Life and Microgravity Sciences and Applications to fund its experiments. In addition, the Mission to Planet Earth program (NASA's Earth Sciences program) provided funds for an externally attached Space Station payload, and the Space Access and Technology program provided funds for technology and commercial payloads for both external and pressurized Space Station deployment.[39] When including all of these sources of funds, as well as the amount allowed in a new contingency account called Russian Cooperation and Program Assurance and some funds in the Construction of Facilities account used for the Space Station, funding for the Station remained fairly steady through 1997 and even rose slightly.

Money problems, however, did not abate. In September 1997, Boeing admitted it was incurring millions of dollars in cost overruns and could have a $600 million overrun at Station completion.[40] NASA also accepted some of the conclusions of the Cost Control Task Force (except for a cost estimate of $24.7 million at completion), chaired by Jay Chabrow, and raised the cost to complete to $22.7 billion.[41] NASA requested an increase of $430 million in the appropriation for FY 1998. Congress responded with $230 million. At the end of the decade, additional funds would be needed for NASA to complete the Space Station, whether they came from Congress or were diverted from other NASA programs.

Table 3–2 shows authorized and appropriated amounts for Research and Development, HSF, and the Space Station from 1989–1998. Table 3–3 shows the programmed amounts for the budget categories included in this chapter. If no programmed amount appears for a particular budget category or for a particular year, there was no amount indicated in the budget documents. Tables 3–4 through 3–43 show the amounts requested by NASA and the programmed amounts. Where the authorization or the appropriation was listed, those amounts are provided. As explained in chapter 1, NASA submits an initial and a revised budget request to Congress before the budget is passed. Where available, both amounts are indicated in the column titled Budget Submission with the two amounts separated by a forward slash. The programmed amount indicates the amount actually spent.

The move toward implementation of "full-cost" accounting, which NASA began with the FY 1997 budget request, aimed to give a more accurate picture of actual project costs. This method of accounting associated all project costs in

[39] "Analysis of Agency Support for International Space Station," *National Aeronautics and Space Administration Fiscal Year 1997 Estimates*, p. SI–2.
[40] Smith, *Space Stations*, 1999, p. CRS–6.
[41] NASA Advisory Council, "Report of the Cost Assessment and Validation Task Force on the International Space Station," April 21, 1998, *http://history.nasa.gov/32999.pdf* (accessed June 12, 2005).

project budgets, regardless of their source. Starting with projected FY 1997 costs, that is, the budget request, NASA showed budget figures using both the traditional method being phased out and the new "full-cost" method. FY 1995 and prior years' budget authority were recalculations reflecting the full cost of all elements associated with a project.[42] Where provided in budget documents, the following tables show an amount for "budget authority" as stated in the FY 1998 budget estimate. In this budget estimate, NASA restated the amounts estimated for the Space Station to include the funds appropriated in FY 1997 and prior years to the current Science, Aeronautics, and Technology, former Construction of Facilities, and former Research and Development appropriations as well as funds appropriated in the HSF appropriation. The amounts from appropriations other than HSF are shown only in the "Space Station-Research" budget category.[43]

The Space Shuttle

This section describes the Space Shuttle system and operations and details of each Shuttle mission between 1989 and 1998. For an overview of the Shuttle's development and a detailed description of events of the prior decade, the reader may consult the *NASA Historical Data Book, Volume V, 1979–1988*.[44] As in the previous chapter, all measurements are given in the unit used in the original reference. Equivalent measurements in alternate units follow in parentheses.

The Space Shuttle system consisted of four main components: an expendable external tank, two reusable solid rocket boosters, a reusable orbiter, and three installed main engines, commonly called the Space Shuttle Main Engines (see Figure 3–15).[45] The structure and systems of the Space Shuttle have remained essentially the same since its inception. Detailed descriptions of its components and systems are available in the *NSTS Shuttle Reference Manual* (1988) and in the *Shuttle Crew Operations Manual*.[46]

---

[42] Budget authority represents the amounts appropriated by Congress in a given fiscal year that provides NASA with the authority to obligate funds. Obligation of funds legally commits NASA to pay contractors and other service providers for materials and services. The ensuing obligations, cost incurrence, and expenditures (outlays) based on the budget authority can occur in a different fiscal year from the year in which Congress provides the budget authority.

[43] "Full-Cost Budgeting," National Aeronautics and Space Administration Fiscal Year 1998 Budget Estimates, pp. SI–6–SI–7.

[44] Judy Rumerman, compiler, *NASA Historical Data Book, Volume V, 1979–1988* (Washington, DC: National Aeronautics and Space Administration Special Publication-4012, 1999), pp. 121–238, 269–358. Also at *http://history.nasa.gov/SP-4012/vol5/cover5.html*.

[45] The external tank, solid rocket boosters, and main engines are described in chapter 2, Launch Systems, of this volume.

[46] *NSTS Reference Manual, 1988*, *http://science.ksc.nasa.gov/shuttle/technology/sts-newsref/stsref-toc.html* (accessed July 6, 2005); *Shuttle Crew Operations Manual*, OI–29, SFOC-FL0884, Rev. B, CPN-3, United Space Alliance (January 13, 2003).

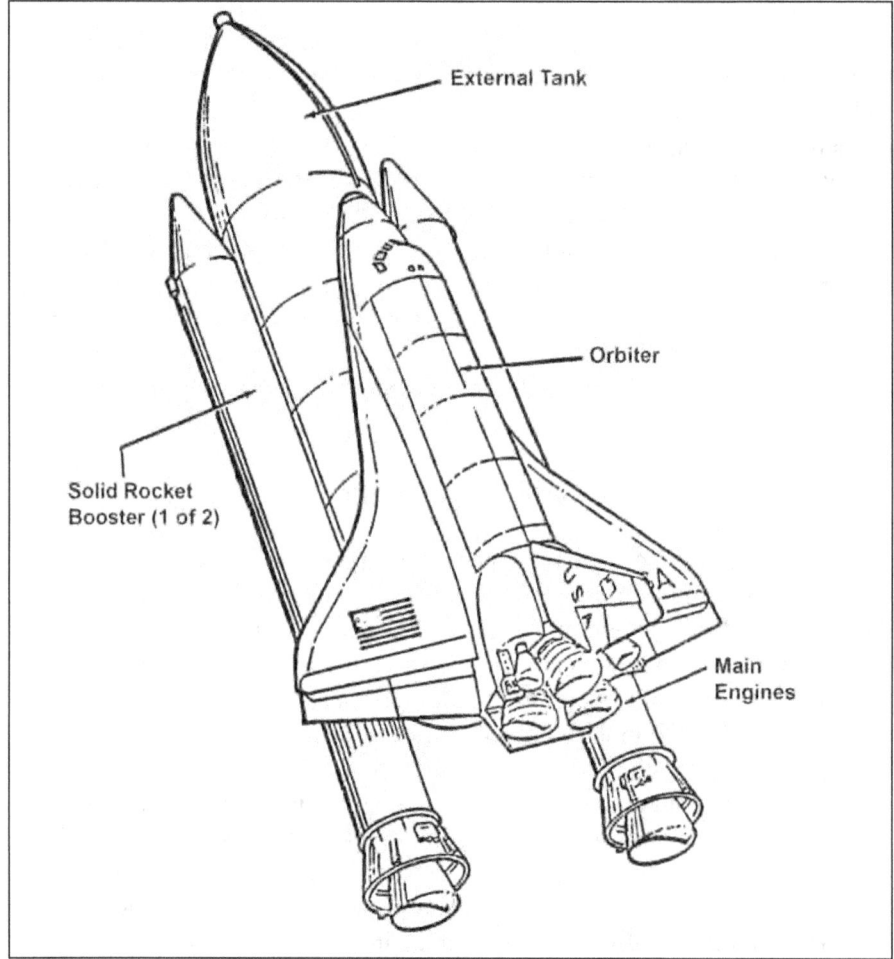

Figure 3–15. Space Shuttle Vehicle Configuration.[47]

The Space Shuttle could perform a variety of missions. These included:

- Delivery of payloads to specified Earth orbits.
- Placement of payloads into parking orbits for subsequent transfer to other orbits or Earth escape trajectories.
- Rendezvous and stationkeeping with detached payloads/space stations.
- Monitoring and checkout of payloads.
- Return of payloads to Earth from a specified orbit.
- Routine and special support to space activities such as sortie missions; rescue; repair; maintenance; servicing; assembly; disassembly; and docking.
- Space Station assembly and operations support.

---

[47] "Space Shuttle Program Description and Requirements Baseline," NSTS 07700, Vol. I, Rev. G, December 17, 1997, pp. 3–15, *http://pbma.hq.nasa.gov/sma/public/Jsc/bp_jsc_44.pdf* (accessed June 28, 2005).

The Shuttle could transport payloads into near-Earth orbit 100 nautical miles to 312 nautical miles (185 kilometers to 578 kilometers) above Earth. Acceleration during ascent never exceeded 3g. On its return to Earth, the orbiter had a crossrange maneuvering capability of about 550 nautical miles (1,019 kilometers) either side of center. The orbiter normally carried crews of up to seven people, although it could carry eight-person flight crews. The usual mission lasted from 4 to 16 days in space.

All Shuttle flights launched from Kennedy Space Center in Cape Canaveral, Florida. The Kennedy and Edwards Air Force bases in California were the primary landing sites. Contingency landing sites were also provided in the event the orbiter needed to return to Earth in an emergency.

On a typical mission, payload bay doors were opened soon after orbit stabilization to allow the orbiter space radiators to dissipate heat. The crew then conducted payload operations from the payload station on the aft flight deck. Upon completion of on-orbit operations, the payload bay doors were closed, and the orbiter was configured for return to Earth. The orbiter returned to Earth by firing the orbital maneuvering system engines to reduce velocity. After reentering Earth's atmosphere, the orbiter glided to its landing at Kennedy Space Center or, if conditions prevented landing at Kennedy, at Edwards Air Force Base. The incorporation of a drag chute and carbon-carbon brakes allowed more missions to land at Kennedy.

Shuttle Orbiter

For most of this decade, the Shuttle orbiter fleet consisted of four vehicles. *Columbia* (OV-102), the first operational orbiter; *Discovery* (OV-103), and *Atlantis* (OV-104) were part of the original orbiter fleet. *Endeavour* (OV-105) replaced the *Challenger* in 1992.

The orbiter was comparable in size and weight to a modern commercial airliner. It had three main engines and two smaller solid orbital maneuvering system engines mounted in the rear that assisted during initial phases of the ascent trajectory. The main engines provided the vehicle acceleration from liftoff to main engine cutoff at a predetermined velocity. In space, the reaction control system engines provided attitude control. Figure 3–16 shows the orbiter's structure.

The orbiter was constructed primarily of aluminum. A thermal protection system made of rigid silica tiles or some other heat-resistant material shielded every part of its external shell and protected it from reentry heat. Tiles covering the upper and forward fuselage sections and the tops of the wings could absorb heat as high as 650°C (1,202°F). Tiles on the underside absorbed temperatures up to 1,260°C (2,300°F). Panels made of reinforced carbon-carbon covered areas that had to withstand temperatures greater than 1,260°C (2,300°F), such as on the nose and leading edges of the wings on reentry.

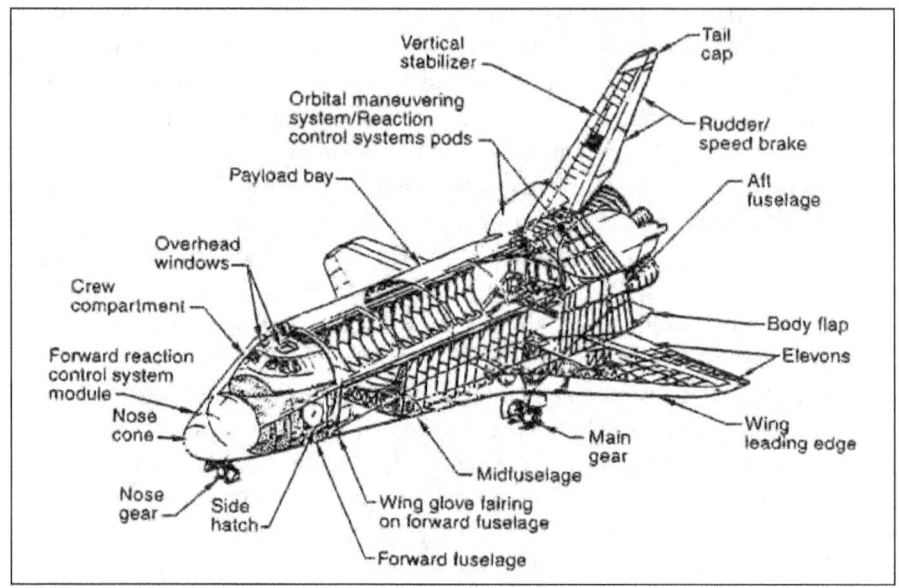

Figure 3–16. Orbiter Structure.[48]

The orbiter structure had nine major sections:

1. The forward fuselage consisting of upper and lower sections that fit clamlike around a pressurized crew compartment
2. Wings
3. Midfuselage
4. Payload bay doors
5. Aft fuselage
6. Forward reaction control system
7. Vertical tail
8. Orbital maneuvering system/reaction control system pods
9. Body flap

Table 3–44 lists nominal orbiter characteristics. The individual mission tables later in this chapter include characteristics for each mission.

Endeavour, NASA's fifth operational orbiter, was the newest addition to the Shuttle fleet. Congress authorized NASA to construct *Endeavour* on August 1, 1987. Table 3–45 lists *Endeavour*'s construction milestones. *Endeavour* was named through a national competition involving students in elementary and secondary schools who were asked to select a name based upon an exploratory or research sea vessel. President George H. W. Bush announced the winning name in May 1989. *Endeavour* entered service on May 7, 1992, on the STS-49 mission.

---

[48] *Shuttle Crew Operations Manual*, p. 1.2–1.

Endeavour incorporated a number of upgrades. They included:

- A 40-foot (12.2 meter)-diameter drag chute that reduced the orbiter's rollout distance by 1,000 feet to 2,000 feet (305 meters to 610 meters).
- An updated avionics system including advanced general purpose computers, improved inertial measurement units and tactical air navigation systems, enhanced master events controllers and multiplexer-demultiplexers, and a solid-state star tracker.
- Improved nosewheel steering mechanisms.
- An improved version of the auxiliary power units providing power to operate the Space Shuttle's hydraulic systems.
- A ground cooling hookup to allow the payload bay to cool the mini-pressurized logistics module.
- Doublers on several wing spars to allow heavier payloads and two wing glove truss tubes having increased wall thickness.

Endeavour was originally equipped as the first extended duration orbiter. This feature was removed during its Orbiter Maintenance Down Period (OMDP) to save weight for ISS missions. During an OMDP, an orbiter is inspected, torn down, overhauled, and upgraded.

Orbiter Upgrades

Many of the improvements incorporated in the *Endeavour* were made to the other orbiters. Some changes, called Category I changes, were required before the return to flight in 1988. Others were made when the orbiters came out of service for regular maintenance and modifications. A number of improvements outfitted the orbiters for visiting the Space Station.

Columbia was the oldest orbiter and the first to undergo a scheduled inspection and retrofit program. It received modifications at the Rocketdyne Division of Rockwell International assembly plant at Palmdale, California, where it had been manufactured, after completion of STS-4, after STS-5, and after STS-9. These modifications added equipment needed to accommodate the PAM to be used for the STS-5 payload and to allow it to accommodate the Spacelab. They also removed the ejection seats, installed Orbiter Experiments Program packages and heads-up displays, and added provisions for GPS navigation, as well as more than 200 other modifications.[49]

On August 10, 1991, after completion of STS-40, *Columbia* returned to Palmdale, California. The spacecraft underwent approximately 50 upgrades there, including the addition of carbon brakes, a drag chute, and improved nose wheel steering; removal of instrumentation used during the test phase of the orbiter; and an enhancement to its thermal protection system. The

---

[49] Dennis R. Jenkins, *Space Shuttle: The History of the National Space Transportation System, The First 100 Missions*, 3rd ed. (Cape Canaveral, FL: Dennis Jenkins, 2001), pp. 435–437.

orbiter returned to Kennedy Space Center in February 1992. On October 8, 1994, *Columbia* went back to Palmdale, California for its first OMDP. Approximately 90 modifications and upgrades were made during a six-month period. Modifications included upgrades to the main landing gear thermal barrier; tire pressure monitoring system, and radiator drive circuitry. Repairs were made to the radiators where micrometeorites had made impacts. Intensive structural inspections took place, and an upgraded corrosion control coating was applied on the wings and rudder. This overhaul left the vehicle in "like-new" condition. *Columbia* was too heavy to fly either to the Russian *Mir* space station or to perform Space Station assembly missions and was not retrofitted for that purpose.

Discovery's first OMDP took place beginning in mid-March 1992 after its return to Kennedy Space Center from Edwards Air Force Base where the STS-42 mission landing took place. A drag chute was installed, and the orbiter received a complete structural inspection and refurbishment of the thermal protection system.[50]

Its next inspection was in 1995. *Discovery* departed from Kennedy on September 27, 1995, arriving at Palmdale, California, to undergo a nine-month OMDP. The vehicle was outfitted with a fifth set of cryogenic tanks, and an external airlock replaced its internal airlock. This gave it the capability to participate in Shuttle-*Mir* docking missions and support missions to the ISS. *Discovery* left Palmdale for its return to Kennedy, riding atop a modified Boeing 747, on June 28, 1996.

Atlantis's first OMDP extended from October 1992 through May 1994 while major work required for *Atlantis* to support missions to *Mir* took place. Modifications included installation of a drag chute, new plumbing lines and electrical connections enabling extended duration missions, improved nosewheel steering, new insulation for the main landing gear doors, more than 800 new heat protection tiles and blankets, and structural modifications to the airframe. *Atlantis* received an Orbiter Docking System, which included both Russian and U.S. hardware.

Atlantis's second OMDP began in November 1997 at Boeing's facility in Palmdale, California, where about 130 modifications were made.[51] Along with detailed nose-to-tail inspections and replacement of dated flight hardware, workers installed thinner and lighter insulation that reduced the orbiter's weight by about 1,000 pounds (454 kilograms), allowing the orbiter to haul heavier cargo into space. An external airlock replaced its internal airlock, freeing up

[50] "Chronology of KSC and KSC Related Events for 1992," Part I at *http://www-lib.ksc.nasa.gov/lib/ archives/chronologies/1992CHRONO1.pdf*, Part II at *http://www-lib.ksc.nasa.gov/lib/archives/chronologies/ 1992CHRONO2.pdf* (accessed July 6, 2005).

[51] "Space Shuttle *Atlantis* Modification Work To Be Performed at Palmdale Facility," *NASA News* Release 97–11, January 16, 1997, *http://www.nasa.gov/lb/centers/johnson/news/releases/1996_1998/97-11.html* (accessed April 28, 2005). Also "Atlantis Scheduled To Return to KSC after 10 Months in Palmdale, CA," *NASA News*, Kennedy Space Center (September 21, 1998), *http://xs4all.nl/~carlkop/atlantis.html* (accessed April 14, 2006). In 1996, Rocketdyne became part of the Boeing Company.

interior space in the middeck and equipping it for ISS docking rather than for docking with *Mir*. The Multifunctional Electronic Display System, or "glass cockpit," replaced the cockpit's four cathode ray tube screens, mechanical gauges, and instruments with full-color, flat-panel displays like those used on modern commercial airliners and military aircraft. The orbiter left Palmdale, California, on September 24, 1998, arriving at Kennedy on September 27.

Endeavour's eight-month OMDP in Palmdale, California, began at the end of July 1996. About 100 modifications were performed. Approximately 10 of those modifications were directly associated with work required to support ISS operations. The most extensive modification was the installation of an external airlock equipped with fluid and power lines to support spacewalks that replaced the original internal airlock. Other modifications included upgrades to the orbiter's power supply system; general purpose computers; the thermal protection system; installation of new, lightweight commander and pilot seats; other weight-saving modifications; and a number of safety and turnaround enhancements.

Payload Accommodations

Shuttle payloads ranged in size from those like the Hubble Space Telescope that weighed thousands of pounds to small payloads weighing less than 60 pounds (27 kilograms). The Shuttle provided several types of payload accommodations including the payload bay and crew compartment, payload carriers, and pressurized modules. For payloads carried in the payload bay, structural supports enabling payloads to withstand the rigors of liftoff and ascent to orbit were provided by main frames below the longeron sills on each side of the bay and by using payload attach fittings placed to suit the payload's dimensions. Large payloads had trunnions that mated directly with the attach fittings. Smaller payloads could be mounted on carriers that fit into the attach fittings. The Shuttle also provided a variety of services including power, thermal control, communications and data handling, and displays and controls for crew interaction, provided through the avionics system.

The Space Shuttle accommodated three basic types of payloads: dedicated, standard, and middeck.[52]

- Dedicated payloads, such as the Spacelab, Hubble Space Telescope, and some DOD payloads, took up the entire cargo-carrying capacity and services of the orbiter. These large payloads occupied the entire payload bay and required the Shuttle's full performance capability.
- Standard payloads were the primary type of Shuttle cargo. Normally, the payload bay could accommodate up to four standard payloads per flight. The avionics system provided power, command, and data services through a standard mixed cargo harness.

---

[52] "Mission Preparation and Prelaunch Operations," *NSTS Shuttle Reference Manual (1988), http://science.ksc.nasa.gov/shuttle/technology/sts-newsref/stsover-prep.html* (accessed July 22, 2005).

- Middeck payloads were small, usually self-contained packages requiring a pressurized environment or direct crew operation. They were stored in compartments that could be as small as 2 cubic feet (.06 cubic meter), allowing the opportunity for limited late stowage and early removal from the Shuttle. This type of payload often consisted of manufacturing-in-space or small life sciences experiments.

Displays and controls for payload operations were located in the aft flight deck, which was in the upper level of the crew compartment. The middeck, located immediately below the flight deck, provided the crew living areas and accommodations for middeck payloads. The orbiter payload bay was approximately 60 feet (18.3 meters) long and 15 feet (4.6 meters) in diameter.

The remote manipulator system (RMS) mechanical arm was mounted along the left side of the payload bay. It was used for payload deployment, retrieval, special handling operations, and other orbiter servicing. The RMS was 50.25 feet (15.3 meters) in length.[53]

Table 3–46 lists some of the Shuttle's various payload accommodations.

Small Payloads

The Shuttle carried a variety of small payloads allowing domestic and international educational, commercial, and government payloads to travel into space. The Small Self-Contained Payload (SSCP) program, popularly known as the GAS program, launched its first payload, G-001, on June 27, 1982, on STS-04. G-001 was built by Utah State University. Through 1998, GAS flew 159 payloads on 35 missions.[54] The GAS program was managed by Goddard Space Flight Center.

Standard GAS containers had volumes of 5 cubic feet (0.15 cubic meter) and 2.5 cubic feet (0.07 cubic meter). The 2.5 cubic-foot (0.07-cubic-meter) container could house payloads weighing up to 100 pounds (90.7 kilograms). The larger container could house payloads up to 200 pounds (45.4 kilograms). The GAS carrier provided limited mechanical and electrical interfaces for self-contained experiments, and the customer needed to provide all required battery, data recording, and sequencing systems.[55]

The Code of Federal Regulation, 14 CFR 1214.9, governed the SSCP program and defined and provided the rules for participating in the program. NASA issued the original SSCP rule in 1980. It established conditions of use,

---

[53] "Space Shuttle System Payload Accommodations, Revision L," NSTS 07700, Vol. XIV (2001), pp. 3-1–3-4, 5-10–5-11, *http://shuttlepayloads.jsc.nasa.gov/data/PayloadDocs/documents/07700/Vol_XIV.pdf* (accessed June 1, 2005).

[54] "GAS Can Experiments," *http://members.fortunecity.com/spaceshuttlealmanac/gascans.htm* (accessed 6 July 2005). "Historical Information, Get Away Special," *http://www.wff.nasa.gov/efpo/ssppo/gas/history.html* (accessed April 26, 2005).

[55] "Space Shuttle System Payload Accommodations, Revision L," NSTS 07700, Vol. XIV (2001), p. 5–11.

reimbursement procedures, and flight scheduling mechanisms for SSCPs flown on the Space Shuttle and ensured equitable allocation of space opportunities to educational, commercial, and U.S. government groups of users.

NASA revised the rule in 1991 and again in 1992, creating 14 CFR 1214.10, "Special Policy on Use of Small Self-Contained Payloads (SSCP) by Domestic Educational Institutions." The revision provided two different pricing structures: an increased standard flight price for commercial and international customers, while the original price remained for domestic educational institutions. On April 23, 1999, NASA revoked both regulations.

In 1995, the SEM program was established to provide students an opportunity to develop experiments not involving complicated engineering. The SEM Carrier System was a self-contained assembly of engineered subsystems functioning together to provide structural support, power, experiment command, and data storage capabilities for microgravity experiments. The system, consisting of a 5-foot (1.5-meter) "canister," contained 10 experiment modules.[56]

The Hitchhiker program became part of the Shuttle Small Payloads Project (SSPP) in 1986. The program expanded GAS capabilities by offering customers power, command, real-time data acquisition and transfer, crew control, and display capability. Hitchhiker customers operated their payloads from the Hitchhiker Control Center at Goddard Space Flight Center using their own ground support equipment (usually a personal computer) to send commands and display data. Users' ground support equipment worked in tandem with Hitchhiker's ground system, the Advanced Carrier Customer Equipment Support System, for communicating with their payloads. Control and monitoring of payloads from remote sites also was used.[57] The first Hitchhiker flight, designated Hitchhiker G-1, took place in 1986 on STS-61-C.

HH-Jr. accommodated experiments requiring power from the orbiter or from internal batteries. A connection to the orbiter enabled the crew to command and check payloads using a laptop computer.

The Hitchhiker carrier system provided electrical power (28 volts DC), command signals, and "downlink" data interfaces. It had provisions for flying payloads along the sidewall of the Space Shuttle payload bay, i.e., the longeron, and on cross-bay carriers/platforms, and provided options for ejecting small spacecraft from the Space Shuttle payload bay. Hitchhiker payloads were contained in mounted canisters attached to mounting plates of various sizes.

---

[56] "Space Experiment Module," Fact Sheet, *http://www.wff.nasa.gov/efpo/ssppo/sem/About/about_facts.html* (archived Web site accessed April 26, 2005).

[57] "Hitchhiker Carrier System," *http://www.wff.nasa.gov/efpo/ssppo/hh/index.html* (archived Web site accessed November 15, 2005).

Spacelab

Spacelab was a non-deployable Shuttle payload that carried investigations from many scientific disciplines, including atmospheric science, solar science, materials science, space plasma physics, the life sciences, and astrophysics. Sometimes a Spacelab mission carried experiments from several disciplines; at other times, it focused on a single discipline. Spacelab fit into the Shuttle orbiter's payload bay. Spacelab's modular structure allowed for a wide range of configurations and objectives and enabled extended experiments to take place in orbit (see Figure 3–17). An integral part of the Space Shuttle system, it was developed jointly by ESA and NASA and designed and produced by ESA.[58]

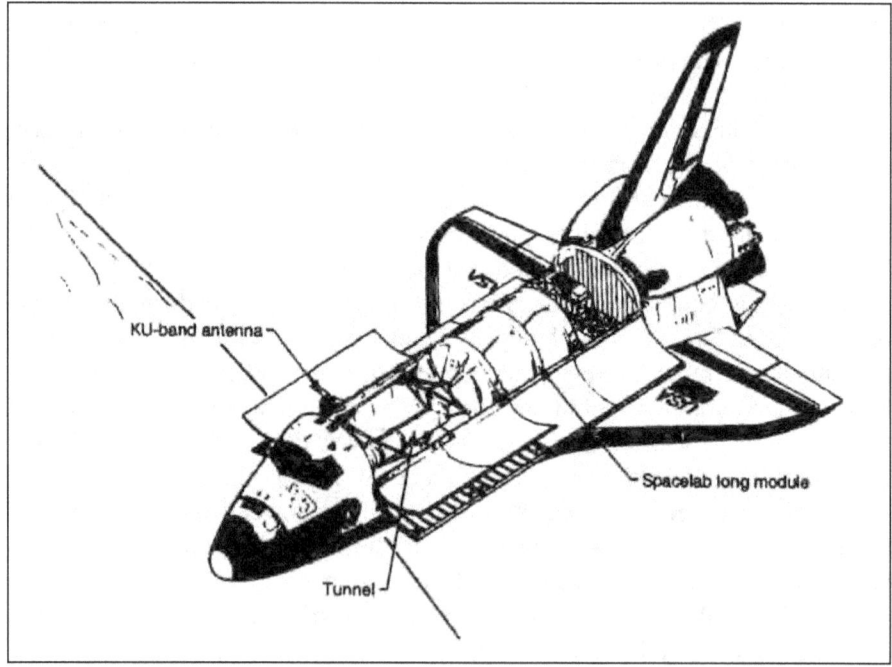

Figure 3–17. Spacelab on Orbit.

Spacelab's four principal components were the pressurized laboratory module, one or more open pallets that exposed materials and equipment to space, a tunnel to gain access to the module, and an instrument pointing system. Electrical power, command and data management, caution and warning, and environmental control and life support systems supported the Spacelab. Figure 3–18 shows Spacelab components. Table 3–47 presents characteristics of the Spacelab module.

---

[58] When Spacelab was first proposed, the ESA was called the European Space Research Organisation (ESRO).

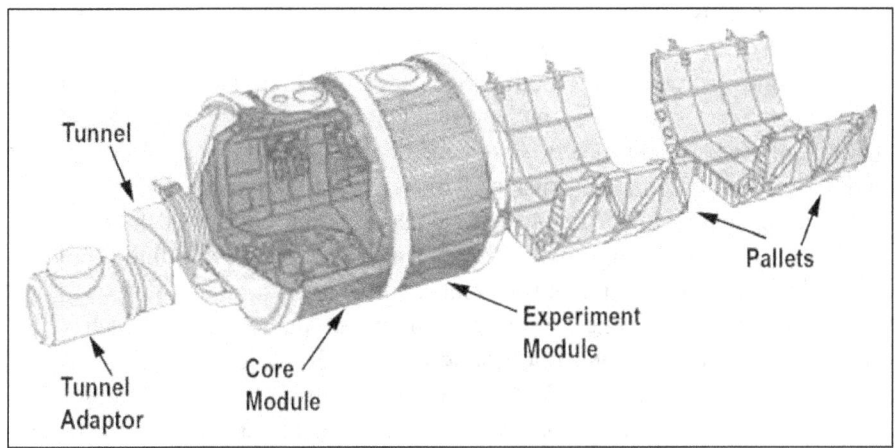

Figure 3–18. Spacelab Components.

The flight crew could control Spacelab experiments from the Spacelab module or from the orbiter's aft flight deck. Experiments located in the module or on the pallets could also be controlled directly from the ground.[59]

The cylindrical pressurized laboratory module had a habitable, shirt-sleeve environment and was available in two segments—a core and an experiment segment—that could be assembled as either a single segment (the core segment) or a double segment (the core and experiment segments, known as the long module). Each segment was 13.5 feet (4.1 meters) in outside diameter and 9 feet (2.7 meters) long. When both segments were assembled with end cones, their maximum outside length was 23 feet (7 meters). The pressurized module was structurally attached to the orbiter payload bay by four attach fittings consisting of three longeron fitting sets (two primary and one stabilizing) and one keel fitting and was covered with passive thermal control insulation. The laboratory equipment was mounted in racks and in other areas. Handrails were mounted in racks, overhead, and on end cones. Foot restraints were also provided on the floor and on rack platforms. Crew objects could be temporarily fastened to Velcro patches throughout the modules.

The core segment (also known as the short module when it was flown alone) contained supporting systems such as data processing equipment and utilities for the module and pallets (if pallets were used with the module). It provided laboratory space with floor-mounted racks and a workbench. When only one segment was needed, the core segment was used. The experiment segment provided more working laboratory space and contained only floor-mounted racks.

---

[59] Marsha R. Torr, "Scientific Achievements of the Spacelab Program: An Overview of the Missions," AIAA 94-4646, AIAA Space Programs and Technologies Conference, September 27–29, 1994 (NASA Goddard Library Electronic Database).

It was flown only in conjunction with the core segment. The modules were designed for a lifetime of 50 missions.[60]

End cones were bolted to both ends of the cylindrical laboratory segments. The truncated cones were 30.8 inches (78.2 centimeters) long; the large end was 161.9 inches (411.2 centimeters) in outside diameter and the small end was 51.2 inches (130 centimeters) in outside diameter. Each cone had three 16.4-inch (41.7-centimeter)-diameter cutouts, two located at the bottom of the cone and one at the top. Feedthrough plates for routing utility cables and lines could be installed in the lower cutouts of both end cones.

The ceiling skin panel of each segment contained a 51.2-inch (130-centimeter)-diameter opening for mounting a viewport adapter assembly. If the assembly was not used, the bolted-down cover plates closed the openings. The Spacelab viewport assembly could be installed in the upper cutout of the aft end cone, and the upper cutout of the forward end cone was for the pressurized module vent and relief valves.

Spacelab pallets were platforms designed for large instruments, experiments requiring direct exposure to space, and systems needing unobstructed or broad fields of view. Experiments could be mounted on the pallets or smaller special support structures if the instruments required exposure to space. For pallet-only missions, the support utilities for the instruments mounted on the pallet were housed in the Spacelab igloo, a temperature-controlled housing providing connections for data gathering, communications, electrical power, and cooling equipment. Vertically attached to the forward end of the first pallet, the igloo was 7.9 feet (2.4 meters) high and 3.6 feet (1.1 meters) in diameter.[61]

Because of the orbiter's center-of-gravity requirements, the Spacelab module had to be installed at the rear end of the orbiter payload bay. Equipment and crew passed through a pressurized tunnel between the crew compartment and the module. The tunnel was cylindrical with an internal unobstructed diameter of 40 inches (101.7 centimeters) and assembled in sections to allow length adjustment for different module configurations. Tunnel lengths of 18.88 feet (5.8 meters) and of 8.72 feet (2.7 meters) could be used. A "joggle" section of the tunnel compensated for the 42.1-inch (1.7-centimeter) vertical offset of the middeck airlock to the module's centerline. There were flexible sections on each end of the tunnel near the orbiter and Spacelab interfaces.

The airlock, tunnel adapter, tunnel, and module were at ambient pressure before launch. The tunnel adapter permitted crew members outfitted for extravehicular activity (EVA) to transfer from the middeck airlock to the payload bay without depressurizing the orbiter crew compartment and module. If an EVA was required, no crew members were permitted in the Spacelab tunnel or module.

---

[60] "Spacelab Module," *http://liftoff.msfc.nasa.gov/Shuttle/spacelab/sl-elements.html* (accessed December 12, 2005).
[61] "Igloo," *http://liftoff.msfc.nasa.gov/Shuttle/spacelab/element-igloo.html* (accessed December 13, 2005).

Some Spacelab mission research required instruments to be pointed with very high accuracy and stability at stars, the Sun, Earth, or other targets of observation. The instrument pointing system provided precision pointing for instruments of diverse sizes and weights up to 15,432 pounds (7,000 kilograms) and could point them to within 2 arc seconds and hold them on target to within 1.2 arc seconds. The system consisted of a three-axis gimbal system mounted on a gimbal support structure and a control system. The control system was based on the inertial reference of a three-axis gyro package and operated by a gimbal-mounted mini-computer.

The Spacelab command and data management system (CDMS) provided a variety of services to Spacelab experiments and subsystems. Most of the CDMS commands were carried out using the computerized system aboard Spacelab, called the data processing assembly (DPA). The DPA formatted telemetry data and transferred the information to the orbiter for transmission, received command data from the orbiter and distributed it to Spacelab subsystems, transferred data from the orbiter to experiments, and distributed timing signals from the orbiter to experiments.[62]

The first Spacelab mission flew in 1983, the last on STS-90 in 1998. The program ended because the experiments performed on Spacelab could now be performed on the Space Station. Table 3–48 lists Spacelab missions from 1989 to 1998.

SPACEHAB

During the 1980s, as directed by legislation and national space policy, the commercial development of space became one of NASA's chief objectives. In the late 1980s, NASA's Office of Commercial Programs identified a significant number of payloads to further this objective, which required a sufficient level of flight activity for their support. In September 1989, a NASA analysis concluded that planned Space Shuttle flights did not offer adequate middeck-class accommodations for these payloads.

In February 1990, NASA initiated the Commercial Middeck Augmentation Module (CMAM) procurement through Johnson Space Center to provide support for these payloads. In November 1990, NASA awarded a five-year contract to SPACEHAB, Inc., of Arlington, Virginia, for the lease of their pressurized modules, the SPACEHAB Space Research Laboratories, to provide additional space by extending the Shuttle orbiter middeck into the Shuttle cargo bay for "crew-tended" payloads. This five-year lease arrangement covered several Shuttle flights and required SPACEHAB, Inc., to provide for the physical and operational integration of the SPACEHAB

---

[62] *Shuttle Crew Operations Manual*, pp. 2.25-1, 2.25-3, 2.25-5, 2.25-10, and 2.25-18.

laboratories into the Space Shuttle orbiters, including experiment and integration services such as safety documentation and crew training.[63]

SPACEHAB contracted with McDonnell Douglas's Huntsville Space Division in Alabama to provide the design, development, and physical integration of two space research laboratories. SPACEHAB also contracted with Alenia Aerospazio of Turin, Italy, to build the laboratories and design and build their passive thermal control systems.[64] These aluminum space research laboratory modules carried commercial and other attached payloads on the Shuttle and were used for *Mir* logistics flights. SPACEHAB unveiled its first module in May 1992; it flew on its first mission, STS-57, in 1993. Table 3–49 lists Shuttle SPACEHAB flights.

The SPACEHAB pressurized laboratory augmented Space Shuttle middeck experiment accommodations and provided Shuttle crew with a place to carry out experiments. It was located in the forward end of the Shuttle orbiter cargo bay and was accessed from the orbiter middeck through a tunnel adapter connected to an airlock. The module contained cooling, power and command, and data provisions in addition to SPACEHAB housekeeping systems (power distribution and control; lighting; fire and smoke detection; fire suppression; atmosphere control; status monitoring and control; and thermal control).

A single module weighed 9,628 pounds (4,367 kilograms), was 9.2 feet (2.8 meters) long, 11.2 feet (3.4 meters) high, and 13.5 feet (4.1 feet) in diameter. It increased pressurized experiment space in the Shuttle orbiter by 1,100 cubic feet (31 cubic meters), quadrupling the working and storage volume available. Environmental control of the laboratory's interior maintained ambient temperatures between 65°F and 80°F (18°C and 27°C) and had a total payload capacity of 3,000 pounds (1,361 kilograms).

The SPACEHAB laboratory could be configured with middeck-type lockers, racks, and/or a logistics transportation system to accommodate a variety of experiments and equipment. It could accommodate up to two SPACEHAB racks, either of which could be a "double rack" or "single rack" configuration. A double rack provided a maximum capacity of 1,250 pounds (567 kilograms) and 45 cubic feet (1.3 cubic meters) of volume, whereas a single rack provided half that capacity. The double-rack was similar in size and design to the racks planned for use in the Space Station.[65] Figure 3–19 shows the dimensions and arrangement of typical SPACEHAB interior configurations.

---

[63] "Space Shuttle Mission STS-57 Press Kit," June 1993, p. 16, *http://www.jsc.nasa.gov/history/shuttle_pk/pk/Flight_056_STS-057_Press_Kit.pdf* (accessed December 2, 2005).

[64] E-mail from Kimberly Campbell, Vice President Corporate Marketing and Communications, SPACEHAB, Inc., December 13, 2005.

[65] "Space Shuttle Mission STS-57 Press Kit," June 1993, pp. 16–17, *http://www.jsc.nasa.gov/history/shuttle_pk/pk/Flight_056_STS-057_Press_Kit.pdf* (accessed December 2, 2005).

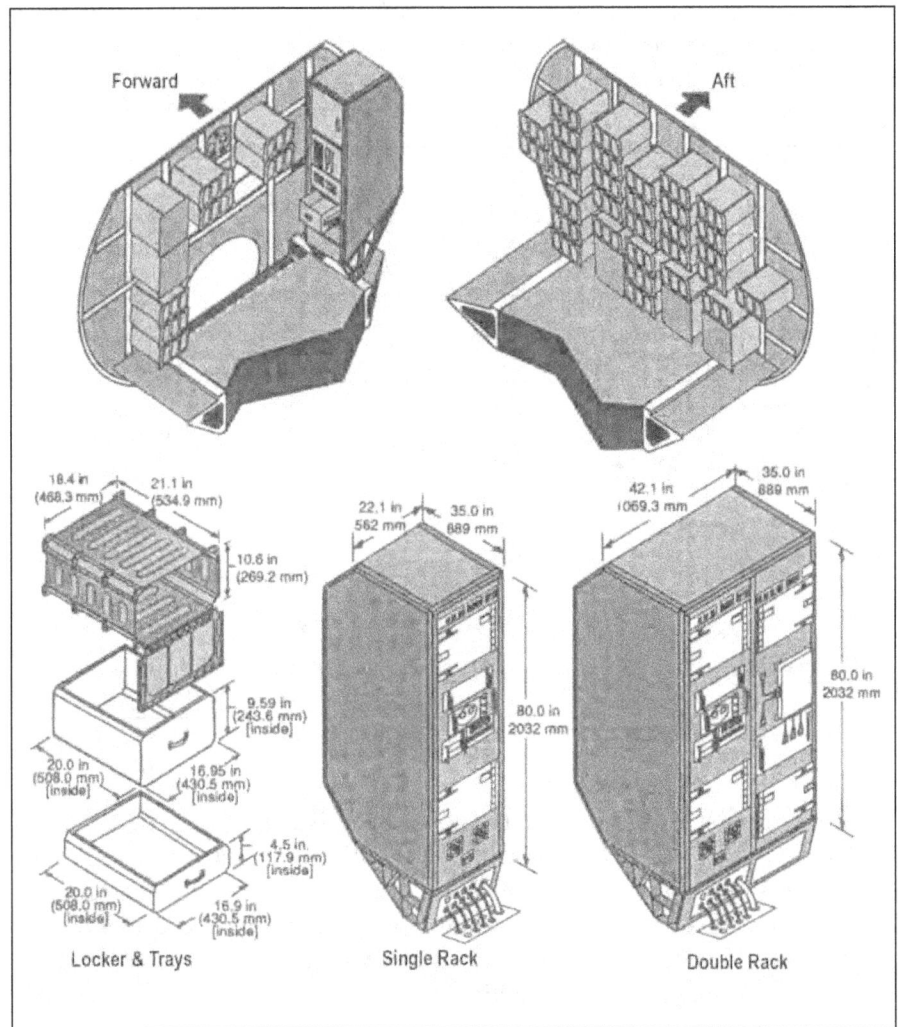

Figure 3–19. Typical SPACEHAB Interior Configurations.

SPACEHAB offered three module configurations to accommodate specific mission requirements. Configuration 1 was connected to the orbiter using a modified Spacelab tunnel adapter and standard orbiter payload support resources from the cabin and payload bay (see Figure 3–20). The SPACEHAB single module-to-orbiter tunnel adapter connection used the Spacelab tunnel adapter, the SPACEHAB transition section, and the Spacelab flex section.

Configuration 2 allowed the SPACEHAB single module to be mounted in a new trunnion location to accommodate the orbiter docking system (ODS). The module was connected to the ODS using a Spacelab flex section, the new Spacelab extension for *Mir*, the SPACEHAB long tunnel segment, the SPACEHAB tunnel segment, and another flex section. All SPACEHAB module subsystems remained the same as in Configuration 1 except for a lower air exchange rate with the orbiter and the addition of two negative pressure relief valves.

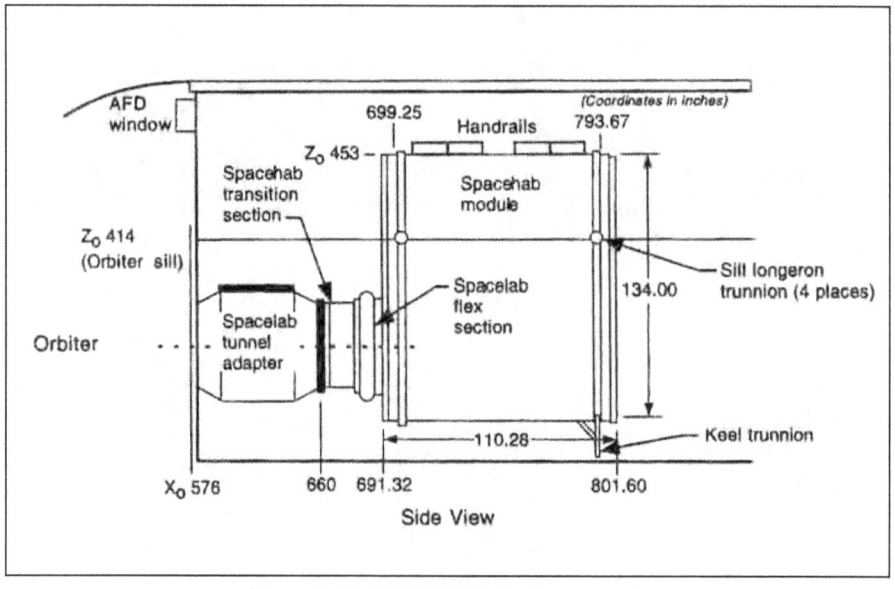

Figure 3–20. SPACEHAB Configuration 1.

Configuration 3 was a double module consisting of one SPACEHAB module and one SPACEHAB module shell joined by an intermediate adapter (see Figure 3–21). This configuration had the same tunnel arrangement and attach points as Configuration 2, except for two trunnions moved farther back to accommodate the additional module. All SPACEHAB module subsystems remained the same as in Configuration 2 except for the addition of a fan and lights in the aft module segment.[66]

Animals in Space

Animals were a valuable part of space life sciences research and flew in space since the earliest days of the space program. All animal experiments aboard the Space Shuttle were housed either in the middeck area or in a laboratory research module specifically configured for the cargo bay. Two types of enclosures were flight-certified for use with on-board animals. Rodent experiments were usually carried in middeck lockers configured with animal enclosure modules that could be loaded onto the Shuttle 12 to 18 hours before launch and removed 3 to 6 hours after landing. Each module contained sufficient food for the duration of the mission and had an on-board water supply.

---

[66] *Shuttle Crew Operations Manual*, pp. 2.24-1–2.24-2.

Animal enclosure modules could not be removed during a flight and were tightly sealed. Daily animal health checks were conducted during flights by opening the locker cover containing the module and pulling the module from its stowage position. The astronaut could observe the animals through the module's transparent cover. These modules were originally developed by General Dynamics for the Student Shuttle Flight Program.

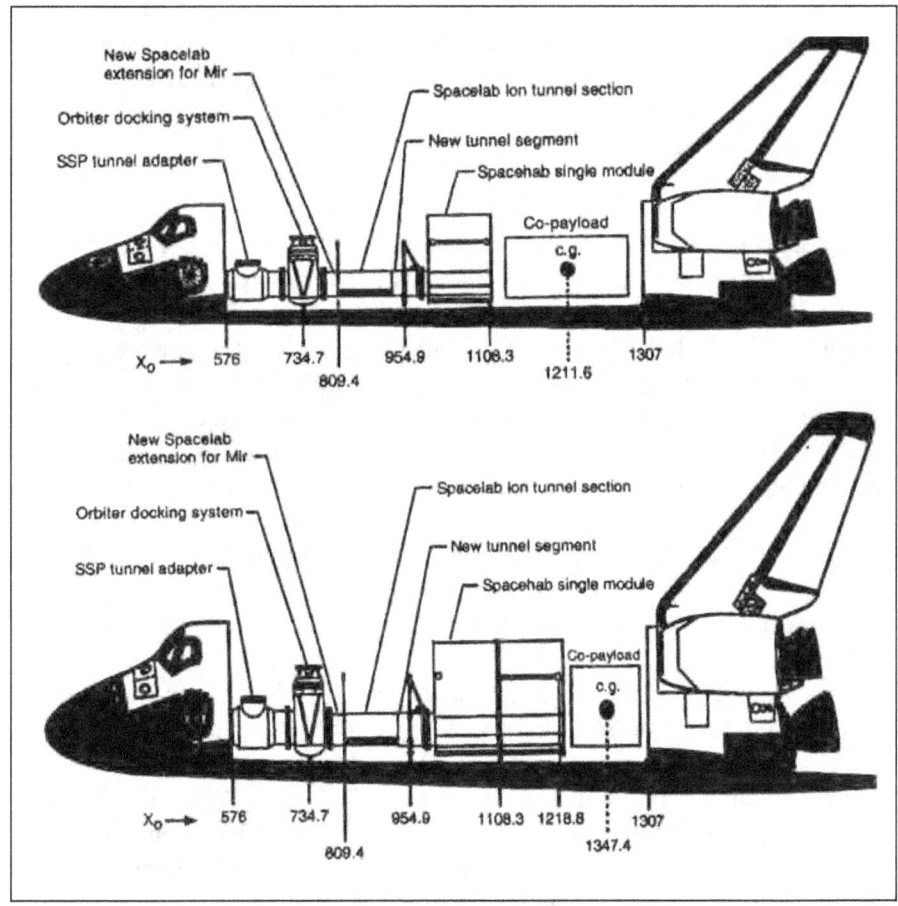

Figure 3–21. SPACEHAB Configurations 2 (top) and 3.

The Spacelab module could be converted into an on-orbit research center providing additional space for rodents and primates. The Research Animal Holding Facility (RAHF) placed into a standard Spacelab double rack with housing space for up to 24 rats or four 1-kilogram (2.2-pound) squirrel monkeys. The facility provided environmental control, food, water, light, and waste management control for the animals. Unlike the sealed animal enclosure module, the animal cages could be removed from the RAHF and transported to a general purpose work area where the animal cages could be opened and the animals removed for tissue or fluid sample collection, administration of specific treatments, or euthanasia and tissue collection. Primates were used only for the Spacelab-3 mission in April 1985, and NASA did not plan to use primates again.[67]

In 1997, NASA issued a document titled "Principles for the Ethical Care and Use of Animals." The document stated three principles to guide the use of animals in research: to use the appropriate species and minimum number of animals required to obtain valid scientific results, to consider the potential societal good and overall ethical value whenever animals were used, and that the minimization of distress, pain, and suffering was a moral imperative.[68]

The Space Shuttle Crew[69]

NASA selects astronauts from a diverse pool of applicants with a wide variety of backgrounds. From the thousands of applications received, only a few are chosen for the intensive astronaut candidate training program.

The first group of astronaut candidates for the Space Shuttle program was chosen in 1978. In July of that year, 35 candidates began a rigorous training and evaluation period at Johnson Space Center to qualify for subsequent assignment for future Space Shuttle flight crews. This group of 20 mission scientist astronauts and 15 pilots completed training and went from astronaut candidate status to astronaut active status in August 1979. Six of the 35 were women and four were minorities. Through 1998, nine additional groups of pilots and mission specialists were added: 19 in 1980; 17 in 1984; 13 in 1985; 15 in 1987; 23 in 1990; 19 in 1992; 19 in 1995; 35 in 1996; and 25 in 1998. In addition, payload specialists, who were individuals other than NASA astronauts chosen to meet specialized requirements, completed the crews. Payload specialists could be from the United States or from other countries. International crew members are indicated in the mission tables that follow along with the agency or country that sponsored them.

Astronauts participating in the Russian Mir program received Russian language training before transferring to the Yuri Gagarin Cosmonaut Training Center for approximately 13 months. Russian language courses continued at the Gagarin Center until the astronaut reached the level required to begin technical training. Russian technical training included theoretical training on Russian vehicle design and systems, EVA training, scientific investigations and experiments, and biomedical training. Four weeks before the Shuttle launch that traveled to *Mir*, the astronaut returned to Johnson Space Center to train and integrate as part of the Shuttle crew.

---

[67] Gary L. Borkowski, William W. Wilfinger, and Philip K. Lane, "Laboratory Animals in Science; Life Sciences Research," *Animal Welfare Information Center Newsletter* 6, no. 2-4 (Winter 1995/1996), *http:// www.nal.usda.gov/awic/newsletters/v6n2/6n2borko.htm* (accessed November 22, 2005).

[68] "NASA Principles for the Ethical Care and Use of Animals," *http://grants.nih.gov/grants/olaw/ references/dc97-2.htm* (accessed November 22, 2005). Also "Care and Use of Animals," NASA Policy Directive (NPD) 8910.1, Effective March 23, 1998 (canceled).

[69] *Astronaut Selection and Training*, Information Summaries, NP-1997-07-006 JSC, July 1997, *http:// spaceflight.nasa.gov/spacenews/factsheets/pdfs/np199707006jsc.pdf* (accessed July 10, 2005).

Commander/Pilot Astronauts

Pilot astronauts served as both Space Shuttle commanders and pilots. During flight, the commander had on-board responsibility for the vehicle, crew, mission success, and safety of the flight. The pilot assisted the commander in controlling and operating the vehicle and might assist in deploying and retrieving satellites using the remote manipulator system mechanical arm.

Mission Specialist Astronauts

Mission specialist astronauts worked with the commander and pilot, and the specialists had overall responsibility for coordinating Shuttle operations in the areas of Shuttle systems, crew activity planning, consumables usage, and experiment/payload operations. Mission specialists were trained in the details of the orbiter on-board systems, as well as the operational characteristics, mission requirements and objectives, and supporting equipment and systems for each of the experiments conducted on their assigned missions. Mission specialists performed EVAs, operated the remote manipulator system, and were responsible for payloads and specific experiment operations.

Payload Commander

The payload commander was an experienced mission specialist who had been designated to represent the NASA Flight Crew Operations Directorate and the Astronaut Office on a Spacelab or complex payload flight. This individual had full authority to work with the payload mission managers to identify and resolve issues associated with payload assignment and integration, training, crew member qualification, and operational constraints.[70]

Payload Specialists

Payload specialists were persons other than NASA astronauts (including foreign nationals) who had specialized on-board duties. They were career scientists or engineers selected by their employer or country for their expertise in conducting a specific experiment or commercial venture on a Space Shuttle mission.[71] They might be added to Shuttle crews if activities having unique requirements were involved and more than the minimum crew size of five were needed. First consideration for additional crew members was given to qualified NASA mission specialists. When payload specialists were required, they were nominated by NASA, the foreign sponsor, or the designated payload sponsor. In the case of NASA or NASA-related payloads, the nominations were based on the recommendations of the appropriate

---

[70] *Shuttle Crew Operations Manual*, p. 2.25-1.
[71] Astronaut Fact Book, NASA Information Summaries, NP-2005-01-001 JSC, January 2005, *http://spaceflight.nasa.gov/spacenews/factsheets/pdfs/astro.pdf* (accessed November 30, 2005).

Investigator Working Group. Although payload specialists were not part of the Astronaut Candidate Program, they were required to have the appropriate education and training related to the payload or experiment. All applicants must meet certain physical requirements and pass NASA space physical examinations with varying standards depending on classification.

Crew Services

Shuttle crew members provided services in three specific areas: EVAs, intravehicular activity, and in-flight maintenance. During EVAs, crew members donned pressurized spacesuits and life support systems, moved outside the protective environment of a spacecraft's pressurized cabin, and performed various payload-related activities in the microgravity environment of space, often outside the payload bay. The current spacesuit, designed for a total maximum duration of 7 hours, provided environmental protection, mobility, life support, and communications. Figure 3–22 shows the extravehicular mobility unit (EMU), or spacesuit.

There were three basic categories of EVA: scheduled, unscheduled, and contingency. A scheduled EVA was any EVA incorporated into the flight plan to complete a specific mission objective, for instance, repairing a satellite or testing equipment. (Figure 3–23 shows the EVA on the first Hubble Space Telescope servicing mission.) A quick-response EVA was a type of scheduled EVA that must be performed within a few hours after discovering a problem. It was usually associated with payload deployment. This type of EVA was prepared for and scheduled before the flight but might not be performed if the problem did not materialize. An unscheduled EVA was conducted to achieve payload operation success or to advance overall mission accomplishments. A contingency EVA was also unscheduled but was needed to ensure safe return of the orbiter and crew.

Even when an EVA was not scheduled, at least two crew members must be prepared to perform a contingency EVA if the situation made it necessary, for example: if payload bay doors failed to close properly and needed manual assistance or if equipment needed to be jettisoned from the orbiter. Beginning in 1998, EVAs were an important part of Space Station assembly. Earlier, U.S. astronauts had participated in spacewalks while on the Russian space station *Mir.* On April 29, 1997, Jerry Linenger became the first American to conduct a spacewalk from a foreign space station and in a non-American-made spacesuit in his 5-hour spacewalk.[72] Table 3–50 lists EVAs performed by U.S. Shuttle crews between 1989 and 1998.

---

[72] "Linenger Increment: A Spacewalk and a Fire, History, Shuttle Flights and *Mir* Increments," *http://spaceflight1.nasa.gov/history/shuttle-Mir/history/h-f-linenger.htm* (accessed July 5, 2005).

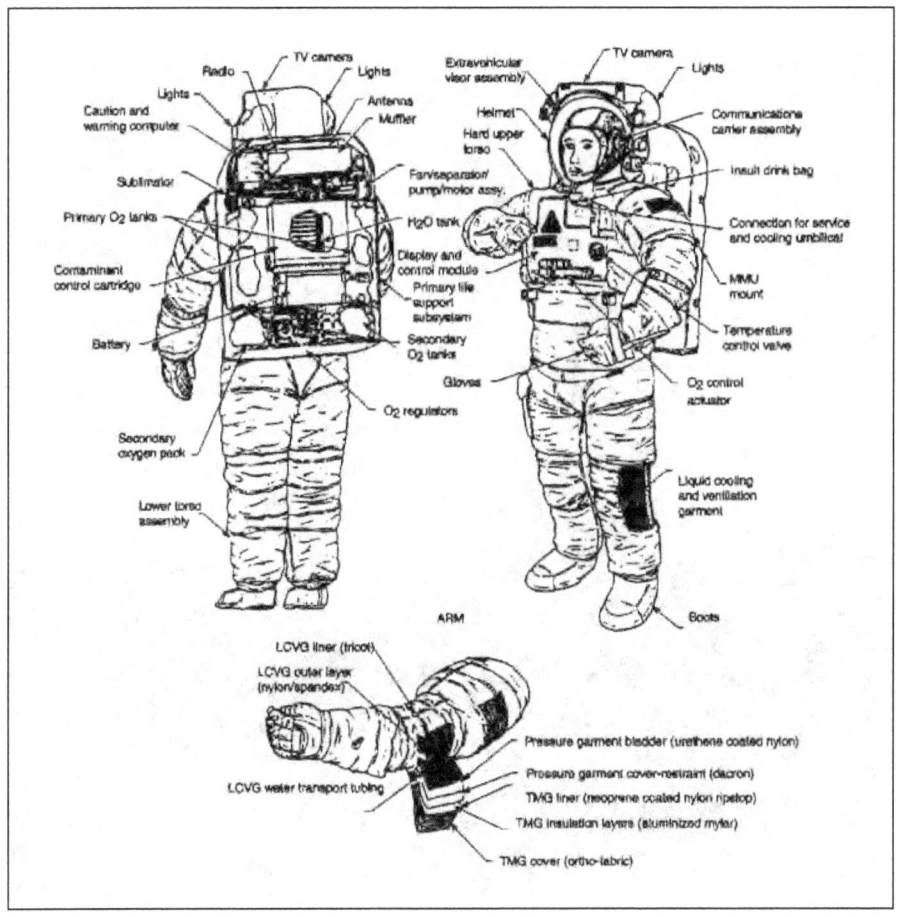

Figure 3–22. Extravehicular Mobility Unit (Spacesuit).[73]

Intravehicular activity (IVA) included crew activities occurring within the orbiter crew compartment or a customer-provided pressurized module such as an attached pressurized module in the payload bay or a free flying module docked with the orbiter. IVA operations included module activation/deactivation, on-orbit operations, and monitoring while hatches were open, allowing free access to the orbiter. Normal operations included IVA activity (other than in-flight maintenance) planned before launch and listed in the mission timeline such as unpacking, assembly, and powering up. Off-nominal operations included performance of backup, malfunction, contingency, or emergency procedures not involving hardware modification or repair. IVA also included all activities in which crew members dressed in spacesuits and using life support systems performed hands-on operations inside a customer-supplied crew module.[74]

---

[73] *Shuttle Crew Operations Manual*, p. 2.11-2.
[74] "Mission Preparation and Prelaunch Operations," *NSTS Shuttle Reference Manual* (1988), *http://science.ksc.nasa.gov/shuttle/technology/sts-newsref/stsover-prep.html#stsover-crewserv* (accessed November 12, 2005).

Figure 3–23. Astronaut F. Story Musgrave is seen anchored on the end of the Remote Manipulator System Arm as he prepares to be elevated to the top of the Hubble Space Telescope to install protective covers on the magnetometers, December 9, 1993. (NASA Photo No. GPN-2000-001085)

In-flight maintenance was any abnormal on-orbit maintenance or repair of a malfunctioning payload conducted by the crew within a pressurized vessel or payload module to keep the payload operable or to return it to operability. In-flight maintenance normally involved removal of payload panels, mating and demating of electrical connectors, or replacement of line replaceable units.[75]

Space Shuttle Abort Modes

Space Shuttle launch abort philosophy is aimed toward safe and intact recovery of the flight crew, orbiter, and its payload. A Shuttle launch scrub or

---

[75] *Space Shuttle Systems Payload Accommodations*, NSTS 07700, Volume XIV, Appendix 9. "System Description and Design Data–Intravehicular Activities," pp. 1-1, 1-2, *http://shuttlepayloads.jsc.nasa.gov/ data/PayloadDocs/documents/07700/App_09.pdf* (accessed November 15, 2005).

abort might occur up to solid rocket booster ignition. Normally, launch scrubs before SSME start were followed by an orderly safing procedure and crew egress, assisted by the closeout crew. A fully fueled Shuttle on the launch pad might present an extremely hazardous situation if toxic vapors, fire, or structural damage were present. A launch abort after SSME start was automatically controlled by the ground launch sequencer. The presence of excess hydrogen was the most serious hazard, resulting in a very dangerous hydrogen fire invisible to the eye. This situation occurred during a launch attempt for STS-41-D in 1984.

Should an abnormal event occur that terminated a flight or prelaunch operation and resulted in substantial damage to the Shuttle and/or injury to personnel, the NASA Test Director would declare a contingency situation. This would alert fire and rescue personnel and put in motion preplanned procedures to minimize further damage and injuries. The NASA Test Director might also initiate action if an emergency condition existed that required immediate action to prevent loss of life or destruction of equipment. In preparation for a potential emergency condition, a hazardous condition might be declared if there was a threat to personnel health or safety. A hazardous condition might develop into an emergency condition.

There were two basic types of ascent abort modes: intact and contingency. Intact aborts were designed to provide a safe return of the orbiter to a planned landing site. Contingency aborts were designed to permit crew survival following more severe failures when an intact abort was not possible. A contingency abort would usually result in a crew bailout.

- Abort-To-Orbit (ATO)—This mode would be chosen if partial loss of main engine thrust occurred late enough to permit reaching a minimal 105-nautical-mile (194.5-kilometer) orbit with orbital maneuvering system engines.
- Abort-Once-Around (AOA)—This mode would be chosen when there was earlier main engine shutdown with the capability to allow one orbit around Earth before landing at Edwards Air Force Base, California; White Sands Space Harbor (Northrup Strip), New Mexico; or the Shuttle Landing Facility at Kennedy Space Center, Florida.
- Transoceanic Abort Landing (TAL)—This mode would be selected when the loss of two main engines midway through powered flight would force a landing at Ben Guerir, Morocco; Moron, Spain; or Banjul, The Gambia.[76]
- Return-To-Launch-Site (RTLS)—This mode would be selected when there was early shutdown of one or more engines, and when there was not enough energy to reach Ben Guerir. It would result in a pitch around and thrust back toward Kennedy Space Center until the Shuttle was within gliding distance of the Shuttle Landing Facility.[77]

---

[76] E-mail from Kyle Herring, NASA Public Affairs Office, Johnson Space Center, November 30, 2005.
[77] *Shuttle Crew Operations Manual*, pp. 6.1-1–6.1-2, 6.2-1.

Since Space Shuttle flights began through 1998, there have been very few aborts. The first on-pad abort-after-ignition occurred on STS-41-D in 1984. STS-51-F experienced both an on-pad abort and an abort-to-orbit during two launch attempts in July 1985. STS-55 experienced an on-pad abort-after-ignition on March 22, 1993, when SSME No. 3 failed to ignite completely.[78]

Space Shuttle Missions

Between 1989 and 1998, 66 Space Shuttle missions flew. The following section describes those missions, presented chronologically. Table 3–51 lists summary data. Tables 3–52 through 3–116 list Shuttle mission characteristics and events. Most mission information was obtained from online Shuttle chronologies and archives.[79] Additional material comes from the press kits for each mission, *U.S. Human Spaceflight* (NASA Monographs in Aerospace History No. 9), and specific pages from the National Space Science Data Center (NSSDC) Master Catalog.[80] Other sources are noted in footnotes beneath the text. Abbreviations relating to crew positions are: CDR–Commander; PLT–Pilot; MC–Mission Commander, MS–Mission Specialist, PC–Payload Commander, and PS–Payload Specialist.

The online Shuttle mission archives generally presented Shuttle altitudes as a single value. However, mission descriptions indicated that altitude often changed during a mission, sometimes for days at a time, to accomplish mission objectives. Unless the change in altitude was especially significant, only the single value presented in the mission archive is noted in the mission tables. The reader can find additional details relating to mission payloads in chapter 4 of this volume, Space Science, and in the next volume of the *NASA Historical Data Book*. Missions shown as "successful" mission were those in which mission objectives were achieved.

STS-29

This mission launched on March 13, 1989, from Kennedy Space Center and landed March 19 at Edwards Air Force Base. The primary payload was the Tracking and Data Relay Satellite-4 (TDRS-4) attached to an Inertial Upper Stage (IUS), which became the third TDRS deployed. After deployment, the IUS propelled the satellite to geosynchronous orbit. See Table 3–52 for further details.

---

[78] Jenkins, pp. 272, 274, 304, 305.
[79] "Mission Chronologies," *http://www-pao.ksc.nasa.gov/kscpao/chron/chrontoc.htm* (accessed May–June 2005); "1999–2004 Shuttle Mission Archives," *http://www.nasa.gov/centers/kennedy/shuttleoperations/ archives/1999-2004.html* (accessed May–June 2005).
[80] Judy A. Rumerman, compiler, *U.S. Human Spaceflight: A Record of Achievement, 1961–1998*, Monographs in Aerospace History, no. 9 (Washington, DC: NASA History Division, July 1998); "Space Shuttle Press Kits," *http://www.jsc.nasa.gov/history/shuttle_pk/shuttle_press.htm* (accessed May–June 2005); NSSDC Master Catalog, *http://nssdc.gsfc.nasa.gov/nmc/sc-query.html* (accessed June–December 2005).

STS-30

This mission launched on May 4, 1989, from Kennedy Space Center and landed May 8 at Edwards Air Force Base. The mission's primary payload was the Magellan/Venus Radar Mapper spacecraft with its attached IUS, which boosted the spacecraft on its proper trajectory for a 15-month journey to Venus. It was the first Shuttle launch of a deep space probe and the first U.S. planetary mission in 11 years. Secondary payloads were the Mesoscale Lightning Experiment (MLE), microgravity research with the Fluids Experiment Apparatus (FEA), and the AMOS experiment. One of the general purpose computers failed on orbit and had to be replaced. It was the first time such an operation was performed while orbiting. See Table 3–53 for further mission details.

STS-28

This classified DOD mission launched on August 8, 1989, from Kennedy Space Center and landed on August 13 at Edwards Air Force Base. See Table 3–54 for further mission details.

STS-34

This mission launched October 18, 1989, from Kennedy Space Center and landed October 23 at Edwards Air Force Base. The Galileo spacecraft was launched on the Shuttle's fifth orbit with a boost from its IUS toward Jupiter by way of Venus. It was the Shuttle's second interplanetary payload.

Also in the payload bay of *Atlantis* was the Shuttle Solar Backscatter Ultraviolet (SSBUV) instrument. The SSBUV provided calibration of backscatter ultraviolet instruments concurrently being flown on free-flying satellites. The SSBUV was contained in two canisters in the payload bay, one holding the SSBUV spectrometer and five supporting optical sensors and a second housing data, command, and power systems. An interconnecting cable provided the communication link between the two canisters. *Atlantis* also carried several secondary payloads involving radiation measurements, polymer morphology, lightning research, microgravity effects on plants, and a student experiment on ice crystal growth in zero gravity. See Table 3–55 for further mission details.

STS-33

This classified DOD mission launched November 22, 1989, from Kennedy Space Center and landed November 27 at Edwards Air Force Base. It was the first night launch since the return to flight. See Table 3–56 for further mission details.

STS-32

This mission launched January 9, 1990, from Kennedy Space Center and landed January 20 at Edwards Air Force Base. Lasting almost 11 days, STS-32 was the longest Shuttle flight to date.

The Long Duration Exposure Facility (LDEF), released into orbit on STS-41-C in 1984, was finally retrieved after nearly six years in space. LDEF was a 14.5-foot by 30-foot (4.4-meter by 9.1-meter) 12-sided array of more than 70 panels designed to obtain data important to designers of spacecraft on the effects of the orbital environment on metals, coatings, and other materials used in constructing spacecraft. It provided an STS-transported, low-cost, reusable, free-flying structure to carry many different science and technology experiments. The LDEF required little or no electric power and data processing while in long-duration spaceflight.

While in space, the LDEF completed 32,422 Earth orbits, allowing investigators to increase their scientific and technological understanding of the space environment and its effects. LDEF experienced one-half of a solar cycle, as it was deployed during a solar minimum and retrieved at a solar maximum. After rendezvousing with the large, cylindrical satellite—one of the most complicated space rendezvous operations ever—the Shuttle crew photographed the LDEF in orbit, grappled it with the remote manipulator system arm, and then stowed it in the cargo bay of *Columbia*. Scientists who examined the LDEF after landing found evidence of erosion and micrometeorite impacts, as expected. By the time LDEF was retrieved, its orbit altitude had decayed to ~175 nautical miles (324 kilometers), and the satellite was a little more than one month away from reentering Earth's atmosphere. Figure 3–24 shows the LDEF.

A SYNCOM DOD communications satellite also was deployed on the mission. See Table 3–57 for further mission details.

STS-36

This classified DOD mission launched February 28, 1990, from Kennedy Space Center and landed March 4 at Edwards Air Force Base. Launch was postponed several times (and then postponed further because of bad weather) because of the illness of the mission commander, John Creighton. The mission was the first time since the Apollo 13 mission in 1970 that a human spaceflight mission had been postponed because of the illness of a crew member.

This flight flew at an inclination orbit of 62 degrees, the highest inclination flown by the Shuttle to date. See Table 3–58 for further mission details.

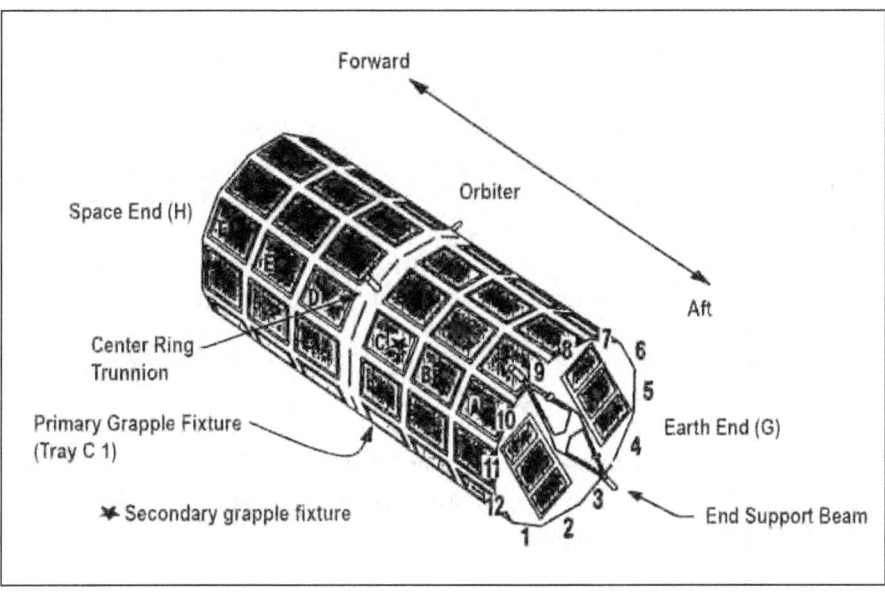

Figure 3–24. The LDEF was retrieved by STS-32 after nearly six years in space.

STS-31

This mission launched April 24, 1990, from Kennedy Space Center and landed April 29 at Edwards Air Force Base. The Hubble Space Telescope, first of the Great Observatories and first large optical telescope to be placed above Earth's atmosphere, was released into orbit by the remote manipulator system arm on the second day of the mission. Because of the need to place the telescope above most of Earth's atmosphere, *Discovery* flew the highest Shuttle orbit to date, reaching an altitude of slightly more than 611 kilometers (330 nautical miles). After the telescope was deployed, the astronauts conducted experiments in crystal growth and monitored the radiation environment aboard the orbiter. See Table 3–59 for further mission details.

STS-41

This mission launched October 6, 1990, from Kennedy Space Center and landed October 10 at Edwards Air Force Base. It was the heaviest payload to date. The deployment of ESA's Ulysses to explore the polar regions of the Sun was the highlight of this four-day mission. Ulysses was released from *Discovery*'s cargo bay on the first day of the mission; on-board rockets were fired to send the spacecraft toward a gravity-assist at Jupiter to observe the polar regions of the Sun. For the first time, a PAM and IUS combined together were used to send the spacecraft into its trajectory. They replaced the canceled Centaur upper stage that had been planned for this mission. After Ulysses's deployment, the astronauts conducted a number of secondary experiments, including measuring atmospheric ozone, studying the effects of atomic oxygen on spacecraft materials, and evaluating a new "hands-off" voice command system in the Shuttle crew cabin.

Also in *Discovery*'s payload bay was the Airborne Electrical Support Equipment, an electrical generating system mounted on the side of the bay to supply power to Ulysses. The INTELSAT Solar Array Coupon, samples of solar array materials mounted on *Discovery*'s remote manipulator system, studied the effects of atomic oxygen wear on solar panels in preparation for a future Shuttle mission to rescue the stranded INTELSAT satellite. See Table 3–60 for further mission details.

STS-38

This classified DOD mission launched November 15, 1990, and landed November 20, 1990, at Kennedy Space Center. See Table 3–61 for further mission details.

STS-35

This mission launched December 2, 1990, from Kennedy Space Center and landed December 10 at Edwards Air Force Base. This mission was the first Shuttle flight dedicated to a single discipline: astrophysics. Using Spacelab pallets with the instrument pointing system and igloo, *Discovery* carried a group of astronomical telescopes called Astro-1 in its cargo bay. The crew included four individuals with doctorates in astronomy: Jeffrey Hoffman, Robert Parker, Samuel Durrance, and Ronald Parise. Despite several hardware malfunctions, the crew observed a wide variety of astronomical targets, from comets to quasars, with particular attention to x-ray and ultraviolet wavelengths. See Table 3–62 for further mission details.

STS-37

This mission launched April 5, 1991, from Kennedy Space Center and landed April 11 at Edwards Air Force Base. The initial landing at Edwards was waved off and rescheduled for the next day at Kennedy. That, too, was waved off because of fog, and the mission landed one orbit later at Edwards.

The Gamma Ray Observatory, the second "Great Observatory," was released by the Shuttle's remote manipulator system arm on the third day of the flight, after astronauts Jerry Ross and Jay Apt made an unscheduled spacewalk to repair an antenna on the spacecraft. The Gamma Ray Observatory was the heaviest science satellite ever launched from the Shuttle (see Figure 3–25).

Later in the mission, Ross and Apt returned to the cargo bay to test Crew and Equipment Translation Aids, rail-mounted mechanical pushcarts planned for use on Space Station *Freedom*. The two spacewalks were the first in more than five years. See Table 3–63 for further mission details.

STS-39

This mission launched April 28 and landed May 6, 1991, at Kennedy Space Center. This was the first unclassified defense-related mission of the Shuttle program. Highlighted by around-the-clock observations, it included experiments sponsored by the U.S. Air Force and the Strategic Defense Initiative Organization. The studies included extensive infrared, ultraviolet, visible, and x-ray observations of the space environment and the Shuttle itself. On-board instruments also returned high-quality images of Earth's aurora. In an experiment related to ballistic missile defense, *Discovery* released a Shuttle Pallet Satellite (SPAS) instrument platform equipped with infrared sensors to fly in formation and observe rocket thruster plumes while the Shuttle performed a complicated series of maneuvers. The satellite was retrieved and returned to Earth at the end of the mission. See Table 3–64 for further mission details.

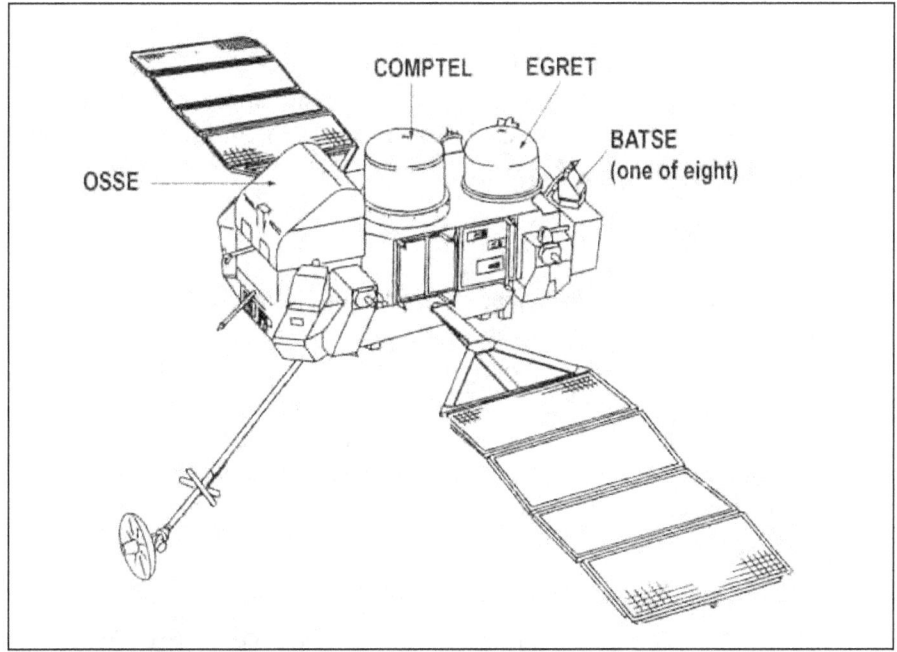

Figure 3–25. The Gamma Ray Observatory, the second "Great Observatory," was the most massive instrument ever launched by the Space Shuttle to date.

STS-40

This mission launched on June 5, 1991, from Kennedy Space Center and landed June 14 at Edwards Air Force Base. The SLS-1 mission was the first mission dedicated entirely to understanding the physiological effects of spaceflight. The crew conducted an extensive series of biomedical experiments during the nine-day mission, and the results were compared with baseline data collected on the ground before and after the flight. In addition to the human subjects, rodents and jellyfish were aboard to test their adaptation to microgravity. See Table 3–65 for further mission details.

STS-43

This mission launched August 2, 1991, and landed August 11, 1991, at Kennedy Space Center. It marked the first scheduled landing at Kennedy's Shuttle Landing Facility since January 1986. The primary payload, the TDRS-5, attached to an IUS, was deployed about 6 hours into flight. The IUS propelled the satellite into geosynchronous orbit as TDRS-5 became the fourth member of the orbiting TDRS cluster. See Table 3–66 for further mission details.

STS-48

This mission launched September 12, 1991, from Kennedy Space Center and landed September 18 at Edwards Air Force Base. The UARS was deployed on the third day of the mission. The 14,500-pound (6,577-kilogram) observatory conducted the most extensive study to date of the upper atmosphere as it investigated the stratosphere, mesosphere, and lower thermosphere. See Table 3–67 for further mission details.

STS-44

This mission launched November 24, 1991, from Kennedy Space Center and landed December 1, 1991, at Edwards Air Force Base. The mission was shortened by three days because one of the orbiter's three inertial measurement units failed.

The unclassified DOD payload included the Defense Support Program (DSP) early warning satellite and attached IUS, which was deployed on the first day of the mission. On-board payloads focused on contamination experiments and medical research. See Table 3–68 for further mission details.

STS-42

This mission launched January 22, 1992, from Kennedy Space Center and landed January 30 at Edwards Air Force Base. The primary payload was the International Microgravity Laboratory-1 (IML-1) using the Spacelab long module. The IML-1 mission was the first in a series of international Shuttle flights dedicated to fundamental life and microgravity sciences research. IML-1 science operations were a cooperative effort between the *Discovery*'s crew in orbit and mission management, scientists, and engineers in a control facility at Marshall Space Flight Center. Though the crew and the ground-based controllers and science teams were separated by many miles, they interacted in much the same way as they would if working side by side. The mission was extended by one day to continue mission work. Figure 3–26 shows STS-42 astronauts in the IML-1 science module.

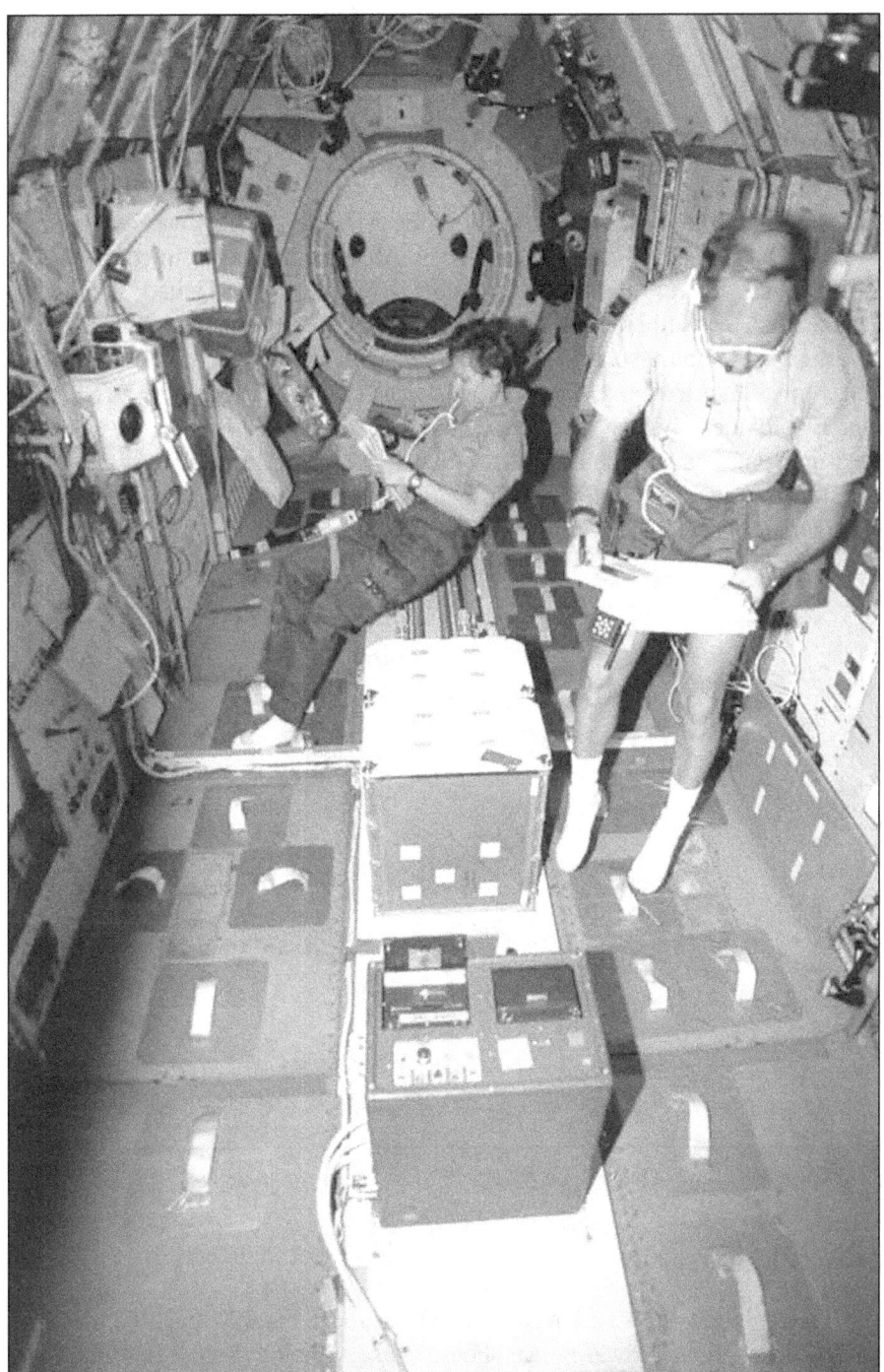

Figure 3–26. Bondar and Thagard work with experiments in the IML-1 Science Module. The two, along with four other NASA astronauts and a second IML-1 payload specialist, spent more than eight days conducting experiments in Earth orbit. Part of the SAMS is in the center foreground. (NASA-MSFC Photo MSFC-9250420)

Other payloads included 12 GAS canisters, a number of middeck payloads, and two SSIP experiments. See Table 3–69 for further details.

STS-45

This mission launched March 24, 1992, and landed April 2 at Kennedy Space Center. It marked the first flight of the ATLAS-1, which was mounted on Spacelab pallets in the orbiter's cargo bay. An international team consisting of the United States, France, Germany, Belgium, the United Kingdom, Switzerland, the Netherlands, and Japan provided 12 instruments performing 13 investigations in atmospheric chemistry, solar radiation, space plasma physics, and ultraviolet astronomy. The ATLAS-1 was co-manifested with the SSBUV, which provided highly calibrated measurements of ozone to fine-tune measurements made by other NASA and NOAA satellites. The mission was extended one day to continue investigations. Figure 3–27 shows the payload configuration. See Table 3–70 for further mission details.

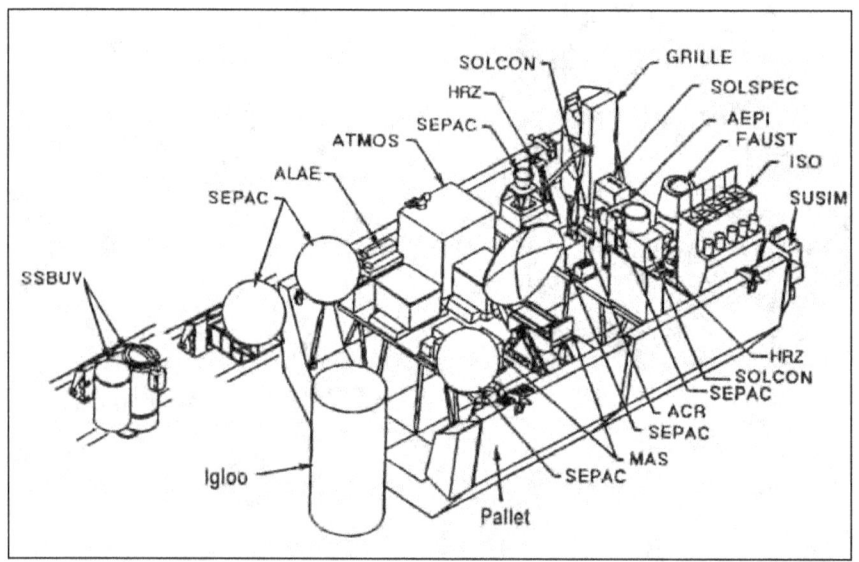

Figure 3–27. ATLAS-1 Payload Configuration.

STS-49

This mission launched May 7, 1992, from Kennedy Space Center and landed May 16 at Edwards Air Force Base. During a mission that was extended by two days, the crew successfully captured and redeployed the INTELSAT VI satellite, which had been in an unusable orbit since the upper stage failed to separate from the second stage of its Titan launch vehicle in

March 1990. Capture of the satellite required three spacewalks and the simultaneous efforts of three spacewalking astronauts as well as the maneuvering skill of the Shuttle commander.

The mission was marked by a number of "firsts." Four spacewalks, the most ever on a single mission, highlighted the first flight of the orbiter *Endeavour*. Two of these were the longest in U.S. spaceflight history to date, the first lasting 8 hours, 29 minutes and the second 7 hours, 45 minutes. The flight also featured the longest spacewalk to date by a female astronaut and was the first spaceflight in which three crew members worked outside the spacecraft at the same time. It also was the first time that astronauts attached a live rocket motor to an orbiting satellite, when they attached a perigee kick motor to the INTELSAT VI satellite, which later boosted it into its proper orbit. This was the first Shuttle mission requiring three astronauts to rendezvous with an orbiting spacecraft.

The crew also practiced assembly techniques for the planned Space Station *Freedom* and tested the new drag chute after orbiter nosegear touchdown at Edwards Air Force Base. See Table 3–71 for further details.

STS-50

This mission launched June 25 and landed July 9, 1992, at Kennedy Space Center. It marked the first use of the Extended Duration Orbiter kit, tanks of liquid oxygen and hydrogen mounted in the payload bay to extend the energy-generating fuel cell's capacity, allowing mission duration to surpass all previous U.S. crewed spaceflights to date with the exception of the three Skylab missions in 1973–1974. The USML-1 made its first flight on this mission. It was the first in a planned series of flights to advance microgravity research efforts in several disciplines. See Table 3–72 for further mission details.

STS-46

This mission launched July 31 and landed August 8, 1992, at Kennedy Space Center. The primary mission objective was the deployment of the ESA's EURECA and the operation of the NASA-Italian TSS, with Italian astronaut Franco Malerba on board the Shuttle. EURECA was the largest satellite produced in Europe. It carried 15 major science experiments, mostly in microgravity sciences.[81] After a delay and a shorter than planned thruster firing, the satellite was successfully boosted to operational orbit.

---

[81] Jenkins, p. 301.

During TSS deployment, the satellite at the end of the tether reached a distance of only 840 feet (256 meters), rather than its planned 12.5 miles (20.1 kilometers) because of a jammed tether line.[82] After additional unsuccessful attempts to free the tether, the satellite was restowed for return to Earth. Figure 3–28 shows the TSS viewed from the orbiter *Atlantis*. See Table 3–73 for further mission details.

Figure 3–28. This Space Shuttle Orbiter Atlantis *(STS-46) on-board photo is a close-up view of the TSS-1 deployment. (NASA-MSFC Photo No. MSFC-9410850)*

STS-47

This mission launched September 12 and landed September 20, 1992, at Kennedy Space Center. It was the first on-time launch since STS-61-B in 1985. Spacelab-J, the first Japanese Spacelab, flew on this flight. The crew included the first African-American woman to fly in space, the first married couple to fly on the same mission, and the first Japanese person to fly on the Space Shuttle. This mission marked the first operational use of the new drag chute, which was deployed before nosegear touchdown. See Table 3–74 for further mission details.

---

[82] Distance of 840 feet (256 meters) of tether deployment was stated in the "STS-46 Mission Chronology," *http://www-pao.ksc.nasa.gov/kscpao/chron/sts-46.htm* (accessed July 6, 2005); The summary of the Investigative Board corroborated that figure, "Report Details Causes of Tethered Satellite Malfunctions," *NASA News* Release 92-196, November 6, 1992, *http://nssdc.gsfc.nasa.gov/space/text/tss-summary.txt* (accessed December 4, 2005); Jenkins, p. 301, and the "STS-46 Mission Archives" at *http:// science.ksc.nasa.gov/shuttle/missions/sts-46/mission-sts-46.html* (accessed December 4, 2005), stated the distance as 860 feet (262 meters).

STS-52

This mission launched October 22 and landed November 1, 1992, at Kennedy Space Center. It deployed the Laser Geodynamic Satellite II (LAGEOS), a joint effort of NASA and the Italian Space Agency. This dense 0.6-meter (2-foot)-diameter sphere was covered by retroflectors to allow study of dynamic motions of Earth's crust using precise laser tracking of the satellite from ground stations around the world. LAGEOS II was deployed on flight day two and boosted into an initial elliptical orbit by the IRIS, flying for the first time. The apogee kick motor later fired to adjust the spacecraft's orbit at an operational altitude of 5,616 kilometers by 5,905 kilometers (3,490 miles by 3,669 miles).[83]

The mission also carried USMP-1, which was activated on the first day of the flight. On-board studies focused on the influence of gravity on basic fluid and solidification processes. See Table 3–75 for further mission details.

STS-53

This mission launched December 2, 1992, from Kennedy Space Center and landed December 9 at Edwards Air Force Base. It was the first flight of *Discovery* after its OMDP. This was the last Shuttle flight for the DOD. *Discovery* deployed a classified payload followed by unclassified flight activities. GAS hardware located in the cargo bay or on the middeck contained or were attached to 10 secondary payloads. See Table 3–76 for further mission details.

STS-54

This mission launched January 13 and landed January 19, 1993, at Kennedy Space Center. The fifth TDRS-6, part of NASA's orbiting communications system, was deployed about 6 hours after liftoff. Figure 3–29 shows the on-orbit configuration.

On the fifth day of the flight, astronauts Mario Runco and Gregory Harbaugh spent almost 5 hours working in the open payload bay, performing a series of EVA tasks to increase NASA's knowledge of working in space. The astronauts tested their abilities to move freely in the cargo bay, climb into foot restraints without using their hands, and simulated carrying large objects in a microgravity environment.

A Hitchhiker experiment, the Diffuse X-ray Spectrometer, collected data on x-ray radiation from stars and galactic gases. See Table 3–77 for further mission details.

---

[83] E-mail from Carey Noll; data provided by LAGEOS science contact Peter Dunn, November 25, 2005.

STS-56

This mission launched April 9, 1993, and landed April 17, 1993, at Kennedy Space Center. The primary payload was the second ATLAS-2, a Spacelab pallet mission that was one element of NASA's Mission to Planet Earth program. The pallet in the payload bay held six instruments, and a seventh was mounted in two GAS canisters.

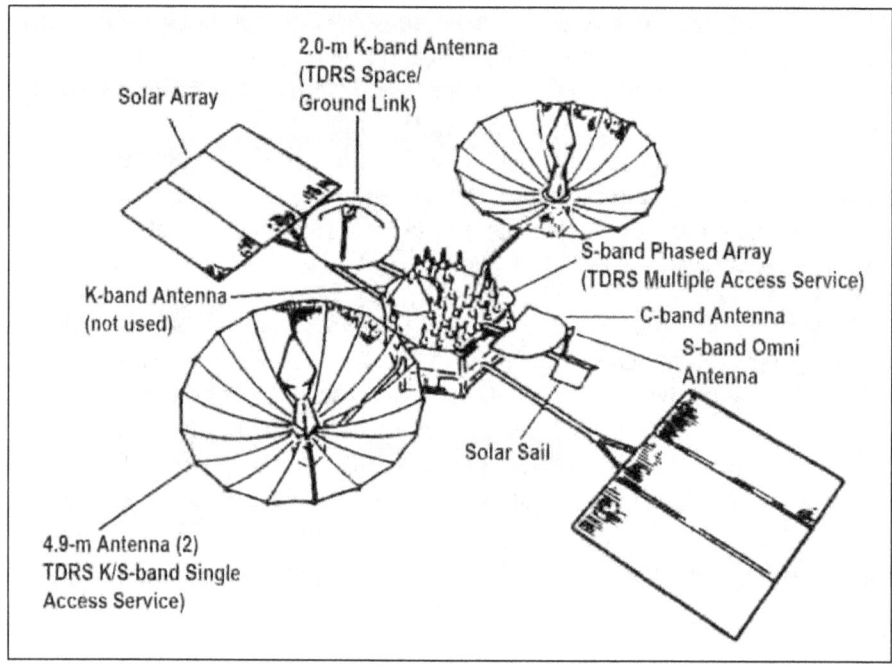

Figure 3–29. TDRS-F (6) On-Orbit Configuration.

The crew used the remote manipulator system arm to deploy the SPARTAN-201 on the second day of the mission. SPARTAN was a free-flying science instrument platform that studied the velocity and acceleration of solar wind and observed the Sun's corona. The collected data was stored on tape for playback after return to Earth. SPARTAN was retrieved on April 13 .

Using the SAREX, the crew also contacted schools around the world and briefly contacted the Russian *Mir* space station, the first contact between the Shuttle and *Mir* using amateur radio equipment. See Table 3–77 for further mission details.

STS-55

This mission launched April 26, 1993, from Kennedy Space Center and landed May 6 at Edwards Air Force Base. This was the last launch scheduled from Pad 39-A at Kennedy Space Center until February 1994 to allow for pad refurbishment and modification. Figure 3-30 shows the STS-55 launch. On May 4, the ninth day of the mission, ground control lost all communication with *Columbia* for about 90 minutes because Mission Control issued an incorrect command.[84]

Figure 3–30. Space Shuttle Columbia *(STS-55) blasts off from Pad 39-A at Kennedy Space Center. This was the last launch from this pad until the next year to allow for pad refurbishment and modification. (NASA-KSC Photo No. KSC-93PC-0626)*

STS-55 was the second German Spacelab mission using the long module, designated D-2. Two crews worked in around-the-clock shifts and conducted approximately 88 experiments relating to materials and life sciences, technology applications, Earth observations, astronomy, and atmospheric physics. The orbiter *Columbia*, the oldest fleet member, passed its 100th day in space on this mission. See Table 3–79 for further mission details.

---

[84] Jenkins, p. 304.

STS-57

This mission launched June 21 and landed July 1, 1993, at Kennedy Space Center. It was the first flight of the commercially developed SPACE-HAB, a laboratory designed to more than double the pressurized workspace for crew-tended experiments.

SPACEHAB's Space Research Laboratory was situated in the forward quarter of the cargo bay. The pressurized laboratory measured approximately 10 feet (3 meters) long and 13.5 feet (4.1 meters) in diameter and contained more than 1,100 cubic feet (31.1 cubic meters) of working volume, enough to house as many as 61 middeck lockers for experiments or a combination of middeck lockers and Space Station racks. Crew members used the modified Spacelab tunnel adapter between the crew compartment and the SPACEHAB laboratory to gain access to the lab once on orbit.

The Space Research Laboratory contained all of the subsystems required to support experiment operations, including environmental controls, command and data handling, electrical power, and thermal control. On this flight, the SPACEHAB laboratory carried payloads from NASA, the U.S. commercial sector, and ESA. The crew operated a total of 22 individual experiments during the mission. Included on the flight were 13 commercial space experiments in materials processing and the effect of spaceflight on human biotechnology: 12 sponsored by the NASA CCDS and one by NASA Langley Research Center. Also on board the SPACEHAB module was an investigation sponsored by the NASA Space Station Freedom Office on closed systems to improve water recycling in the future Space Station environment.

Rendezvous and retrieval of the more than 9,000-pound (4,082-kilogram) EURECA-1 scientific satellite took place on flight day four. (See Figure 3–31 for the EURECA mission scenario.) On flight day five, astronauts David Low and Peter Wisoff spent part of a 5-hour, 55-minute EVA manually stowing the antennae, which would not respond to ground commands. The satellite had been deployed on the STS-46 mission in 1992. The crew spent the remainder of the EVA using the robot arm to complete activities associated with mass handling, mass fine alignment, and high torque. During the mission, the crew also spoke with President William J. Clinton. See Table 3–80 for further mission details.

STS-51

This mission launched on September 12 and landed on September 22, 1993, at Kennedy Space Center. The ACTS was deployed on this mission. The attached Transfer Orbit Stage booster was used for the first time to propel this communications technology spacecraft to geosynchronous transfer orbit.

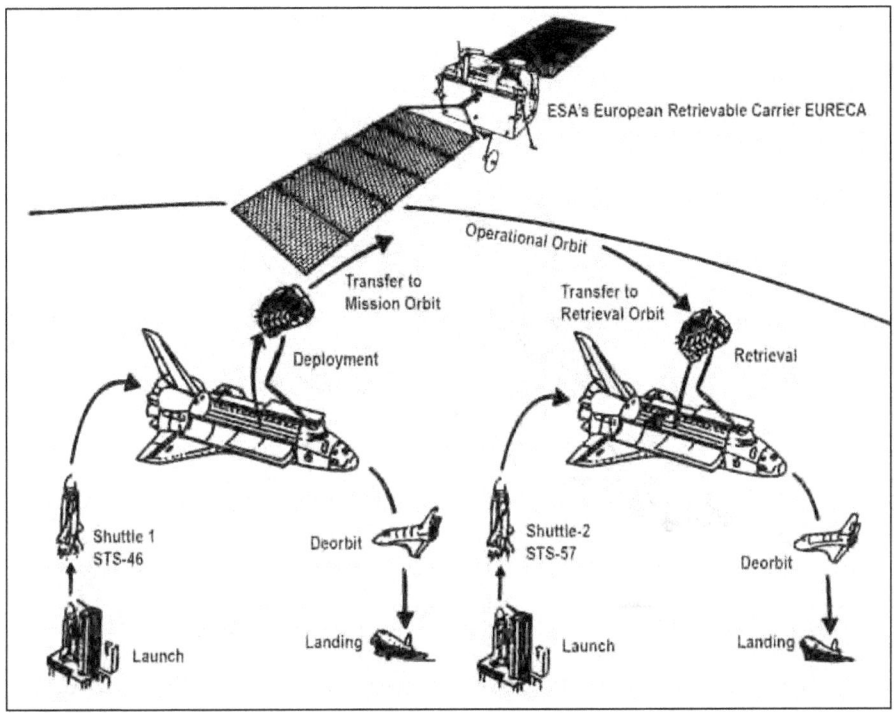

Figure 3–31. EURECA Mission Scenario. EURECA was deployed on STS-46, transferred to operational orbit, and was retrieved on the STS-57 Mission and brought back to Earth.

The second primary payload, the ORFEUS-SPAS, first in a series of ASTRO-SPAS astronomical missions, was also deployed. The joint German–U.S. astrophysics payload was controlled from the SPAS Payload Operations Control Center at Kennedy Space Center, the first time a Shuttle payload was managed from Florida. An IMAX camera mounted on SPAS recorded extensive footage of the orbiter for the first time. The crew also used the IMAX handheld camera to take out-the-window shots of the SPAS operations. After six days spent collecting data, the remote manipulator system arm retrieved the satellite and returned it to the orbiter payload bay. Figure 3–32 shows the position of the ORFEUS-SPAS and ACTS/TOS payloads in the orbiter. Figure 3–33 shows the ORFEUS-SPAS configuration.

Mission specialists James Newman and Carl Walz also performed a spacewalk that lasted 7 hours, 5 minutes, 28 seconds. Last in a series of generic spacewalks begun earlier in the year, the spacewalk's objective was to evaluate tools, tethers, and foot restraint platforms for the upcoming Hubble Space Telescope servicing mission. The findings reassured the designers and planners of the mission that their preparations were sound. See Table 3–81 for further mission details.

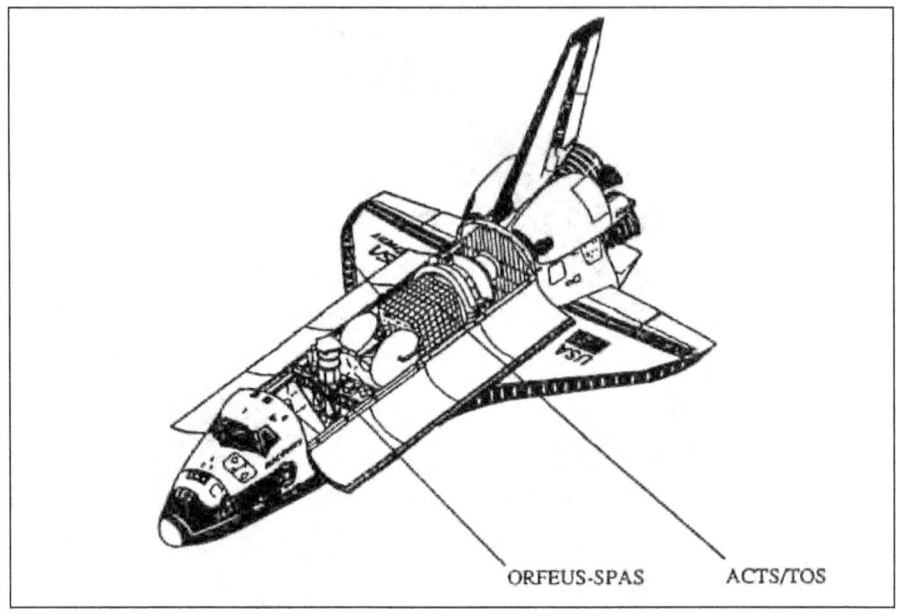

Figure 3–32. ORFEUS-SPAS and ACTS/TOS in the Bay of Discovery *on STS-51.*

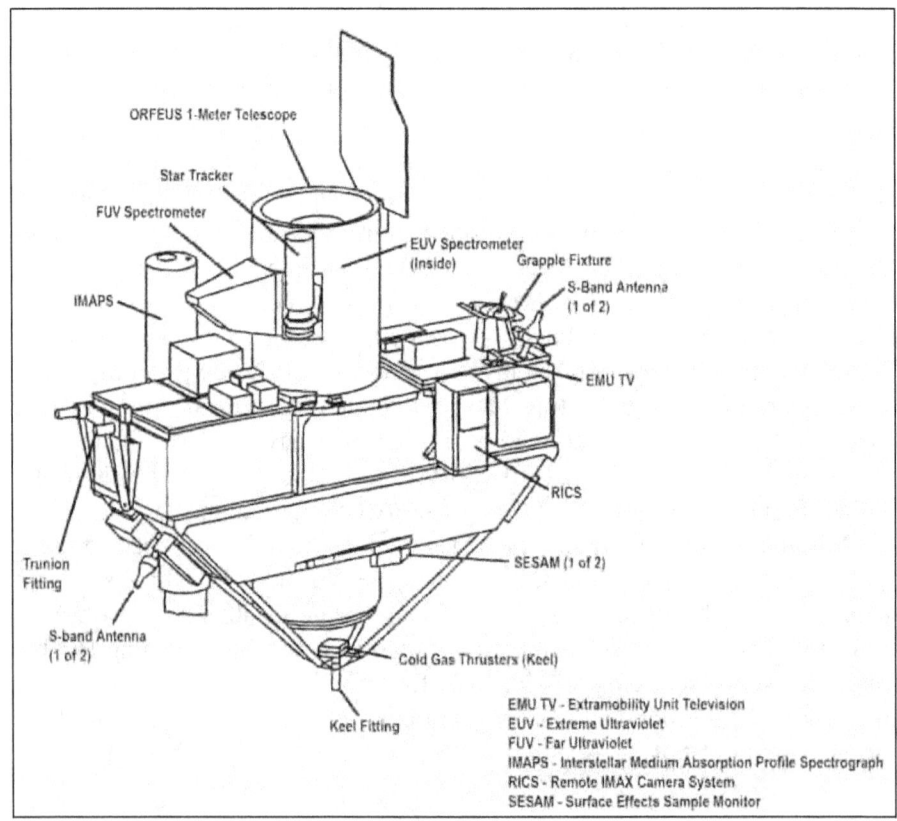

Figure 3–33. ORFEUS-SPAS Configuration.

STS-58

This mission launched October 18, 1993, from Kennedy Space Center and landed November 1 at Edwards Air Force Base. This was the longest Shuttle flight to date.

STS-58 was the second dedicated Spacelab Life Sciences mission and the second use of the extended duration orbiter. The crew conducted 14-neurovestibular, cardiovascular, cardiopulmonary, metabolic, and musculoskeletal medical experiments. Eight of the experiments centered on the crew, and another six focused on 48 rodents carried on board. With the completion of her fourth spaceflight, astronaut Shannon Lucid accumulated the most flight time for a female astronaut on the Shuttle, 838 hours. See Table 2–82 for further mission details.

STS-61

This mission launched December 2 and landed December 13, 1993, at Kennedy Space Center. This was the first Hubble Space Telescope servicing mission, one of the most challenging and complex human spaceflight missions ever attempted. During a record five back-to-back spacewalks totaling 35 hours, 28 minutes, two teams of astronauts completed the first servicing of the Hubble Space Telescope, updating instruments, correcting the spherical aberration clouding the telescope's vision, and replacing faulty gyroscopes. Both the handheld and cargo bay IMAX cameras captured coverage of the EVAs. Footage from the cameras was used in the 2001 movie *Destiny in Space*. See Table 3–83 for further mission details.

STS-60

This mission launched February 3 and landed February 11, 1994, at Kennedy Space Center. The first Shuttle flight of 1994 marked the first flight of a Russian cosmonaut, Sergei Krikalev, on the U.S. Space Shuttle—part of the Implementing Agreement on NASA/Russian Space Agency Cooperation in HSF, an international agreement between the two countries on human spaceflight. The mission also was the second flight of the SPACEHAB pressurized module and carried the 100th GAS payload to fly in space. STS-60 flew four GAS experiments as well as three other payloads on the GAS Bridge.

Discovery also carried the WSF, an attempt to grow innovative semiconductor film materials for use in advanced electronics while in the near vacuum of space. The 12-foot (3.7-meter)-diameter parabolic-shaped WSF included a communications and avionics system, solar cells and batteries, and a propulsion thruster. It was to be deployed by the remote manipulator arm and fly in formation with *Discovery* at a distance of up to 46 statute miles (74 kilometers) from the orbiter for 56 hours. The remote manipulator arm was supposed to retrieve the WSF from space. However, after two unsuccessful attempts to deploy the facility, it was decided that for the remainder of the mission, all WSF operations would take place at the end of the remote manipulator system and there would be no WSF free-flying operations. See Table 3–84 for further mission details.

STS-62

This mission launched March 4 and landed March 18, 1994, at Kennedy Space Center. The primary payloads were the USMP-2 and the OAST-2 suite of experiments. USMP-2 included five experiments investigating materials processing and crystal growth in microgravity. OAST's six experiments focused on space technology and spaceflight. Both payloads were located in the payload bay, activated by crew members, and operated by teams on the ground.

The USMP-2 experiments were conducted early in the mission. Later, to facilitate the OAST-2 experiments, *Columbia*'s orbit was lowered about 20 nautical miles (37 kilometers). The crew also conducted a number of biomedical activities aimed at better understanding and countering the effects of prolonged spaceflight. See Table 3–85 for further mission details.

STS-59

This mission launched April 9, 1994, from Kennedy Space Center and landed April 20 at Edwards Air Force Base. The SRL-1 was this mission's primary payload. It gathered data on Earth and the effect of humans on the planet's carbon, water, and energy cycles. SRL-1 was located in the Shuttle's payload bay, activated by crew members, and operated by teams on the ground. SRL-1 included an atmospheric instrument called the Measurement of Air Pollution from Satellites (MAPS), the Spaceborne Imaging Radar-C (SIR-C), and the X-band Synthetic Aperture Radar (X-SAR). Figure 3–34 shows the location of the X-SAR panels and the SIR-C-band and L-band panels on the payload bay pallet. The German Space Agency and the Italian Space Agency provided the X-SAR. More than 400 sites were imaged, including 19 primary observation sites (supersites) in Brazil, Michigan, North Carolina, and Central Europe. The total area covered was 25.6 million square miles (~50 million square kilometers).[85] Thirteen countries were represented in the project with 49 principal investigators and more than 100 scientists, coordinated by the Jet Propulsion Laboratory (JPL). Roughly 65 hours of data were collected.[86] The MAPS experiment measured the global distribution of carbon monoxide in the troposphere, or lower atmosphere.

---

[85] "SIR-C/X-SAR Flight 1 Statistics," JPL Fact Sheet, *http://southport.jpl.nasa.gov/sir-c/getting_data/ missions_stats.html* (accessed December 7, 2005). Also e-mails from Bruce Chapman, JPL, December 7, 2005.
[86] "SIR-C/X-SAR Flight 1 Statistics," JPL Fact Sheet.

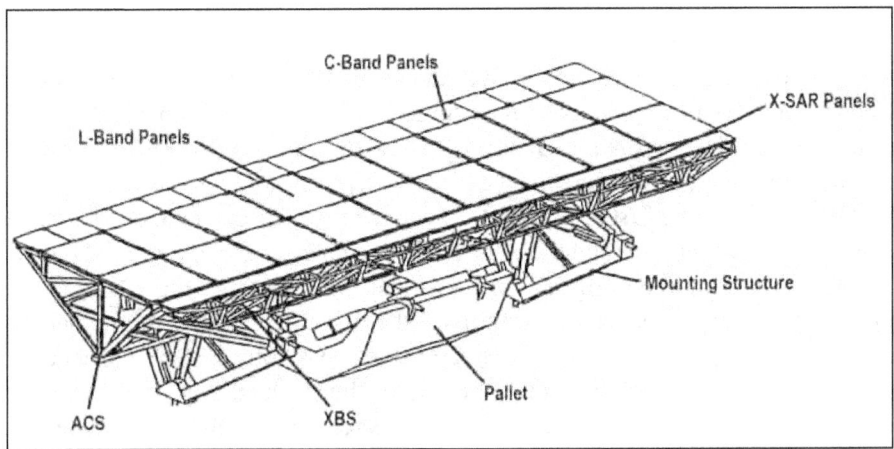

Figure 3–34. The SIR and X-SAR located on a Spacelab pallet in the Shuttle's payload bay on STS-59.

This was the first flight test of an improved thermal protection tile. Known as Toughened Uni-Piece Fibrous Insulation (TUFI), the new tile material was an advanced version of the material protecting the Space Shuttle from the intense heat that built up as it reentered Earth's atmosphere. On this mission, six tiles located on the triangular carrier panel between and below two of the main engines sustained no damage.[87] Figure 3–35 shows the location of the various payloads on *Endeavour.* See Table 3–86 for further mission details.

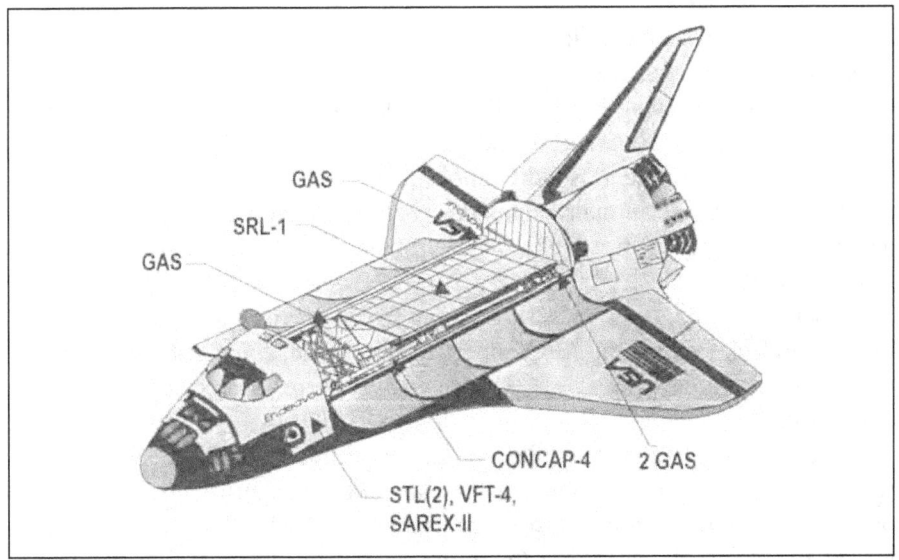

Figure 3–35. Payloads on Endeavour *(STS-59).*

[87] "STS-59 Shuttle Mission Report," June 1994, NSTS-08291, NASA-TM-110527, p. 25, *http://ntrs.nasa.gov/archive/nasa/casi.ntrs.nasa.gov/19950016676_1995116676.pdf* (accessed July 7, 2005).

STS-65

This mission launched July 8 and landed July 23, 1994, at Kennedy Space Center. This was *Columbia*'s last mission before its scheduled modification and refurbishment. The first female Japanese astronaut, Chiaki Naito-Mukai, flew on this mission. She set a record for the longest flight by a female astronaut. This flight also marked the first time that liftoff and reentry were captured on videotape from the crew cabin. The flight was the longest to date, lasting 14 days, 18 hours.

The IML-2 was the primary payload. The IML-2 carried more than twice the number of experiments and facilities as the first IML mission. More than 80 experiments, representing more than 200 scientists from six space agencies, were located in the IML Spacelab module in the payload bay. Two teams of crew members performed round-the-clock research on the behavior of materials and life in near weightlessness.

Fifty of the experiments related to life sciences, including bioprocessing, space biology, human physiology, and radiation biology. Some of the equipment used for these investigations had flown on previous Spacelab flights, such as ESA's Biorack, making its third flight. The IML-2 Biorack housed 19 experiments featuring chemicals and biological samples such as bacteria; mammalian and human cells; isolated tissues and eggs; sea urchin larvae; fruit flies; and plant seedlings.

DARA provided the Slow Rotating Centrifuge Microscope (NIZEMI), a slow-rotating centrifuge that allowed study of how organisms react to different gravity levels. Samples studied included jellyfish and plants. For the first time, researchers could determine how organisms reacted to forces one and one-half times Earth's gravity.

Nearly 30 experiments in materials processing were conducted with nine different types of science facilities. DARA provided the Electromagnetic Containerless Processing Facility (TEMPUS), flying for the first time on IML-2, to allow study of the solidification of materials from the liquid state in a containerless environment. Solidification phenomena were of great interest to science and also used in many industrial processes. Science teams detected for the first time a phase in a nickel-niobium sample that was masked by other forces on Earth.

Another facility, the ESA's APCF, was flying for the second time. Housed in two middeck lockers, the APCF operated autonomously after being activated on the first flight day. Some 5,000 video images were made of crystals grown during flight.

The mission further advanced the concept of telescience, where researchers on the ground could monitor in real time experiments on board the orbiter. The flight set a new record of more than 25,000 payload commands issued from Spacelab Mission Operations Control at Huntsville, Alabama. Figure 3–36 shows the layout of the IML-2 module racks. See Table 3–87 for further mission details.

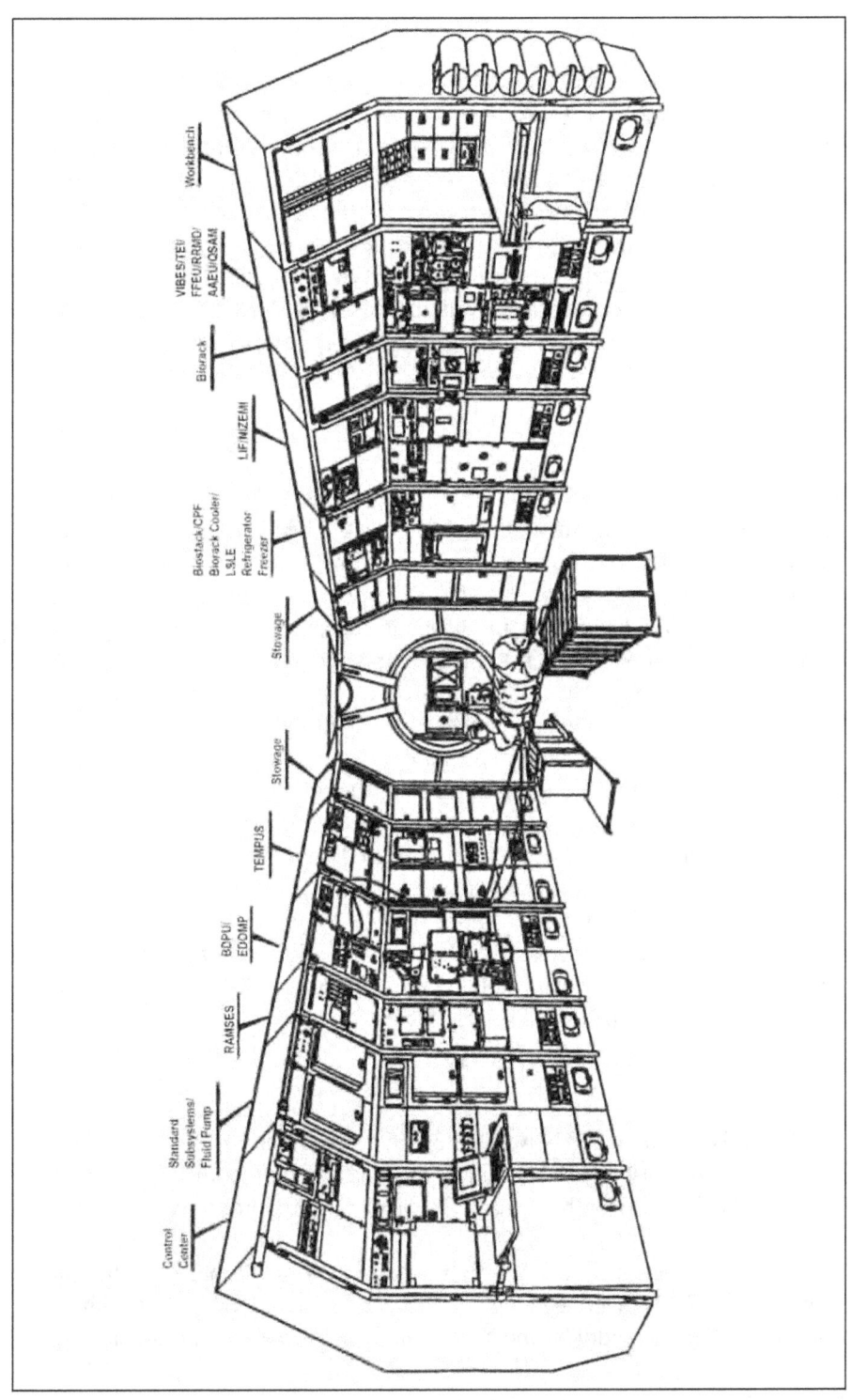

Figure 3–36. IML-2 Module Racks.

STS-64

This mission launched September 9, 1994, from Kennedy Space Center and landed September 20 at Edwards Air Force Base. STS-64 marked the first flight of the LIDAR LITE, which used laser optical radar for the first time to perform atmospheric research as part of NASA's Mission to Planet Earth program. The LITE operated for 53 hours and yielded more than 43 hours of high-rate data. Sixty-five groups from 20 countries made validation measurements with ground-based and aircraft instruments to verify LITE data. During the mission, the crew also released and retrieved the SPARTAN-201 satellite using the remote manipulator system arm. See Table 3–88 for further mission details.

STS-68

This mission launched September 30, 1994, from Kennedy Space Center and landed October 11 at Edwards Air Force Base. The mission set another duration record, lasting more than 16.5 days. The SRL-2, part of NASA's Mission to Planet Earth, flew for the second time in the same year. It gathered data on Earth and the effect of humans on the planet's carbon, water, and energy cycles. Flying the laboratory in different seasons allowed investigators to compare observations between the two flights, which took place in mid-April and at the end of September. The mission also tested the ability of SRL-2's imaging radar to distinguish between changes caused by human-induced phenomena, such as oil spills, and naturally occurring events. The mission demonstrated the maneuvering capability of the orbiter as the crew piloted the *Endeavour* to within 30 feet (9.1 meters) of where it had flown during the first SRL mission on STS-59. The total area covered on this mission was 32 million square miles (roughly 83 million square kilometers).[88]

Five GAS payloads were among the other cargo bay payloads. They included two canisters from the U.S. Postal Service that held 500,000 commemorative stamps honoring the 25th anniversary of Apollo 11. See Table 3–89 for further mission details.

STS-66

This mission launched November 3, 1994, from Kennedy Space Center and landed November 14 at Edwards Air Force Base. The landing was diverted to Edwards Air Force Base because of Tropical Storm Gordon, which prevented landing in Florida.

ATLAS-3, the third ATLAS flight, sat on a Spacelab pallet in the Shuttle cargo bay and collected data about the Sun's energy output, the chemical makeup of Earth's middle atmosphere, and how these factors affected global

---

[88] "SIR-C/X Flight 2 Statistics," *http://southport.jpl.nasa.gov/sir-c/getting_data/missions_stats.html* (accessed December 7, 2005).

ozone levels. The second primary payload, the CRISTA-SPAS, was released on the second day of the mission and retrieved with the Shuttle's remote manipulator system arm. This payload continued the joint NASA-German Space Agency series of scientific missions. CRISTA-SPAS flew at a distance of about 25 miles to 44 miles (40 kilometers to 70 kilometers) behind the Shuttle and collected data for more than eight days before being retrieved and returned to the cargo bay. See Table 3–90 for further mission details.

STS-63

This mission launched February 3 and landed February 11, 1995, at Kennedy Space Center. On this flight, Eileen Collins became the first female to serve as a Shuttle pilot.

STS-63 had special importance as a precursor and dress rehearsal for the Shuttle missions that would rendezvous and dock with the Russian space station *Mir*. After flying to and "stationkeeping" at 400 feet (122 meters) from *Mir*, *Discovery* approached to 37 feet (11 meters) before backing off to 400 feet (122 meters) and performing a fly-around. The six-person Shuttle crew included Vladimir Titov, the second Russian cosmonaut to fly on the Space Shuttle. (Figure 3-37 shows the *Mir* space station as seen from *Discovery*.) Crew members Bernard Harris, Jr., and C. Michael Foale performed a spacewalk away from the payload bay to test spacesuit modifications intended to keep spacewalkers warmer and to demonstrate large-object handling techniques. The mass-handling part of the EVA was curtailed when the astronauts became very cold. Harris became the first African-American to walk in space.

The mission also deployed SPARTAN-204, a free-flying spacecraft that made astronomical observations in the far ultraviolet spectrum (see Figure 3–38).

The SPACEHAB module flew for the third time with an array of technological, biological, and other scientific experiments. SPACEHAB introduced two new system features to reduce the demands on crew time. The first was a video switch allowing one camcorder to transmit images to the ground at the same time another unit collected a digital image on a freeze frame and sent it down independently of other orbiter video downlink operations. The second, an enhanced experiment data interface with the SPACEHAB telemetry system, allowed an experimenter with a standard RS232 computer interface to tie directly into the system and send continuous information down to the ground, off-loading this task from the crew and enhancing ground controller monitoring of experiment status. The SPACEHAB laboratory on this mission had two 12-inch (30.5-centimeter)-diameter windows with a NASA docking camera to assist in *Mir* proximity operations. See Table 3–91 for further mission details.

Figure 3–37. Russia's Mir *Space Station during rendezvous operations with the Space Shuttle*
Discovery. *Docked at the bottom of the* Mir *facility is a Soyuz vehicle. On STS-63,* Discovery
approached Mir, *flew around the Russian Space Station, and then backed off. This provided
practice for future docking missions. (NASA Photo STS063-712-017)*

STS-67

This mission launched March 2, 1995, from Kennedy Space Center and
landed March 18 at Edwards Air Force Base. The mission set a duration record
of more than 16.5 days.

Astro-2 was the second mission using the Spacelab instrument pointing
system and igloo/pallet to conduct astronomical observations and obtain
scientific data on astronomical objects in the ultraviolet regions of the spectrum.
The Spacelab's three telescopes made observations in complementary regions
of the spectrum and gathered data that would add to scientists' understanding of
the universe's history and the origins of stars. Figure 3–39 shows the Astro-2
suite of instruments.

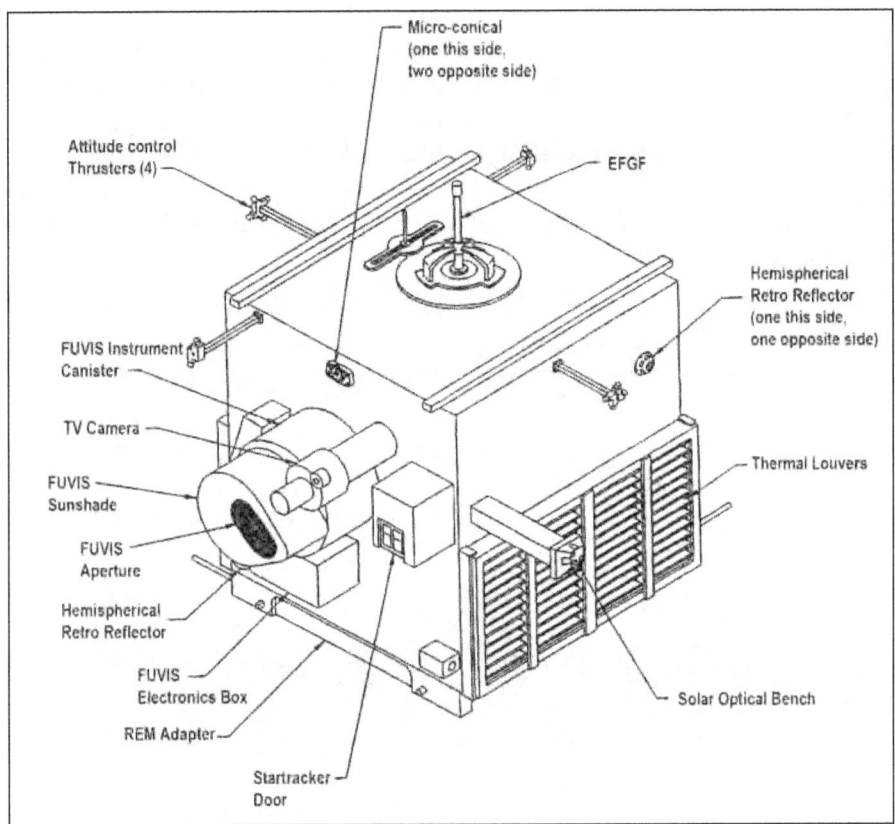

Figure 3–38. SPARTAN-204 was a free-flying spacecraft that observed the far ultraviolet spectrum. Weighing some 2,500 pounds (1,134 kilograms), it was to give the astronauts practice handling heavy loads in the cold, nighttime space environment in preparation for Space Station assembly. But both astronauts reported they were becoming very cold, and the mass handling part of the mission was curtailed

STS-67 was the first advertised Shuttle mission connected to the Internet. Users of more than 200,000 computers from 59 countries logged on to the Astro-2 home page at Marshall Space Flight Center. More than 2.4 million requests were recorded during the mission, many answered by the crew in-orbit. See Table 3–92 for further mission details.

STS-71

This mission launched June 27 and landed July 7, 1995, at Kennedy Space Center. This flight marked the 100th U.S. human spaceflight and was the first in a series of Shuttle flights that docked with the Russian space station *Mir*. After docking on flight day three, *Mir* and *Atlantis* remained joined for five days. The seven-person Shuttle crew included two Russian cosmonauts who remained on *Mir* after *Atlantis* returned to Earth. Two other cosmonauts and the U.S.

astronaut Norman Thagard, who had flown to *Mir* aboard the Russian Soyuz spacecraft in March 1995, returned to Earth on *Atlantis* after more than 100 days in space. To ease their return to gravity, the three lay on their backs on specially designed seats installed in the orbiter's middeck. The returning crew of eight equaled the largest crew to fly on the Shuttle. The mission demonstrated the successful operation of the Russian-designed docking system, which was based on concepts used during the Apollo-Soyuz Test Project in 1975. See Table 3–93 for further mission details.

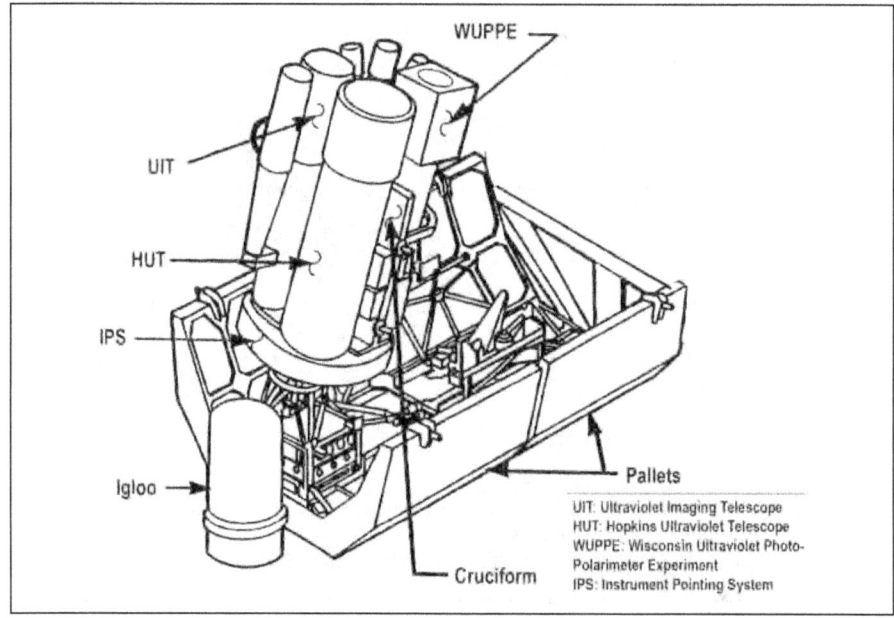

Figure 3–39. Astro-2 Suite of Instruments.

STS-70

This mission launched July 13 and landed July 22, 1995, from Kennedy Space Center. The TDRS-7 deployment marked completion of NASA's TDRS system, which provided communication, tracking, telemetry, data acquisition, and command services to the Shuttle and other low orbital spacecraft missions from geosynchronous orbit. STS-70 also marked the first flight of the new Block I Space Shuttle Main Engine. The engine featured a new high-pressure, liquid oxygen turbopump, two-duct powerhead, baffleless main injector, single-coil heat exchanger, and start sequence modifications that increased its stability and safety. See Table 3–94 for further mission details.

STS-69

This mission launched September 7 and landed September 18, 1995, at Kennedy Space Center. The Shuttle deployed the WSF-2, which, flying separately from the Shuttle, produced an "ultravacuum" in its wake and allowed experimentation in the production of advanced, thin film semiconductor materials. The WSF-2, deployed on flight day five, became the first spacecraft to maneuver itself away from the orbiter (rather than the other way around) by firing a small cold gas nitrogen thruster to move away from *Endeavour.*

The SPARTAN 201-03 also was deployed and retrieved. The SPARTAN's primary objective was to study the outer atmosphere of the Sun and its transition into the solar wind that constantly flows past Earth. The timing of the SPARTAN flight was intended to coincide with the passage of the Ulysses spacecraft over the Sun's north polar region to expand the range of data being collected about the origins of the solar wind.

During the spacewalk on this mission, which lasted 6 hours, 46 minutes, astronauts James Voss and Michael Gernhardt evaluated the thermal improvements made to their EVA suits and reported that they remained comfortable. They also tested a variety of tools and techniques perhaps necessary for ISS assembly. The spacewalk was the 30th EVA of the Shuttle program.

STS-69 also was the second flight of a "dog crew," a flight crew tradition that began on STS-53, on which both Walker and Voss flew. As Dog Crew II, each STS-69 astronaut adopted a dogtag or nickname: Walker was Red Dog; Cockrell was Cujo; Voss, Dog Face; Newman, Pluto; and Gernhardt, Under Dog. See Table 3–95 for further mission details.

STS-73

This mission launched October 20 and landed November 5, 1995, at Kennedy Space Center. USML-2, the second United States Microgravity Laboratory, was the primary payload. Some of the experiments resulted from the outcome of investigations on USML-1, which flew aboard *Columbia* on STS-50. The research during USML-2 concentrated on the same overall areas as USML-1, and many experiments flew for the second time. Research was conducted in five areas: fluid physics, materials science, biotechnology, combustion science, and commercial space processing. Two teams of crew members worked around-the-clock in the 23-foot (7-meter) Spacelab module located in *Columbia*'s payload bay.

The crew took time out from Spacelab work to tape the ceremonial first pitch for Game 5 of the Major League Baseball World Series, marking the first time the thrower was not actually in the ballpark for the pitch. This was the second longest Shuttle flight to date. See Table 3–96 for further mission details.

STS-74

This mission launched November 12 and landed November 20, 1995, at Kennedy Space Center. It was the second in a series of dockings with *Mir*. The mission marked the first time that astronauts from ESA, Canada, Russia, and the United States were in space on the same complex at one time.

Unlike the first docking flight during which a crew exchange took place, the second docking focused on delivery of equipment to *Mir*. The primary payload of the mission was the Russian-built Docking Module (DM), designed to become a permanent extension on *Mir* to afford better clearances for Shuttle-*Mir* linkups. Two solar arrays were stowed on the DM for later transfer to *Mir* by spacewalking cosmonauts. See Table 3–97 for further mission details.

STS-72

This mission launched January 11 and landed January 20, 1996, at Kennedy Space Center. The crew of STS-72 captured and returned to Earth a Japanese microgravity research spacecraft, the Space Flyer Unit, which had been launched by Japan in March 1995. The mission also deployed and retrieved the OAST-Flyer spacecraft, the seventh in a series of missions aboard reusable, free-flying SPARTAN carriers. The flight also included two spacewalks to test hardware and tools to be used during ISS assembly. See Table 3–98 for further mission details.

STS-75

The mission launched February 22 and landed March 9, 1996, at Kennedy Space Center. The mission was the 50th Shuttle flight since NASA's return to flight following the *Challenger* accident and the 75th Shuttle flight. The mission was a reflight of the TSS (see STS-46). The tether broke three days into the mission, just short of its full deployment length, resulting in the loss of the Italian satellite.

The other primary Shuttle payload was USMP-3, the third United States Microgravity Payload. The payload included U.S. and international experiments, all of which had flown at least once before. See Table 3–99 for further mission details.

STS-76

This mission launched March 22, 1996, from Kennedy Space Center and landed March 31 at Edwards Air Force Base. The mission, Shuttle-*Mir* Mission 3, featured the third docking of the Space Shuttle *Atlantis* and the Russian space station *Mir*. Docking occurred between the ODS in the forward area of the *Atlantis* payload bay and the DM installed during STS-74 on *Mir*'s Kristall module docking port. The mission included a spacewalk, logistics operations, and scientific research. About 1,500 pounds (680 kilograms) of water and two tons of scientific equipment, logistical material, and resupply items were transferred from *Atlantis* to *Mir*, including a gyrodyne, transformer, batteries, food, water, film, and clothing. Experiment samples and miscellaneous equipment were brought to *Atlantis* from *Mir*. Astronaut Shannon Lucid, the second U.S. astronaut and the first U.S. woman to live on the Russian space station, began what turned out to be a marathon stay on *Mir* of four and one-half months, eclipsing the previous record set by Norman Thagard.

STS-76 marked the first flight of a SPACEHAB pressurized module to support Shuttle-*Mir* dockings. The single module served primarily as stowage area for a large supply of equipment slated for transfer to the Space Station. It also carried ESA's Biorack experiment rack for on-orbit research.

This mission experienced an unusual anomaly on the orbiter's ride back from Edwards Air Force Base to Kennedy Space Center after the Shuttle flight. A warning light for an engine on the Shuttle Carrier Aircraft 747 indicated an engine fire. The plane returned to Edwards Air Force Base, and the engine was replaced before the journey recommenced. See Table 3–100 for further mission details.

STS-77

This mission launched May 19 and landed May 29, 1996, at Kennedy Space Center. This was the first flight that used three Block I main engines and the first Shuttle mission controlled from the new Mission Control Center at Johnson Space Center. The new facility replaced the Apollo-era complex that had been used for previous Shuttle missions.

The mission was highlighted by four rendezvous activities with two different payloads: deployment and retrieval of the Passive Aerodynamically Stabilized Magnetically Damped Satellite (PAMS), one of four Technology Experiments for Advancing Missions in Space (TEAMS), and of the SPARTAN 207/Inflatable Antenna Experiment (IAE) satellite. During its 90-minute mission, the IAE tested the performance of a large inflatable antenna, laying the groundwork for future technology development on inflatable space structures. At the end of the mission, the crew jettisoned the antenna structure and stowed the spacecraft.

The six-person *Endeavour* crew also performed microgravity research aboard the SPACEHAB module. The single module carried almost 3,000 pounds (1,361 kilograms) of experiments and support equipment for 12 commercial space product development payloads in the areas of biotechnology, electronic materials, polymers, and agriculture. One of the additional payloads, the Commercial Float Zone Facility, was an international collaboration between the United States, Canada, and Germany. See Table 3–101 for further mission details.

STS-78

This mission launched June 20 and landed July 7, 1996, at Kennedy Space Center. Five space agencies (NASA, ESA, the French Space Agency—CNES, the CSA, and the Italian Space Agency) and research scientists from 10 countries worked together on the LMS Spacelab, which built on previous Shuttle Spacelab flights dedicated to life sciences and microgravity investigations. More than 40 experiments were flown and grouped into the areas of life sciences, which included the following: human physiology and space biology; microgravity science (including basic fluid physics investigations and advanced semiconductor and metal alloy materials processing); and medical research in protein crystal growth. The investigations focused on the effects of long-duration spaceflight on human physiology, and crew members conducted the types of experiments that would fly on the ISS. LMS investigations were conducted via the most extensive telescience to date. Investigators were located at four remote European and four remote U.S. locations, similar to what would happen with the ISS. The mission also made extensive use of video imaging to help crew members perform in-flight maintenance procedures on experiment hardware. This was the longest Shuttle flight flown, lasting almost 17 days. See Table 3–102 for further mission details.

STS-79

This mission launched September 16 and landed September 26, 1996, at Kennedy Space Center. On this mission, astronaut Shannon Lucid set the world's women's and U.S. records for length of time in space: 188 days and 5 hours. The mission was the fourth Shuttle docking with the *Mir* space station and the first exchange of U.S. crew aboard a Russian spacecraft. Lucid returned to Earth on *Atlantis*, and astronaut John Blaha replaced her on *Mir* for a planned four-month stay.

The mission also marked the second flight of the SPACEHAB module in support of Shuttle-*Mir* activities and the first flight of the SPACEHAB double module configuration. During five days of mated operations, the two crews transferred more than 4,000 pounds (1,814 kilograms) of supplies to *Mir*, including logistics, food, and water generated by orbiter fuel cells. Three experiments also were transferred: Biotechnology System (BTS) for study of cartilage development; Material in Devices as Superconductors (MIDAS) to measure electrical properties of high-temperature superconductor materials; and CGBA, which contained several smaller experiments including self-contained aquatic systems. About 2,000 pounds (907 kilograms) of experiment samples and equipment were transferred from *Mir* to *Atlantis*. The total logistical transfer to and from *Mir* of more than 6,000 pounds (2,722 kilograms) was the most extensive to date. See Table 3–103 for further mission details.

STS-80

This mission launched November 19 and landed December 7, 1996, at Kennedy Space Center. This was the third flight of the WSF, which had flown on STS-60 and STS-69. It was the second flight of the German-built ORFEUS-SPAS-2. Both the WSF and ORFEUS-SPAS were deployed and retrieved during the mission, making it the first time that two satellites were flying freely at the same time. ORFEUS-SPAS II was the third flight to use the German-built ASTRO-SPAS science satellite. The 1-meter (3.1-foot)-diameter ORFEUS-Telescope with the Far Ultraviolet (FUV) Spectrograph and the Extreme Ultraviolet (EUV) Spectrograph comprised the main payload attached to the ASTRO-SPAS framework. The Interstellar Medium Absorption Profile Spectrograph (IMAPS) was a separate instrument, IMAPS operated independently of the ORFEUS telescope. Another science payload was the Surface Effects Sample Monitor (SESAM), a passive carrier for state-of-the-art optical surfaces and potential future detector materials. The SESAM investigated the impact of the space environment on materials and surfaces in different phases of a Space Shuttle mission, from launch to orbit phase to reentry into Earth's atmosphere.

Two planned 6-hour EVAs were canceled because of a jammed outer airlock hatch. This flight again broke the record for the longest Shuttle flight, lasting slightly more than 17.5 days. See Table 3–104 for further mission details.

STS-81

This mission launched January 12 and landed January 22, 1997, at Kennedy Space Center. STS-81 was the fifth of nine planned missions to *Mir* and the second involving an exchange of U.S. astronauts. Astronaut Jerry Linenger replaced astronaut John Blaha aboard *Mir* after Blaha spent 118 days on *Mir* and 128 days in space. *Atlantis* carried the SPACEHAB double module, which provided additional middeck locker space for experiments. While the vehicles were docked, crews transferred nearly 6,000 pounds (2,722 kilograms) of logistics to *Mir*, including approximately 1,600 pounds (726 kilograms) of water, 1,138 pounds (516 kilograms) of U.S. science equipment, and 2,206 pounds (1,000 kilograms) of Russian logistical equipment. About 2,400 pounds (1,089 kilograms) of materials returned from *Mir* to Earth on *Atlantis*. See Table 3–105 for further mission details.

STS-82

This mission launched February 11 and landed February 21, 1997, at Kennedy Space Center. It was the second in a series of planned servicing missions to the Hubble Space Telescope. The orbiter's robot arm captured the Hubble Space Telescope so it could be serviced, and two teams of astronauts performed five spacewalks. The crew took more than 150 crew aids and tools on the mission, ranging from a simple bag for carrying some of the smaller tools to sophisticated battery-operated power tools. See Table 3–106 for further mission details.

STS-83

This mission launched April 4 and landed April 8, 1997, at Kennedy Space Center. This mission lasted only 4 days and returned to Earth 12 days early because of a problem with one of the fuel cells that provided electricity and water to the orbiter. The MSL-1 was rescheduled for STS-94. See Table 3–107 for further mission details.

STS-84

This mission launched May 15 and landed May 24, 1997, at Kennedy Space Center. This was the sixth docking with the *Mir* space station and the third involving an exchange of U.S. astronauts. Astronaut J. Michael Foale replaced astronaut Jerry Linenger, who had been in space 132 days. The mission resupplied materials for experiments to be performed aboard *Mir* and returned experiment samples and data to Earth. Altogether nearly 249 items were moved between the two spacecraft, with nearly 1,000 pounds (565 kilograms) of water moved to *Mir*, for a total of nearly 7,500 pounds (3,402 kilograms) of water, experiment samples, supplies, and hardware. See Table 3–108 for further mission details.

STS-94

This mission launched July 1 and landed July 17, 1997, at Kennedy Space Center. It was the reflight of MSL-1, which had flown on STS-83. The mission involved the same vehicle, crew, and experiment activities as planned on the earlier MSL-1 mission. The crew maintained 24-hour/two-shift operations. Using the Spacelab module as a testbed, the MSL-1 tested some of the hardware, facilities, and procedures that would be used on the ISS. The 33 investigations also yielded new knowledge in the fields of combustion, biotechnology, and materials processing. Scientists from NASA, ESA, the German Space Agency, and the National Space Development Agency of Japan contributed the 25 primary experiments, 4 glovebox investigations, and 4 accelerometer studies on MSL-1. A record number of commands—more than 35,000—were sent from the Spacelab Mission Operations Control Center at Marshall Space Flight Center to the MSL-1. See Table 3–109 for further mission details.

STS-85

This mission launched August 7 and landed August 19, 1997, at Kennedy Space Center. The CRISTA-SPAS-2 was the primary payload. It was deployed and, after more than 200 hours of free flight, was retrieved using *Discovery*'s robot arm. (See Figure 3–40 for a drawing of the SPAS-2.) CRISTA-SPAS-2 was the fourth in a series of cooperative ventures between the German Space Agency and NASA. This was the satellite's second flight. The satellite consisted of three telescopes and four spectrometers. The three CRISTA telescopes collected 38 full atmospheric profiles of the middle atmosphere. Two other instruments mounted on the SPAS also studied Earth's atmosphere. The MAHRSI obtained new vertical profile data on the distribution of hydroxyl (OH) and nitric oxide in the mesosphere and upper stratosphere conditions under very different (both seasonal and diurnal) from its previous flight on STS-66. The SESAM carried state-of-the-art optical surfaces to study the impact of the atomic oxygen and space environment on materials and services. Twenty-two sounding rockets and 40 balloons were launched to provide correlating data.

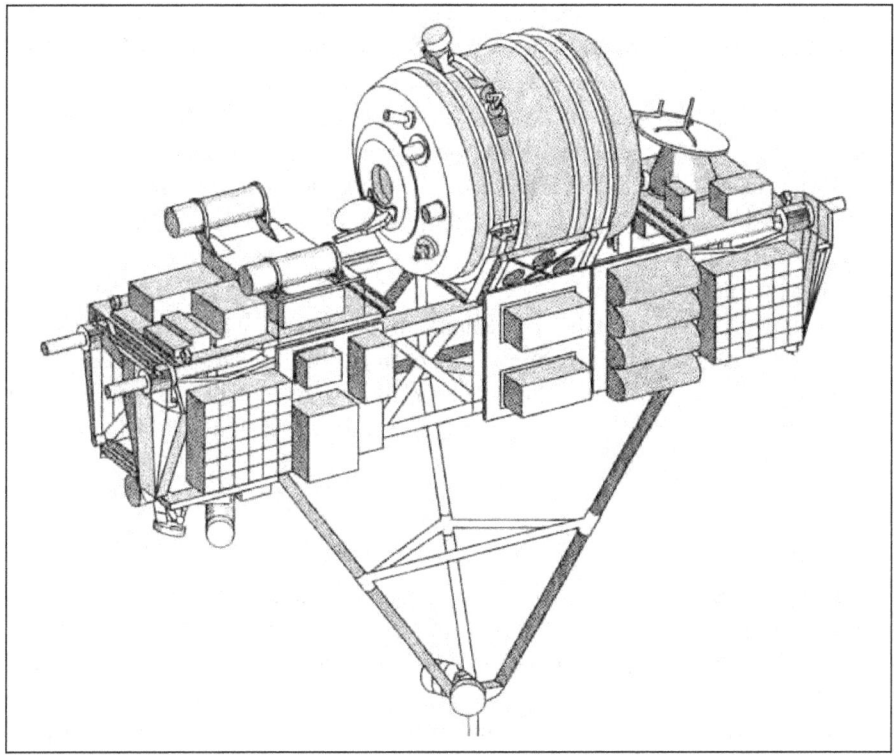

Figure 3–40. The SPAS-2 was a German-built, reusable free-flying vehicle that could be deployed and retrieved by the Space Shuttle's Remote Manipulator System. The original SPAS was used on STS-7 with materials processing and defense-related sensor payloads. The SPAS-2 was used on ORFEUS-SPAS and CRISTA-SPAS missions.

The Technology Applications and Science experiments, Manipulator Flight Demonstration supplied by Japan, and international Extreme Ultraviolet Hitchhiker were other mission payloads. The crew also worked with the Orbiter Space Vision System (OSVS), which would be used during ISS assembly. The OSVS featured a series of dots strategically placed on various payload and vehicle structures that permitted precise alignment and pointing capability. See Table 3–110 for further mission details.

STS-86

This mission launched September 25 and landed October 6, 1997, at Kennedy Space Center. It was the seventh docking between *Atlantis* and the *Mir* space station and the fourth exchange of U.S. astronauts. U.S. astronaut J. Michael Foale returned to Earth aboard *Atlantis* after a stay of 134 days on *Mir* and 145 days in space. His stay on *Mir* was the second longest spaceflight in U.S. history behind Shannon Lucid's 188-day flight in 1996. Foale was replaced by David Wolf.

The first joint U.S.-Russian EVA during a Shuttle flight took place on this mission. During a 5-hour, 1-minute spacewalk on October 1, Vladimir Titov and Scott Parazynski affixed a 121-pound (55-kilogram) Solar Array Cap to the docking module for future use by *Mir* crew members to seal off the suspected leak in Spektr's hull.[89] Parazynski and Titov also retrieved four MEEPs from the outside of *Mir* and tested several components of the SAFER jet packs.

Atlantis carried the SPACEHAB double module to support the transfer of logistics and supplies to *Mir* and the return of experiment hardware and specimens to Earth (see Figure 3–41). More than 4 tons (3,628 kilograms) of materials were transferred between SPACEHAB and *Mir*, including approximately 1,700 pounds (771 kilograms) of water; experiment hardware for ISS Risk Mitigation Experiments to monitor the *Mir* for crew health and safety; a gyrodone; batteries; three pressurization units with breathing air; an attitude control computer; and many other items. See Table 3–111 for further mission details.

STS-87

This mission launched November 19 and landed December 5, 1997, at Kennedy Space Center. It was the first time since 1992 that eight Shuttle flights were conducted in one year. The mission carried the USMP-4 and the SPARTAN 201-04 satellite as the primary payloads. It included experiments that studied how the weightless environment of space affected various physical processes. During this mission, payload specialist Leonid Kadenyuk became the first Ukrainian to fly aboard the Space Shuttle. Six minutes into the climb to orbit, *Columbia*'s computers commanded the orbiter to roll from an inverted position under its fuel tank to a "heads-up" position to provide early communications access to the TDRS system. That enabled NASA to phase out the Bermuda tracking station to save costs to the Shuttle program.

---

[89] The accident that caused the leak is described later in this chapter.

Figure 3–41. The SPACEHAB double module is lifted into the payload changeout room at Launch Pad 39-A for insertion into the payload bay of Atlantis. *On STS-86, about 3-1/2 tons (3,175 kilograms) of science and logistical equipment and supplies were exchanged between* Atlantis *and* Mir. *(NASA Photo No. KSC-97PC-1340)*

An unexpected event occurred when the attitude control system aboard the free-flying SPARTAN solar research satellite malfunctioned, causing the satellite to rotate outside the Shuttle. After unsuccessful attempts to capture the satellite using the orbiter's mechanical arm, crew members performed an unscheduled spacewalk lasting 7 hours, 43 minutes, successfully recapturing the satellite and lowering it onto its berth in the payload bay manually. The anomaly prevented all planned research on SPARTAN from being performed. A second spacewalk lasting 7 hours, 33 minutes tested a crane to be used for constructing the ISS and a free-flying camera to monitor conditions outside the Station without requiring EVAs. See Table 3–112 for further mission details.

STS-89

This mission launched January 22 and landed January 31, 1998, at Kennedy Space Center. The eighth *Mir*-Shuttle linkup and the fifth crew exchange took place. Astronaut David Wolf, who had been on *Mir* since September 1997 and had spent 128 days in space, was replaced by astronaut Andrew Thomas. In addition to using the SPACEHAB Logistics Double Module to supply *Mir* with more than 8,000 pounds (3,629 kilograms) of scientific equipment, logistical hardware, and water, the mission recovered the Optical Properties Monitor from *Mir*. This important experiment exposed material samples composed mostly of optical instruments and coatings to space conditions. See Table 3–113 for further mission details.

STS-90

This mission launched April 17 and landed May 3, 1998, at Kennedy Space Center. This was the 23rd and final Spacelab module flight, which had spanned the prior 15 years. The key science focused on Neurolab, a set of investigations relating to the effects of microgravity on the nervous system. The experiments studied vestibular system adaptation and space adaptation syndrome, adaptation of the central nervous system and the pathways that control the ability to sense location and orientation in the absence of gravity, and the effect of microgravity on a developing nervous system (Figure 3–42).

The mission was a joint venture of six space agencies and seven U.S. research agencies. Investigator teams from nine countries conducted 31 studies in the microgravity environment of space. The agencies participating in this mission included six institutes of the National Institutes of Health, the National Science Foundation, and the Office of Naval Research, as well as the space agencies of Canada, France, Germany, Japan, and the ESA. See Table 3–114 for further mission details.

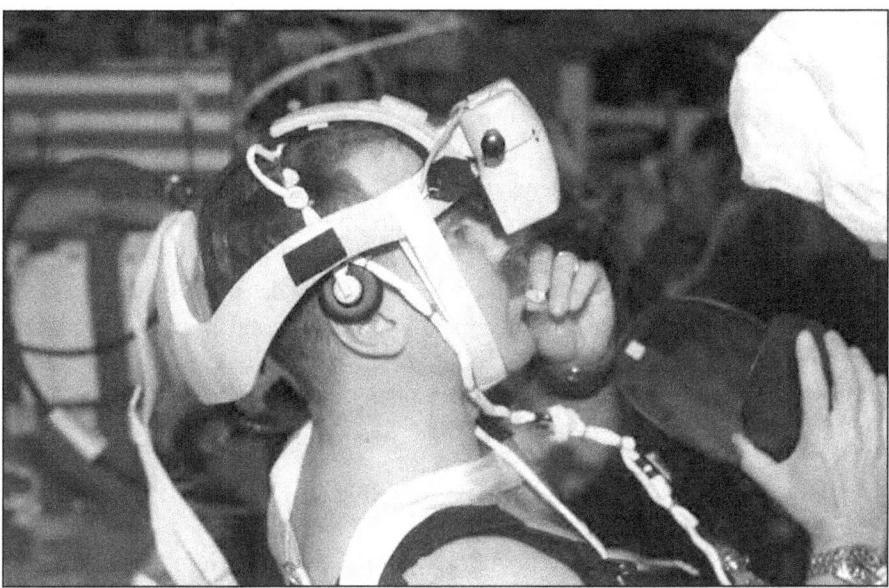

Figure 3–42. This Electronic Still Camera (ESC) image shows Dafydd R. "Dave" Williams, Mission Specialist, working with the Virtual Environment Generator (VEG), in the Neurolab on board Columbia, *on April 20, 1998. The VEG was used to discover how the balance between visual and vestibular cues shifts toward the visual system in weightlessness. The VEG was a head-mounted display that showed computer-generated virtual reality scenes generated by a three-dimensional graphics computer. (NASA Photo No. STS90-E-5041)*

STS-91

This mission launched June 2 and landed June 12, 1998, at Kennedy Space Center. It was the ninth and last *Mir* docking mission. It was the first docking mission for *Discovery*. Astronaut Andrew Thomas returned to Earth after completing 130 days of living and working on *Mir*. No U.S. astronaut was delivered to *Mir*. Thomas' transfer ended a total of 907 days spent by seven U.S. astronauts aboard the Russian space station as long-duration crew members.

Discovery carried the single SPACEHAB module in its payload bay. The module housed experiments performed by the astronauts and served as a cargo carrier for the items transferred to *Mir* and returned to Earth.[90] During the docked phase of STS-91, astronauts and cosmonauts transferred more than 1,100 pounds (500 kilograms) of water, and almost 4,700 pounds (2,132 kilograms) of cargo experiments and supplies were exchanged between the two spacecraft.

---

[90] *Mir* remained in orbit until March 23, 2001, when it returned to Earth after 86,331 total orbits. Five of *Mir's* modules were still pressurized at the time of deorbit and burst into flame as fragments fell into the South Pacific Ocean as ground controllers had planned. (Roger D. Launius, *Space Stations, Base Camps to the Stars* (Washington, DC: Smithsonian Institution, 2003), pp. 172–173. Between the final Shuttle-*Mir* docking and June 2000, the station remained crewed by Russian cosmonauts. In January 2001, a Progress cargo vehicle was launched in preparation for its March docking and deorbit of *Mir*. "Mir Chronicles," *http://www.russianspaceweb.com/mir_chronology.html* (accessed November 29, 2005). Also "Mir Space Station Observing," *http://satobs.org/mir.html* (accessed November 29, 2005).

STS-91 also carried into space the Alpha Magnetic Spectrometer (AMS) Investigation. The objectives of this investigation were to search for anti-matter and dark matter in space and to study astrophysics.

The mission was the first use of the super lightweight external tank (SLWT). This new tank was the same size, 154 feet long and 27 feet in diameter, (47 meters by 8.2 meters) as the external tank used on previous Shuttle launches but 7,500 pounds (3,401 kilograms) lighter. It was made of an aluminum lithium alloy, and the structural design had been improved, making the SWLT 30 percent stronger and 5 percent less dense. The walls of the redesigned hydrogen tank were machined in an orthogonal, waffle-like pattern, providing more strength and stability than the previous design. These improvements made additional payload capacity available to the ISS. See Table 3–115 for further mission details.

STS-95

This mission launched October 29 and landed November 7, 1998, at Kennedy Space Center. The mission conducted a variety of science experiments in the pressurized SPACEHAB module, deployed and retrieved the SPARTAN free-flyer payload, and carried out operations with the Hubble Space Telescope Orbiting Systems Test (HOST) and the IEH payloads in the payload bay. This mission was dubbed "the John Glenn Mission" because of its famous crew member. The scientific research mission returned space pioneer John Glenn to orbit 36 years, 8 months, and 9 days after he became the first American to orbit Earth. A battery of tests on Glenn and Pedro Duque furthered research on how the absence of gravity affected balance, perception, immune system response, bone and muscle density, metabolism and blood flow, and sleep.

The HOST provided a unique opportunity to test key pieces of new Hubble Space Telescope hardware before installation on future servicing missions. By flying the Shuttle in an orbit similar to Hubble Space Telescope's, the HOST allowed engineers to determine how the new equipment would perform on the telescope. HOST engineers monitored the effects of radiation on Hubble Space Telescope's new hardware, including an advanced computer, digital data recorder, and cryogenic cooling system. All the new technologies on the HOST mission performed as expected.

The SPARTAN spacecraft investigated physical conditions and processes of the hot outer layers of the Sun's atmosphere, or solar corona. While deployed from the Shuttle, SPARTAN gathered measurements of the solar corona and solar wind. This was a reflight of the SPARTAN payload flown on STS-87 that had developed problems soon after deployment from the Shuttle. See Table 3–116 for further mission details.

STS-88

This mission, the last in the 1989–1998 decade, launched December 4 and landed December 15, 1998, at Kennedy Space Center. This mission marked the start of ISS assembly when the U.S. module Unity mated with the Russian Zarya module that had been launched by a Proton rocket on November 20.[91] Astronauts Jerry Ross and James Newman conducted three spacewalks to attach cables, connectors, and hand rails. The two modules were powered up after the astronauts' entry.

Ross and Newman met other EVA objectives as they tested a SAFER unit, a self-rescue device to aid a spacewalker who becomes separated from the spacecraft during an EVA. They also nudged two undeployed antennas on Zarya into position; removed launch restraint pins on Unity's four hatchways for mating future additions of Station modules and truss structures; installed a sunshade over Unity's two data relay boxes to protect them from harsh sunlight; stowed a tool bag on Unity and disconnected umbilicals used for the mating procedure with Zarya; installed a handrail on Zarya; and made a detailed photographic survey of the Station.

Astronauts completed assembly of an early S-band communications system that allowed flight controllers in Houston to send commands to Unity's systems and keep tabs on the health of the Station, and conducted a successful test of the videoconferencing capability of the early communications system that the first permanent crew would use. Astronauts Sergei Krikalev and Nancy Currie also replaced a faulty unit in Zarya.

A new spacewalk record was established as Ross completed his seventh walk, totaling 44 hours, 9 minutes. Newman moved into third place with four EVAs totaling 28 hours, 27 minutes. See Table 3–117 for further mission details.

Space Station

Overview[92]

The Space Station is the largest and most complex international scientific project in history. Space Station development began in 1984 with President Ronald Reagan's call to create a permanent human presence in space. Called *Freedom* in its early planning stages, *Alpha* in 1993, and finally the ISS, assembly began in 1998 with the launch of the first two components, the Unity and Zarya modules. Led by the United States, the ISS has drawn upon the scientific and technological resources of 16 nations: Canada, Japan, Russia, 11 ESA member nations, and Brazil. The United States is responsible for

---

[91] "Zayra" means "sunrise." See the section describing Space Station development later in this chapter for more details of this mission.

[92] Most material in the overview came from "The International Space Station: An Overview," *NASA Facts,* IS-1999-06-ISS022, June 1999, *http://spaceflight.nasa.gov/spacenews/factsheets/pdfs/issovw.pdf* (accessed June 28, 2005).

developing and ultimately operating the major elements and systems aboard the Station. Beginning in 1993, Russia has been a prime partner in Space Station development, contributing both Space Station elements and knowledge gleaned from years of long-duration spaceflight.

The completed ISS, as configured in 1999, will have a mass of about 1 million pounds (453,592 kilograms), more than four times as large as the Russian *Mir* space station. It will measure about 360 feet (110 meters) across and 290 feet long (88 meters), with almost an acre of solar panels to provide electrical power to six laboratories. The first two ISS modules, the Russian-launched Zarya control module and the U.S.-launched Unity connecting module, were assembled in orbit in late 1998. This orbiting two-module complex had a mass of more than 74,000 pounds (33,566 kilograms) and measured 76 feet long (23 meters) with a 78-foot (23.8-meter) wingspan of the solar arrays. The Station's internal pressurized volume was 4,635 cubic feet (131.2 cubic meters). By early 1999, about 500,000 pounds (226,796 kilograms) of Station components had been built at factories around the world.

The ISS orbits at an altitude of 250 statute miles (402 kilometers) with an inclination of 51.6 degrees. This orbit allows launch vehicles of all the international partners to reach the Station, providing the capability to deliver crews and supplies. The orbit also allows excellent Earth observations with coverage of 85 percent of the globe and overflight of 95 percent of the population.

The program was organized into three phases since it became the ISS. The first phase of the ISS, the Shuttle-*Mir* program, began in 1995 and involved more than two years of continuous stays by U.S. astronauts on *Mir* and nine Shuttle-*Mir* docking missions. Seven U.S. astronauts spent a cumulative total of 32 months aboard *Mir* with 28 months of continuous occupancy since March 1996. By contrast, it took the U.S. Space Shuttle fleet more than 12 years and 60 flights to achieve an accumulated one year in orbit.

The knowledge and experience NASA gained through the Shuttle-*Mir* program could not have been achieved in any other way. NASA acquired valuable skills in international crew training activities; operating an international space program; and meeting the challenges of long-duration spaceflight for astronauts and ground controllers. Dealing with the real-time challenges encountered during Shuttle-*Mir* missions also resulted in unprecedented cooperation and trust between members of the U.S. and Russian space programs that has enhanced ISS development.

Many of the research programs planned for the ISS benefit from longer times in space. It is envisioned that research in the Station's six laboratories will lead to discoveries in medicine, materials, and fundamental science that will benefit people around the world. Through its research and technology, the ISS also will serve as an indispensable step in preparing for future human space exploration.

See Table 3–131 for a chronology of Space Station development.

ISS Partners

A worldwide team consisting of the United States, Canada, ESA, Japan, Russia, Italy, and Brazil is providing components for the ISS.

The United States, through NASA, is the initiator, integrator, and leader of the ISS effort. The United States is contributing the truss structures making up the Station's framework; four pairs of large solar arrays; three connecting modules, or nodes, with ports for spacecraft and for passage to other ISS elements; a cupola; an unpressurized logistics carrier; and an airlock accommodating U.S. and Russian spacesuits. NASA is also furnishing laboratory, habitation, and centrifuge accommodation modules.

NASA's integrated services include thermal control; power; environmental control and life support; communications, tracking, and data handling services; guidance, navigation, and control; and crew health maintenance as well as ground operations and launch site processing facilities.

Canada's CSA is providing the Mobile Servicing System, a 55-foot (16.8-meter), 125-ton (113,398-kilogram)-capacity robotic arm called the Space Station Remote Manipulator System (SSRMS), as well as a 12-foot (3.7-meter) Special Purpose Dexterous Manipulator (SPDM) arm. The Mobil Servicing System will aid in ISS assembly and maintenance. Canada will also supply the Space Vision System, a Shuttle-tested advanced camera to assist astronauts in viewing the SSRMS.

The European Space Agency comprises Belgium, Denmark, France, Germany, Italy, the Netherlands, Norway, Spain, Sweden, Switzerland, and the United Kingdom. The ESA is providing the Columbus Orbital Facility to be launched on the Ariane 5 expendable launch vehicle and the Automated Transfer Vehicle. The ESA was cooperating on development of the X-38 Crew Return Vehicle (canceled in 2002).

The Columbus Orbital Facility will carry 10 refrigerator-size racks for holding experiments, half of them European research projects. The Automated Transfer Vehicle will be used for logistics and propellant resupply as well as for reboost of the ISS.

Japan's National Space Development Agency is providing the Japanese Experiment Module. This experiment module houses the pressurized module, Exposed Facility, a remote manipulator system, and an Experiment Logistic Module. The pressurized module comprises a laboratory to accommodate 10 racks for holding experiments. The Exposed Facility is an external platform for up to 10 unpressurized experiments in the vacuum of space.

The 32-foot (9.8-meter) remote manipulator system will be used for servicing the Exposed Facility system and for changing payloads. The Experiment Logistic Module will be used for pressurized and unpressurized logistics resupply missions.

The Russian Space Agency is supplying about one-third the mass of the ISS in the form of a service module, Universal Docking Module, Science Power Platform, Docking Compartment, and research modules. The service module provides early living quarters for ISS crews, while the Universal Docking Module provides docking for both Russian and U.S. space vehicles. The Russian Space Agency provides crew transfers on the Soyuz and logistics resupply, Station reboosting, and orientation adjustments with its Progress and other vehicles. Russia built the first ISS element launched into orbit, the U.S.-funded Zarya.

Italy is participating as part of the ESA as well as independently providing three Multi-Purpose Logistics Modules through the ASI. The modules will be used on the Shuttle to carry pressurized cargo and payloads to the ISS. The structural design of the modules forms the basis for the design of the ESA's Columbus Orbital Facility. The agency also will supply Nodes 2 and 3 to NASA.

The Instituto Nacional de Pesquisas Espaciais (INPE) in San Jose dos Campos, Brazil, will provide six items under the direction of the Brazilian Space Agency, Agencia Espacial Brasileira (AEB). These constitute attachment devices and a pallet on which experiments and equipment will ride in Shuttle missions to the ISS. Brazil's Technology Experiment Facility will provide long-term space exposure for selected experiments, while Window Observation Research Facility 2 will be devoted to observation and remote sensing development.

Background

In January 1984, in his State of the Union address, President Ronald Reagan called for NASA "to develop a permanently manned space station and to do it within a decade."[93] From that day on, NASA committed to building a Space Station, then with a 1994 completion date on the calendar. The Agency created the necessary organizational structure and began work on the baseline concept. NASA stated in April 1988 that "the Space Station is essential if the United States is to maintain preeminence in key areas of civil space activities during the 1990s and beyond."[94] By the end of 1988, President Ronald Reagan had named the orbiting structure *Freedom*; and NASA had formed an international partnership with nine European nations; Canada; and soon Japan; as well as their respective space agencies; the ESA; the CSA; and Japan's NASDA. These alliances pledged cooperation during the detailed design, development, and operation and utilization phases of the Space Station program and agreed to provide the components of this modular orbiting laboratory.

---

[93] "State of the Union Message, January 25, 1984," *Public Papers of the Presidents of the United States: Ronald Reagan, 1984* (Washington, DC: Government Printing Office, 1986), pp. 87–95.

[94] Office of Space Station, National Aeronautics and Space Administration, "Space Station Capital Development Plan, Fiscal Year 1989," Submitted to the Committee on Science, Space and Technology, U.S. House of Representatives and the Committee on Commerce, Science and Transportation, U.S. Senate, April 1988, p. 1.

At the same time, NASA chose its prime contractor team of Boeing Aerospace, McDonnell Douglas, General Electric, and the Rocketdyne Division of Rockwell International for the implementation and execution phases, hardware development, and advanced design. NASA awarded four 10-year contracts with a total value of approximately $6.7 billion to correspond to the four "work packages" centered at four NASA Centers: Marshall Space Flight Center, Johnson Space Center, Goddard Space Flight Center, and Lewis Research Center. The Station was considered a facility that would "allow evolution in keeping with the needs of Station users and the long-term goals of the United States."[95]

Since 1987, the Station had been planned for completion in two phases. Phase I, known as the "revised baseline configuration," included a single horizontal boom, U.S. laboratory and habitat modules, accommodation for attached payloads, U.S. and European polar platforms, 75 kilowatts of photovoltaic power, European and Japanese laboratory modules, the Canadian Mobile Servicing System, and provisions for evolution. An earlier structural configuration, consisting of a dual keel with additional accommodations for attached payloads, had been moved to Phase II, known as the "enhanced configuration." Under the 1987 plan, the dual keel would be added only when support requirements of the attached payloads exceeded the capacity of the original solar panel truss. NASA had changed from the earlier single-phase, dual-keel plan to two phases because planners doubted that the Space Shuttle could schedule enough flights within the available time to deliver the truss elements needed for a dual keel.[96] Phase II also included an additional 50 kilowatts of power from the solar dynamic system, the satellite servicing facility, and the U.S. co-orbiting laboratory satellite.[97] NASA also decided to give the Station a "man-tended" status in Phase I and upgrade to "permanent habitability" after Phase I was complete.[98] Phase II was initially planned for sometime after the 20th assembly flight in early 1998, but it never received funds and was postponed indefinitely. Table 3–118 shows the Phase I and II contractors, tasks, and contract values.

Space Station Freedom

Space Station *Freedom* development did not proceed smoothly. Congress balked at the rising cost, which by 1987 had grown by more than 80 percent to $14.5 billion in FY 1984 dollars.[99] NASA also slipped the goal of a permanently

---

[95] "Fact Sheet, Presidential Directive on National Space Policy, February 11, 1988," (actual policy statement was classified), *http://www.fas.org/spp/military/docops/national/policy88.htm* (accessed March 1, 2005).

[96] David M. Harland and John E. Catchpole, *Creating the International Space Station* (Chichester, UK: Springer-Praxis Books, 2002), p. 118.

[97] U.S. General Accounting Office, *Space Station: NASA's Search for Design, Cost, and Schedule Stability Continues*, GAO/NSIAD-91-125, March 1991, p. 23, *http://archives.gao.gov/d21t9/143481.pdf* (accessed June 1, 2005).

[98] Man-tended referred to short-term occupation of the Station while a Shuttle orbiter or Soyuz was docked at the Station. Permanent habitability meant a continuous presence on the Station.

[99] Launius, pp. 134–136.

occupied Space Station, as stated by President Ronald Reagan, to 1995 and the assembly completion date from 1994 to early 1997.[100] Support by the science community also was uneven as some questioned the value of the Station for scientific research and worried that money spent on *Freedom* would reduce the amount available for other scientific pursuits. Consequently, Space Station *Freedom* was redesigned several times in an effort to reduce the price and streamline construction.

Congress kept tight control over Space Station funding, insisting that NASA request funds annually rather than appropriate funds the project could use over several years. This forced NASA to repeatedly justify the Station's cost, causing increased friction between NASA and Congress. In 1988, when introducing the FY 1990 budget request, NASA Administrator James Fletcher stated that the proposed cost was as low as possible and there was no room for further reductions. Congress still was unwilling to provide adequate funding, and Fletcher resigned on April 8, 1989. Dale Meyers, Fletcher's deputy, briefly became Administrator until the President appointed Richard Truly later in the year. In June, James Odom, Space Station program head only since March 1988, retired, and Truly appointed former astronaut William Lenoir to the job. Lenoir was also given the task of working out the consolidation of the Office of Space Station and Office of Space Flight, the organization managing the Space Shuttle.

As of April 1989, Space Station *Freedom* was planned as a 476-foot (145-meter) main truss assembly. Components included:

- A U.S. laboratory module
- A habitation module that would allow a continuous human presence
- A European attached pressurized module
- A Japanese experiment module
- Four resource nodes
- One standard and one hyperbaric airlock
- A logistics carrier
- A flight telerobotic servicer
- A Canadian mobile servicing system
- Attached payload accommodations equipment
- A propulsion assembly

The laboratories would provide for extensive science, applications, and technology development. There would also be provisions for external attached payloads, and three additional free-flyer spacecraft would be provided.[101]

---

[100] U.S. General Accounting Office, *Space Station: NASA's Search for Design, Cost, and Schedule Stability Continues*, GAO/NSIAD-91-125, March 1991, p. 21, *http://archive.gao.gov/d21t9/143481.pdf* (accessed 1 June 2005).

[101] National Aeronautics and Space Administration, Office of Space Station, *Space Station Freedom Capital Development Plan, Fiscal Year 1990*, April 1989, pp. 12–27 (NASA History Office Folder 009508).

This configuration did not survive. First, in June 1989, NASA began overhauling the Space Station assembly sequence. It decided to rely only on the current Space Shuttle capabilities for lifting and assembling Station components. The Agency abandoned the possibility of a Shuttle advanced solid rocket motor to increase the Shuttle's carrying capacity as well as the availability of an orbital maneuvering vehicle or a Shuttle-C vehicle.[102] Phase II was postponed indefinitely, and the polar platform transferred from the Space Station program to NASA's Office of Space Science and Applications for use in Earth observation studies.

Next, in July 1989, NASA formed a Configuration Budget Review team headed by W. Ray Hook of Langley Research Center. The team established three control boards that meticulously reviewed the program to develop preliminary options for ways the program could exist within the severe budget constraints threatened by Congress. The team presented these options to NASA Space Station management and the international partners.[103]

Based on this major program review, NASA announced late in 1989 a "rephasing" of the program to meet an anticipated budget cut of nearly $300 million for FY 1990 and "to reduce technical, schedule, and cost risk."[104] (Congress had in the meantime passed a FY 1990 funding bill in October 1989 to fund the Space Station program at $1.8 billion, $250 million less than the administration's $2.05 billion budget request.) Major program modifications included:

- Swapping the hydrogen-oxygen attitude control thrusters for more conventional hydrazine thrusters requiring little development cost.
- Eliminating a completely closed-loop environmental system that recycled *Freedom*'s water and air supply, instead shipping supplies to *Freedom* from Earth on the Shuttle.
- Rearranging the module layout to eliminate several interconnecting node modules.
- Eliminating two airlocks that would have provided redundancy and storage space for spacesuits and other EVA equipment and replacing them with a single airlock with hyperbaric treatment capability for treating decompression sickness.
- Eliminating two deployable booms that would have held the propulsion system and the communications and tracking antennas and adding two truss bays below the standard truss to hold that equipment, one on each end.

---

[102] The orbital maneuvering vehicle was planned as a reusable, remotely controlled free-flying "space tug." The Shuttle-C vehicle was a heavy-duty, uncrewed, Shuttle-like hauler of cargo.
[103] Harland and Catchpole, p. 122.
[104] "NASA Officials Make Some Changes to Space Station Freedom To Reduce Risk," *Station Break* 2, no. 1 (January 1990) (NASA History Office Folder 009522).

- Reducing the number of attachment fixtures for external payloads and utilities.[105]
- Scrapping development of new high-pressure spacesuits, leaving the crew with existing suits used for Shuttle EVA.[106]

These changes kept the first element launch scheduled for March 1995 but delayed the assembly complete date by 18 months. Figure 3–43 shows the proposed configuration for the completed Space Station as of 1991.

The various modules had been determined early in *Freedom*'s development. These remained in place throughout the various modifications that followed. The module cluster consisted of the U.S. laboratory and habitation modules, the ESA's attached pressurized module, and the Japanese Experiment Module.

The U.S. laboratory module (see Figure 3–44) was a pressurized, shirt-sleeve laboratory containing racks to house experiments and *Freedom*'s systems. The racks were a standard size (approximately as large as a refrigerator) to simplify replacement and for commonality with the racks in the other laboratory modules to be delivered later in the assembly. Of the 24 racks in the U.S. lab, 15 were allotted to the users to perform research and development activities. The remaining nine racks were for *Freedom*'s systems, such as the environmental control system and guidance and navigation systems. The U.S. habitation module (see Figure 3–45) served as living quarters for *Freedom*'s crew. It provided room for relaxation, personal hygiene, and exercise, as well as on-board medical facilities.

The attached pressurized module from the ESA (see Figure 3–46) and the Japanese Experiment Module laboratories (see Figure 3–47) provided 20 and 10 user racks, respectively, enabling investigations into material properties, fluid dynamics, and the behavior of living organisms in a weightless environment. The attached pressurized module provided a shirt-sleeve environment for astronauts and was equipped with power supply, thermal control, environmental control and life support, and data handling systems. The Japanese Experiment Module also provided a pressurized shirt-sleeve environment for astronauts and was equipped with power supply, thermal control, environmental control and life support, and data handling systems. An external platform, called the Exposed Facility (see Figure 3–48) would be attached to the rear of the Japanese Experiment Module. This facility provided additional attach ports for external payloads. The Japanese Experiment Module's robotic arm could replace or service payloads on the Exposed Facility.

The SSRMS and the SPDM were Canada's contribution to Space Station *Freedom* (see Figure 3–49). The SPDM had two 6-foot (1.8-meter) robotic arms for delicate tasks, such as connecting and disconnecting utilities,

---

[105] Billie Deason, "Budget-Minded Changes Alter Freedom Plans," *Space News Roundup*, NASA Johnson Space Center (April 27, 1990): 3 (NASA History Office Folder 009523).
[106] Robert Zimmerman, *Leaving Earth; Space Stations, Rival Superpowers, and the Quest for Interplanetary Travel* (Washington, DC: Joseph Henry Press, 2003), p. 222.

exchanging orbital replacement units, and assisting in Space Station assembly, maintenance, and repair activities. The SSRMS and the SPDM, together with the mobile transporter, made up the Mobile Servicing System. This system had lighting and video capabilities to assist astronauts in remote handling and visual inspection of payloads.

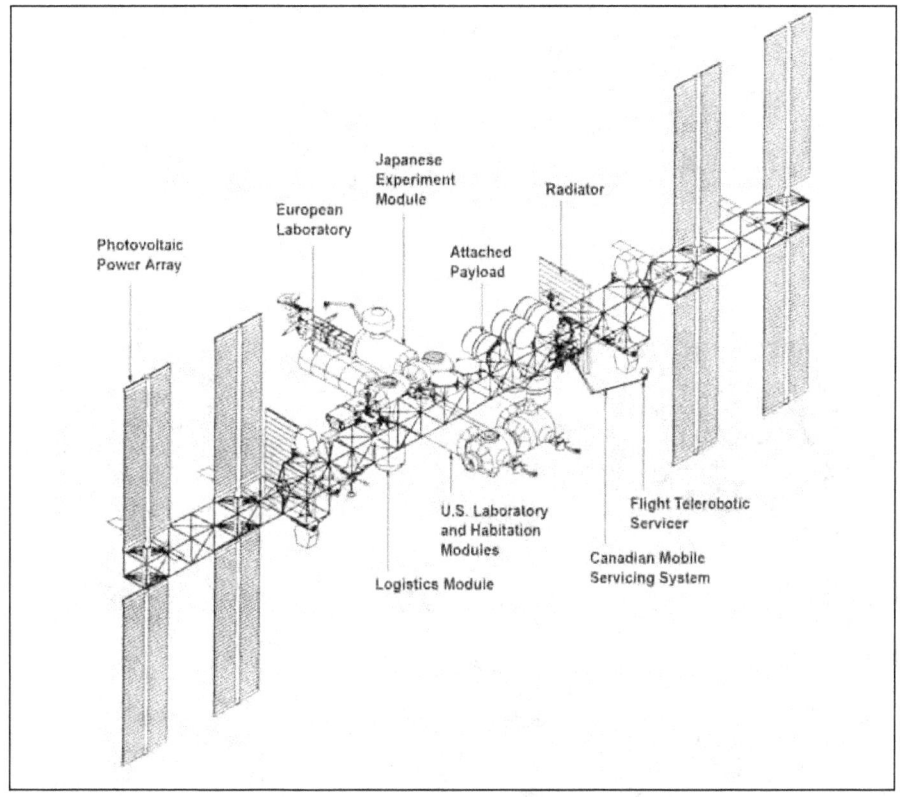

Figure 3–43. Space Station Freedom *Configuration, 1991. (NASA History Office Folder 009524)*

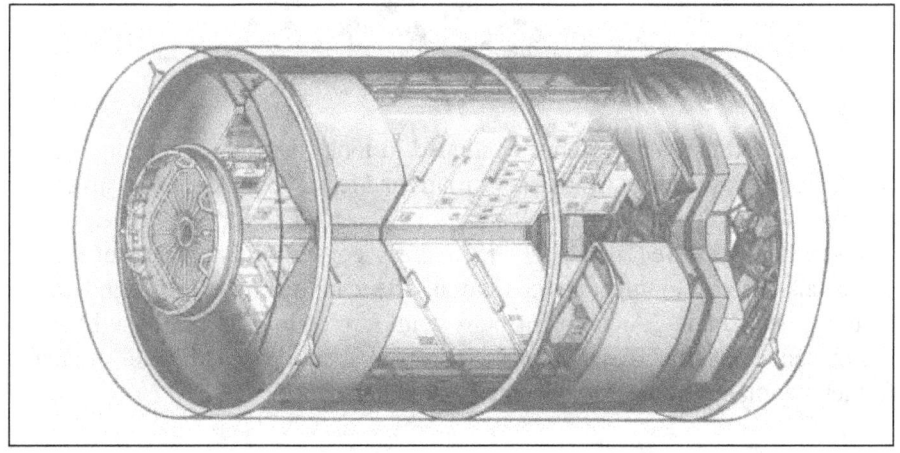

Figure 3–44. U.S. Laboratory Module.

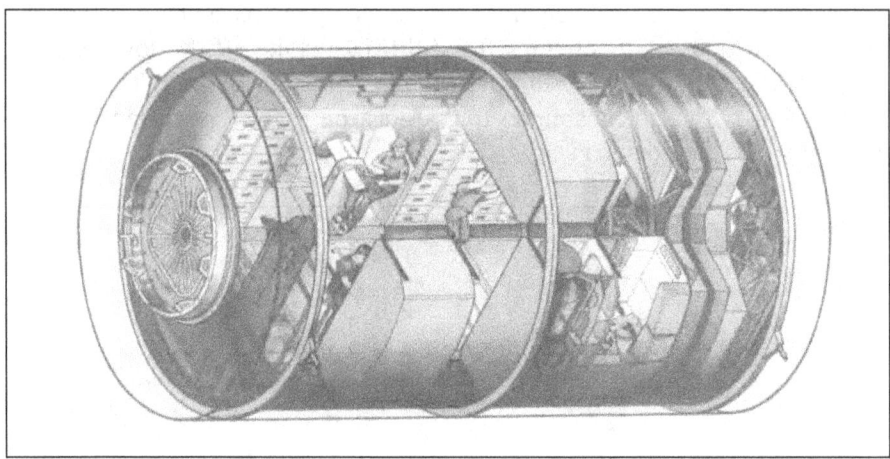

Figure 3–45. U.S. Habitation Module.

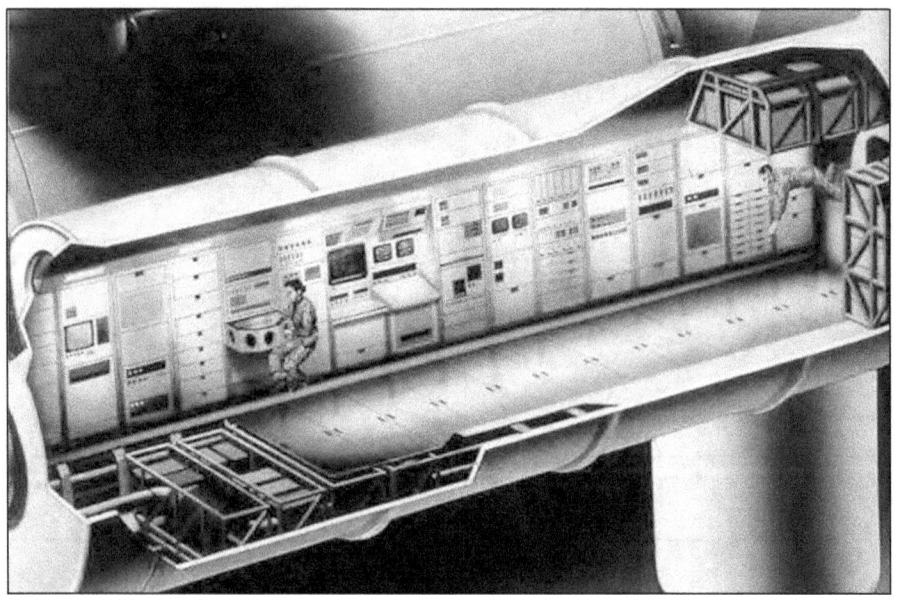

Figure 3–46. The ESA Attached Pressurized Module.

NASA and the White House continued to look for ways to reconfigure the Station to cut costs, forming advisory groups and teams to look at alternatives and propose recommendations. In January 1990, NASA formed the External Maintenance Task Team (EMTT) to address concerns about the number of spacewalks needed to maintain the Station. The team recommended significantly reducing annual EVA. A complementary team formed in June 1990, the External Maintenance Solutions Team, addressed problems raised by the EMTT and made further recommendations for reducing EVA maintenance.[107]

---

[107] *Space Station Freedom Media Handbook,* 1992, pp.18–19.

In the fall of 1990, the White House formed the Advisory Committee on the Future of the U.S. Space Program, chaired by Norman Augustine, to "assess alternative approaches and make recommendations for implementing future civil space goals" and to advise the NASA Administrator on overall approaches to implement the U.S. space program. The Committee had 120 days to make "a serious no-holds-barred" review of the space program and recommend improvements.[108] The Committee recommended that "steps should be taken to reduce the Station's size and complexity, permit greater end-to-end testing prior to launch, reduce transportation requirements, reduce extra-vehicular assembly and maintenance, and, where it can be done without affecting safety, reduce cost." The Committee also recommended revamping the program to emphasize life sciences and human space operations, including microgravity research as appropriate. Although Congress had given NASA only 90 days to "implement a revised space station design and assembly sequence," the Committee stated that this might prove inadequate and as much time as needed should be taken. The Committee also strongly recommended the immediate availability of a crew rescue vehicle.[109]

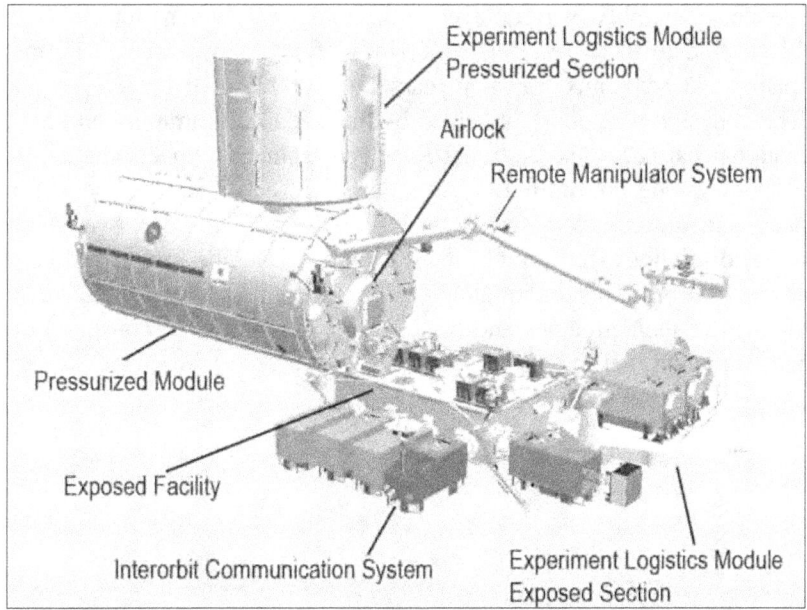

Figure 3–47. Japanese Experiment Module.

NASA seriously considered the reviews, recommendations, and direction from Congress, delivering to Congress in March 1991 a restructuring report laying out an extensively redesigned Station. It had a $30 billion price tag

---

[108] Harland and Catchpole, p. 124.
[109] Advisory Committee on the Future of the U.S. Space Program, "Report of the Advisory Committee on the Future of the U.S. Space Program, December 1990," *http://www.hq.nasa.gov/office/pao/History/augustine/racfup1.htm* (accessed March 15, 2005).

(including launches) but was smaller, easier to assemble in orbit, and would require fewer Shuttle flights to build.[110] The number of crew members on board was reduced from eight to four and the flight telerobotic servicer (FTS) was moved from the Space Station program to NASA's Office of Aeronautics, Exploration, and Technology. One large solar panel on *Freedom* was eliminated, reducing the panel's power from 75 kilowatts to 35 kilowatts. Following Congress's instructions for components that could be built in stages, the main truss was shortened to 353 feet (108 meters) and modified so it could be pre-integrated and tested with all subsystems before launch, reducing the EVA time needed to build and maintain the Station. The U.S. laboratory and habitation modules were shortened by 40 percent and also could be built, preassembled, and checked out on the ground. Because of the shortened truss, the facilities for large attached payloads were no longer needed and were canceled, although the hardpoints on the truss would still be used for small payloads. The cancellation of the FTS and the attached payload accommodation equipment eliminated Work Package 3, and NASA terminated its contract with GE Astro.[111] The schedule was also rephased. The first element launch was moved to early 1996. Man-tended capability was delayed until mid-1997, when docked Shuttles would be able to use *Freedom* for periods of up to two weeks. Permanent occupation was postponed for three years until 2000.[112]

This redesign was poorly received by the science community and NASA's international partners. The National Research Council's Space Studies Board stated that the redesign did not "meet the basic research requirements" for life sciences and microgravity research and applications, "the two principal scientific disciplines for which it is intended."[113] The modifications also displeased the Station's international partners both because of the delay in deployment of their modules and because they had not been consulted on the changes, a violation of their agreements.

[110] *Space Station Freedom Media Handbook*, 1992, p. 19. Also Marcia S. Smith, Congressional Research Service, testimony to the Science Committee, U.S. House of Representatives, "NASA's Space Station Program: Evolution and Current Status," 4 April 2001, *http://www.spaceref.com/news/viewsr.htm?pid-2562* (accessed June 3, 2005); also *http://www.house.gov/science/full/apr04/smith.htm* (accessed June 7, 2005).

[111] "Goddard Announces Contract Termination," *NASA News* Release 91-27, February 15, 1991, *ftp://ftp.hq.nasa.gov/pub/pao/pressrel/1991/91-027.txt* (accessed March 22, 2005). The contract for the payload accommodation equipment also included two polar platforms to be used for on-orbit research as part of the EOS. A new contract for these items was drawn up with GE Astro.

[112] Peter Bond, *The Continuing Story of the International Space Station* (London: Springer-Praxis Books, 2002), p. 114.

[113] National Research Council, Commission on Physical Sciences, Mathematics, and Applications, "Space Studies Board Position on Proposed Redesign of Space Station Freedom," March 29, 1991 (NASA History Office Folder 009524).

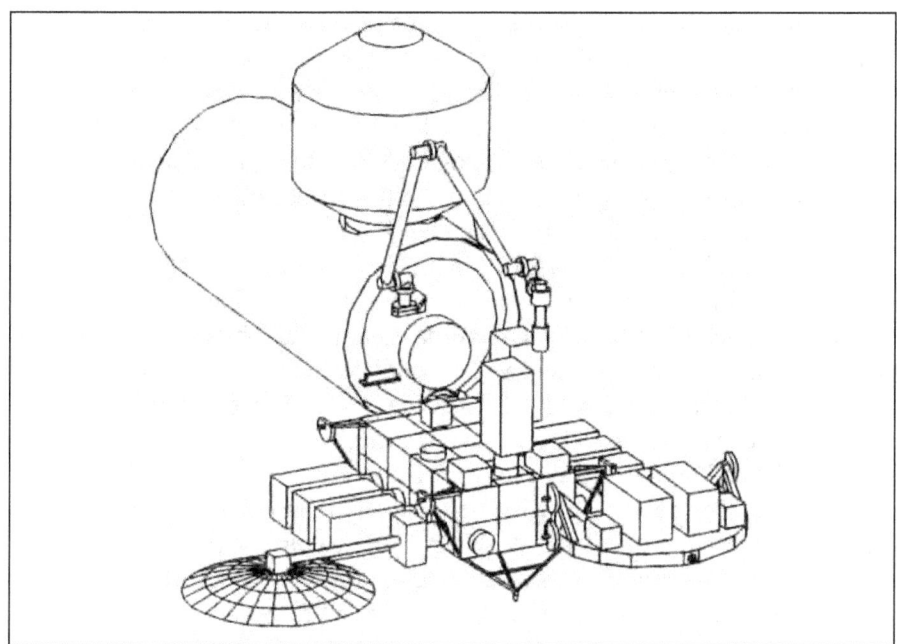

Figure 3–48. Japanese Experiment Module Exposed Facility.

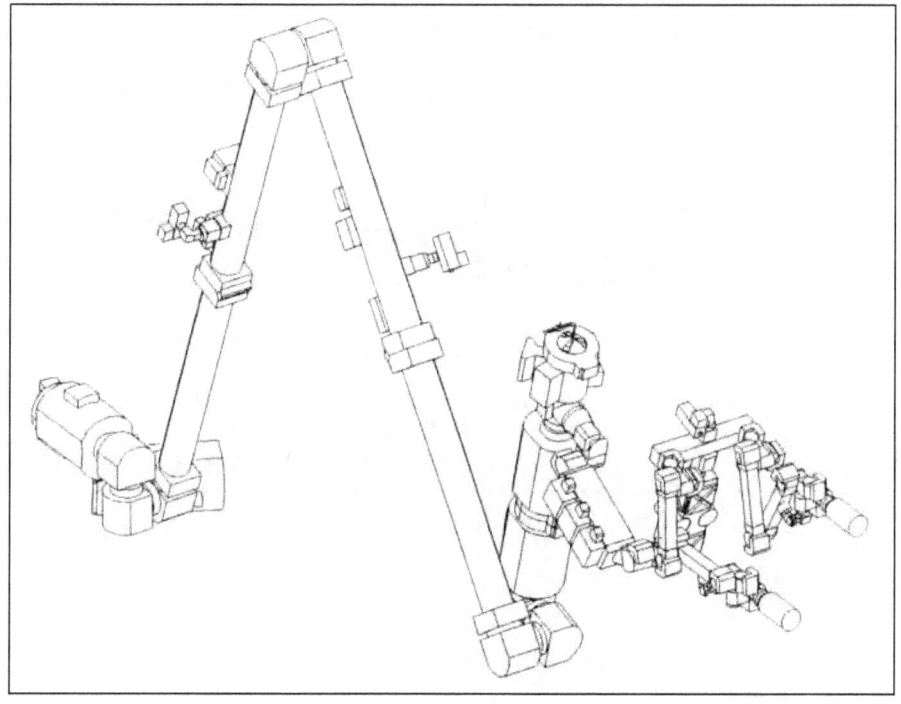

Figure 3–49. Canada's SPDM attached to the SSRMS.

Nevertheless, Vice President Dan Quayle and the National Space Council endorsed the report on March 21, 1991. After the House Appropriations Committee recommended cutting off all funding for *Freedom* and canceling it, the Senate, on September 27, 1991, agreed to the House bill and granted NASA its full FY 1992 funding request of $2,028,900,000 for Space Station *Freedom*. President George H. W. Bush signed the bill on December 9.[114] This vote to cancel the program was the first of many that the Station survived.

In spring of 1992, the NASA Space Station Freedom Office issued a strategic plan for the program.[115] The plan presented a "vision of what *Freedom* will accomplish, as well as its mission, goals, and objectives." The plan described the three-phase process with a separate "man-tended capability" and a "permanently manned capability." (See Figures 3–50 and 3–51). During the man-tended capability period, the crew would remain on-board *Freedom* only while the Space Shuttle was docked, returning to Earth with the Shuttle after each mission. The beginning of the permanently manned capability would be marked by the addition of the Assisted Crew Return Vehicle, to be added during 1999. Figure 3–52 shows the progression from Stage 1, first element launch, through Stage 6, man-tended capability, to Stage 17, permanently manned capability, as envisioned in July 1992. Assembly would take approximately four years, beginning in the fall of 1995 with the first element launch, and would require 18 mission build flights during that period to transport *Freedom*'s components into orbit.[116]

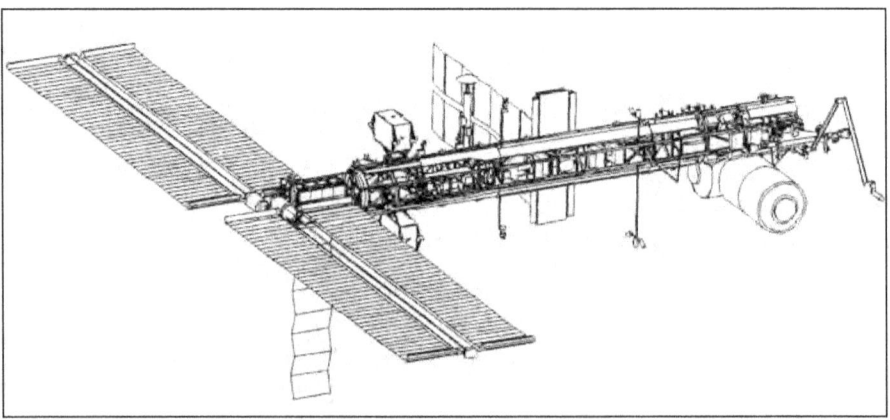

Figure 3–50. Man-Tended Capability, 1992.[117]

[114] *National Aeronautics and Space Administration Authorization Act, Fiscal Year 1992*, Public Law 102-195, 102nd Congress, 1st sess. (December 9, 1991).
[115] *NASA Space Station Freedom Strategic Plan 1992*, undated. (NASA History Office Folder 16941).
[116] *Space Station Freedom User's Guide*, August 1992, pp. 2-1–2-2. (NASA History Office Folder 009554).
[117] *NASA Space Station Freedom Strategic Plan 1992*, p. 9.

The Station would orbit from 335 kilometers (208 nautical miles) to 460 kilometers (285 nautical miles) above Earth at a 28.5-degree inclination. An orbit around Earth would take approximately 90 minutes. Table 3–119 lists Space Station *Freedom* characteristics as of May 1992.

Russian Involvement

While NASA and Congress were embroiled in budget battles and restructuring of the Space Station, the leaders of the Soviet Union and United States were discussing cooperation in space. Early in July 1991, soon after dissolution of the Warsaw Pact, Vice President Dan Quayle and Oleg Shishkin, minister of General Machine Building in the Soviet Union, met to discuss a venture in which the United States and Soviet Union could cooperatively use *Mir* for human spaceflight missions.[118] On July 31, at a summit meeting in Moscow, President George H.W. Bush and Soviet President Mikhail Gorbachev signaled their growing cordiality by signing an agreement for an astronaut to visit *Mir* and a cosmonaut to fly on the Space Shuttle. The two also discussed Russia's desire to enter the commercial space launch market.[119] In December, Gorbachev resigned after an unsuccessful coup staged by hard-liners and the disintegration of the Soviet Union. Boris Yeltsin became the head of the new Russian Federation.

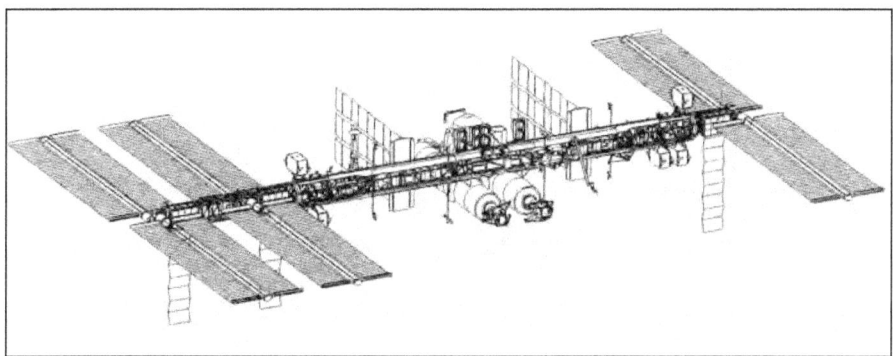

Figure 3–51. Permanently Manned Capability, 1992.

Soon after, in February 1992, President George H. W. Bush asked NASA Administrator Truly to resign. On April 1, Daniel Goldin assumed NASA's helm, inheriting a program that was behind schedule and over cost.[120] At roughly the same time, Yeltsin created the civilian Russian Space Agency headed by

---

[118] Launius, p. 152.

[119] John M. Logsdon, "Appendix B: The Evolution of U.S.-Russian Cooperation in Human Space Flight," in John M. Logsdon and James R. Millar, eds., *U.S.-Russian Cooperation in Human Space Flight: Assessing the Impacts* (Washington, DC: Institute for European, Russian and Eurasian Studies, The George Washington University, 2001), *http://www.gwu.edu/~spi/usrusappb.html* (accessed June 3, 2005).

[120] W. Henry Lambright, *Transforming Government: Dan Goldin and the Remaking of NASA*, The PricewaterhouseCoopers Endowment for The Business of Government, March 2001, p. 19.

Yuri Koptev. The two new agency heads met informally in Washington, DC, to discuss possibilities for cooperation. This meeting was followed by a summit between President George H. W. Bush and Yeltsin on June 17, 1992, in which the two agreed "to give consideration to" a joint mission. The two leaders signed the "Agreement Between the United States of America and the Russian Federation Concerning Cooperation in the Exploration and Use of Outer Space for Peaceful Purposes." The cooperation would include a "Space Shuttle and *Mir* Space Station mission involving U.S. astronauts and Russian cosmonauts." The leaders also agreed to a Shuttle flight by Russian cosmonauts in 1993, a flight on a long-duration mission on *Mir* by a U.S. astronaut in 1994, and a docking mission between the Shuttle and *Mir* in 1995.[121]

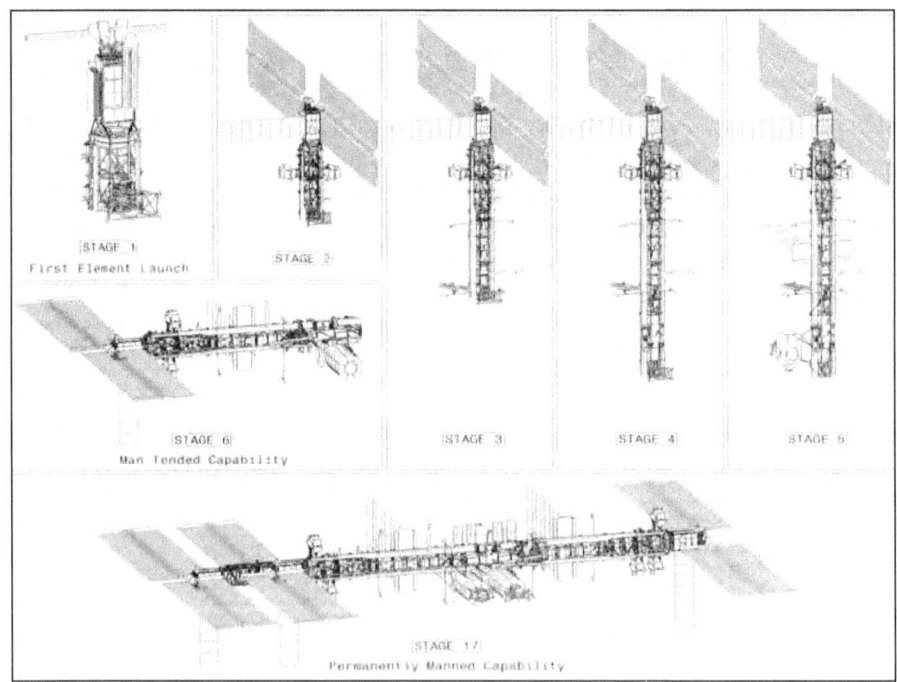

Figure 3–52. Space Station Freedom *Assembly Stages as Envisioned in July 1992, Showing the progression from Stage 1, First Element Launch, through Stage 6, Man-Tended Capability, to Stage 17, Permanently Manned Capability. (Grumman)*

On June 18, 1992, Russia and the United States formally signed a new U.S.-Russian Space Cooperation Agreement and ratified the first contract between NASA and the Russian aerospace firm NPO-Energia, a quasi-independent industrial conglomerate that ran the *Mir* space station. The agreement called for "a rendezvous [and] docking mission between the *Mir* and the Space Shuttle in 1994 or 1995"; "detailed technical studies of the

---

[121] "How 'Phase 1' Started," Shuttle-*Mir* Background, *http://spaceflight.nasa.gov/history/shuttle-mir/history/h-b-start.htm (*accessed June 3, 2005).

possible use of [Russian] space technology" for U.S. missions, including Space Station *Freedom*; and "steps to encourage private companies to expand their search for new commercial space business."[122] The one-year contract with NPO-Energia, valued at $1 million, was to study applications of Russian space technology to the Space Station Freedom program. NASA also expressed interest in the potential use of the Soyuz as a crew rescue vehicle for Space Station *Freedom*, of Russia's automated rendezvous and docking system known as Androgynous Peripheral Docking Assembly being used with *Mir*, and of the *Mir* for long lead-time life sciences experiments in support of the Space Station Freedom program.[123]

On October 5, 1992, NASA and the Russian Space Agency signed an "Implementing Agreement Between the National Aeronautics and Space Administration of the United States of America and the Russian Space Agency of the Russian Federation on HSF Cooperation." This agreement detailed the cooperation that had been called for in the June 1992 agreement and the necessary legal and other provisions associated with the cooperation. Particulars included an exchange of cosmonauts and astronauts on each other's spacecraft, with U.S. astronauts delivered to *Mir* by Soyuz, spending more than 90 days there, and returning on the Shuttle, Russian cosmonauts on *Mir* being "changed out" on the same Shuttle flight that would deliver a U.S. astronaut; and evaluation of the Russian Androgynous Peripheral Docking Assembly. The joint effort was named "the Shuttle-*Mir* Program."[124] The United States later proposed expanding the program to include more docking missions between the Shuttle and *Mir*, increasing the presence of U.S. astronauts on *Mir* to a maximum period of two years, and delivering up to two tons of hardware to the U.S. Space Station on Russian modules.

Redesign and Space Station Alpha[125]

In January 1993, William J. Clinton was inaugurated as President. One of his goals was reducing the federal deficit. A NASA assessment early in the year revealed that *Freedom* was $1 billion over budget.[126] The Office of Management and Budget warned Goldin that the President planned to cut NASA's budget and perhaps terminate Space Station *Freedom*. Goldin argued for the necessity of the Station to NASA's mission and existence.[127] President William J. Clinton reconsidered, and rather than cancel the program, directed NASA to redesign the Station and produce a configuration that reduced costs while still providing meaningful international participation as well as the "essential resources to

---

[122] Logsdon and Millar, Appendix B.
[123] "NASA Ratifies First Contract with Russian Space Program," *NASA News* Release 92-91, June 18, 1992, *ftp://ftp.hq.nasa.gov/pub/pao/pressrel/1992/92-091.txt* (accessed June 3, 2005).
[124] Logsdon and Millar, Appendix B.
[125] Advisory Committee on the Redesign of the Space Station, "Final Report to the President," June 10, 1993, pp. 1–3, 21–24, 34, 40–41.
[126] Smith, *Space Stations*, 1999, p. CRS-3.
[127] Lambright, p. 17.

advance the nation's scientific and technology development capabilities in space."[128] Consequently, on March 9, 1993, the President formally directed NASA to undertake a "rapid and far-reaching redesign of the Station" with a goal of significantly reducing development, operations, and utilization costs." The aim of this redesign was to cut the cost from the planned $14.4 billion to an administration goal of $9 billion and reduce the complexity of the current design and program while still achieving the goals for long-duration scientific research. The President directed NASA to give him several design options with various costs and capabilities.

At the request of the U.S. Office of Science and Technology Policy, the Redesign Team was to consider options at three cumulative-cost levels: $5 billion, $7 billion, and President William J. Clinton's ceiling of $9 billion. The cost of each option for fiscal years 1994 through 1998 was to accommodate the international partners and cover total expenditures for the Station, including development, operations, utilization, Shuttle integration, facilities, research operations, and transition cost. The Station Redesign Team, led initially by Dr. Joseph Shea and subsequently by Col. Bryan O'Connor, first met on March 10, 1993. Over approximately three months, the Station Redesign Team developed three options.

An advisory committee, chaired by vice presidential appointee and MIT president, Dr. Charles Vest, beginning in April 1993, assessed each option, looking at technical and scientific capability, accuracy of projected costs, and structure of management and operations. The committee made a number of observations.[129] All three options had a firm requirement for an assured crew return capability—a space "lifeboat" or "parachute." The advisory committee noted that the United States was not currently developing such a vehicle, but that the Russian Soyuz spacecraft was considered a viable contender. The committee recommended changing the Station's inclination to 51.6 degrees to allow use of the Soyuz. White House guidelines included considering Russian participation and use of *Mir*, although later clarification from the White House emphasized that the redesign effort was not to focus on "present or future Russian capabilities."[130]

Redesign Options

Option A was a modular buildup using many *Freedom* systems. Option A eliminated the two U.S. nodes, simplifying the pressurized volumes. Many of the subsystems, including data management, software, electrical power, thermal systems, and pressurized modules, were also simplified. Option A contained two "sub-options," one with a Lockheed Bus-1 spacecraft for navigation and propulsion, the second without it. The Station would be 100

---

[128] National Aeronautics and Space Administration, "Space Station Redesign Team Final Report to the Advisory Committee on the Redesign of the Space Station," June 1993, p. 259.
[129] The Advisory Committee was also called the Vest Panel.
[130] Logsdon and Millar, Appendix B.

feet (30 meters) shorter than the original design. Permanent human capability would be achieved in September 2000 after 16 Shuttle flights. The total cost of Option A was $17 billion.

Option B was derived most closely from Space Station *Freedom*. Except for minor changes, the phasing of capabilities and subsystems remained the same. Option B offered two advantages: 1) mature hardware, hardware mostly designed already with prototypes tested, and 2) the design of the baseline Station had evolved after years of engineering review and iteration with the research community. Option B used an evolutionary approach. The Option B Station was larger than the current design and would require 20 Shuttle flights to achieve an international permanent human capability in December 2001, and a greater number of EVAs. Option B's total cost was $19.7 billion.

Option C was a single-launch core Station and deviated most from the original design. All basic systems of this option would be checked out before launch, and it would be operational as soon the astronauts arrived. It had the largest inhabited volume and number of experiment racks. Because few of Option C's systems were mounted on the outside of the Station, less EVA maintenance was required, and therefore more crew time was available for research. This option placed a pressured module, derived from Space Shuttle components, in orbit with a single launch. Seven Shuttle flights would add international modules, and a permanent human capability would begin early in 2001. The total cost was $15.5 billion.

The advisory committee noted that none of the redesign options met the White House goal of completing development by the end of October 1998. The committee concluded, though, that Option A reached its human-tended configuration by that date. None of the options met the targets of $5 billion, $7 billion, or $9 billion. Even so, the proposed options, the committee believed, would still save from $6 billion to $10 billion when compared to the current anticipated cost of Space Station *Freedom* while permitting the development of a "very capable station."[131]

The advisory committee determined that Options A and C were "most deserving of further consideration." The international partners however, the report stated, expressed "strong reservations" about Option C based on this option's "relative lack of maturity and programmatic uncertainties." The committee also endorsed the Redesign Team's recommendation of a single prime contractor responsible for total system integration, including cost, schedule, and performance, and the establishment of a single NASA management team combining project and program levels into a dedicated program office and locating this core management team at a host Center.[132]

---

[131] Advisory Committee on the Redesign of the Space Station, "Final Report to the President" (June 10, 1993), p. 40.
[132] Advisory Committee on the Redesign of the Space Station, "Final Report to the President" (June 10, 1993), pp. 1–3, 7, 34, 40.

On June 17, 1993, President William J. Clinton announced his selection of "a reduced cost, scaled-down version of the original Space Station *Freedom*." Called "*Alpha*," this was a hybrid of two options with a $10.5 billion price tag over FYs 1994–1998 and a total cost of $17.4 billion.[133] *Alpha* had four phases: "1) photo-voltaic (PV) power station on-orbit for increased power to a docked orbiter/spacelab; 2) human tended capability (adding a U.S. Laboratory); 3) international human tended (adding an additional PV array and international modules); and 4) permanent human capability (adding a third PV array, the U.S. habitat module, and two Russian Soyuz capsules)."[134] The President also directed NASA to develop an implementation plan by September 1993 that included plans to con-tinue and expand international participation to take advantage of political developments arising from the end of the Cold War.

The President's endorsement of the new Space Station failed to protect the project from attacks by Congress. In June, a vote to cancel the Station was defeated by only one vote. A week later, another bid to cancel the program failed by 24 votes. Furthermore, scientists continued to say that the new design had even fewer science benefits than before. With the end of the Cold War, the Station's political benefits had also evaporated.[135] Nevertheless, NASA moved ahead with the program. A Space Station Transition Team worked through July and August to refine Option A. On August 17, Goldin named Johnson Space Center as the host Center for the new Space Station program, reporting directly to NASA Headquarters, and Boeing Defense and Space Group as the prime contractor.[136] The change subordinated the other prime contractors, Grumman, McDonnell Douglas, and the Rocketdyne Division of Rockwell International, to Boeing and moved the program office from Reston, Virginia, to Houston, Texas, along with approximately 1,000 government and contractor jobs.

On September 7, President William J. Clinton formally chose the small, four-person *Alpha* Station approved in June. *Alpha* essentially merged the U.S. Space Station *Freedom* and the Russian *Mir-2* into a new Space Station, international in scope. Congress and the Administration agreed to a fixed annual budget of $2.1 billion and a total cap of $17.4 billion. This was below the required annual peak of $2.8 billion identified in the redesign. To manage with the allotted funds, NASA revised the assembly plans and slipped the scheduled permanent habitability capability date to September 2003.[137]

---

[133] Lambright, p. 18. Also Smith, *Space Stations*, 1999, p. CRS- 3.
[134] Statement of the President, June 17, 1993 (NASA History Office Folder no. 009576). Also Launius, p. 178.
[135] Marcus Lindroos, "International Space Station (ISS) Plan," *Space Stations and Manned Spaceflight in the 1980s and 90s*, April 5, 2002, *http://www.abo.fi/~mlindroo/Station/Slides/sld061.htm* (accessed June 6, 2005).
[136] "Space Station Host Center and Prime Contractor Announced," *NASA News* Release 93–148, August 17, 1993, *ftp://ftp.hq.nasa.gov/pub/pao/pressrel/1993/93-148.txt* (accessed May 23, 2005).
[137] Launius, p. 179.

Throughout the redesign process, President William J. Clinton worked to develop closer ties with the new Russian government. During April 3–4, 1993, President William J. Clinton and Vice President Albert A. Gore met with Russian leaders at a summit in Vancouver, Canada, with the goal of furthering cooperation in space. President William J. Clinton invited Russia to participate in the new Station, and Russian President Yeltsin agreed. This summit resulted in "a comprehensive strategy of cooperation to promote democracy, security, and peace" and establishment of the "United States-Russian Commission on technological cooperation in the areas of energy and space" working group headed by Albert A. Gore and Russian Prime Minister Viktor Chernomyrdin.

The United States-Russian Commission met on September 1–2, 1993.[138] One result from the meeting was agreement on a three-phase structure leading to a complete Space Station. The first phase, from 1994 to 1997, was the Shuttle-*Mir* program. It included up to 10 Shuttle flights to *Mir* as well as stays on *Mir* by U.S. astronauts. The second phase, from 1998 to 2000, would enable the Station to support three people. It included building the Station's core and an interface to the Shuttle and would involve the United States, Russia, and Canada. Russia would be paid $400 million as "compensation for services" during phases 1 and 2. The third phase, from 2000 to 2004, would complete the Station's assembly with European, Russian, and Japanese components in place.

On November 1, 1993, Goldin and Russian Space Agency Director Yuri Koptev signed an "Addendum to Program Implementation Plan" for Space Station *Alpha*. The plan described the overall concept of the relationship between NASA and the Russian Space Agency, the components and operations, and science and technology utilization during the three phases. It also laid out program management and financial management roles and responsibilities. It noted that "Russia will become a full international partner in the Space Station."[139] President William J. Clinton, however, was concerned about Russia's plan to sell missile technology to India. At a November 29 top-level White House meeting, an agreement was reached that Russia would be a new partner—"the primary partner," the Station would be designated the ISS, and Russia would cancel its planned sale of missile technology to India and receive $100 million annually from NASA to compensate for the canceled missile sale.[140]

For the most part, discussions between Russia and the United States had not involved the other Station partners, although they had been kept informed of progress, nor had they formally been asked to approve Russian participation in the program as a partner. On October 16, 1993, the United States met with its partners in Paris, France to formally inform them of its intent to invite Russia to join the Space Station program. On November 7, the partners jointly met with the Russian Space Agency to review the details of the November 1 addendum.

---

[138] Logsdon and Millar, Appendix B.
[139] "Addendum to Program Implementation Plan," *Alpha* Station, November 1, 1993 (NASA History Office Folder 009576).
[140] Lambright, p. 18.

Finally, on December 6, 1993, in Washington, DC, the original Space Station partners decided to formally invite the Russian Federation to join the partnership. Over the next four years, the United States and partners worked to revise the Station intergovernmental agreements and memoranda of understanding to accommodate the Russian Federation. All the partners except Japan signed the new agreements on January 29, 1998.[141]

The Gore-Chernomyrdin Commission met again during December 16–17 in Moscow, Russia. There, Prime Minister Chernomyrdin announced that Russia had accepted the invitation to join the ISS program. Goldin and Koptev signed a protocol that expanded the terms of the 1992 HSF Cooperation agreement, detailing the activities that were to span the next decade and result in a completed Space Station. The two agencies agreed to up to 10 Shuttle flights to *Mir* with astronauts spending a total of 24 months on board the Station, a program of scientific and technological research, and the upgrade and extension of the *Mir* lifetime to the period 1995–1997. The protocol named some of the specific Shuttle missions for joint *Mir*-Shuttle activities.[142] Russia was to provide 12 hardware construction launches and six to eight utilization and resupply flights a year aboard Russian boosters.[143] Finally, Albert A. Gore and Chernomyrdin signed a "Joint Statement on Space Station Cooperation" describing the steps needed to formally bring Russia into the ISS partnership. It also noted that NASA and the Russian Space Agency had "agreed to contractual arrangements for up to $400 million through 1997 to facilitate the Shuttle-*Mir* program, joint technology developments, and the international Space Station."[144] This agreement ended a longstanding NASA practice that cooperative programs must not involve an exchange of funds.

The Shuttle-Mir Program

Russia has had more experience with long-duration spaceflight than any other nation, using the country's Soyuz spacecraft to ferry cosmonauts to and from Salyut space stations. The earliest Salyuts were equipped only for short stays, but beginning with its second-generation Salyut stations, the Soviet Union

---

[141] John M. Logsdon, *Together in Orbit: The Origins of International Participation in the Space Station*, Monographs in Aerospace History, no. 11 (Washington, DC: National Aeronautics and Space Administration, 1988), pp. 42–43.

[142] "How Phase 1 Started," *http://spaceflight.nasa.gov/history/shuttle-mir/history/h-b-start.htm* (accessed July 14, 2006); Also "Protocol to the Implementing Agreement between the National Aeronautics and Space Administration of the United States of America and the Russian Space Agency of the Russian Federation on Human Spaceflight Cooperation," December 16, 1993. Cited and quoted in Launius, pp. 153–155.

[143] "Russia Joins Station Effort, Will Get $1 Billion Over Life of Project," *Aerospace Daily* (December 17, 1993): 441 (NASA History Office Folder 009576).

[144] "NASA and Russian Space Agency Sign Agreement for Additional Space Shuttle/*Mir* Missions," *NASA News* Release 93–222, 16 December 1993, *ftp://ftp.hq.nasa.gov/pub/pao/pressrel/1993/93-222.txt* (accessed June 6, 2005). Also "U.S.-Russian Joint Commission on Economic and Technological Cooperation: Joint Statement on Space Station Cooperation," December 16, 1993 (NASA History Office Folder 17040).

began sending crews into space for extended periods. Russia also used modified uncrewed Soyuz spacecraft, called Progress, to carry food, propellant, and supplies to these orbiting outposts.

The final Salyut space station, *Salyut 7*, was abandoned in 1986 and reentered Earth's atmosphere over Argentina in 1991. The *Mir* space station replaced the Salyut. This third-generation space station was the world's first permanent space station, orbiting Earth since a Proton booster sent its core into space on February 20, 1986. The first *Mir* crew arrived in March 1986, and several Russian crews have spent extended periods on board *Mir*, sometimes for more than a year. Space travelers from other countries have also visited *Mir*.

Mir's modular design allowed several different vehicles or modules to be docked together (see Figure 3–53). Kvant-1 was added to the core module in 1987. This module housed the first set of six gyroscopes, instruments for astrophysical observations, and an experimental unit for electrophoresis. *Mir* also received an additional deployable solar panel.[145] Kvant-2, added in 1989, carried an EVA airlock, solar arrays, and life support equipment. Kristall, weighing 19.6 tons (17,781 kilograms), was added in 1990. This module carried scientific equipment, retractable solar arrays, and a docking node equipped with a special androgynous docking mechanism designed to receive spacecraft weighing up to 100 tons (90,718 kilograms).[146]

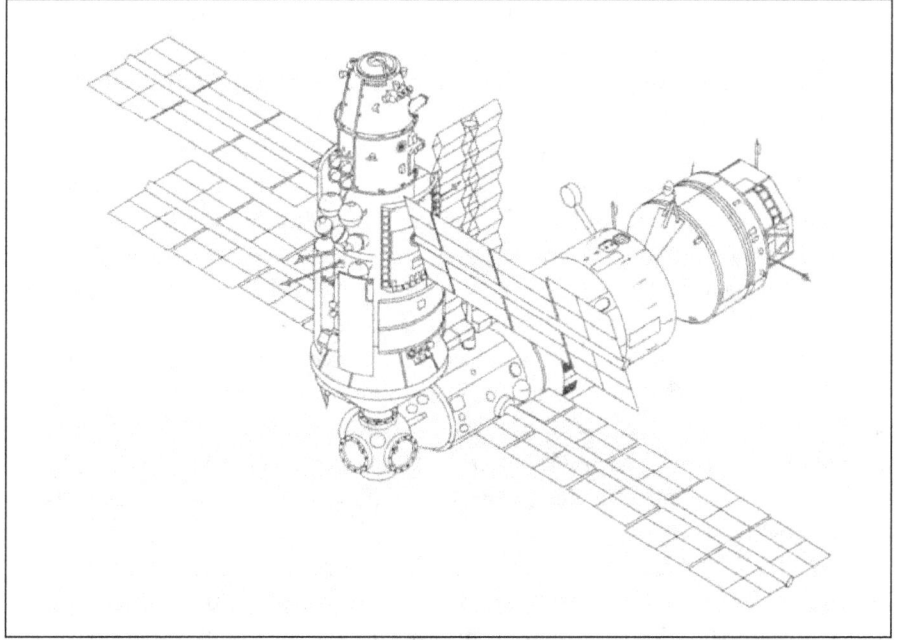

Figure 3–53. Mir *Space Station, 1989, with Base Block, Center; Kvant-1 Module, Right; and Kvant-2 Module, Top.*

---

[145] "Kvant-1 Module," *http://www.russianspaceweb.com/Mir_kvant.html* (accessed June 8, 2005).
[146] Launius, p. 146.

The next *Mir* module to be installed, Spektr, was originally designed for military experiments but had been grounded for years after the intended launch date because of financial problems in the former Soviet Union. It was rescued in the mid-1990s with the advent of U.S.–Russian cooperation and was refurbished for its new role—to house experiments for the Shuttle-*Mir* program. Spektr was finally launched on a Russian Proton rocket on May 20, 1995, and was berthed at *Mir*'s radial port opposite Kvant-2 after Kristall was moved out of the way. The module carried four solar arrays and scientific equipment, including more than 1,600 pounds (726 kilograms) of U.S. equipment. Earth observation was the focus of scientific study for this module, specifically natural resources and atmosphere.

Piroda was the last science module added to *Mir*. It docked to *Mir* on April 26, 1996. Piroda's primary purpose was to add Earth remote sensing capability. Along with remote sensing equipment, Piroda carried hardware for materials processing and meteorological and ionospheric research and equipment for U.S., French, and German experiments.

The Shuttle-*Mir* program that would span three years was the first phase of the cooperative program leading to construction of the ISS. The program used the U.S. Space Shuttle and the Russian *Mir* to provide experience to American and Russian crews and to conduct early joint scientific research. The program objectives were to: 1) learn to work with an international partner; 2) reduce risks associated with developing and assembling a Space Station; 3) gain operational experience for NASA on long-duration missions; and 4) conduct life science, microgravity, and environmental research.[147] The program involved launching the Shuttle to take cargo to and from *Mir* and leaving U.S. astronauts aboard *Mir* for four to five months.

The program began February 3, 1994, on STS-63 when cosmonaut Sergei Krikalev became the first Russian to fly on a U.S. spacecraft to join his American colleagues on the Space Shuttle *Discovery*. The same mission demonstrated a close rendezvous between *Discovery* and *Mir*. The next year, on June 27, 1995, STS-71 collected Norman Thagard from *Mir*, who had spent 115 days on the Space Station after arriving aboard a Russian Soyuz TM-21 spacecraft. He was the first American to visit *Mir*. On this mission, the Space Shuttle *Atlantis* for the first time docked with *Mir* using the androgynous unit on the Kristall module that had been delivered to *Mir* in 1990. The photos below show a rendition of *Atlantis* docked to *Mir* (see Figure 3–54) and the two vehicles connected as photographed by a *Mir* crew member in the Soyuz (see Figure 3–55). Table 3–120 lists all Shuttle-*Mir* flights.

On STS-74 in November 1995, *Atlantis* delivered and permanently attached the new Androgynous Peripheral Docking Assembly to Kristall's androgynous docking unit. This docking module improved clearance between *Atlantis* and

---

[147] Frank L. Culbertson, Jr., "Phase 1; Shuttle-Mir Program Overview," May 12, 1997 (NASA History Office Folder 15522). Also George C. Nield and Pavel Mikhailovich Vorobiev, ed., "Phase 1 Program Joint Report," NASA Special Publications 1000-6108/ (In English), National Aeronautics and Space Administration, January 1999, p. 3. (NASA History Office Folder 16480).

Mir's solar arrays on later docking flights. During the STS-74 flight, the Shuttle crew used the orbiter's remote manipulator system robot arm to hoist the docking module from the payload bay and berth its bottom androgynous unit atop *Atlantis*'s docking system. *Atlantis* then docked to Kristall. When *Atlantis* undocked from the docking module, the docking module remained permanently connected to Kristall.[148]

Figure 3–54. A technical rendition of the Space Shuttle Atlantis *docked to the Kristall Module of* Mir. *This configuration shows the STS-71/Mir Expedition 18 completed in June 1995. The Russian-developed Androgynous Peripheral Docking System linked the orbiter to the Kristall Module. (NASA Photo No. S-93-46073)*

---

[148] "International Space Station: Russian Space Stations," NASA Facts, ISS-1997-06-004JSC, International Space Station, January 1997, *http://spaceflight.nasa.gov/history/shuttle-mir/references/documents/russian.pdf* (accessed June 7, 2005). Also "STS-74," *http://science.ksc.nasa.gov/shuttle/missions/sts-74/mission-sts-74.html* (accessed June 10, 2005).

Figure 3–55. Undocking of Space Shuttle Atlantis *and* Mir *Space Station on STS-71. A* Mir *cosmonaut took this photo of* Atlantis *connected to Russia's* Mir *from a stationkeeping Soyuz on July 4, 1995. (NASA-MSFC Photo No. MSFC-9704176)*

A milestone occurred in 1997 when U.S. astronaut Jerry Linenger participated in the first U.S.-Russian EVA. On April 29, Linenger and *Mir* Commander Vasily Tsibliev conducted a 5-hour EVA to attach a monitor to the outside of the Station. The Optical Properties Monitor was to remain on *Mir* for nine months, studying the effects of the space environment on optical properties, such as mirrors used in telescopes.

In the midst of Shuttle flights to *Mir*, two serious accidents and a number of system problems on *Mir* raised doubts about the safety of *Mir* for U.S. crews and the reliability of the Russian equipment.[149] A fire on February 24, 1997,

---

[149] Launius, p. 166.

ignited in the Kvant-1 module when an oxygen canister malfunctioned. Of considerable size, the flames blocked access to one of the Soyuz spacecraft serving as a rescue vehicle. Although the fire burnt for only about 90 seconds, it filled the Station with sooty smoke, forcing the crew to wear masks and goggles until the area was cleared and it was certain there was no health hazard. It took a day of Station cleaning before the crew could return to their science mission.

In the months after the fire, the aging *Mir* experienced a number of systems failures and anomalies affecting such things as oxygen generation; carbon dioxide levels; temperature inside the habitable elements; crew exposure to ethylene glycol; power levels; power availability; air quality; and attitude control. The crew spent considerable time making repairs and keeping *Mir* habitable.

A life-threatening incident occurred on June 25, 1997, with astronaut J. Michael Foale aboard *Mir*. A Progress resupply vehicle loaded with garbage from *Mir* ran into the Spektr module while *Mir-23* Commander Vasily Tsibliev was attempting a test manual docking of the Progress using remote controls. The Progress flew off course, and the crew was unable to regain control of the tumbling cargo ship before it struck a solar panel on Spektr, destroying it. The Progress then bounced off the module, breaching the hull and buckling a radiator. Seconds later, a hissing sound alerted the crew to escaping oxygen, which was quickly traced to the Spektr module, now depressurizing, that the Progress had punctured. Crew members cut the cables leading into the Spektr, which sealed off the Spektr from the rest of the Station, and repressurized the remaining modules, leaving Foale's personal effects and several NASA science experiments inside the sealed-off area. For two days, the crew operated without power, which forced the shutdown of a number of key systems, including the oxygen generators and carbon dioxide scrubbers.

Meanwhile, the gyrodones which kept *Mir* in the proper attitude failed, destabilizing the vehicle and sending it into a spin that required firing the *Mir* engines to stop. Two weeks later, on July 7, another Progress vehicle brought supplies and repair materials to *Mir*. A fly-around of *Mir* and a 6-hour EVA on September 6, 1997 by Foale and *Mir* Commander Anatoly Solovyev to inspect damage to the Spektr module determined the location of the puncture on the module's hull.[150] The *Mir* crew pumped air into the module, and the Shuttle crew observed that the leak seemed to be located at the base of a damaged solar panel. The crew worked for months to return the damaged solar array to use, install a modified hatch so power lines could be routed while still keeping Spektr sealed, and restore damaged systems.[151] Because it was uncertain whether Spektr might again experience depressurization, even with repairs to the module, it remained sealed off and the scientific equipment in the module was lost.[152]

---

[150] Launius, p. 169.
[151] Launius, pp. 167–169.
[152] Marcia S. Smith, "The Shuttle-Mir Program: Testimony Before the U.S. House of Representatives Committee on Science," September 18, 1997, *http://www.house.gov/science/smith_9_18.html* (accessed June 7, 2005).

The collision prompted Congress to call on NASA to conduct a safety review of *Mir* before allowing any more astronauts to visit the Station. Some, concerned with the safety of American crews and the advancing age of *Mir*, as well as Russia's ability to meet its obligations, demanded an end to the United States program with Russia. A Task Force Red Team, led by Maj. General Ralph Jacobson, conducted a safety assessment of *Mir* to decide whether to allow a long-duration stay of astronaut David Wolf on *Mir*. The Task Force recommended to Administrator Goldin that "it was safe to launch Dave Wolf to *Mir* on STS-86 and continue U.S. presence on *Mir* . . ."[153] It also reaffirmed the conclusions of NASA's internal reviews to proceed with plans to exchange U.S. astronauts on *Mir*. A. Thomas Young conducted an additional external assessment and endorsed the safety process.[154] Goldin decided to continue the program even though some members of Congress and the NASA Inspector General opposed it. The Shuttle-*Mir* program concluded with no further crises.

International Space Station

Background
The ISS evolved from the U.S. Space Station Freedom program and the Russian *Mir* space station program. Approximately 75 percent of the hardware created for *Freedom* provided by the United States and its international partners was incorporated into the ISS design, (see Table 3–121). When complete, the ISS will be the largest artificial structure ever to orbit Earth.

Development
Space Station *Freedom* was formally terminated on February 1, 1994, when NASA and contractor officials from Boeing, McDonnell Douglas, and the Rocketdyne Division of Rockwell International signed documents marking the end of the *Freedom* work package contracts. This consolidated responsibility for the design, development, and integration of the program under a single prime contract with Boeing Defense and Space Group. NASA and Boeing signed a major modification to the November 15, 1993, letter contract between the two parties, changing Boeing's scope of work from a transitional contract to a hardware design and development contract. Work on the components named in the work packages would continue with McDonnell Douglas and Rocketdyne as subcontractors to Boeing.[155]

[153] Lt. Gen. Thomas P. Stafford, "Statement before the Committee on Science, U.S. House of Representatives, May 6, 1998," NASA Advisory Council Task Force on the Shuttle-*Mir* Rendezvous and Docking Missions and Task Force on International Space Station Operational Readiness, *http://www.house.gov/science/stafford_05-06.htm* (accessed June 8, 2005).
[154] "Panels Give Astronaut a 'Go' for Launch to *Mir*," *NASA News* Release 97-214, September 25, 1997, *ftp://ftp.hq.nasa.gov/pub/pao/pressrel/1997/97-214.txt* (accessed June 8, 2005).
[155] "NASA Marks Space Station Milestone," *NASA News* Release 94-014, February 2, 1994, *http://www.nasa.gov/centers/johnson/news/releases/1993_1995/94-014.html* (accessed May 24, 2005).

There were frequent revisions to the ISS assembly schedule. In March 1994, the "preliminary" schedule in the November 1993 Implementation Plan was revised, and the first Shuttle launch was moved from July 1997 to December. The completion date slipped from October 2001 to June 2002. At the time, congressional critics expressed doubts about the cost and schedule savings that Russia's participation would provide and repeatedly introduced motions to cancel the program. The 103rd Congress, which met in 1993 and 1994, defeated five attempts to terminate the Space Station program in NASA funding bills and three other attempts in broader legislation.[156] NASA defended its actions by stating that the schedule slip resulted from the need to stay within the $2.1 billion annual budget ceiling. Table 3–122 lists the assembly schedule as of April 1994.

The most serious problems came from the financial and political circumstances of NASA's partners. The ESA stopped development of its Hermes spaceplane in 1993 and removed the attached pressurized module and free-flying platform from its list of contributions to the Station, leaving only its scaled down Columbus laboratory. Canada trimmed $400 million from its $1 billion contribution to the Station. Some of Russia's own launches were delayed because of lack of funds for rockets. Political unrest and instability in the splintered country resulted in dropped communication with *Mir* and damage to ground facilities. Although Russia's contributions were supposed to be "enhancing" rather than "enabling," the country's contributions were essential. The Station could not function without Russia's critical elements, which included the FGB, reboost and refueling, a service module, a power mast, and Soyuz spacecraft for emergency return. To counter charges that Russia would not carry out its commitments, NASA declared to Congress that, if given funds, it would buy, rather than lease, the FGB from its manufacturer, Khrunichev. Other backups were identified in case Russia did not meet its commitments.[157] The weak spot, NASA admitted, in the "critical path" was the service module, which the Russians were to develop as their principal contribution. NASA had no alternative to that element.

The ISS System Design Review, held in March 1994, was a major technical milestone. The Review confirmed the validity of the baseline configuration, schedule, and cost of the completed ISS. The ISS would operate at an altitude of approximately 240 nautical miles (444 kilometers) and would orbit at a 51.6-degree inclination (the *Mir* inclination) to offer better Earth observation opportunities. It would have six crew members and 33 standard user racks for science operations.

As summarized in the System Design Review, planned assembly was to begin with the launch of the Russian FGB in November 1997. A docking compartment would be added before the first U.S. launch in December 1997.

---

[156] Smith, *Space Stations,* IB93017, 1996, *http://www.fas.org/spp/civil/crs/93-017.htm#legn* (accessed June 25, 2005).

[157] Bond, pp. 125–127.

The Russian service module was to be added to the Station in January 1998 followed by the universal docking module and the science power platform. The U.S. laboratory module would be launched on the third U.S. flight in May 1998. It would mark the beginning of human-tended science operations.

The Canadian-built robotic arm would be launched on the next flight in June 1998, and the addition of the Soyuz transfer vehicle in August 1998 would allow for extended on-orbit operations. The Japanese Experiment Module would be launched in early 2000, and the ESA laboratory module would be added in June 2001. Assembly would be complete in June 2002. The sequence provided for 13 Russian and 16 U.S. assembly flights. Use of the Ariane 5 launcher to lift the European module to the Station was added to the technical baseline. The U.S. contribution to the ISS, as stated in the System Design Review, was estimated at $17.4 billion from FY 1994 until assembly was complete in 2002.[158]

In April 1994, soon after the System Design Review ended, the heads of the various ISS agencies met in Washington to endorse the successful review and reaffirm their commitment to bringing Russia into the program as soon as possible.

Despite the successful review and the administration's support, criticism from Congress continued, and Congress introduced bills into the 1994 budget cycle to terminate the Station. But a bipartisan coalition of House legislators on June 29 defeated the motion 278-155. This vote was considered a signal that legislators felt that NASA was "getting its act together."[159] On August 3 the Senate rejected a similar motion to cancel the Station.[160]

In July 1994, the Space Station Control Board, which included representatives from NASA, the international partners, and Boeing, approved a revised assembly sequence (see Table 3–123). The new schedule substituted a U.S.-built solar array for a planned Russian-built array because of uncertainties whether the Russian array would be ready early enough in Station construction. This array fit between the Russian service module and the FGB. It would also provide more power to researchers during Phase 2. The U.S. truss would be attached temporarily to a small truss on top of the U.S. node and moved to a permanent position later in the assembly. The revised schedule moved launch of the third Station element, Russia's service module, from January to May 1998. The Board also agreed with U.S. plans to purchase the FGB from Khrunichev to assure its availability when ISS assembly began and the ESA plans to launch its laboratory module on an Ariane expendable launch vehicle rather than the Shuttle.[161]

---

[158] "Space Station System Design Review Completed," *NASA News* Release 94-53, March 24, 1994, *ftp:// ftp.hq.nasa.gov/pub/pao/pressrel/1994/94-053.txt* (accessed June 8, 2005).

[159] "New Coalition of Lawmakers Gives Space Station Resounding Victory," *Aerospace Daily* 171, no. 1 (July 1, 1994): 1. (NASA History Office Folder 009577)

[160] "Goldin Hails Solid Senate Vote on Space Station," *NASA News* Release 94-127, August 3, 1994, *ftp:// ftp.hq.nasa.gov/pub/pao/pressrel/1994/94-127.txt* (accessed December 4, 2005).

[161] "Station Control Board Ratifies Improved Assembly Sequence," *NASA News* Release 94-117, July 15, 1994, *ftp://ftp.hq.nasa.gov/pub/pao/pressrel/1994/94-117.txt* (accessed June 8, 2005). Also "Worries Over Russian Readiness Led to Station Schedule Shuffle," *Aerospace Daily* 171 (July 18, 1994): 87 (NASA History Office Folder 009577).

On August 31, 1994, NASA and Boeing agreed on key elements of the prime contract for the ISS. For the first time, NASA and Boeing concurred on the scope of work, program schedule, cost ceiling, and fee arrangement by fiscal year and on the completion and established contractual terms and conditions. NASA and Boeing hoped that the final contract would be in place before the end of the year.[162]

At the end of September, ISS managers released another updated assembly plan. This sequence incorporated early provisions for a centrifuge to augment the Station's science capabilities, allowed for earlier construction of Russia's Solar Power Platform in the late 1998 to mid-1999 timeframe, and meshed the latest weight estimates for Station components with Space Shuttle launch commitments. The change provided the Russian portion of the ISS with power and eliminated the need to transfer U.S. power to the Russian modules.[163]

The program to modify the Shuttle to increase lift capability, needed because of the launch to a higher inclination, was, according to a 1995 General Accounting Office (GAO) report, "challenging" and had a "questionable" schedule, "particularly in a declining budget environment." The assembly schedule continued to be complicated. It might be impossible, the GAO said, for the Shuttle to meet the demanding ISS assembly schedule. The GAO recommended that NASA obtain an independent review to assess the Agency's plans for increasing the Shuttle's lift capability, identify the associated risks, and weigh the costs and benefits of the tight scheduling of Shuttle flights for ISS assembly.[164] The program also had to overcome another attempt by the House of Representatives in a July 1995 vote to cut off funding for the program, ending ISS construction, and a similar Senate motion in September.[165]

ISS specifications and assembly schedules, as published by NASA, changed from 1994 through 1998. Updated assembly schedules were issued in September 1994, 1996, 1997, and 1998, each with later dates for assembling the various components.[166] Station mass also increased significantly from 831,000 pounds (376,935 kilograms) in 1994 to 924,000 pounds (419,119 kilograms) in 1996, and 1,015,000 pounds (460,396 kilograms) in 1998.[167]

---

[162] NASA had selected Boeing as prime contractor in September 1993, and the two had signed a letter agreement in November. "NASA and Boeing Reach Agreement on Space Station Contract," *NASA News* Release 94-144, September 1, 1994, *ftp://ftp/hq.nasa.gov/pub/pao/pressrel/1994/94-144.txt* (accessed June 8, 2005).

[163] "Space Station Managers Release Updated Assembly Plan," *NASA News* Release 94-164, September 30, 1994, *ftp://ftp.hq.nasa.gov/pub/pao/pressrel/1994/94-164.txt* (accessed December 2, 2005).

[164] U.S. General Accounting Office, *Space Shuttle: Declining Budget and Tight Schedule Could Jeopardize Space Station Support* GAO/NSIAD-95-171, July 1995, pp. 1–2, 10, *http://www.gao.gov/archive/1995/ns95171.pdf* (accessed June 11, 2005).

[165] Patrice Hill, "House Democrats Fail in Effort To Kill Space Construction," *The Washington Times,* July 28, 1995. A6; "Senator Tries To Kill Space Station," *UPI,* September 26, 1995, *NASA Earlybird News,* NASA Public Affairs Office, News and Information Branch (NASA History Office Folder 16936).

[166] National Aeronautics and Space Administration "International Space Station (ISS) Phase I-III Overview," (undated, c. April 1995) (NASA History Office Folder 16936).

[167] "International Space Station: Assembly Complete With Shuttle," National Aeronautics and Space Administration Fact Sheet HqL-408, September 1994; "International Space Station: Assembly Complete," National Aeronautics and Administration Fact Sheet HqL-426, January 1996; "International Space Station Pocket Information Card," National Aeronautics and Space Administration, June 1998 (NASA History Office Folder 17083).

Construction

Construction of Space Station components and systems progressed while the Shuttle-*Mir* flights were occurring, although not without challenges and problems. The first major ISS event of 1995 occurred January 13, 1995, when NASA and Boeing signed a $5.63 billion contract to manage the building of the core Station, including two nodes, an airlock, and laboratory and habitation modules, as well as their integration. The contract also called for the design and development of the Station. With its other responsibilities, Boeing was directed to interact with NASA's international partners to ensure the compatibility of all the components. Soon after, NASA reached an agreement with the Russian Space Agency to purchase the FGB, the first ISS element. The two agencies signed a protocol on February 5 in Houston. Texas, reflecting the contract terms negotiated by Boeing subcontractor Lockheed Missiles & Space Co. and Khrunichev, the Russian manufacturer of the FGB. The agreement called for the design, development, manufacturing, test, and delivery of the FGB initially at a price of $190 million. By the time the contract was signed on August 15, the cost had risen to $210 million.[168]

In May 1995, the ISS completed a series of tests to evaluate elements of its Water Recovery System and its ability to remove bacteria, fungi, and live viruses from the water supply. It was the first time its ability to remove viral particles was assessed. Designers intended to recycle the Station's water supply once it was occupied. By mid-September, the United States had produced 54,000 pounds (24,494 kilograms) of ISS hardware, with nearly 80,000 pounds (36,287 kilograms) estimated to be produced by the end of the year. The international partners had manufactured a total of more than 60,000 pounds (27,216 kilograms). By the end of the summer, estimates predicted that Boeing alone would have built almost 41,000 pounds (18,597 kilograms) of ISS hardware, including pressurized aluminum modules where the Station crew would work, and the payload racks to house systems and experiments. Subcontractor McDonnell Douglas had delivered about 5,000 pounds (2,268 kilograms) of qualification and flight hardware. Rocketdyne had built about one-third of its hardware, including about 30 percent of the solar cells needed for the entire program—more than 75,000 solar cells according to Rocketdyne's program manager. Rocketdyne had also provided photovoltaic modules for a Russian-assembled replacement solar array that the Shuttle *Atlantis* would deliver to *Mir*. Astronauts were also well into their training for EVAs.

By the end of September, Boeing had successfully completed the main structure of the U.S. laboratory module. The structure consisted of three cylindrical sections, two bulkheads, and the hatch openings through which the astronauts would enter and exit. Also completed were critical design reviews on

[168] "NASA/Russian Space Agency Reach Agreement on Key Station Element," *NASA News* Release 95-13, February 8, 1995 (NASA History Office Folder 16936). Also Launius, pp. 181–182. "Boeing, Khrunichev Sign Contract for Space Station Element," *NASA News* Release 95-138, August 15, 1995, *ftp:// ftp.hq.nasa.gov/pub/pao/pressrel/1995/95-138.txt* (accessed June 8, 2005).

the communications and tracking systems, as well as demonstrations showing full compatibility between the ISS's S-band subsystem and NASA's TDRS System that the Shuttle used for communications and tracking.[169]

In January 1996, the exteriors of the U.S. Station modules were completed. One module was to house astronauts on board the ISS. Two nodes, a laboratory module and an airlock, were also completed. In May 1996, the air purification system passed a major test at Marshall Space Flight Center. The month-long test evaluated the air purification system's ability to control carbon dioxide, oxygen, and air pressure inside the living and laboratory quarters.[170] The next month, Rocketdyne successfully conducted tests in the neutral buoyancy simulator on a mockup of a truss that would house the communications and tracking, attitude stabilization, thermal control, and electrical power distribution systems.[171] In November, the first U.S. module, Node 1, successfully completed its final pressure test at the Boeing plant in Huntsville, Alabama. Node 1 was shipped to Kennedy Space Center in June 1997.

International Contributions

Although the ESA was scheduled to contribute the Columbus laboratory module, by mid-June 1995 the ESA still had not reached agreement over the size and scope of its involvement in the Space Station.[172] On October 18, the ESA Council met in Toulouse, France, and approved the program "European Participation in the International Space Station Alpha." The program incorporated a number of cutbacks from earlier plans because of financial constraints. The approved program consisted of the following:

- Columbus laboratory development and launch, a module permanently attached to the ISS for conducting scientific experiments, research, and development.
- The Automated Transfer Vehicle (ATV), a logistics vehicle launched by an Ariane 5 for carrying research and system equipment, gases, and propellant to the ISS, and removing trash from the Station.
- Station utilization preparation and astronaut-related activities.

---

[169] "Space Station Completes Major Life Support System Tests," *NASA News* Release 95-61, May 3, 1995, *ftp://ftp.hq.nasa.gov/pub/pao/pressrel/1995/95-61.txt* (accessed June 8, 2005); Frank Morring, Jr., "Space Station: Contractors Say Project Well Underway; Schedule Critical," *Focus, a Supplement to Aerospace Daily* (May 19, 1995): p. 278 (NASA History Office Folder 16936); "U.S. Structure for International Space Station Completed," *NASA News* Release 95-161, September 26, 1995, *ftp://ftp.hq.nasa.gov/pub/pao/ pressrel/1995/95-161.txt* (accessed June 8, 2005); Dave Cooling, "Research Outpost Beyond the Sky," *IEEE Spectrum* (October 1995):.28-33 (NASA History Office Folder 16936).

[170] "Space Station Air Purification System Completes Major Test," *NASA News*, Marshall Space Flight Center Release 96-96, May 10, 1996, *ftp://ftp.hq.nasa.gov/pub/pao/pressrel/1996/96-96.txt* (accessed June 14, 2005).

[171] "Space Station Truss Tested in Neutral Buoyancy Simulator," *NASA News*, Marshall Space Flight Center Release 96-121, June 13, 1996, *ftp://ftp.hq.nasa.gov/pub/pao/pressrel/1996/96-121.txt* (accessed June 14, 2005).

[172] Peter B. deSelding, "ESA's Role in Space Station Still Shaky," *SpaceNews* 6 (June 19–25, 1995): p. 1 (NASA History Office Folder 16936).

- Studies of a European Crew Transport Vehicle (CTV), leading to involvement in the X-38 demonstrator and possible participation in the Crew Return Vehicle.
- Exploitation of the results of the Atmospheric Reentry Demonstration (developed under the Hermes program) for the ATV and CTV.

Soon after the Toulouse conference, the ESA and its prime contractor, Daimler Benz Aerospace, signed a contract to undertake Columbus laboratory development using a consortium of European subcontractors.[173] Table 3–124 lists the Columbus laboratory characteristics.

The ESA intended to use an Ariane 5 to launch Columbus. But on June 4, 1996, the first Ariane 5 launch failed, destroying the launch vehicle and its payload. The ESA decided against using the Ariane to launch Columbus, and NASA agreed to launch Columbus on the Shuttle. In return, Alenia Aerospazio, an Italian space company under contract to the Italian Space Agency, would supply the second and third nodes of the ISS, saving NASA the cost of building them.[174] The Italian Space Agency also agreed to provide three pressurized Multi-Purpose Logistics Modules. With the ability to be attached to both the Station and the Shuttle and with components to provide some life support, the modules would serve both as "moving vans" by carrying equipment, experiments, and supplies between the ISS and the Shuttle, and as attached Station modules. While traveling between the ISS and Earth, these modules would be isolated, and crew members could not enter them from the Shuttle cabin. This would retain the Station environment.

Construction of the first Italian module, named Leonardo, began in April 1996 at the Alenia Aerospazio factory in Turin, Italy. A special Beluga cargo aircraft delivered the module to Kennedy Space Center from Italy in August 1998, with launch planned for 2001. The cylindrical module was approximately 21 feet (6.4 meters) long and 15 feet (4.6 meters) in diameter. The module weighed almost 4.5 tons (4,082 kilograms) and could carry up to 10 tons (371,946 kilograms) of cargo packed into 16 equipment racks. Two more multipurpose modules, named Raffaello and Donnatello, were planned for later Shuttle flights.[175] Figure 3–56 shows Leonardo being processed at Kennedy Space Center.

[173] European Space Agency, *Columbus: Europe's Laboratory on the International Space Station*, BR-144, October 1999, pp. 5–7.

[174] Harland and Catchpole, pp. 190–91, 196. Also "Space Station Assembly Elements: U.S. Node 2," http://spaceflight.nasa.gov/station/assembly/elements/node2/index.html (accessed June 15, 2005).

[175] "Space Station Assembly: Multi-Purpose Logistics Modules," *http://www.nasa.gov/mission_pages/station/structure/elements/mplm.html* (accessed June 20, 2005); "Leonardo Module: A 'Moving Van' for the International Space Station," *NASA Facts*, Johnson Space Center, IS-1998-10-ISS021-JSC, November 1998, *http://spaceflight.nasa.gov/spacenews/factsheets/pdfs/mplm.pdf* (accessed June 15, 2005).

Figure 3–56. Processing of Leonardo, the first multi-purpose logistics module, takes place at Kennedy Space Center on December 3, 1998. The module was one of three from Italy's Alenia Aerospazio. Leonardo will be operated by NASA and supported by the Italian Space Agency. (NASA-KSC Photo No. KSC-98PC-0892)

On October 14, 1997, NASA and the Brazilian Space Agency (AEB), a new international partner, signed an implementing arrangement providing for the design, development, operation, and use of Brazilian-developed flight equipment and payloads for the ISS. In exchange for AEB-supplied equipment and support, NASA would give Brazil access to NASA ISS facilities on orbit and a flight opportunity for one Brazilian astronaut.[176]

Problems with Russia

Although construction was progressing on the Russian FGB, and the module would be assembled and ready for testing in December 1996, Russia's persistent lack of funds was causing a major program crisis.[177] In December 1995, the Russian Space Agency announced that the Russian government owed FGB manufacturer Khrunichev money for 1995 work and, unless the Russian government released the funds needed to work on the FGB and the service module, it would be unable to meet the FGB's launch date and unable to build the service module, both essential components. On March 27, 1996, NASA Administrator Goldin stated that he would give Russia one month or six weeks to get "stalled . . . effort moving again." But by July, Khrunichev had received

---

[176] "NASA Signs International Space Station Agreement with Brazil," *NASA News* Release 97-233, October 14, 1997, *ftp://ftp.hq.nasa.gov/pub/pao/pressrel/1997/97-233.txt* (accessed June 14, 2005).

[177] "Station's First Module Assembled; Ready for Testing," *NASA News* Release 96-253, December 9, 1996 (NASA History Office Folder 17083).

only a letter as a guarantee in seeking a loan to fund work on the service module, which was now acknowledged to be "months" behind schedule. By late September, it seemed unlikely that the service module would be ready to launch in April 1998. At the end of 1996, the Russian Space Agency acknowledged that the service module would have to be delayed still further to December 1998 because the promised funds had not arrived.[178]

In the meantime, early in 1997, NASA allocated $100 million to Lockheed to initiate development of the Interim Control Module as a backup, based on the propulsion module of a classified military satellite. The Interim Control Module could provide propulsion until the service module became available, although the interim module would require substantial modifications and would cost time and money. At this point, Russia was seriously in danger of being dropped from the program. Although promises were forthcoming, money was not. In April, NASA and the Russian Space Agency formally agreed to slip launch of the FGB from November 1997 to mid-1998, 11 months later than originally planned, and to launch the Interim Control Module if the service module could not be launched later in 1998. NASA also stated that it would devote "equal attention" to contingency planning. On April 9, 1997, NASA announced that the ISS's on-orbit assembly was slipped to "no later than" October 1998.[179] On April 11, the Russian government arranged for bank loans to Energia by the end of May. Khrunichev soon resumed work on the service module, and NASA expressed "cautious optimism" that the ISS was back on track.

Cost and Schedule Problems

On May 15, 1997, the Space Station Control Board released a new assembly schedule, Revision C.[180] According to this revision, the FGB would launch in June 1998, eight months later than earlier planned; the U.S. node would launch in July; and the service module would launch in December (see Table 3–125). A Shuttle flight was added as a contingency to send up the Interim Control Module in December 1998 if delivery of the service module slipped into 1999.[181] On May 31 at a meeting in Tokyo, the heads of the five participating space agencies accepted the revised schedule. NASA also requested that Congress create a new Russian Program Assurance budget category to finance construction of the Interim Control Module and other contingency options.

---

[178] Harland and Catchpole, pp. 191–194.
[179] "NASA Revises International Space Station Schedule," *NASA News* Release 97-65, April 9, 1997, *ftp:// ftp.hq.nasa.gov/pub/pao/pressrel/1997/97-065.txt* (accessed June 13, 2005).
[180] "Assembly Sequence, 5/15/97 Rev C," National Aeronautics and Space Administration, International Space Station (NASA History Office Folder 11613); "Space Station Control Board Approves New Assembly Schedule," *NASA News* Release 97-98, May 15, 1997, *ftp://ftp.hq.nasa.gov/pub/pao/pressrel/ 1997/97-098.txt* (accessed June 13, 2005).
[181] Harland and Catchpole, pp. 191–197.

A September 1997 meeting in Houston, Texas, of representatives of all the ISS partners formally approved Revision C of the assembly and launch schedule that had received preliminary approval in May. The first U.S.-built Station element, Node 1, was scheduled to launch on STS-88, the first Shuttle assembly mission, in July 1998. In June 1997, Node 1 had been shipped from Alabama to Kennedy Space Center to begin launch preparations. The schedule called for launch of the ESA's Columbus in October 2002. The Russian service module was scheduled for a December 1998 launch. At an earlier General Designer's Review, the Russian Space Agency had assured NASA that it could meet the scheduled launch date. The first ISS element, the Functional Cargo Block, was "on track" for a June 1998 launch. The module had completed manufacturing at Khrunichev on September 15 and had been moved to the RSC-Energia facilities for further testing.[182] However, in November it was announced that, because of manufacturing problems, the module had fallen two months further behind schedule and the time was unlikely to be recovered.[183]

Cost continued to be a problem. From 1996 to 1997, cost overruns on the project increased at an alarming pace. In 1997, the GAO released two reports before the Senate Subcommittee on Science, Technology, and Space reporting on the program's cost and schedule status and estimated cost at completion. The June report stated that Russia's inability to meet its financial responsibilities had resulted in a projected eight-month delay in launching the service module. The GAO report also said that cost control problems under the prime contract had "steadily worsened." Since April 1996, the cost overrun "more than tripled" to $291 million.[184] On September 18, 1997, the GAO released a second report updating the status of the ISS prime contractor's cost and schedule performance. The situation, the GAO now claimed, had continued to worsen from a cost overrun of $291 million in April 1997 to a cost overrun of $355 million as of July 1997.[185] In September 1997, NASA and Boeing revealed that Boeing's prime contract would have at least a $600 million overrun at completion, and NASA needed $430 million more than expected in FY 1998.[186] In November, prime contractor Boeing admitted to a House panel that Boeing's costs were millions of dollars over the company's contract amount.

---

[182] "Control Board Reports International Space Station Launch on Target, Finalizes Assembly Sequence," *NASA News* Release 97-222, October 1, 1997, *ftp://ftp.hq.nasa.gov/pub/pao/pressrel/1997/97-222.txt* (accessed June 14, 2005).

[183] "NASA Says Russian Service Module Two Months Behind Schedule," *Aerospace Daily* 184 (November 6, 1997): p. 201 (NASA History Office Folder 16949).

[184] U.S. General Accounting Office, "Space Station: Cost Control Problems Continue To Worsen," Testimony Before the Senate Subcommittee on Science, Technology, and Space, June 18, 1997, *http://www.gao.gov/archive/1997/ns97177t.pdf* (accessed June 15, 2005).

[185] U.S. General Accounting Office, "Space Station: Deteriorating Cost and Schedule Performance under the Prime Contract," Testimony of Allen Li Before the Senate Subcommittee on Science, Technology, and Space, GAO/T-NSIAD-97-262, September 18, 1997, p. 1, *http://www.gao.gov/archive/1997/ns97262t.pdf* (accessed June 15, 2005).

[186] Smith, *Space Stations*, IB93017, May 16, 2001, CRS Issue Brief for Congress, *http://www.ncseonline.org/NLE/CRSreports/Science/st-58.cfm* (accessed March 2, 2005). Also Catchpole, pp. 197–198.

In September 1997, Administrator Goldin requested that the NASA Advisory Council establish a "cost control task force . . . on the International Space Station . . . to conduct a prompt, independent, and thorough analysis" of the . . . "factors that affect cost growth and control . . . " The task force, chaired by Jay Chabrow, delivered its report to the Advisory Council on April 15, 1998. Although the report credited the program with having made "notable and reasonable progress," the report estimated that NASA would need an extra $7 billion, increasing the 1993 estimate of $17.4 billion to $24.8 billion, and up to three more years to complete the project. The report attributed the cost overrun to "program size, complexity, and ambitious schedule goals . . . beyond that which could be reasonably achieved within . . . the $17.4 billion total cap."[187] Russian participation also was a "major threat" to the program. Rather than the anticipated $1 billion cost savings to the United States from Russia's provision of the FGB and an Assured Crew Return Vehicle, Russia's economic situation had negated most of the savings, depleted a major part of program reserves, and caused schedule slips.[188]

In NASA's response to the Task Force report released on June 15, 1998, NASA identified approximately $1.4 billion in additional costs. The response also noted that after the report's release, the assembly schedule was changed once more to accommodate a four-month service module schedule slip (Revision D). Consequently, the first-element launch (the FGB) was moved to November 1998 with ISS assembly complete by January 2004.[189] On May 31, representatives of all ISS partners agreed to officially target a November 1998 launch for the first ISS component and to revise launch target dates for the remainder of the assembly plan. The partners set an April 1999 target launch date for the service module, and the first ISS crew would be launched aboard a Soyuz spacecraft in the summer of 1999 to begin a five-month stay on the Station.[190] Table 3–126 shows a partial assembly sequence as of May 1998.

January 29, 1998, marked an important ISS milestone. On that day, senior government officials from 15 countries met in Washington, DC, and signed agreements establishing the framework for cooperation among the partners for the design, development, operation, and utilization of the ISS. The "Space Station Intergovernmental Agreement," signed by Canada, 11 member states of the ESA, the Russian Federation, Japan, and the United States, established "a long term international cooperative framework on the basis of genuine partnership . . . of a permanently inhabited civil Space Station for peaceful

[187] NASA Advisory Council, "Report of the Cost Assessment and Validation Task Force on the International Space Station," Apri 21, 1998, *http://history.nasa.gov/32999.pdf* (accessed June 12, 2005).
[188] "Report of the Cost Assessment and Validation Task Force on the International Space Station," April 21, 1998. *http://history.nasa.gov/32999.pdf* (accessed July 18, 2006).
[189] "International Space Station (ISS) Response to the Cost Assessment and Validation Task Force on the ISS," June 15, 1998, *http://spaceflight.nasa.gov/spacenews/releases/1998/cav_response.pdf* (accessed June 15, 2005).
[190] "International Space Station Partners Adjust Target Dates for First Launches, Revise Other Station Assembly Launches," *NASA News* Release 98-93, June 1, 1998, *ftp://ftp.hq.nasa.gov/pub/pao/pressrel/1998/98-093.txt* (accessed June 21, 2005).

purposes, in accordance with international law." (See Figure 3–57 for the commemorative plaque issued in honor of the signing.) NASA Administrator Goldin also signed three bilateral memoranda of understanding (MOU) with the heads of the Russian Space Agency, ESA, and the Canadian Space Agency on January 29, and with the government of Japan on February 24. These MOUs described in detail the roles and responsibilities of the agencies in the design, development, operation, and utilization of the ISS. They spelled out the management structure and interfaces necessary to ensure the effective operation and utilization of the ISS. These new agreements superseded previous ISS agreements signed in 1988 among the United States, Europe, Japan, and Canada, reflecting changes to the program resulting from Russian participation in the program and the 1993 design changes.[191]

On-Orbit Assembly Begins

On-orbit assembly of the ISS began November 20 with the launch of Russia's Functional Cargo Block (renamed Zarya, meaning "sunrise") to orbit by a Russian Proton rocket from the Baikonur Cosmodrome in Kazakhstan. Work on this U.S.-funded module had begun in 1994 and was completed in 1998.[192] Zarya provided the ISS's initial propulsion and power. Characteristics of Zarya are listed in Table 3–127.

The first U.S. component of the ISS, Node 1, was launched on December 3, 1998, on mission STS-88 (see Figure 3–58 for the module inside the *Endeavour*'s payload bay). The module was named "Unity" to commemorate the joining of ISS modules from Russia and the United States and to honor the spirit of international cooperation and achievement in building the Station. Unity provided six docking ports for the attachment of other modules. Unity also provided external attachment points for the truss and internal storage and pressurized access between modules. Two PMAs connected Unity to *Endeavour* at one end and to Zarya at the other. The PMAs also provided passageways for crew, equipment, and supplies. PMA-1 connected to Zarya using an Androgynous Peripheral Attach System similar to the Russian docking system used for Shuttle-*Mir* docking. The other end of the PMA connected to Unity using a Passive Common Berthing Mechanism. PMA-2 linked the Shuttle to Unity using an Androgynous Peripheral Attach System provided with a hatch that had an 8-inch (20-centimeter) viewport. The other end attached to Unity using a Passive Common Berthing Mechanism. Figure 3–59 shows interior and exterior views of Unity. Table 3–128 lists Unity's characteristics.

---

[191] "Partners Sign ISS Agreements," National Aeronautics and Space Administration, *http:// spaceflight1.nasa.gov/station/reference/partners/special/iss_aggrements/* (accessed June 21, 2005). Also "Space Station Agreements," ESA, *http://www.spaceflight.esa.int/users/file.cfm?filename=fac-iss-la-ssa* (accessed June 21, 2005).

[192] The module was built by the Khrunichev State Research and Production Space Center in Moscow under a subcontract to The Boeing Company for NASA. "Space Station Assembly Elements: Zarya Control Module," *http://spaceflight.nasa.gov/station/assembly/elements/fgb/index.html* (accessed December 3. 2005).

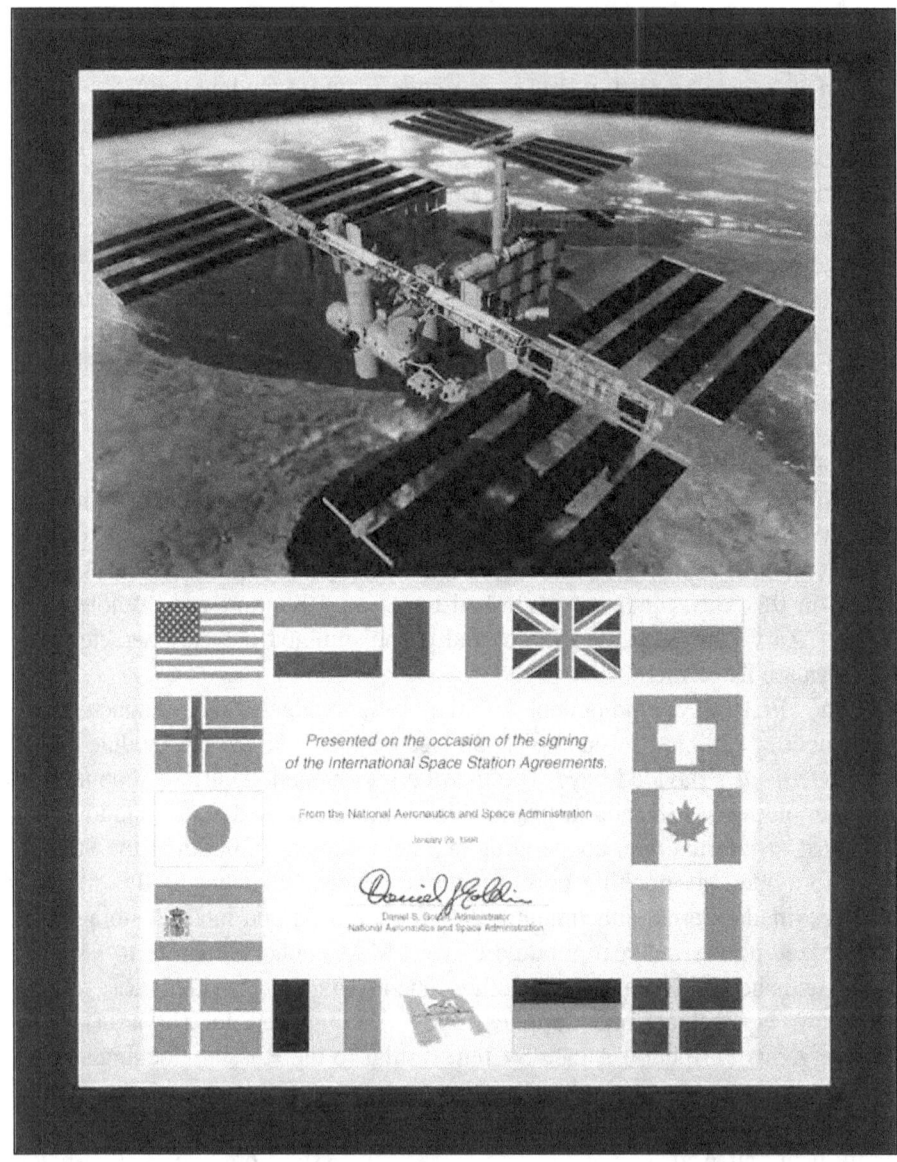

*Figure 3–57. This commemorative plaque was presented on the signing of the International Space Station Agreements (*http://spaceflight.nasa.gov/station/reference/partners/special/ iss_aggrements/ISS_Agreements_lo.jpg*).*

On December 5, the 12.8-ton (11,612-kilogram) Unity connecting module was first attached to *Endeavour*'s docking system. On December 6, using the Shuttle's 50-foot (15.2-meter) robot arm, Zarya was captured from orbit, and the two units docked. Figure 3–60 shows the Canadian-built remote manipulator system maneuvering astronauts Newman and Ross into position to work on the Unity module. Figure 3–61 shows the two modules docked together.

On December 7, 9, and 12, astronauts Ross and Newman conducted three spacewalks to:

- Attach PMA-1 to Zarya.
- Test a SAFER unit, a self-rescue device should a spacewalker become separated from the spacecraft during an EVA.
- Nudge two undeployed antennas on Zarya into position.
- Remove launch restraint pins on Unity's four hatchways for mating future ISS modules and truss structures.
- Install a sunshade over Unity's two data relay boxes to protect them against harsh sunlight.
- Stow a tool bag on Unity.
- Disconnect umbilicals used for the mating procedure with Zarya.
- Install a handrail on Zarya.
- Make a detailed photographic survey of the Station.

The astronauts completed assembly of an early S-band communications system allowing flight controllers in Houston, Texas, to send commands to Unity's systems and monitor the Space Station's health. The astronauts also conducted a successful test of the videoconferencing capability of the early communications system that the first permanent crew would use. Mission Specialists Krikalev from Russia and Currie also replaced a faulty unit in Zarya.

Unity and Zarya were successfully engaged at 9:48 p.m. on December 6, and Unity came to life at 10:49 p.m. on December 7. At 2:54 p.m. on December 10, history was made as Shuttle Commander Cabana and Krikalev floated into the ISS together, followed by the rest of the crew. At 4:12 p.m., Cabana and Krikalev opened the hatch to Zarya and entered. On December 11, at 5:41 p.m., Cabana and Krikalev closed the hatch to Zarya, and they closed the door to Unity at 7:26 p.m. The ISS flew free at 3:25 p.m. on December 13, as Shuttle Pilot Sturckow separated *Endeavour* from the ISS. Orbital events relating to the Zarya and Unity missions are listed in Table 3–129.

Laboratory Accommodations
The laboratories provided by the United States and the international partners were to focus on six major research disciplines: microgravity science, life science, space science, Earth science, engineering research and technology, and space product development. Table 3–130 provides an overview of the ISS science laboratories as of early 1999.

Figure 3–58. The Unity Module Inside the Payload Bay of
Space Shuttle Endeavour, *November 19, 1998.*

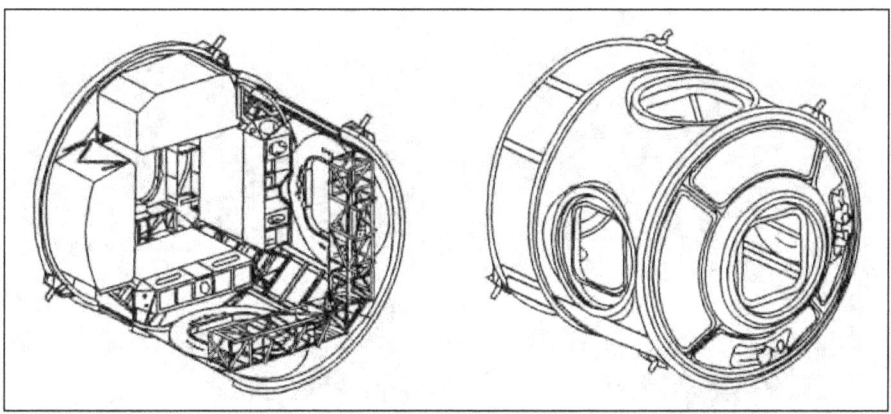

*Figure 3–59. Diagram of Interior and Exterior of Unity Connecting Module.
(NASA Photo No. 98PC-1731)*

*Figure 3–60. In December 1998, the crew of the STS-88 Mission began construction of the ISS,
joining the U.S.-built Unity Node to the Russian-built Zarya Module. The crew used a large-
format IMAX camera to take this photo, which shows astronauts Newman (left) and Ross
maneuvering into position to continue work on Unity. (NASA Photo No. S99-03771)*

Figure 3–61. Unity and Zarya Modules. This photograph, taken during the STS-88 Mission, shows the connected Unity Module (Node 1) and Zarya (the Functional Cargo Block) after Unity's release from Endeavour's *cargo bay. (NASA-MSFC Photo No. 0100335)*

Crew Return Vehicle

One of the requirements for the Space Station was a vehicle for returning crews to Earth in an emergency. In 1987, NASA Administrator James Fletcher requested $3 million from Congress for a study of a Crew Emergency Return Vehicle to be delivered to the Station by the Shuttle and used as a "lifeboat" to return stranded crew members. In October 1989, NASA issued a request for proposals for the renamed "Assured Crew Return Vehicle." Langley Research Center proposed a Crew Rescue Vehicle (CRV), called the HL-20, that could carry a crew of eight and would be carried by the VentureStar reusable launch vehicle. The HL-20 proved too expensive, and NASA instead awarded $1.5 million contracts to Lockheed and Rockwell International in 1990 to refine their concepts for a "lifting body" vehicle that would evolve into the X-38. At the time, a 1992 start was planned for hardware development.[193]

---

[193] Harland and Catchpole, p. 119.

Figure 3–62. The complete ISS as envisioned in 1997 superimposed over the Straits of Magellan and the Mediterranean Sea. The drawing in Figure 3–63 (next page) shows the complete ISS with its components and the contribution of each component.

The X-38 project began in 1995 at Johnson Space Center using data from past lifting-body programs and the U.S. Army's Guided Precision Delivery Systems from Yuma Proving Grounds. The design closely resembled the X-24 wingless lifting body concept tested at Dryden Flight Research Center between 1969 and 1971. The vehicle would be able to return up to seven ISS crew members to Earth. In early 1996, a contract was awarded to Scaled Composites to construct two atmospheric test vehicles. Scaled Composites delivered the first vehicle, the V131, to Johnson Space Center in September 1996, where it was outfitted for its initial flight tests at Dryden Flight Research Center. The second vehicle, the V132, was delivered to Johnson Space Center in December 1996.

The test vehicles were shells made of composite materials such as fiberglass and graphite epoxy and strengthened with steel and aluminum at stress points. The test vehicle weights ranged from 15,000 pounds (6,804 kilograms) to about 25,000 pounds (11,340 kilograms). The prototypes were 23.5 feet (7.2 meters) long, 11.6 feet (3.5 meters) wide, and 8.4 feet (2.6 meters) high, approximately 80 percent the size of the proposed full-size CRV. The vehicles landed on skids, similar to the X-15 research aircraft, instead of wheels. The second test vehicle, the V132, carried a full flight control system, including electro-mechanical control surface actuators similar to those planned for the production CRV.

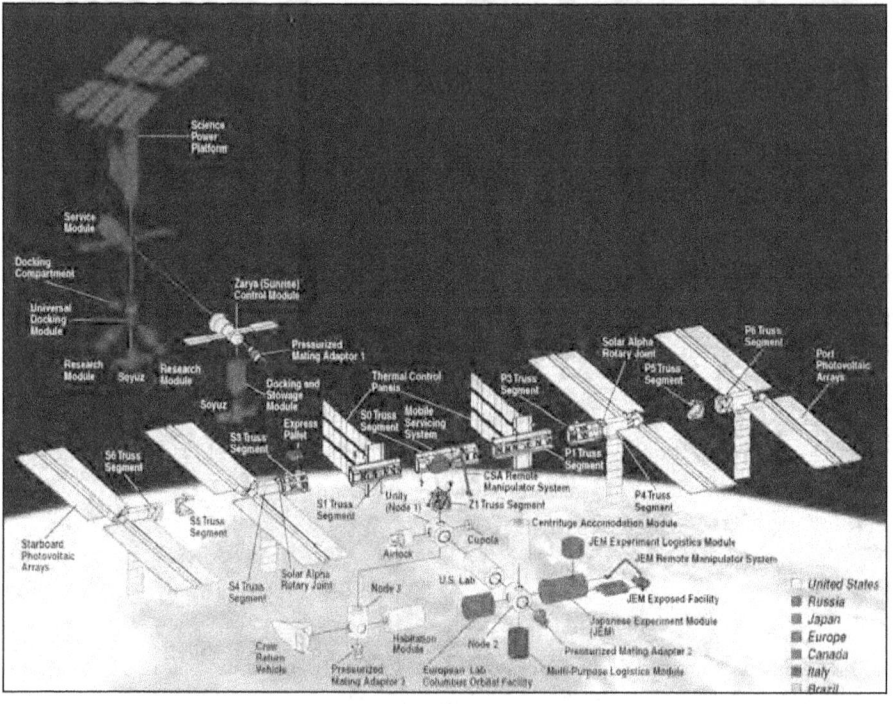

Figure 3–63. This drawing shows the completed ISS and contributions by each of the international partners, 1998. (NASA Photo No. S99-01389)

The unpiloted, captive-carry flight tests of the test airframes attached to B-52 aircraft began in July 1997 at Dryden Flight Research Center to study their aerodynamics while attached to the aircraft's wing pylon.[194] The first free-flight drop tests took place on March 12, 1998, and continued into 1999. The tests included use of a parafoil spanning 121.5 feet (37 meters) with an area of 5,500 square feet (511 square meters). These flight tests studied launch characteristics and assessed the operation of the parachute from deployment of the small drogue through reefing of the main parafoil and landing (see Figure 3–64, which shows the X-38 descending at the end of its first free flight on March 12, 1998). Drop tests used Navstar GPS signals for guidance. A production X-38 would weigh 20,000 pounds (9,072 kilograms), with its deorbit engine weighing 5,000 pounds (2,268 kilograms). The X-38 program was canceled in 2002 due to budget pressures associated with the ISS.[195]

---

[194] This B-52 was the same aircraft used for the X-15 program.

[195] "X-38," NASA Fact Sheets, Dryden Flight Research Center, *http://www.nasa.gov/centers/dryden/news/ FactSheets/FS-038-DFRC.html* (accessed May 24, 2005). Also Mark Lindroos, *Space Stations and Manned Spaceflight in the 1980s and 90s, http://www.abo.fi/~mlindroo/Station/Slides/index.htm,* "X-38 Crew Rescue Vehicle," *http://www.abo.fi/~mlindroo/Station/Slides/sld054.htm,* "HL-20 Crew Rescue Vehicle," *http://www.abo.fi/~mlindroo/Station/Slides/sld053.htm,* "NASA Assured Crew Return Vehicle," and *http://www.abo.fi/~mlindroo/Station/Slides/sld052.htm* (accessed May 23, 2005).

*Figure 3–64. The X-38 descends under its steerable parafoil over the California desert during its first free flight at Dryden Research Center, March 12, 1998.
(NASA-DFRC Photo No. EC98-44452-2)*

Table 3–1. Percent of NASA's R&D or HSF Budget
Allocated for Space Station

Year (Fiscal)	Space Station Appropriation (in thousands of dollars)	Percent of R&D or HSF Appropriation
1989	900,000	21
1990	1,800,000	33
1991	1,900,000	34
1992	2,929,000	32
1993	2,100,000	30
1994	2,100,000	27
1995	2,100,000	38
1996	2,144,000[a]	39
1997	1,840,200[b]	34
1998	2,351,300	42

[a] Marcia S. Smith, *Space Stations,* (Washington, DC: Congressional Research Service, The Library of Congress, 1999), p. CRS-10. Neither the appropriations bill (*Making Appropriations for Fiscal Year 1996 To Make a Further Downpayment Toward a Balanced Budget, and for Other Purposes,* Public Law 104-134, 104th Congress, 1st sess, [April 26, 1996]), nor Conference Report H. Rept.104-537 for FY 1996, provided any figure at all for the Space Station.

[b] Authorized amount; no amount for the Space Station specified in appropriations bill.

Table 3–2. Authorized/Appropriated Budget (FY 1989–FY 1998)
(in thousands of dollars)

	1989	1990	1991	1992	1993	1994
Research and Development Appropriation[a]	4,191,700	5,366,050	5,600,000	6,413,800	7,089,300	7,509,300
Space Station Authorization	900,000	1,800,000	2,907,000[b]	2,028,900[c]	2,100,000	1,900,000[d]
Space Station Appropriation[e]	900,000	1,800,000[f]	1,900,000[g]	2,029,000[h]	2,100,000	2,100,000[i]

	1995	1996	1997	1998
HSF Appropriation[j]	5,573,900	5,456,600	5,362,900	5,506,500
Space Station Authorization	2,120,900[k]	2,121,000[l]	1,840,200	2,121,300[m]
Space Station Appropriation[n]	2,100,000[o]	2,144,000[p]	1,800,000	2,351,300[q]
Russian Cooperation Authorization	150,100	100,000[r]	100,000	—[s]

[a] Authorized and appropriated amounts for individual life sciences and microgravity science categories were not included in budget bills, so they cannot be included in this table.

[b] House multiyear authorization bill was "laid aside." Senate multiyear authorization bill (S.916) was agreed to by House. Bill did not go to President for signature.

[c] Of this amount, $18 million was authorized for an Assured Crew Return Vehicle.

[d] Bill was passed by House and sent to the Senate, but the Senate never acted on it and there was no bill passed and signed by the President.

[e] From annual appropriations bills.

[f] Of this amount, $750 million was not to be available until June 1, 1990.

[g] Not in H.R. 5158. Added in Conference Committee, October 25, 1990, and signed into law, Public Law 101-507.

[h] Amount was specified in Conference Report, October 2, 1991, and was not included in text of appropriations bill. A stated appropriated amount that was greater than the authorized amount was most likely due to rounding in the appropriations budget document.

[i] Appropriated "space station activities, including payloads," as stated in Conference Report on H.R. 2491, House of Representative, October 4, 1993. Of this amount, no more than $160,000,000 million was to be available for termination costs connected with Space Station *Freedom* contracts, no more than $172,000,000 million was to be for Space Station operations and utilization capability development, and no more than $99,000,000 million was to be for supporting development. Of the total amount appropriated for the Space Station, not more than $1,100,000,000 billion was to be made available before March 31, 1994. Not more than $100,000,000 million was to be used to support cooperative space ventures between the United States and Russia, of which no more than $50,000,000 million was to be only for space transportation capability development activities and $50,000,000 million was to be only for space science activities other than life sciences.

j   Space Shuttle appropriations are included in chapter 2, Launch Systems, of this volume. The HSF appropriation began to be used for the Space Station and Space Shuttle budgeted amounts except for a small amount allocated to the Space Station through the Science, Aeronautics and Technology appropriation.

k   An amendment to the authorization bill required the NASA Administrator to submit a five-year program plan that showed "the total amount of estimated expenditures and proposed appropriations necessary to support the programs, projects, and activities…." The authorization bill was passed by both the House and Senate but was not sent to the President for signature. The authorization bill was passed October 5, 1994, after passage of the appropriations bill (September 28, 1994).

l   The *International Space Station Authorization Act of 1995*, (H.R. 1601, 104th Congress, 1st sess.) authorized appropriations for the ISS for FY 1996 "and all subsequent fiscal years…through fiscal year 2002…." It specified that the total amount authorized not exceed \$13,141,000,000 billion, and that amount was "to remain available until expended, for complete development and assembly of, and to provide for initial operations…of the International Space Station." The bill authorizing Space Station appropriations was separate from the bill authorizing all other NASA appropriations. The bill was referred to the Senate Commerce Committee but did not become law.

m   The authorization bill passed by the House was referred to the Senate Committee where it remained.

n   From annual appropriations bills.

o   The amount of \$2.1 billion appropriated for the Space Station was stated in the Conference Report on H.R. 4624, Departments of Veterans Affairs and Housing and Urban Development, and Independent Agencies Appropriations Act of 1995 (House of Representatives–September 12, 1994) that accompanied the bill signed by the President. The amount was not directly stated in the appropriations bill. This Congress, the 103rd, defeated five attempts to terminate the Space Station program in NASA funding bills, and three other attempts in broader legislation. The amount of \$2.113 billion was stated in Marcia S. Smith, *Space Stations*, IB93017, updated December 12, 1996, CRS Issue Brief, *http://www.fas.org/spp/civil/crs/93-017.htm#legn* (accessed June 25, 2005).

p   Smith, *Space Stations*, 1999, p. CRS-10. Neither the appropriations bill (*Making Appropriations for Fiscal Year 1996 To Make a Further Downpayment Toward a Balanced Budget, and for Other Purposes*, Public Law. 104-134, 104th Congress, 1st sess. (April 26, 1996)) nor conference report H. Rept.104-537 for FY 1996 gives any figure at all for the Space Station.

q   The appropriations act stated that only \$1.5 billion should be made available before March 31, 1998. The amount of \$2,351,300,000 billion for the Space Station was introduced in House Report 105-297. This exceeded the authorized ceiling of \$2.1 billion by \$230 million but was below the \$430 million requested by NASA. It was agreed that the appropriation would include "\$80,000,000 from funds in the mission support account identified by the Agency (\$25,000,000 from the Tracking and Data Relay Satellite (TDRS), \$20,000,000 from environmental programs, \$30,000,000 from Research Operations Support, and \$5,000,000 from facilities), \$100,000,000 in addition to the Agency's request, and \$50,000,000 by reallocation from within the amounts requested in the Human Space Flight account."

r   The authorization for Russian Cooperation was in a separate bill from the Space Station authorization. H.R. 2405, 1st session, 104th Congress, "National Aeronautics and Space Administration Authorization Act of 1995" was passed by the House and sent to the Senate. No action was taken by the Senate.

s   Budget category not shown in authorization or appropriation bill.

Table 3–3. Programmed Budget (FY 1989–1998) (thousands of dollars)[a]

	1989	1990	1991	1992	1993	1994	1995	1996	1997	1998
Research and Development					HSF					
Spacelab[b]	87,600	93,700	129,300	99,200	112,800	125,500	90,000	86,700	40,100	9,100
Space Station (Total)	900,000	1,749,623	1,900,000	—	2,162,000	1,939,200	1,889,600	2,143,600	2,148,600	2,331,300
U.S./Russian Cooperative Program and Program Assurance	—			—	—	70,800	50,100	—		—
Russian Space Agency Contract Support/U.S.-Russian Cooperation			—	—		100,000	100,000	100,000	100,000	—
Mir Support		—	—	—	—	70,800	50,100	—	—	
Russian Program Assurance		—	—	—	—	—	—	—	200,000	110,000
Space Station Development	842,000	1,661,223	1,790,700	1,996,745	2,125,000	1,918,200	1,749,400	1,746,200	1,809,900	1,604,800
Development–Management and Integration	187,700	—	—	—	—	—	—	—	—	—

Table 3–3. Programmed Budget (FY 1989–1998) (thousands of dollars)[a] (Continued)

	1989	1990	1991	1992	1993	1994	1995	1996	1997	1998
Development–Pressurized Modules	155,500	—	371,200	—	—	—	—	—	—	—
Development–Assembly Hardware/Subsystems	267,200	—	731,200	—	—	—	—	—	—	—
Development–Platforms and Servicing	51,200	—	2,800	—	—	—	—	—	—	—
Development–Power Systems	124,000	—	292,800	—	—	—	—	—	—	—
Development–Operations/Utilization Capability	56,400	—	149,800	—	—	—	—	—	—	—
Development–Flight Hardware	—	—	—	—	2,085,500	1,609,700	1,319,900	1,468,900	1,540,700	1,461,000
Development–Test, Manufacturing and Assembly	—	—	—	—	—	99,000	91,900	73,500	95,700	97,400
Development–Operations Capability and Construction	—	—	—	—	—	151,000	169,800	112,600	115,700	—

Table 3–3. Programmed Budget (FY 1989–1998) (thousands of dollars)[a] (Continued)

	1989	1990	1991	1992	1993	1994	1995	1996	1997	1998
Development–Transportation Support	—	—	—	—	25,700	58,500	117,600	63,500	55,700	45,500
Development–Flight Technology Demonstrations	—	—	—	—	—	—	30,000	12,900	2,100	900
Development–Operations Capability and Construction	—	—	—	—	13,800	—	20,200	14,800	—	—
Assured Crew Return Vehicle	—	—	—	6,000	7,000	—	—	—	—	—
Flight Telerobotic System/Servicer	46,000	79,400	—	—	—	—	—	—	—	—
Space Station Utilization	—	—	—	—	30,000	21,000	31,300	—	—	—
Space Station Operations	—	—	—	—	—	—	108,900	120,000	142,600	500,200
Shuttle/Spacelab Payload Mission Management and Integration	—	—	—	—	94,100	108,700	102,300	53,600	24,200	—

Table 3–3. Programmed Budget (FY 1989–1998) (thousands of dollars)[a] (Continued)

	1989	1990	1991	1992	1993	1994	1995	1996	1997	1998
Space Station Integration Planning and Attached Payloads	8,000	9,000	3,000	—	—	—	—	—	—	—
Space Station Research	—	—	—	—	—	—	—	277,400	196,100	226,300
Space Station Transition Definition/Advanced Programs	12,000	9,000	3,000	—	—	—	—	—	—	—
Life Sciences	79,100	106,051	137,400	157,650	139,500	186,800	140,500	109,600	—	—
Life Sciences Flight Experiments	—	—	—	94,700	—	—	—	—	—	—
Human Spaceflight and Systems Engineering	28,600	40,678	58,300	—	—	—	—	—	—	—
Space Biological Sciences	10,100	21,067	22,800	—	—	—	—	—	—	—
Life Sciences Research and Analysis	38,200	40,306	44,800	50,700	52,900	55,100	50,700	55,200	—	—

Table 3–3. Programmed Budget (FY 1989–1998) (thousands of dollars)[a] (Continued)

	1989	1990	1991	1992	1993	1994	1995	1996	1997	1998
Life and Microgravity Sciences (Total)	—	—	—	157,650	407,500	—	467,400	304,200	243,700	—
Centrifuge	—	—	—	—	5,500	—	—	—	—	—
Search for Extraterrestrial Intelligence	2,200	4,000	11,500	12,250	—	—	—	—	—	—
Life Sciences Flight Program	—	—	—	—	81,100	—	89,800	54,400	—	—
Advanced Human Support Technology	—	—	—	—	—	—	—	—	19,700	—
Biomedical Research and Countermeasures Program	—	—	—	—	—	—	—	—	44,100	—
Gravitational Biology and Ecology Program	—	—	—	—	—	—	—	—	33,600	—

[a] Empty cells indicate that no programmed amounts were shown in the annual budget. See the individual budget tables that follow for additional details.

[b] Included in the Space Transportation Capability Development budget category.

Table 3–4. Spacelab Funding History (in thousands of dollars)

Year (Fiscal)	Submission	Programmed
1989	80,400/88,600	87,600
1990	98,900/95,600	93,700
1991	130,700/129,300	129,300
1992	150,200/96,000	99,200
1993	122,600/114,459	112,800
1994	139,900/125,500	125,500
1995	92,300/98,600	90,000
1996	97,000/86,700	86,700
1997	62,400/50,300	40,100
1998	14,200/11,900	9,100

Table 3–5. Space Station (Total) Funding History (in thousands of dollars)

Year (Fiscal)	Submission	Authorization	Appropriation	Programmed	Budget Authority (Full Cost)
1989	967,400/900,000	900,000	900,000	900,000	n/a
1990	2,050,200/1,749,623	1,800,000	1,800,000	1,749,623	n/a
1991	2,451,000/1,900,000[a]	2,907,000	1,900,000	1,900,000	n/a
1992	2,028,900/2,028,900	2,028,900	2,029,000	—[b]	n/a
1993	2,250,000/2,122,467	2,100,000	2,100,000	2,162,000	n/a
1994[c]	—/1,937,000	1,900,000	2,100,000	1,939,200	2,106,000
1995	1,889,600/1,889,600	1,889,600	2,100,000	1,889,600	2,112,900
1996	1,833,600/1,863,600	2,121,000	2,144,000[d]	2,143,600	2,143,600
1997	1,802,000/2,148,600	1,840,200	1,800,000	2,148,600	2,148,600
1998	2,121,300/2,501,300	2,121,300	2,351,300	2,331,300	2,121,300

[a]   Congress reduced the FY 1991 funding requested for the Space Station by $551.0 million. A study to restructure the program was incomplete and did not allow for sufficient definition of requirements to develop detailed estimates.

[b]   Program was being restructured and no programmed amount was shown.

[c]   Space Station Freedom program was budgeted within the Office of Space Systems Development.

[d]   Smith, Space Stations, 1999, p. CRS-10. Neither the appropriations bill (Making Appropriations for Fiscal Year 1996 To Make a Further Downpayment Toward a Balanced Budget, and for Other Purposes, Public Law 104-134, 104th Congress, 1st sess, (April 26, 1996)), nor conference report H. Rept.104-537, gives any figure at all for the Space Station for FY 1996.

Table 3–6. *U.S./Russian Cooperative Program Funding History (in thousands of dollars)*

Year (Fiscal)	Submission	Authorization	Programmed	Budget Authority (Full Cost)
1994	—[a]	—	70,800	70,800
1995	—/50,100	50,100	50,100	50,100
1996	29,200/29,200	100,000	—	29,200
1997	38,200/50,000	100,000/100,000	—[b]	38,200

[a] Budget category not established at time of budget submission.
[b] Budget line item was discontinued. New budget line item, U.S./Russian Cooperation and Program Assurance, was established.

Table 3–7. Russian Space Agency Contract Support[a] Funding History
(in thousands of dollars)

Year (Fiscal)	Submission	Programmed
1994	—	100,000
1995	—[b]/100,000	100,000
1996	100,000/100,000	100,000
1997	100,000/100,000	300,000[bc]
1998	—/50,000	—

[a] Changed to U.S./Russian Cooperation and Program Assurance budget category in FY 1998 budget.
[b] Budget category not established at time of initial budget submission.
[c] Consisted of $100,000,000 million from the disestablished budget category of U.S./Russian Cooperative Program and $200,000,000 million reallocated from elsewhere within the HSF account.

Table 3–8. Mir *Support Funding History*
(in thousands of dollars)

Year (Fiscal)	Submission	Programmed
1994	—[a]	70,800
1995	—/50,100	50,100
1996	29,200/29,200	—
1997	38,200/—	—

[a] Budget category not established at time of budget submission.

Table 3–9. Russian Program Assurance Funding History
(in thousands of dollars)

Year (Fiscal)	Submission	Programmed
1997	—[a]	200,000
1998	—/50,000	110,000

[a] No budget category at time of budget submission.

Table 3–10. Space Station Development Funding History
(in thousands of dollars)

Year (Fiscal)	Submission	Programmed	Budget Authority (Full Cost)
1989	935,400/842,000	842,000	n/a
1990	1,970,200/1,661,223	1,661,223	n/a
1991[a]	2,299,800/—[b]	1,790,700	n/a
1992	—[c]/2,022,900	1,996,745	n/a
1993	2,200,000/2,115,467	2,125,000[d]	n/a
1994	—[e]/1,911,000	1,918,200	1,918,200
1995	1,662,000[f]/ 1,752,400[g]	1,749,400	1,749,400
1996	1,612,800[h]/1,696,200	1,746,200	1,746,200
1997	1,513,200/1,766,300	1,809,900	1,766,300
1998	1,386,100/1,789,900	1,604,800[i]	1,386,100

[a] The distribution by program element (Development, Flight Telerobotic Servicer, Operations, and Advanced Programs) for the FY 1991 revised estimate and the FY 1992 budget estimate were under review, pending the preliminary results of the 90-day study to restructure the Space Station program, directed by the conference report accompanying the FY 1991 *Departments of Veterans Affairs and Housing and Urban Development, and Independent Agencies Appropriations Act, 1991,* Public Law 101-507, 101st Congress, 2nd sess, (November 5, 1990).
[b] No revised budget submitted shown.
[c] The FY 1992 budget estimate was submitted before completion of the Space Station restructuring activity, and no project estimates were available.
[d] Included $13,800,000 million for construction of facilities.
[e] No initial budget estimate shown for this category.
[f] Included $20,200,000 million for construction of facilities.
[g] Included $20,200,000 million for construction of facilities.
[h] Included $14,800,000 million for construction of facilities.
[i] Budget category was renamed "Vehicle."

Table 3–11. Development–Management and
Integration Funding History (in thousands of dollars)

Year (Fiscal)	Submission	Programmed
1989	169,400/187,700	187,700
1990	230,200/198,258	—[a]
1991	248,000/—	—

[a] No programmed amount shown.

Table 3–12. Development–Pressurized Modules[a] Funding History (in thousands of dollars)

Year (Fiscal)	Submission	Programmed
1989	188,000/155,500	155,500
1990	366,000/303,900	—
1991	522,100/—	371,200
1992	—/433,500	—
1993	448,400/—	—[b]

[a] Consisted of Work Package 1, managed by Marshall Space Flight Center.
[b] No programmed amounts shown. Work packages were restructured into other budget categories with restructuring of program.

Table 3–13. Development–Assembly Hardware/Subsystems[a] Funding History (in thousands of dollars)

Year (Fiscal)	Submission	Programmed
1989	288,000/263,200	267,200
1990	762,000/666,300	—[b]
1991	872,600/—	731,200
1992	—/764,600	—
1993	766,200/—	—[c]

[a] Consisted of Work Package 2, managed by Johnson Space Center.
[b] No programmed amounts shown.
[c] No programmed amounts shown.

Table 3–14. Development–Platforms and Servicing[a] Funding History (in thousands of dollars)

Year (Fiscal)	Submission	Programmed
1989	56,000/51,200	51,200
1990	130,000/107,500	—
1991	34,100/—[b]	2,800[c]

[a] Consisted of Work Package 3, managed by Goddard Space Flight Center.
[b] All activities associated with this budget category were eliminated in the 1991 restructuring.
[c] Costs associated with termination of contracts.

Table 3–15. Development–Power Systems[a] Funding History (in thousands of dollars)

Year (Fiscal)	Submission	Programmed
1989	154,000/120,000	124,000
1990	298,000/249,925	—[b]
1991	—/367,900	292,800
1992	—/306,500	—[c]
1993	350,400	—

[a] Consisted of Work Package 4, managed by Lewis Research Center.
[b] No programmed amount shown.
[c] No programmed amount shown.

Table 3–16. Development–Operations/Utilization Capability Funding History (in thousands of dollars)

Year (Fiscal)	Submission	Programmed
1989	80,000/64,400	56,400
1990	184,000/135,340	—[a]
1991	255,100	149,800
1992	—/253,600	—[b]
1993	377,100/—	—

[a] No programmed amount shown.
[b] No programmed amount shown.

Table 3–17. Development–Flight Hardware Funding History (in thousands of dollars)

Year (Fiscal)	Submission	Programmed
1993	—[a]	2,085,500
1994	—/1,642,400	1,609,700
1995	1,127,000/1,319,900	1,319,900
1996	1,277,200/1,413,300	1,468,900
1997	1,244,400/1,480,500	1,540,700
1998	1,157,900/1,529,000	1,461,000

[a] Budget category introduced with the redesigned Space Station. Budget categories were restructured.

*Table 3–18. Development–Test, Manufacturing and Assembly
Funding History (in thousands of dollars)*

Year (Fiscal)	Submission	Programmed
1994	—ᵃ/87,600	99,000
1995	117,000/94,900	91,900
1996	90,300/68,600	73,500
1997	78,200/97,300	95,700
1998	93,600/97,400	97,400

ᵃ Budget category introduced with the redesigned Space Station, not at time of initial budget submission. Budget categories were restructured.

*Table 3–19. Development–Operations Capability and Construction
Funding History (in thousands of dollars)*

Year (Fiscal)	Submission	Programmed
1994	—ᵃ/151,000	151,000
1995	257,800/169,800	169,800
1996	137,100/117,100	112,600
1997	111,300/130,700	115,700
1998	85,400/115,100	—

ᵃ Budget category introduced with the redesigned Space Station, not at time of initial budget submission. Budget categories were restructured.

*Table 3–20. Development–Transportation Support
Funding History (in thousands of dollars)*

Year (Fiscal)	Submission	Programmed
1993	—ᵃ	25,700
1994	—/30,000	58,500
1995	100,000/117,600	117,600
1996	83,000/74,100	63,500
1997	76,100/55,700	55,700
1998	47,800/47,000	45,500

ᵃ Budget category introduced with the redesigned Space Station. Budget categories were restructured.

Table 3–21. Development–Flight Technology Demonstrations Funding History (in thousands of dollars)

Year (Fiscal)	Submission	Programmed
1994	—a	—
1995	40,000/30,000	30,000
1996	10,400/8,300	12,900
1997	3,200/2,100	2,100
1998	1,400/1,400	900

[a] Budget category was introduced with the redesigned Space Station. Budget categories were restructured.

Table 3–22. Development–Operations Capability and Construction Funding History (in thousands of dollars)

Year (Fiscal)	Submission	Programmed
1993	—a	13,800
1994	—/—	—
1995	20,200/20,200	20,200
1996	14,800/14,800	14,800b

[a] Budget category introduced with the redesigned Space Station. Budget categories were restructured. Included in Construction of Facilities appropriation.
[b] Included with Operations beginning with FY 1997.

Table 3–23. Shuttle/Spacelab Payload Mission Management and Integration Funding History (in thousands of dollars)a

Year (Fiscal)	Submission	Programmed
1993b	101,100/94,018	94,100
1994	117,700c/111,500	108,700
1995	112,400/113,900	102,300
1996	85,400/77,600	53,600
1997	54,400/24,200	24,200
1998	6,900/4,900d	—e

[a] This category included funds to manage the mission planning, integration, and execution of all NASA Spacelab and attached Shuttle payloads.
[b] Transferred to OLMSA program.
[c] Included in OLMSA budget.
[d] Changed to Mission Integration Function in OLMSA realignment of budget categories that occurred with FY 1999 congressional budget submission (and revisions to FY 1998 budget submission).
[e] No programmed funds in this budget category. Included with OLMSA funds.

Table 3–24. Space Station Integration Planning and Attached Payloads Funding History (in thousands of dollars)

Year (Fiscal)	Submission	Programmed
1989	8,000	8,000
1990	23,000/4,975	4,975
1991	15,000/3,000	3,000
1992	—[a]	—[b]

[a] Functions associated with Space Station Integration Planning were transferred to the Materials Processing budget category beginning in FY 1992.
[b] No programmed funds in this budget category.

Table 3–25. Assured Crew Return Vehicle Funding History (in thousands of dollars)

Year (Fiscal)	Submission	Programmed
1992	—[a]	6,000
1993	15,000/7,000	7,000
1994	—/5,000	—
1995	—[b]	—

[a] Budget category not established at time of budget submission.
[b] The redesigned Space Station was to use an Assured Crew Return Vehicle based on a Soyuz vehicle and launched on a Russian booster for rescue and crew rotation. The Soyuz Assured Crew Return Vehicle was a Russian element of the Space Station and required no U.S. funding in FY 1995.

Table 3–26. Flight Telerobotic System/Servicer Funding History (in thousands of dollars)

Year (Fiscal)	Submission	Programmed
1989	20,000/46,000	46,000
1990	15,000[a]/79,400[b]	79,400
1991	106,300/—[c]	—

[a] NASA was actively pursuing approaches to encourage the private sector to invest in the Flight Telerobotic Servicer (FTS). The requested funding was to provide for supporting development activities.
[b] After consideration of industry responses, a decision was made that the FTS was not a viable candidate for full commercial development. The increased budget estimate was consistent with the decision to provide for a NASA procurement through a prime contractor.
[c] All activities associated with this budget category were eliminated in the 1991 restructuring.

Table 3–27. Space Station Utilization Support
Funding History (in thousands of dollars)

Year (Fiscal)	Submission	Programmed
1993	—a	30,000
1994	—/21,000	21,000
1995	96,600/28,300	31,300
1996	67,900/47,400	—b
1997	72,100/—	—

a Budget category not established at time of budget submission.
b Budget category disestablished; included with Space Station Research budget category beginning in FY 1997.

Table 3–28. Space Station Operations Funding History
(in thousands of dollars)

Year (Fiscal)	Submission	Programmed	Budget Authority (Full Cost)
1990	25,000/—a	—	n/a
1991	8,900/—b	—	n/a
1992	—	—	n/a
1993	35,000/—	—	n/a
1994	—	—	n/a
1995c	131,000/108,900	108,900	108,900
1996	152,900/120,000	120,000	120,000
1997	216,700/177,600	142,600	177,600
1998	490,100/490,100	500,200d	490,100

a Deletion of the requested amount reflected a delay due to program rephasing associated with the rebaselining activities of the configuration baseline review, which indicated that FY 1990 resources would not be required to meet the revised program milestones.
b Amounts designated for Space Station Operations were deferred or canceled as the Space Station schedule slipped.
c Space Station Operations budget category included vehicle operations and ground and transportation operations.
d Included construction of facilities.

*Table 3–29. Space Station Research[a] Funding History
(in thousands of dollars)*

Year (Fiscal)	Submission	Programmed	Budget Authority (Full Cost)[b]
1994	—[c]	—[d]	187,800
1995	—	—	254,600
1996	—	277,400	277,400
1997	—/204,700	196,100	204,700
1998	245,100/221,300	226,300	245,100

[a] Included *Mir* research and support, utilization support, research facilities, science utilization (HSF), and science utilization (Construction of Facilities).

[b] Reflected amounts used for Space Station activities from Science, Aeronautics, and Technology appropriation and from Research and Development and Construction of Facilities appropriations.

[c] Budget category not established.

[d] No programmed amount in HSF appropriation. Full cost budget authority included Space Station Research amounts from other appropriation categories (Science, Aeronautics, and Technology and former appropriation categories of Research and Development and Construction of Facilities).

*Table 3–30. Space Station Transition Definition/Advanced Programs
Funding History (in thousands of dollars)*

Year (Fiscal)	Submission	Programmed
1989	12,000/12,000	12,000
1990	25,000/9,000[a]	9,000
1991	36,000	3,000

[a] Name of budget category was changed to Advanced Programs, consisting of advanced system studies, advanced development, and support for human exploration.

Table 3–31. Life Sciences Funding History
(in thousands of dollars)

Year (Fiscal)	Submission	Programmed
1989	101,700	79,100
1990	124,200/106,278	106,051
1991	163,000/138,000	137,400
1992	183,900/145,800	157,650
1993	177,200/140,550	139,500
1994	143,900/188,200[a]	186,800
1995	145,600/140,700	140,500
1996	—[b]/136,400	109,600
1997	144,300/97,400	—[c]
1998	85,500/[d]	—[e]

[a] Realignment of budget categories. Life Sciences under the Office of Space Science moved to OLMSA. OLMSA had programs for Life Sciences and Microgravity Science Research, Shuttle/Spacelab Payload, and Mission Management and Integration, and corresponding budget categories.
[b] Became subcategory under OLMSA. Included Research and Analysis and Flight Program budget categories. No initial submission for this budget category.
[c] No programmed amount shown.
[d] Realigned to Advanced Human Support Technology Program, Biomedical Research and Countermeasures Program, and Gravitational Biology and Ecology Program.
[e] No programmed amount shown.

Table 3–32. Life Sciences Flight Experiments
Funding History (in thousands of dollars)

Year (Fiscal)	Submission	Programmed
1989	54,500/—[a]	—
1990	—	—
1991	—	—
1992	—	94,700
1993	89,700[b]/81,089	—
1994	94,700/—[c]	—

[a] Budget category appeared in initial submission for FY 1989 budget but not in revised budget or in programmed amount. This budget category did not reappear in the budget until the FY 1993 budget submission (which also listed the FY 1992 programmed amount).
[b] Moved to Space Applications Microgravity Flight Experiments.
[c] Budget category disestablished. No submission or programmed amount shown.

Table 3–33. Human Spaceflight and Systems Engineering
Funding History (in thousands of dollars)

Year (Fiscal)	Submission	Programmed
1989	—[a]	28,600
1990	42,800/40,678	40,678
1991	71,000/58,300	58,300
1992	58,600/70,100[b]	—
1993	71,400/—[c]	—

[a] Budget category not established at time of budget submission.
[b] Included all Spacelab flight program activities.
[c] Budget category disestablished. No submitted or programmed amount for this category shown.

Table 3–34. Space Biological Sciences
Funding History (in thousands of dollars)

Year (Fiscal)	Submission	Programmed
1989	—[a]	10,100
1990	27,600/21,200	21,067
1991	32,000/22,800	22,800
1992	31,100/14,600	—[b]
1993	18,300/—	—

[a] Budget category not established at time of budget submission.
[b] Budget category disestablished. No further funds requested or programmed.

Table 3–35. Life Sciences Research and Analysis
Funding History (in thousands of dollars)

Year (Fiscal)	Submission	Programmed
1989	47,200/38,200	38,200
1990	47,000/40,400	40,306
1991	47,900/44,800	44,800
1992	64,700/47,600	50,700
1993[a]	55,600/53,940	52,900
1994	49,200/55,100	55,100
1995	51,900/50,700	50,700
1996[b]	50,400/55,200	55,200
1997	49,800/58,000	—[c]
1998	50,000/—	—

[a] Moved to Space Applications Microgravity Research and Analysis.
[b] Became Research and Analysis dealing specifically with Life Sciences programs under OLMSA.
[c] Budget category disestablished. No funds requested or programmed for this budget category.

Table 3–36. Lifesat/Radiation Biology Initiative History
(in thousands of dollars)

Year (Fiscal)	Submission	Programmed
1992	15,000/—[a]	—

[a] All funding for the Lifesat program was deleted per congressional direction. The program was additionally reduced by $5 million as part of the congressionally-directed general reduction to space science and applications.

Table 3–37. Life and Microgravity Sciences (Total)
Funding History (in thousands of dollars)[a]

Year (Fiscal)	Submission	Programmed	Budget Authority (Full Cost)[b]
1992	—[c]	157,650[d]	n/a
1993	177,200/140,550	407,500	n/a
1994	351,000/515,300[e]	—[f]	96,000
1995	470,900/—	467,400	158,200
1996	—/488,500	304,200	210,800
1997	498,500/243,700	243,700	267,800
1998	214,200/214,200	—[g]	345,000

[a] OLMSA combined several of the budget categories formerly from the Life Sciences budget category within the Office of Space Science and Applications, specifically from the Life Sciences and Materials Programs together with their supporting Spacelab management function. OLMSA consisted of Life Sciences, Microgravity Science Research, Shuttle/Spacelab Payload Mission Management and Integration, and Space Station Payload Facilities budget categories. Materials Processing, previously funded under Space Applications, was renamed Microgravity Research, and remained a distinct element within the new structure. The addition of the Shuttle/Spacelab Payload Mission Management and Integration, which was transferred from Physics and Astronomy, served to consolidate the on-orbit research in these disciplines together with their associated space access infrastructure.

[b] Did not include all Life and Microgravity Sciences activities. Included only items related to the Space Station. From FY 1994–FY 1998, these totaled: Space Station Facilities ($694,700,000 million), Life Sciences and Aerospace Medicine ($122,900,000 million), Microgravity Research ($158,500,000 million), and STS (Space Shuttle)/Spacelab Mission Management ($83,700,000 million).

[c] Office of Life and Microgravity Sciences and corresponding budget not established at this time.

[d] Former funding structure.

[e] Reflected new funding structure.

[f] No programmed amount shown.

[g] No programmed amount shown.

Table 3–38. Centrifuge Funding History (in thousands of dollars)

Year (Fiscal)	Submission	Programmed
1993	18,400/5,521	5,500[a]
1994	—[b]	—

[a] Centrifuge budget category was removed from Space Science budget.

[b] No funds were included in the FY 1994 request for the centrifuge facility. Funding plans were being reevaluated in accordance with NASA's reexamination of plans for the Space Station.

Table 3–39. Search for Extraterrestrial Intelligence
Funding History (in thousands of dollars)

Year (Fiscal)	Submission	Programmed
1989	—[a]	2,200[b]
1990	6,800/4,000	4,000
1991	12,100/12,100	11,500
1992	14,500/13,500	12,250
1993	13,500/—[c]	—

[a] Budget category not established at time of budget submission.
[b] Funded as part of Research and Analysis budget category.
[c] The Search for Extraterrestrial Intelligence (SETI) program was deleted from the Space Life Science program. Technology developed under the program was incorporated in the Towards Other Planetary System (TOPS) program in the Planetary Exploration Research and Analysis program, in accordance with congressional direction. Per congressional direction, funding was terminated for SETI within the Life Sciences program.

Table 3–40. Life Sciences Flight Program
Funding History (in thousands of dollars)

Year (Fiscal)	Submission	Programmed
1993	—[a]	81,100
1994	—[b]/133,100	—[c]
1995	93,700/—[d]	89,800
1996[e]	—[f]/81,200	54,400
1997	56,400/39,400	—[g]

[a] Budget category was not included in FY 1993 budget submission.
[b] No initial budget submission shown for this budget category.
[c] No programmed amount shown.
[d] Budget category not listed in revised budget submission.
[e] Became Research and Analysis budget category dealing specifically with Life Sciences programs under OLMSA.
[f] No initial budget submission for this budget category.
[g] No programmed amount listed.

Table 3–41. Advanced Human Support Technology Program[a]
Funding History (in thousands of dollars)

Year (Fiscal)	Submission	Programmed
1997	—	19,700
1998	—[b]/17,900	—[c]

[a] Formerly Life Sciences budget category under the Office of Space Science.
[b] Budget category did not appear until revised budget request was submitted.
[c] No programmed amount shown.

Table 3–42. Biomedical Research and Countermeasures Program[a]
Funding History (in thousands of dollars)

Year (Fiscal)	Submission	Programmed
1997	—	44,100
1998	—/40,600	—

[a] Formerly Life Sciences budget category under the Office of Space Science.

Table 3–43. Gravitational Biology and Ecology Program[a]
Funding History (in thousands of dollars)

Year (Fiscal)	Submission	Programmed
1997	—	33,600
1998	—/30,000	—

[a] Formerly Life Sciences budget category under the Office of Space Science.

Table 3–44. Orbiter Characteristics

Component	Characteristics
Length	37.2 m (122.2 ft)
Height	~17 m (56.7 ft)
Vertical stabilizer	8 m (26.2 ft)
Wingspan	23.8 m (78.1 ft)
Body flap	
Area	12.6 sq m (135.6 sq ft)
Width	6.1 m (20 ft)
Aft fuselage	
Length	5.5 m (18 ft)
Width	6.7 m (22 ft)
Height	6.1 m (20 ft)
Mid fuselage	
Length	18.3 m (60 ft)
Width	5.2 m (17.1 ft)
Height	4.0 m (13.1 ft)
Airlock	
Inside diameter	160 cm (5.2 ft)
Length	211 cm (6.9 ft)
Minimum clearance	91.4 cm (3 ft)
Opening capacity	46 cm by 46 cm by 127 cm (1.5 ft by 1.5 ft by 4.2 ft)
Payload bay	4.6 m by 18.3 m (15 ft by 60 ft)
Forward fuselage crew cabin	71.5 cu m (2,525 cu ft)
Payload bay doors	
Length	18.3 m (60 ft)
Diameter	4.6 m (15.1 ft)
Surface area	148.6 sq m
Weight	1,480 kg (3,263 lb)
Wing	
Length	18.3 m (60 ft)
Maximum thickness	1.5 m (4.9 ft)
Elevons	4.2 m and 3.8 m (13.8 ft and 12.5 ft)
Tread width	6.9 m (22.7 ft)
Structure type	Semi-monocoque
Structure material	Aluminum
Gross takeoff weight	Variable depending on payload and on-board consumables
Nominal landing weight	Variable
Inert weight (approx.)	74,844 kg (165,003 lb)

Table 3–44. Orbiter Characteristics (Continued)

Component	Characteristics
Main engines	
Number	3
Average thrust (104%)	1,752 kN (393,800 lb) at sea level
Nominal burn time	522 seconds

Table 3–45. Endeavour *Construction Milestones*

Date	Milestone
February 15, 1982	Begins structural assembly of crew module
July 31, 1987	Contract awarded to Rockwell
September 28, 1987	Begins structural assembly of aft-fuselage
December 22, 1987	Wings arrive at Palmdale, California, from Grumman
August 1, 1987	Final assembly begins
July 6, 1990	Final assembly completed
April 25, 1991	Rollout from Palmdale, California
May 7, 1991	Delivery to Kennedy Space Center
April 6, 1992	Flight readiness firing
May 7, 1992	First flight (STS-49)

Table 3–46. Space Shuttle Payload Accommodation[a]

Location/Accommodation	Description
Payload Bay	
Attached	Attached payloads were generally large payloads (14.5 ft/4.4 m maximum diameter) mounted directly to the payload bay attach fittings on an "across the bay" structure.
Deployable/Retrieval	Deployable/retrievable payloads were offered the same basic set of services as attached payloads with extensions allowing for mate/demate with attach hardware, remote command, and control, etc.
Sidewall	The sidewall payloads mounted to the orbiter's sidewall.
Payload Carriers	
Spacelab Pallet	Spacelab pallets were U-shaped platforms for mounting payloads. The pallets had hard points for mounting heavy equipment. Each pallet could hold up to 3 tons (2,722 kg) if the weight was evenly distributed. Each pallet was 13 ft (3.9 m) wide and 10 ft (3 m) long.
Mission Peculiar Equipment Support Structure (MPESS)	The MPESS was an A-frame structure spanning the width of the payload bay. Payloads could be mounted on the top and sides of the structure.
GAS	The GAS carrier system accommodated payloads in canisters mounted in the Shuttle payload bay on the sidewall or on a cross-bay truss structure.
Hitchhiker	The Hitchhiker carrier was intended for payloads requiring power, data, and command services. Hitchhiker provided real-time data transfer for experimenters and crew control/display capability.
Hitchhiker-Jr.	The Hitchhiker-Jr. (HH-Jr.) carrier provided mechanical and electrical interfaces similar to the GAS carrier but had avionics to monitor carrier and payload functions and power services.

Table 3–46. Space Shuttle Payload Accommodation[a] (Continued)

Location/Accommodation	Description
SPARTAN	SPARTAN was a reusable, three-axis stabilized, free-flying carrier providing extended mission flight opportunities for a variety of scientific studies in low-Earth orbit. SPARTAN was taken into orbit by the Space Shuttle, deployed, and operated via ground commands. The satellite was retrieved either on the same Shuttle mission or on a later mission and returned to the ground for reuse.
Space Experiment Module (SEM)	SEM was a canister assembly providing self-contained structure, power, command, and data storage capabilities for microgravity experiments.
Crew Compartment	
Middeck	The middeck offered accommodations in a pressurized environment for payloads that could be stowed within a middeck locker or mounted on an adapter plate that replaced one or more lockers.
Pressurized Modules	
Spacelab	Spacelab modules added significant "shirtsleeve" workspace and laboratory facilities to the Space Shuttle. (See following sections.)
SPACEHAB	SPACEHAB modules also added significant "shirtsleeve" workspace and laboratory facilities to the Space Shuttle. (See following sections.)

[a] "Payload Accommodations and Services," *http://shuttlepayloads.jsc.nasa.gov/flying/accommodations/accommodations.htm* (accessed July 12, 2005).

Table 3–47. Spacelab Module Characteristics[a]

Item	Characteristic
Diameter	4.06 m (13.3 ft)
Module length (1 segment)	2.70 m (8.9 ft)
Module shell material	2219-T851 aluminum
Electrical power	28 VDC +/- 4 VDC
Internal ambient temperature	18°-27°C (64.4°-80.6°F)
Humidity	30%-70% relative humidity
Air leakage	1.3 kg/day max (2.9 lb)
Payload mass	4,500 kg (9,921 lb) (long module)
Payload volume	22.2 cu m (784 cu ft) (long module)
Electrical power to payload	3.9 kW continuous 6.5 kW peak
Other features of payload	Optical window Airlock

[a] E. Vallerani, "Pressurised Module Elements from Spacelab to Columbus," in *Spacelab, 1983–1993; Ten Years Experience in Cooperative Manned Space Activities*; Proceedings from the CEAS European Forum, October 1993, Florence, Italy (Washington, DC: American Institute of Aeronautics and Astronautics, 1993), p. 16.

Table 3–48. Spacelab Missions (1989–1998[a])

Mission	Date	Purpose	Configuration
Astro-1/STS-35	December 2, 1990	Astronomy	Igloo plus 2 pallets
Spacelab Life Sciences (SLS)-1/STS–40	June 5, 1991	Space life sciences	Long module
International Microgravity Laboratory (IML)-1/STS-42	January 22, 1992	Microgravity studies	Long module
Atmospheric Laboratory for Applications and Science (ATLAS)-1/STS-45	March 24, 1992	Atmospheric studies	Igloo plus 2 pallets
United States Microgravity Laboratory (USML)-1/STS-50	June 25, 1992	Microgravity studies	Long module/ extended duration orbiter
Spacelab J1/STS-47	September 12, 1992	Microgravity and life sciences	Long module
ATLAS-2/STS-56	April 8, 1993	Atmospheric studies	Igloo plus 1 pallet
Spacelab D2/STS-55	April 26, 1993	Microgravity studies	Long module plus U.S. Microgravity Laboratory
SLS 2 LM/STS-58	October 18, 1993	Life sciences	Long module/ extended duration orbiter
IML-2/STS-65	July 8, 1994	Microgravity	Long module/ extended duration orbiter
ATLAS-3/STS-66	November 3, 1994	Atmospheric Physics	Igloo plus 2 pallets
Astro-2/STS-67	March 2, 1995	Astronomy	Igloo plus 2 pallets, extended duration orbiter
Spacelab-*Mir* LM/ STS-71	June 27, 1995	Life sciences	Long module
USML-2/STS-73	October 20, 1995	Microgravity	Long module/ extended duration orbiter
Life and Microgravity Spacelab (LMS) 1/ STS-78	June 20, 1996	Life and microgravity sciences	Long module/ extended duration orbiter
Microgravity Science Laboratory (MSL)-1/ STS-83[b]	April 4, 1997	Materials sciences	Long module/ extended duration orbiter

Table 3–48. Spacelab Missions (1989–1998ᵃ) (Continued)

Mission	Date	Purpose	Configuration
MSL-1R/STS-94 (reflight of MSL-1)	July 1, 1997	Materials sciences	Long module/ extended duration orbiter
Neurolab/STS-90	April 17, 1998	Neurobiological life sciences	Long module/ extended duration orbiter

ᵃ "Spacelab," European Space Agency, *http://www.esa.int/esapub/achievements/Sc72s4.pdf* (accessed July 22, 2005).
ᵇ Shortened mission due to concerns about one of the three fuel cells. Reflown on STS-94.

Table 3–49. SPACEHAB Missionsᵃ

Mission	Date	Payload
STS-57	June 21, 1993	SPACEHAB Module
STS-60	February 3, 1994	SPACEHAB Module
STS-63	February 3, 1995	SPACEHAB Module
STS-76	March 22, 1996	SPACEHAB Module
STS-77	May 19, 1996	SPACEHAB Module
STS-79	September 16, 1996	Logistics Double Module
STS-81	January 12, 1997	Logistics Double Module
STS-84	May 15, 1997	Logistics Double Module
STS-86	September 25, 1997	Logistics Double Module
STS-89	January 22, 1998	Logistics Double Module
STS-91	June 2, 1998	Logistics Single Module/ SPACEHAB Universal Communications System
STS-95	October 29, 1998	SPACEHAB Module

ᵃ "Past Missions," SPACEHAB, *http://spacehab.com/missions/past_shi.htm* (accessed July 5, 2005).

Table 3–50. Space Shuttle Extravehicular Activity (1989–1998[a])

Mission	Date	Astronaut	Individual EVA Time per Astronaut	Cumulative Time in Space	Description
STS-37	April 7, 1991	Ross and Apt	4 hr, 26 min	20 hr, 26 min	Deploy jammed Gamma Ray Observatory high-gain antenna
	April 8, 1991	Ross and Apt	5 hr, 47 min		Test Crew and Equipment Translation Aid cart and other EVA equipment
STS-49	May 10, 1992	Thuot and Hieb	3 hr, 43 min	50 hr, 52 min	Unsuccessful attempt to retrieve INTELSAT VI satellite and install perigee kick motor
	May 11, 1992	Thuot and Hieb	5 hr, 30 min		Unsuccessful attempt to retrieve INTELSAT VI satellite and install perigee kick motor
	May 13, 1992	Thuot, Hieb, and Akers	8 hr, 29 min		Retrieve INTELSAT VI satellite and install perigee kick motor
	May 14, 1992	Thornton and Akers	7 hr, 44 min		Test equipment for Space Station Freedom program (assembly of Station by EVA Methods experiment)
STS-54	January 17, 1993	Runco and Harbaugh	4 hr, 28 min	8 hr, 56 min	ISS preparation (Detailed Test Objective)
STS-57	June 25, 1993	Low and Wisoff	5 hr, 50 min	11 hr, 40 min	Hubble Space Telescope preparation (Detailed Test Objective), European Retrievable Carrier (EURECA) antenna stow

Table 3–50. Space Shuttle Extravehicular Activity (1989–1998ᵃ) (Continued)

Mission	Date	Astronaut	Individual EVA Time per Astronaut	Cumulative Time in Space	Description
STS-51	September 16, 1993	Newman and Walz	7 hr, 5 min	14 hr, 10 min	Hubble Space Telescope preparation (Detailed Test Objective)
STS-61	December 4, 1993	Musgrave and Hoffman	7 hr, 54 min	70 hr, 56 min	First Hubble Space Telescope servicing; prepare worksite, change gyroscopes, fuse plugs
	December 5, 1993	Akers and Thornton	6 hr, 36 min		Replace Hubble Space Telescope solar arrays
STS-61	December 6, 1993	Musgrave and Hoffman	6 hr, 47 min		Replace Wide Field and Planetary Camera (WF/PC) with Wide Field Planetary Camera-2 (WFPC-2)
	December 7, 1993	Akers and Thornton	6 hr, 50 min	70 hr, 56 min	Install Corrective Optics Space Telescope Axial Replacement (COSTAR) system
	December 8, 1993	Musgrave and Hoffman	7 hr, 21 min		Replace solar array drive electronics
STS-64	September 16, 1994	Lee and Meade	6 hr, 51 min	13 hr, 42 min	Simplified Aid for EVA Rescue (SAFER) test
STS-63	February 9, 1995	Foale and Harris	4 hr, 39 min	9 hr, 18 min	EVA Development Flight Test (EDFT) (SPARTAN Mass Handling)
STS-69	September 16, 1995	Voss and Gernhardt	6 hr, 46 min	13 hr, 32 min	EDFT (task board with ISS EVA interfaces)

Table 3–50. Space Shuttle Extravehicular Activity (1989–1998ᵃ) (Continued)

Mission	Date	Astronaut	Individual EVA Time per Astronaut	Cumulative Time in Space	Description
STS-72	January 15, 1996	Chiao and Barry	6 hr, 9 min	26 hr, 6 min	EDFT (ISS assembly and maintenance hardware)
	January 17, 1996	Chiao and Scott	6 hr, 54 min		EDFT (ISS assembly and maintenance hardware); test EMU thermal modifications
STS-76	March 27, 1996	Clifford and Godwin	6 hr, 2 min	12 hr, 4 min	EDFT (*Mir* environmental effects payload)
STS-82[b]	February 13, 1997	Smith and Lee	6 hr, 42 min	66 hr, 22 min	Second Hubble Space Telescope servicing. Replace Faint Object Spectrograph (FOS) with Near Infrared Camera and Multi-Object Spectrometer (NICMOS), replace Goddard High Resolution Spectrograph (GHRS) with Space Telescope Imaging Spectrograph (STIS)
	February 14, 1997	Harbaugh and Tanner	7 hr, 27 min		Replace fine guidance sensor, tape recorder, install improve electronics
	February 15, 1997	Smith and Lee	7 hr, 11 min		Replace data interface unit, reaction wheel assembly
	February 16, 1997	Harbaugh and Tanner	6 hr, 34 min		Replace SADE, magnetometer cover, thermal blankets
	February 17, 1997	Smith and Lee	5 hr, 17 min		Install thermal blanket patches

Table 3–50. Space Shuttle Extravehicular Activity (1989–1998[a]) (Continued)

Mission	Date	Astronaut	Individual EVA Time per Astronaut	Cumulative Time in Space	Description
STS-87[c]	November 24, 1997	Scott and Doi	7 hr, 43 min	12 hr, 43 min	Rescue SPARTAN
	December 3, 1997	Scott and Doi	5 hr, 0 min		ISS preparation
STS-88[d]	December 7, 1998	Ross and Newman	7 hr, 21 min	42 hr, 44 min	ISS assembly
	December 8, 1998	Ross and Newman	7 hr, 2 min		
	December 12, 1998	Ross and Newman	6 hr, 59 min		

[a] David S.F. Portree and Robert M. Treviño, *Walking to Olympus: An EVA Chronology*, Monographs in Aerospace History, no. 7 (Washington, DC: National Aeronautics and Space Administration, 1997), pp. 80–126, Available at *http://spaceflight.nasa.gov/spacenews/factsheets/pdfs/EVACron.pdf*; David M. Harland, *The Story of the Space Shuttle*, (London: Springer, 2004), pp. 188–189; *Shuttle Crew Operations Manual*, p. 2-11-1.

[b] "STS-82 Mission Chronology," *http://www-pao.ksc.nasa.gov/kscpao/chron/sts-82.htm* (accessed July 5, 2005).

[c] "STS-87 Shuttle Mission Archive," *http://www-pao.ksc.nasa.gov/kscpao/shuttle/missions/sts-87/mission-sts-87.html* (accessed July 5, 2005).

[d] "STS-88 Shuttle Mission Archive," *http://www-pao.ksc.nasa.gov/kscpao/shuttle/missions/sts-88/mission-sts-88.html* (accessed July 5, 2005).

Table 3–51. Space Shuttle Missions Summary (1989–1998)

Flt No.	Mission/Orbiter	Dates	Crew	Major Payloads
28	STS-29/*Discovery*	March 13, 1989 – March 18, 1989	CDR: Michael L. Coats PLT: John E. Blaha MS: James F. Buchli, Robert C. Springer, James P. Bagian	NASA Payload Deployed: Tracking and Data Relay Satellite-D (4)
29	STS-30/*Atlantis*	May 4, 1989 – May 8, 1989	CDR: David M. Walker PLT: Ronald J. Grabe MS: Mark C. Lee, Norman E. Thagard, Mary L. Cleave	NASA Payload Deployed: Magellan
30	STS-28/*Columbia*	August 8, 1989 – August 13, 1989	CDR: Brewster H. Shaw, Jr. PLT: Richard N. Richards MS: James C. Adamson, David C. Leestma, Mark N. Brown	NASA Payload Deployed: None Other Government Payload Deployed: DOD SDS-2 (USA 40)[a] and USA-41[b]
31	STS-34/*Atlantis*	October 18, 1989 – October 23, 1989	CDR: Donald E. Williams PLT: Michael J. McCulley MS: Shannon W. Lucid, Franklin Chang-Diaz, Ellen S. Baker	NASA Payload Deployed: Galileo
32	STS-33/*Discovery*	November 22, 1989 – November 27, 1989	CDR: Frederick D. Gregory PLT: John E. Blaha MS: Manley L. Carter, Jr., F. Story Musgrave, Kathryn C. Thornton	NASA Payload Deployed: None Other Government Payload Deployed: DOD satellite[c]

Table 3–51. Space Shuttle Missions Summary (1989–1998) (Continued)

Flt No.	Mission/Orbiter	Dates	Crew	Major Payloads
33	STS-32/Columbia	January 9, 1990 – January 20, 1990	CDR: Daniel C. Brandenstein PLT: James D. Wetherbee MS: Bonnie J. Dunbar, Marsha S. Ivins, G. David Low	NASA Payload Deployed: None Long Duration Exposure Facility (LDEF) retrieved (was deployed on STS-41-C) Other Government Payload Deployed: DOD SYNCOM IV-5 (LEASAT F5)
34	STS-36/Atlantis	February 28, 1990 – March 4, 1990	CDR: John O. Creighton PLT: John H. Casper MS: David C. Hilmers, Richard M. Mullane, Pierre J. Thuot	NASA Payload Deployed: None Other Government Payload Deployed: DOD KH 11-10 (AFP 731)[d]
35	STS-31/Discovery	April 24, 1990 – April 28, 1990	CDR: Loren J. Shriver PLT: Charles F. Bolden, Jr. MS: Steven A. Hawley, Kathryn D. Sullivan, Bruce McCandless, II	NASA Payload Deployed: Hubble Space Telescope
36	STS-41/Discovery	October 6, 1990 – October 10, 1990	CDR: Richard N. Richards PLT: Robert D. Cabana MS: Bruce E. Melnick, Thomas D. Akers, William M. Shepherd	NASA Payload Deployed: Ulysses
37	STS-38/Atlantis	November 15, 1990 – November 20, 1990	CDR: Richard O. Covey PLT: Frank L. Culbertson, Jr. MS: Carle J. Meade, Robert C. Springer, Charles D. Gemar	NASA Payload Deployed: None Other Government Payload Deployed: DOD electronics intelligence satellite USA 67[e]

Table 3–51. Space Shuttle Missions Summary (1989–1998) (Continued)

Flt No.	Mission/Orbiter	Dates	Crew	Major Payloads
38	STS-35/*Columbia*	December 2, 1990 – December 10, 1990	CDR: Vance D. Brand PLT: Guy S. Gardner MS: John M. Lounge, Jeffrey A. Hoffman, Robert A.R. Parker PS: Ronald A. Parise, Samuel T. Durrance	NASA Payload Deployed: None Carried Astro-1 observatory
39	STS-37/*Atlantis*	April 5, 1991 – April 11, 1991	CDR: Steven R. Nagel PLT: Kenneth D. Cameron MS: Linda M. Godwin, Jerry L. Ross, Jay Apt	NASA Payload Deployed: Gamma Ray Observatory
40	STS-39/*Discovery*	April 28, 1991 – May 6, 1991	CDR: Michael L. Coats PLT: L. Blaine Hammond, Jr. MS: Gregory J. Harbaugh, Donald McMonagle, Guion S. Bluford, Jr., Charles Lacy Veach, Richard J. Hieb	NASA Payload Deployed: None Shuttle Pallet Satellite instrument platform released and retrieved
41	STS-40/*Columbia*	June 5, 1991 – June 14, 1991	CDR: Bryan D. O'Connor PLT: Sidney M. Gutierrez MS: James P. Bagian, Tamara E. Jernigan, Margaret Rhea Seddon PS: F. Drew Gaffney, Millie Hughes-Fulford	NASA Payload Deployed: None Carried SLS-1 using Spacelab pallets with instrument pointing system and igloo

Table 3–51. Space Shuttle Missions Summary (1989–1998) (Continued)

Flt No.	Mission/Orbiter	Dates	Crew	Major Payloads
42	STS-43/*Atlantis*	August 2, 1991 – August 11, 1991	CDR: John E. Blaha PLT: Michael A. Baker MS: Shannon W. Lucid, James C. Adamson, G. David Low	NASA Payload Deployed: Tracking and Data Relay Satellite-5 (TDRS-5)
43	STS-48/*Discovery*	September 12, 1991 – September 18, 1991	CDR: John O. Creighton PLT: Kenneth S. Reightler, Jr. MS: James F. Buchli, Charles D. Gemar, Mark N. Brown	NASA Payload Deployed: Upper Atmosphere Research Satellite (UARS)
44	STS-44/*Atlantis*	November 24, 1991 – December 1, 1991	CDR: Frederick D. Gregory PLT: Terence T. Henricks MS: F. Story Musgrave, Mario Runco, Jr., James S. Voss PS: Thomas J. Hennen	NASA Payload Deployed: None Other Government Payloads Deployed: Defense Support Program satellite DSP F16 (USA 75)f
45	STS-42/*Discovery*	January 22, 1992 – January 30, 1992	CDR: Ronald J. Grabe PLT: Stephen S. Oswald MS: Norman E. Thagard, David C. Hilmers, William F. Readdy PS: Roberta L. Bondar, Ulf D. Merbold	NASA Payload Deployed: None Carried International Microgravity Laboratory-1 (IML-1) using Spacelab long module

Table 3–51. Space Shuttle Missions Summary (1989–1998) (Continued)

Flt No.	Mission/Orbiter	Dates	Crew	Major Payloads
46	STS-45/*Atlantis*	March 24, 1992 – April 2, 1992	CDR: Charles F. Bolden, Jr. PLT: Brian Duffy MS: Kathryn D. Sullivan, David C. Leestma, C. Michael Foale PS: Byron K. Lichtenberg, Dirk D. Frimout	NASA Payload Deployed: None Carried ATLAS-1 on Spacelab pallets
47	STS-49/*Endeavour*	May 7, 1992 – May 16, 1992	CDR: Daniel C. Brandenstein PLT: Kevin P. Chilton MS: Pierre J. Thuot, Kathryn C. Thornton, Richard J. Hieb, Thomas D. Akers, Bruce E. Melnick	NASA Payload Deployed: None Commercial Payload: INTELSAT VI
48	STS-50/*Columbia*	June 25, 1992 – July 9, 1992	CDR: Richard N. Richards PLT: Kenneth D. Bowersox PC: Bonnie J. Dunbar MS: Ellen S. Baker, Carl J. Meade PS: Lawrence J. DeLucas, Eugene H. Trinh	NASA Payload Deployed: None Carried U.S. Microgravity Laboratory-1 (USML-1) Spacelab module
49	STS-46/*Atlantis*	July 31, 1992 – August 8, 1992	CDR: Loren J. Shriver PLT: Andrew M. Allen PC: Jeffrey A. Hoffman MS: Franklin R. Chang-Diaz, Claude Nicollier, Marsha S. Ivins PS: Franco Malerba	NASA-Italian Space Agency Payload: Tethered Satellite System (TSS)-1 ESA Payload: EURECA

Table 3–51. Space Shuttle Missions Summary (1989–1998) (Continued)

Flt No.	Mission/Orbiter	Dates	Crew	Major Payloads
50	STS-47/*Endeavour*	September 12, 1992– September 20, 1992	CDR: Robert L. Gibson PLT: Curtis L. Brown, Jr. PC: Mark C. Lee MS: Jerome Apt, N. Jan Davis, Mae C. Jemison PS: Mamoru Mohri	NASA Payload Deployed: None Carried Japanese Spacelab-J using Spacelab long module
51	STS-52/*Columbia*	October 22, 1992– November 1, 1992	CDR: James D. Wetherbee PLT: Michael A. Baker MS: Charles Lacy Veach, William M. Shepherd, Tamara E. Jernigan PS: Steven G. MacLean	NASA-Italian Space Agency Deployed Payload: Laser Geodynamic Satellite II (LAGEOS)/Italian Research Interim Stage (IRIS) Carried U.S. Microgravity Payload (USMP-1)
52	STS-53/*Discovery*	December 2, 1992– December 9, 1992	CDR: David M. Walker PLT: Robert D. Cabana MS: Guion S. Bluford, Jr., James S. Voss, Michael R. Clifford	NASA Payload Deployed: None

Other Government Payload: DOD SDS-2 (USA 89)g |
| 53 | STS-54/*Endeavour* | January 13, 1993 – January 19, 1993 | CDR: John H. Casper PLT: Donald R. McMonagle MS: Mario Runco, Jr., Gregory J. Harbaugh, Susan J. Helms | NASA Payload Deployed: TDRS-6 |
| 54 | STS-56/*Discovery* | April 8, 1993 – April 17, 1993 | CDR: Kenneth D. Cameron PLT: Stephen S. Oswald MS: C. Michael Foale, Kenneth D. Cockrell, Ellen Ochoa | NASA Payload Deployed and Retrieved: SPARTAN 201 Carried ATLAS-2 Spacelab using Spacelab pallet and igloo |

Table 3–51. Space Shuttle Missions Summary (1989–1998) (Continued)

Flt No.	Mission/Orbiter	Dates	Crew	Major Payloads
55	STS-55/*Columbia*	April 26, 1993 – May 6, 1993	CDR: Steven R. Nagel PLT: Terence T. Henricks MS: Jerry L. Ross, Charles J. Precourt, Bernard A. Harris, Jr. PS: Ulrich Walter, Hans W. Schlegel	NASA Payload Deployed: None Carried German Spacelab D2 using long module
56	STS-57/*Endeavour*	June 21, 1993 – July 1, 1993	CDR: Ronald J. Grabe PLT: Brian Duffy MS: G. David Low, Nancy J. Sherlock (Currie), Peter J.K. Wisoff, Janice E. Voss	NASA Payload Deployed: None Retrieved EURECA; Carried SPACEHAB 01 research module
57	STS-51/*Discovery*	September 12, 1993 – September 22, 1993	CDR: Frank L. Culbertson, Jr. PLT: William F. Readdy MS: James. H. Newman, Daniel W. Bursch, Carl E. Walz	NASA Payload Deployed: Advanced Communications Technology Satellite (ACTS); NASA-German Payload Deployed: Orbiting and Retrievable Far and Extreme Ultraviolet Spectrograph-Shuttle Pallet Satellite (ORFEUS-SPAS)
58	STS-58/*Columbia*	October 18, 1993 – November 1, 1993	CDR: John E. Blaha PLT: Richard A. Searfoss MS: Margaret Rhea Seddon, William S. McArthur, Jr., David A. Wolf, Shannon W. Lucid PS: Martin J. Fettman	NASA Payload Deployed: None Carried SLS-2 long module

Table 3–51. Space Shuttle Missions Summary (1989–1998) (Continued)

Flt No.	Mission/Orbiter	Dates	Crew	Major Payloads
59	STS-61/*Endeavour*	December 2, 1993 – December 13, 1993	**CDR:** Richard O. Covey **PLT:** Kenneth D. Bowersox **MS:** Kathryn C. Thornton, Claude Nicollier, Jeffrey A. Hoffman, F. Story Musgrave, Thomas D. Akers	NASA Payload Retrieved and Redeployed: First Hubble Space Telescope Servicing Mission
60	STS-60/*Discovery*	February 3, 1994 – February 11, 1994	**CDR:** Charles F. Bolden, Jr. **PLT:** Kenneth S. Reightler, Jr. **MS:** N. Jan Davis, Ronald M. Sega, Franklin R. Chang-Diaz, Sergei K. Krikalev	NASA Payload Deployed: Wake Shield Facility (WSF)–attempt to deploy failed; Carried SPACEHAB 02 research module
61	STS-62/*Columbia*	March 4, 1994 – March 18, 1994	**CDR:** John H. Casper **PLT:** Andrew M. Allen **MS:** Pierre J. Thuot, Charles D. Gemar, Marsha S. Ivins	NASA Payload Deployed: None Carried USMP-2 and Office of Aeronautics and Space Technology (OAST)-2
62	STS-59/*Endeavour*	April 9, 1994 – April 20, 1994	**CDR:** Sidney M. Gutierrez. **PLT:** Kevin P. Chilton **MS:** Jerome Apt, Michael R. Clifford, Thomas D. Jones **PC:** Linda M. Godwin	NASA Payload Deployed: None Carried Space Radar Laboratory (SRL-1)
63	STS-65/*Columbia*	July 8, 1994 – July 23, 1994	**CDR:** Robert D. Cabana **PLT:** James D. Halsell, Jr. **MS:** Richard J. Hieb, Carl E. Walz, Leroy Chiao, Donald A. Thomas **PS:** Chiaki Naito-Mukai	NASA Payload Deployed: None Carried IML-2 Spacelab long module

Table 3–51. Space Shuttle Missions Summary (1989–1998) (Continued)

Flt No.	Mission/Orbiter	Dates	Crew	Major Payloads
64	STS-64/Discovery	September 9, 1994 – September 20, 1994	CDR: Richard N. Richards PLT: L. Blaine Hammond, Jr. MS: Jerry M. Linenger, Susan J. Helms, Carl J. Meade, Mark C. Lee	NASA Payload Deployed and Retrieved: SPARTAN-201 Carried Light Detection and Ranging (LIDAR) in Space Technology Experiment (LITE)
65	STS-68/Endeavour	September 30, 1994 – October 11, 1994	CDR: Michael A. Baker PLT: Terrence W. Wilcutt MS: Steven L. Smith, Daniel W. Bursch, Peter J.K. Wisoff PC: Thomas D. Jones	NASA Payload Deployed: None Carried SRL-2
66	STS-66/Atlantis	November 3, 1994 – November 14, 1994	CDR: Donald R. McMonagle PLT: Curtis L. Brown, Jr. MS: Joseph R. Tanner, Jean-Francois Clervoy, Scott E. Parazynski PC: Ellen Ochoa	NASA-German Space Agency Payload Deployed and Retrieved: Cryogenic Infrared Spectrometers and Telescopes for the Atmosphere-Shuttle Pallet Satellite (CRISTA-SPAS) Carried ATLAS-3 Spacelab
67	STS-63/Discovery	February 3, 1995 – February 11, 1995	CDR: James D. Wetherbee PLT: Eileen M. Collins MS: C. Michael Foale, Janice Voss, Vladimir G. Titov PC: Bernard A. Harris, Jr.	NASA Payload Deployed and Retrieved: SPARTAN-204 Carried SPACEHAB 03 research module

Table 3–51. Space Shuttle Missions Summary (1989–1998) (Continued)

Flt No.	Mission/Orbiter	Dates	Crew	Major Payloads
68	STS-67/*Endeavour*	March 2, 1995 – March 18, 1995	CDR: Stephen S. Oswald PLT: William G. Gregory MS: John M. Grunsfeld, Wendy B. Lawrence PC: Tamara E. Jernigan PS: Samuel T. Durrance, Ronald A. Parise	NASA Payload Deployed: None Carried Astro-2 on Spacelab module
69	STS-71/*Atlantis*	June 27, 1995 – July 7, 1995	CDR: Robert L. Gibson PLT: Charles J. Precourt MS: Gregory J. Harbaugh, Bonnie J. Dunbar PC: Ellen S. Baker	NASA Payload Deployed: None First Shuttle-*Mir* docking
70	STS-70/*Discovery*	July 13, 1995 – July 22, 1995	CDR: Terence T. Henricks PLT: Kevin R. Kregel MS: Nancy J. Sherlock (Currie), Donald A. Thomas, Mary Ellen Weber	NASA Payload Deployed: TDRS-7
71	STS-69/*Endeavour*	September 7, 1995 – September 18, 1995	CDR: David M. Walker PLT: Kenneth D. Cockrell MS: James H. Newman, Michael L. Gernhardt PC: James S. Voss	NASA Payload Deployed and Retrieved: WSF-2
72	STS-73/*Columbia*	October 20, 1995 – November 5, 1995	CDR: Kenneth D. Bowersox PLT: Kent V. Rominger MS: Catherine G. Coleman, Michael E. Lopez-Alegria PC: Kathryn C. Thornton PS: Fred W. Leslie, Albert Sacco, Jr.	NASA Payload Deployed: None Carried USML-2 Spacelab long module

Table 3–51. Space Shuttle Missions Summary (1989–1998) (Continued)

Flt No.	Mission/Orbiter	Dates	Crew	Major Payloads
73	STS-74/*Atlantis*	November 12, 1995 – November 20, 1995	CDR: Kenneth D. Cameron PLT: James D. Halsell, Jr. MS: Chris A. Hadfield, Jerry L. Ross, William S. McArthur, Jr.	NASA Payload Deployed: None Second Shuttle-*Mir* docking
74	STS-72/*Endeavour*	January 11, 1996 – January 20, 1996	CDR: Brian Duffy PLT: Brent W. Jett, Jr. MS: Leroy Chiao, Winston E. Scott, Koichi Wakata, Daniel T. Barry	NASA Payload Deployed and Retrieved: SPARTAN-OAST Flyer Retrieved Japanese Space Flyer Unit
75	STS-75/*Columbia*	February 22, 1996 – March 9, 1996	CDR: Andrew M. Allen PLT: Scott J. Horowitz MS: Jeffrey A. Hoffman, Maurizio Cheli, Claude Nicollier PC: Franklin R. Chang-Diaz PS: Umberto Guidoni	NASA-Italian Space Agency Payload Deployed: Tethered Satellite System (TSS)-1R Carried USMP-3
76	STS-76/*Atlantis*	March 22, 1996 – March 31, 1996	CDR: Kevin P. Chilton PLT: Richard A. Searfoss MS: Ronald M. Sega, Michael R. Clifford, Linda M. Godwin, Shannon W. Lucid (to *Mir*)	NASA Payload Deployed: None Third Shuttle-*Mir* docking Carried SPACEHAB Single Module
77	STS-77/*Endeavour*	May 19, 1996 – May 29, 1996	CDR: John H. Casper PLT: Curtis L. Brown, Jr. MS: Andrew S.W. Thomas, Daniel W. Bursch, Mario Runco, Jr., Marc Garneau	NASA Payload Deployed and Retrieved: SPARTAN-207 carrying Inflatable Antenna Experiment Carried SPACEHAB research module

Table 3–51. Space Shuttle Missions Summary (1989–1998) (Continued)

Flt No.	Mission/Orbiter	Dates	Crew	Major Payloads
78	STS-78/*Columbia*	June 20, 1996 – July 7, 1996	CDR: Terence T. Henricks PLT: Kevin R. Kregel MS: Richard M. Linnehan, Charles E. Brady, Jr. PC: Susan J. Helms PS: Jean-Jacques Favier, Robert Brent Thirsk	NASA Payload Deployed: None Carried Life and Microgravity Spacelab long module
79	STS-79/*Atlantis*	September 16, 1996 – September 26, 1996	CDR: William F. Readdy PLT: Terence W. Wilcutt MS: Jerome Apt, Thomas D. Akers, Carl E. Walz, John E. Blaha (to *Mir*), Shannon W. Lucid (returned from *Mir*)	NASA Payload Deployed: None Fourth Shuttle-*Mir* docking Carried SPACEHAB-05 Double Module
80	STS-80/*Columbia*	November 19, 1996 – December 7, 1996	CDR: Kenneth D. Cockrell PLT: Kent V. Rominger MS: Thomas D. Jones, F. Story Musgrave PC: Tamara E. Jernigan	NASA-German Space Agency Payload Deployed and Retrieved: ORFEUS- SPAS II, Wake Shield Facility-3
81	STS-81/*Atlantis*	January 12, 1997 – January 22, 1997	CDR: Michael A. Baker PLT: Brent W. Jett, Jr. MS: Peter J.K. Wisoff, John M. Grunsfeld, Marsha S. Ivins, Jerry M. Linenger (to *Mir*), John E. Blaha (returned from *Mir*)	NASA Payload Deployed: None Fifth Shuttle-*Mir* docking Carried SPACEHAB Double Module

Table 3–51. Space Shuttle Missions Summary (1989–1998) (Continued)

Flt No.	Mission/Orbiter	Dates	Crew	Major Payloads
82	STS-82/*Discovery*	February 11, 1997 – February 21, 1997	CDR: Kenneth D. Bowersox PLT: Scott J. Horowitz MS: Joseph R. Tanner, Steven A. Hawley, Gregory J. Harbaugh, Steven L. Smith, PC: Mark C. Lee	NASA Payload Retrieved and Redeployed: Second Hubble Space Telescope Servicing Mission
83	STS-83/*Columbia*	April 4, 1997 – April 8, 1997	CDR: James D. Halsell, Jr. PLT: Susan L. Still MS: Michael L. Gernhardt, Donald A. Thomas PC: Janice E. Voss PS: Roger K. Crouch, Gregory T. Linteris	NASA Payload Deployed: None Carried MSL-1 Spacelab long module
84	STS-84/*Atlantis*	May 15, 1997 – May 24, 1997	CDR: Charles J. Precourt PLT: Eileen M. Collins MS: Carlos I. Noriega, Edward Tsang Lu, Elena V. Kondakova, C. Michael Foale (to *Mir*), Jerry M. Linenger (returned from *Mir*) PC: Jean-Francois Clervoy	NASA Payload Deployed: None Sixth Shuttle-*Mir* docking Carried SPACEHAB double module
85	STS-94/*Columbia*	July 1, 1997 – July 17, 1997	CDR: James D. Halsell, Jr. PLT: Susan L. Still MS: Michael L. Gernhardt, Donald A. Thomas PC: Janice E. Voss PS: Roger K. Crouch, Gregory T. Linteris	NASA Payload Deployed: None Carried MSL-1 Spacelab (reflight of STS-83)

Table 3–51. Space Shuttle Missions Summary (1989–1998) (Continued)

Flt No.	Mission/Orbiter	Dates	Crew	Major Payloads
86	STS-85/Discovery	August 7, 1997 – August 19, 1997	CDR: Curtis L. Brown, Jr. PLT: Kent V. Rominger. MS: Robert L. Curbeam, Jr., Stephen K. Robinson PC: N. Jan Davis PS: Bjarni V. Tryggvason	Deployed and Retrieved NASA-German Space Agency Payload: CRISTA-SPAS-II
87	STS-86/Atlantis	September 25, 1997 – October 6, 1997	CDR: James D. Wetherbee PLT: Michael J. Bloomfield MS: Vladimir G. Titov, Scott E. Parazynski, Jean-Loup J.M. Chretien, Wendy B. Lawrence, David A. Wolf (to Mir), C. Michael Foale (return from Mir)	NASA Payload Deployed: None Seventh Shuttle-Mir docking Carried SPACEHAB double module
88	STS-87/Columbia	November 19, 1997 – December 5, 1997	CDR: Kevin R. Kregel PLT: Steven W. Lindsey MS: Kalpana Chawla, Winston E. Scott, Takao Doi PS: Leonid K. Kadenyuk	NASA Payload Deployed and Retrieved: SPARTAN-201-04 Carried USMP-4
89	STS-89/Endeavour	January 22, 1998 – January 31, 1998	CDR: Terrence W. Wilcutt PLT: Joe Frank Edwards, Jr. MS: James F. Reilly, Michael P. Anderson, Salizhan Shakirovich Sharipov, Andrew S.W. Thomas, David A. Wolf PC: Bonnie J. Dunbar	NASA Payload Deployed: None Eighth Shuttle-Mir linkup

Table 3–51. Space Shuttle Missions Summary (1989–1998) (Continued)

Flt No.	Mission/Orbiter	Dates	Crew	Major Payloads
90	STS-90/*Columbia*	April 17, 1998 – May 3, 1998	CDR: Richard A. Searfoss PLT: Scott D. Altman MS: Kathryn P. Hire, Dafydd Rhys Williams PC: Richard M. Linnehan PS: Jay C. Buckey, James A. Pawelczyk	NASA Payload Deployed: None Carried Neurolab Spacelab
91	STS-91/*Discovery*	June 2, 1998 – June 12, 1998	CDR: Charles J. Precourt PLT: Dominic L. Pudwill-Gorie MS: Franklin R. Chang-Diaz, Wendy B. Lawrence, Janet Lynn Kavandi, Valery Victorovitch Ryumin, Andrew Thomas (returned from *Mir*)	NASA Payload Deployed: None Ninth (final) Shuttle-*Mir* docking Carried SPACEHAB double module
92	STS-95/*Discovery*	October 29, 1998 – November 7, 1998	CDR: Curtis L. Brown, Jr. PLT: Steven W. Lindsey MS: Scott E. Parazynski, Pedro Duque PC: Stephen K. Robinson PS: Chiaki Mukai, Sen. John H. Glenn, Jr.	NASA Payload Deployed and Retrieved: SPARTAN-201 Carried SPACEHAB single research module Hubble Space Telescope HOST mission
93	STS-88/*Endeavour*	December 4, 1998 – December 15, 1998	CDR: Robert D. Cabana PLT: Frederick R. Sturckow MS: Jerry L. Ross, Nancy J. Currie, James H. Newman, Sergei K. Krikalev	NASA Payload Deployed: Space Station Unity module First ISS assembly mission

---

[a] "USA 40," NSSDC Master Catalog, Spacecraft, *http://nssdc.gsfc.nasa.gov/database/MasterCatalog?sc=1989-061B* (accessed December 1, 2005). The Federation of American Scientists lists this payload as USA 40 - SDS B1, *http://www.fas.org/spp/military/program/list.htm* (accessed July 6, 2005).

[b] "USA 41," *http://nssdc.gsfc.nasa.gov/database/MasterCatalog?sc=1989-061C* (accessed December 22, 2005).

c   No satellite was listed for this Shuttle launch in NSSDC Master Catalog. The Federation of American Scientists lists this payload as USA 48–Magnum 2. Also listed as USA 48–Magnum 2 in the Launch Log of Jonathan's Space Report, *http://planet4589.org/space/log/space/log/launchlog.txt* (accessed November 30, 2005).

d   "KH 11-10," *http://nssdc.gsfc.nasa.gov/database/MasterCatalog?sc=1990-019B* (accessed November 30, 2005) Listed as USA 53 AFP-731 MISTY in the Launch Log of Jonathan's Space Report.

e   "USA 67,"*http://nssdc.gsfc.nasa.gov/database/MasterCatalog?sc=1990-097B* (accessed November 30, 2005). Federation of American Scientists lists this payload as USA 67 - SDS B-2. Listed as USA 67 QUASAR 2 in the Launch Log of Jonathan's Space Report.

f   "USA 75," *http://nssdc.gsfc.nasa.gov/database/MasterCatalog?sc=1991-080B* (accessed November 30, 2005). Also listed in the Federation of American Scientists and the Launch Log of Jonathan's Space Report.

g   "USA 89," *http://nssdc.gsfc.nasa.gov/database/MasterCatalog?sc=1992-086B* (accessed November 30, 2005). Federation of American Scientists lists this payload as SDS B-3.

Table 3–52. STS-29 Mission Characteristics

Vehicle	OV-103
	Discovery
Crew	CDR: Michael L. Coats
	PLT: John E. Blaha
	MS: James F. Buchli, Robert C. Springer, James P. Bagian
Launch	March 13, 1989, 9:57:00 a.m., EST, Kennedy Space Center, Pad 39-B. Launch manifested for February 18 was reassessed for late February/early March launch to replace suspect liquid oxygen turbopumps on *Discovery*'s three main engines and faulty master events controller. Launch on March 13 was delayed 1 hour, 50 minutes due to morning ground fog and upper winds.
Orbital Altitude & Inclination	184 nmi (341 km), 28.45 deg
Launch Weight (lb/kg)	256,357 /116,282
Landing & Postlanding Operations	March 18, 1989, 6:35:51 a.m., PST, Runway 22, Edwards Air Force Base. Orbiter returned to Kennedy Space Center March 24, 1989.
Rollout Distance (ft/m)	9,339/2,847
Rollout Time (seconds)	51
Mission Duration	119 hr, 38 min, 52 sec
Landed Revolution No.	79
Mission Support	Spaceflight Tracking and Data Network (STDN)
Primary Objective	Launch of Tracking and Data Relay Satellite-D
Deployed Satellites	Tracking and Data Relay Satellite-D (TDRS-D/IUS), designated TDRS-4.
Experiments	Orbiter Experiments Autonomous Supporting Instrumentation System (OASIS-I): Recorded environmental data for the Shuttle's primary payloads. It was configured to interface with TDRS-D (4). OASIS-I measured thermal, acoustic, vibration, stress, and acceleration parameters of the satellite during ascent and other phases of the mission.

Table 3–52. STS-29 Mission Characteristics (Continued)

Experiments	Space Station Heat Pipe Advanced Radiator Element (SHARE): SHARE was an attached payload with the TDRS-D/IUS satellite initiated by the Office of Aeronautics and Space Technology (OAST) and managed by the Office of Space Station. The OSSA performed mission implementation and integration. SHARE was a part of NASA's Thermal Energy Management Processes (TEMP) program, supporting the development of various two-phase heat transport systems for use on Space Station *Freedom* or other future space missions. SHARE was designed to test and measure the thermal performance of a single high-capacity heat pipe radiator element in space and to demonstrate whether a natural process could serve as a dependable, durable cooling system for Space Station *Freedom*. Although a faulty manifold design resulted in early termination of the experiment, significant data with respect to the design of spaceborne heat pipes was obtained.
	Air Force Maui Optical System (AMOS) Calibration Test: The tests allowed ground-based electro-optical sensors located on Mt. Haleakala, Maui, Hawaii, to collect imagery and signature data of the orbiter during cooperative overflights. The scientific observations made of the orbiter while performing reaction control system thruster firings, water dumps, or payload bay light activation were used to support the calibration of the AMOS sensors and the validation of spacecraft
	Chromosome and Plant Cell Division in Space Experiment (CHROMEX): Sponsored by the State University of New York at Stony Brook. The CHROMEX was designed to determine whether the roots of a plant developed similarly in microgravity and on Earth. One objective was to test whether the normal rate, frequency, and patterning of cell division in the root tip could be sustained in microgravity. Another objective was to determine whether the fidelity of chromosome partitioning was maintained during and after flight. The CHROMEX used shoots of cell culture-derived daylily (*Hemerocallis* cv. Autumn Blaze) and both tissue cultured and seedling clones of haplopappus (*Haplopappus gracilis*).[a]

Table 3–52. STS-29 Mission Characteristics (Continued)

Experiments	Protein Crystal Growth (PCG): A joint project of NASA's Office of Commercial Programs (OCP) and OSSA, in conjunction with the Center for Macromolecular Crystallography, a NASA OCP-sponsored Center for the Commercial Development of Space (CCDS) at the University of Alabama at Birmingham. Scientific methods and commercial potential were investigated for growing large, high-quality protein crystals in microgravity. Proteins were processed at a "cold" temperature.
	IMAX Corporation Camera Experiment (IMAX): A 70-mm motion picture camera system photographed Earth with particular emphasis on environmentally threatened areas.
	Shuttle Student Involvement Project (SSIP) 82-8—Effects of Weightlessness in Spaceflight on the Healing of Bone Fractures: The project was designed to determine if the environmental effects of spaceflight inhibit bone healing. The subjects were four rats with a small piece of bone removed by a veterinarian from a non-weight-bearing bone. A similar group of rats remained on Earth as a control group. The effects of zero gravity on the origin, development, and differentiation of osteoblasts (bone cells) and their production of callus were studied.
	SSIP 83-9—Chicken Embryo Development in Space: This experiment studied the effects of the space environment on the development of chicken embryos. In the experiment, 32 eggs—16 fertilized two days before launch and 16 fertilized nine days before launch—were placed in an incubator to see if any changes in the developing embryo could be attributed to weightlessness. An identical group of 32 eggs remained on Earth as a control group. On return to Earth, half of each group of eggs were opened and examined to identify any differences in cartilage, bone and digit structures, muscle system, nervous system, facial structure, and internal organs. The second half (16 from spaceflight, 16 from control) were hatched 21 days after fertilization. All embryos fertilized two days before being carried into orbit had died.
Get Away Specials	None
Mission Results	Successful

[a] "CHROMEX-01/STS-29: Life Sciences Objectives," *http://lifesci.arc.nasa.gov/lis2/Chapter4_Programs/ CHROMEX/CHROMEX_01.html* (accessed December 1, 2005).

Table 3–53. STS-30 Mission Characteristics

Vehicle	OV-104
	Atlantis
Crew	CDR: David M. Walker
	PLT: Ronald J. Grabe
	MS: Mark C. Lee, Norman E. Thagard, Mary L. Cleave
Launch	May 4, 1989, 2:46:59 p.m. EDT, Kennedy Space Center, Pad 39-B. The April 28 launch was scrubbed at T-31 seconds because of a problem with the liquid hydrogen recirculation pump on the No. 1 main engine and a vapor leak in the 4-in (10-cm) liquid hydrogen recirculation line between the orbiter and the external tank. Repairs were made and launch reset for May 4. Liftoff was delayed until the last 5 minutes of a 64-minute window opening at 1:48 a.m. EDT due to cloud cover and high winds at the Kennedy Space Center Shuttle runway, violating return-to-launch-site limits.
Orbital Altitude & Inclination	184 nmi (341 km), 28.85 deg
Launch Weight (lb/kg)	261,118/118,441
Landing & Postlanding Operations	May 8, 1989, 12:43:26 p.m. PDT, Runway 22, Edwards Air Force Base. Orbiter returned to Kennedy Space Center May 15, 1989.
Rollout Distance (ft/m)	10,295/3,138
Rollout Time (seconds)	64
Mission Duration	96 hr, 57 min, 31 sec
Landed Revolution No.	64
Mission Support	STDN
Primary Objective	Launch Magellan
Deployed Satellites	Magellan/IUS
Experiments	FEA: This modular microgravity chemistry/physics laboratory was used to process samples of indium in a float-zone mode. It examined the application of floating zone processes and their effects on crystal quality. The apparatus consisted of a heating element that moved along a track to melt and recrystallize the sample, which was sealed in a Pyrex tube. The super 8-mm camera and computer provided inflight monitoring and data for postflight analysis of each test.
	MLE: This experiment was used to observe and record the visual characteristics of large-scale lightning as seen from space using the on-board cargo bay TV and 35-mm cameras.
	AMOS Calibration Test: See STS-29.
Get Away Specials	None
Mission Results	Successful

Table 3–54. STS-28 Mission Characteristics

Vehicle	OV-102 *Columbia*
Crew	CDR: Brewster H. Shaw, Jr.
	PLT: Richard N. Richards
	MS: James C. Adamson, David C. Leestma, Mark N. Brown
Launch	August 8, 1989, 8:37:00 a.m., EDT, Kennedy Space Center, Pad 39-B. Liftoff occurred during a classified launch window within a launch period extending from 7:30 a.m. to 11:30 a.m., EDT.
Orbital Altitude & Inclination	191 statute mi[a] (166 nmi/307 km), 57 deg
Launch Weight (lb/kg)	Classified
Landing & Postlanding Operations	August 13, 1989, 6:37:08 a.m., PDT, Runway 17, Edwards Air Force Base. Orbiter returned to Kennedy Space Center August 21, 1989.
Rollout Distance (ft/m)	6,015/1,833
Rollout Time (seconds)	46
Mission Duration	121 hr, 0 min, 9 sec
Landed Revolution No.	80
Mission Support	STDN
Primary Objective	DOD mission
Deployed Satellites	SDS 2 (USA 40) and USA 41 military satellites[b]
Experiments[c]	Cosmic Ray Upset Experiment (CRUX)-B
	Interim Operational Contamination Monitor (IOCM): The OCM was an automatic operation system for the measurement of contamination in the payload bay for the entire mission. The IOCM continuously measured collected particulate and molecular mass at preprogrammed collection surface temperatures.
	Multi-Purpose Experiment Canister (MPEC): MPEC was a modified GAS canister containing an ejectable, classified U.S. Air Force Space Systems Division experiment.
	AMOS Calibration Test: See STS-29.
	Cloud Logic to Optimize Use of Defense Systems (CLOUDS)-1A: This experiment involved photographic sequences of cloud fields to correlate space data and ground data simultaneously and develop functions quantifying the relationship between apparent cloud cover and the viewing angle for various cloud formations.
	Radiation Monitoring Equipment (RME)–III: This experiment measured ionizing radiation in the orbiter during sequenced time intervals and digitally stored the resulting data.

Table 3–54. STS-28 Mission Characteristics (Continued)

Experiments[d]	Shuttle Activation Monitor (SAM): The SAM measured the amounts of gamma rays in the Shuttle's crew cabin.
	Visual Function Tester (VFT)-2: This experiment was a biomedical study to determine the effects of microgravity on human visual performance. The experiment examined the vision of crew members exposed to microgravity with regard to contrast thresholds, directional perception, and pattern sensitivities.
	Aerodynamic Coefficient Identification Package (ACIP): The package instrumentation included triaxial sets of linear accelerometers, angular accelerometers, and angular rate gyros, which sensed the orbiter's motions during flight. ACIP provided the vehicle motion data that was used in conjunction with the Shuttle Entry Air Data System (SEADS) environmental information for determining aerodynamic characteristics below about 300,000 ft altitude.
Get Away Specials[e]	G-0335 Customer: Naval Postgraduate School No information submitted on payload.
	G-0341 Customer: DOD Space Test Program No information submitted on payload.
Mission Results	Successful
Remarks	DOD mission

[a] Jenkins, p. 294. Classified mission. No altitude given in NASA sources.

[b] "USA 40," NSSDC Master Catalog, Spacecraft, *http://nssdc.gsfc.nasa.gov/database/MasterCatalog?sc =1989-061B* (accessed December 1, 2005). The Federation of American Scientists lists this payload as USA 40–SDS B1, *http://www.fas.org/spp/military/program/list.htm* (accessed July 6, 2005); "USA 41," *http://nssdc.gsfc.nasa.gov/database/ScQuery* (accessed December 22, 2005).

[c] Experiments were "acknowledged payloads" by the U.S. Air Force according to Jenkins, p. 296.

[d] Experiments were "acknowledged payloads" by the U.S. Air Force according to Jenkins, p. 296.

[e] *The First 100 GAS Payloads*, NASA Goddard Space Flight Center, Shuttle Small Payloads Project, pp. 108–109.

Table 3–55. STS-34 Mission Characteristics

Vehicle	OV-104 *Atlantis*
Crew	CDR: Donald E. Williams PLT: Michael J. McCulley MS: Shannon W. Lucid, Franklin Chang-Diaz, Ellen S. Baker
Launch	October 18, 1989, 12:53:40 p.m., EDT, Kennedy Space Center, Pad 39-B. Launch set for October 12 was rescheduled due to a faulty main engine controller on the No. 2 main engine. Launch set for October 17 was rescheduled due to weather constraints for a return-to-launch-site landing at the Kennedy Space Center Shuttle Landing Facility.
Orbital Altitude & Inclination	185 nmi (342 km), 34.3 deg
Launch Weight (lb/kg)	257,569/116,831
Landing & Postlanding Operations	October 23, 1989, 9:33:00 a.m., PDT, Runway 23, Edwards Air Force Base. The orbiter returned to Kennedy Space Center October 29, 1989.
Rollout Distance (ft/m)	9,677/2,950
Rollout Time (seconds)	61
Mission Duration	119 hr, 39 min, 24 sec
Landed Revolution No.	79
Mission Support	STDN
Primary Objective	Launch of Galileo
Deployed Satellites	Galileo/IUS
Experiments	SSBUV Instrument: This experiment compared the observations of several ozone-measuring instruments aboard NOAA satellites and other Earth-monitoring spacecraft being flown. SSBUV data was used to check the calibration of the ozone sounders on these spacecraft to ensure the most accurate readings possible for detecting atmospheric ozone trends and verify the accuracy of the data set of atmospheric ozone and solar irradiance data. The SSBUV used the Space Shuttle's orbital flight path to assess instrument performance by directly comparing data from identical instruments aboard orbiting satellites as the Shuttle and the satellite passed over the same Earth location within a 1-hour window. These orbital coincidences could occur 17 times per day.

Table 3–55. STS-34 Mission Characteristics (Continued)

Experiments	Growth Hormone Concentration and Distribution (GHCD) in Plants: This experiment studied the effects of microgravity on the concentration, turnover properties, and behavior of plant growth hormone (Auxin) in corn shoot tissue. There were four plant canisters: two placed into a gaseous nitrogen freezer, and two were undisturbed.
	IMAX Cargo-Bay Camera: This 70-mm motion picture camera system photographed Galileo deployment and various Earth features.
	Student Experiment (SE-82-15)—Zero Gravity Growth of Ice Crystals from Supercooled Water With Relation to Temperature: This student experiment observed the formation of ice crystals at various supercooled temperatures below 0°C (32°F) in a gravity-free environment for high-altitude meteorology and researched the relationships between water vapor saturation and crystal geometry to better understand the absence of gravity.
	Polymer Morphology (PM-1): This self-contained payload examined the effects of microgravity on the processing of polymers.
	MLE: See STS-30
	Sensor Technology Experiment (STEX): This dosimetry experiment consisted of a self-contained, battery-powered, automated sensor with an internal solid state memory. The STEX Investigated natural background radiation and the radiation from the Radioisotope Thermoelectric Generators (RTG).
Experiments	AMOS: See STS-29
Get Away Specials	None
Mission Results	Successful

Table 3–56. STS-33 Mission Characteristics

Vehicle	OV-103 *Discovery*
Crew	CDR: Frederick D. Gregory PLT: John E. Blaha MS: Manley L. Carter, Jr., F. Story Musgrave, Kathryn C. Thornton
Launch	November 22, 1989, 7:23:30 p.m. EST, Kennedy Space Center, Pad 39-B. Launch set for November 20 was rescheduled to allow the changeout of suspect integrated electronics assemblies on the solid rocket boosters.
Orbital Altitude & Inclination	302 nmi (559 km), 28.45 deg
Launch Weight (lb/kg)	Classified
Landing & Postlanding Operations	November 27, 1989, 4:30:16 p.m. PST, Runway 4, Edwards Air Force Base. Orbiter returned to Kennedy Space Center December 4, 1989.
Rollout Distance (ft/m)	7,764/2,366
Rollout Time (seconds)	46
Mission Duration	120 hr, 6 min, 49 sec
Landed Revolution No.	79
Mission Support	STDN
Primary Objective	DOD mission
Deployed Satellites	Classified military satellite[a]
Experiments	AMOS: See STS-29 Auroral Photography Experiment (APE)-B: The APE-B experiment photographed the airglow aurora, auroral optical effects, Shuttle glow phenomenon, and thruster emissions. CLOUDS-I: See STS-28 RME-III: See STS-28 VFT-1: The crew members executed a variety of head movements with eyes both opened and closed, and they reported on sensations of movement. The crew measured the sensory effects accompanying observation of shifting red lights (light-emitting diodes) on a visual target display.
Get Away Specials	Reflight of SD-301, Cosmic Ray Induced Error Rate in Memory Chips (CRUX Cosmic Ray Upset Experiment) Unofficial "classified" GAS.
Mission Results	Successful
Remarks	DOD mission

[a] No satellite listed for this Shuttle launch in NSSDC Master Catalog. The Federation of American Scientists lists this payload as USA 48–Magnum 2. Also listed as USA 48–Magnum 2 in the "Launch Log of Jonathan's Space Report," *http://planet4589.org/space/log/launchlog.txt* (accessed November 30, 2005).

Table 3–57. STS-32 Mission Characteristics

Vehicle	OV-102
	Columbia
Crew	CDR: Daniel C. Brandenstein
	PLT: James D. Wetherbee
	MS: Bonnie J. Dunbar, Marsha S. Ivins, G. David Low
Launch	January 9, 1990, 7:35:00 a.m., EST, Kennedy Space Center, Pad 39-A. Launch scheduled for December 18, 1989, was postponed to complete and verify modifications to Pad A, being used for the first time since January 1986. The January 8, 1990, launch was scrubbed due to weather conditions. This was the first use of Mobile Launch Platform for the Shuttle.
Orbital Altitude & Inclination	178 nmi (330 km), 28.5 deg
Launch Weight (lb/kg)	255,994/116,117
Landing & Postlanding Operations	January 20, 1990, 1:35:37 a.m., PST, Runway 22, Edwards Air Force Base. Orbiter returned to Kennedy Space Center January 26, 1990.
Rollout Distance (ft/m)	10,731/3,271
Rollout Time (seconds)	64
Mission Duration	261 hr, 0 min, 37 sec
Landed Revolution No.	171
Mission Support	STDN
Primary Objective	Deployment of SYNCOM IV-F5 and retrieval of LDEF
Deployed Satellites	Defense communications satellite SYNCOM IV-F5 (also known as LEASAT 5)
Experiments	Characterization of Neurospora Circadian Rhythms (CNCR): This experiment, sponsored by NASA's OSSA Life Sciences Division, determined if Neurospora (a pink bread mold) circadian rhythms persisted in microgravity by eliminating most of the exogenously derived environment cues from Earth. The experiment provided data on endogenously-driven biological clocks that could be applied to other organisms.
	Interim Operational Contamination Monitor (IOCM): See STS-28.
	PCG: These were a continuing series of approximately 120 different PCG experiments conducted simultaneously using as many as 24 different proteins. See STS-29.
	FEA: The apparatus was used to conduct the Microgravity Disturbances Experiment. See STS-30.
	American Flight Echocardiograph (AFE): The AFE obtained data on the inflight cardiovascular dynamics of the space adaptation process to help develop optimal countermeasures to debilitating effects for reasons of personal and operational safety.

Table 3–57. STS-32 Mission Characteristics (Continued)

Experiments	Latitude/Longitude Locator (L3): The crew conducted tests to determine the accuracy and usability of the L3 system in finding the latitude and longitude of known ground sites. Consisted of a modified Hasselblad camera equipped with a wide-angle 40-mm lens, a camera-computer interface developed by Johnson Space Center engineers, and a Graphics Retrieval and Information Display (GRID) 1139 Compass Computer. The crew photographed the same target twice at an interval of approximately 15 seconds and fed information to the GRID computer, which computed two possible locations. The crew, by knowing if the target was north or south of the flight path, could determine which of the two locations was correct and the target's latitude and longitude.
	MLE: See STS-30.
	IMAX: This 70-mm motion picture camera system, mounted in the payload bay, was used to photograph various Earth features.
	AMOS Calibration Test: See STS-29.
	Shuttle Infrared Leeside Temperature Sensing (SILTS): Mounted on *Columbia*'s vertical tail, SILTS consisted of a cylindrical housing of approximately 20 in (50.8 cm) diameter capped at the leading edge by a hemispherical dome. Mounted inside the dome was an infrared camera that obtained high-resolution infrared imagery of the upper (leeside) surfaces of *Columbia*'s port (left) wing and fuselage during entry. The images provided detailed temperature maps at the surface of the leeside thermal protection materials and indicated the degree of aerodynamic heating of the surface in flight. *Columbia*'s computer activated SILTS at about 400,000 ft (121,920 m) and terminated SILTS after the orbiter passed through the period of significant aerodynamic heating.[a]
	SEADS: The SEADS nosecap on the *Columbia* contained 14 penetration assemblies, each containing a small hole to sense the nosecap surface air pressure. Measurement of the pressure levels and distribution allowed postflight determination of vehicle attitude and atmospheric density during entry. SEADS operated at an altitude range of 280,000 ft (85,344 m) to landing.[b]
Get Away Specials	None
Mission Results	Successful

[a] "STS-32 Press Information, December 1989," p. 54, *http://www.jsc.nasa.gov/history/shuttle_pk/mrk/ FLIGHT_033-STS-032_MRK.pdf* (accessed December 20, 2005).
[b] "STS-32 Press Information, December 1989," p. 56.

Table 3–58. STS-36 Mission Characteristics

Vehicle	OV-104
	Atlantis
Crew	CDR: John O. Creighton
	PLT: John H. Casper
	MS: David C. Hilmers, Richard M. Mullane, Pierre J. Thuot
Launch	February 28, 1990, 2:50:22 a.m., EST, Kennedy Space Center, Pad 39-A. Launch set for February 22 was postponed to February 23, February 24, and February 25 due to the illness of the crew commander and weather conditions. Launch set for February 25 was scrubbed due to malfunction of the range safety computer. Launch set for February 26 was scrubbed due to weather conditions (Note: the external tank was loaded only for launch attempts on February 25, February 26, and launch on February 28). Launch February 28 was set for a classified window lying within a launch period from midnight to 4 a.m. EST.
Orbital Altitude & Inclination	132 nmi (244 km), 62 deg
Launch Weight (lb/kg)	Classified
Landing & Postlanding Operations	March 4, 1990, 10:08:44 a.m., PST, Runway 23, Edwards Air Force Base. Orbiter returned to Kennedy Space Center on March 13, 1990.
Rollout Distance (ft/m)	7,900/2,408
Rollout Time (seconds)	53
Mission Duration	106 hr, 18 min, 23 sec
Landed Revolution No.	72
Mission Support	STDN
Primary Objective	DOD mission
Deployed Satellites	KH-11-10 (AFP-731)[a]
Experiments	RME-III-12: See STS-28.
	VFT-1: See STS-33.
	VFT-2: See STS-28.
Get Away Specials	None
Mission Results	Successful
Remarks	DOD mission

[a] "KH 11-10," NSSDC Master Catalog Display: Spacecraft, *http://nssdc.gsfc.nasa.gov/database/MasterCatalog?sc=1990-019B* (accessed November 29, 2005). Listed as "USA 53, AFP-731 MISTY" in the Launch Log of *Jonathan's Space Report.*

Table 3–59. STS-31 Mission Characteristics

Vehicle	OV-103 *Discovery*
Crew	CDR: Loren J. Shriver
	PLT: Charles F. Bolden, Jr.
	MS: Steven A. Hawley, Kathryn D. Sullivan, Bruce McCandless, II
Launch	April 24, 1990, 8:33:51 a.m. EDT, Kennedy Space Center, Pad 39-B. Launch scheduled for April 18, then April 12, then moved back to April 10 following the Flight Readiness Review. This was the first time that a date set at a Flight Readiness Review was before that shown on previous planning schedules. April 10 launch was scrubbed at T-4 minutes due to a faulty valve in auxiliary power unit (APU) No. 1. The APU was replaced and payload batteries were recharged. The countdown was briefly halted again at T-31 seconds on April 24, because computer software failed to shut down a fuel valve line on ground support equipment. Engineers ordered the valve to shut and countdown continued.
Orbital Altitude & Inclination	330 nmi (611 km) at HST deployment, 28.45 deg
Launch Weight (lb/kg)	249,109/112,994
Landing & Postlanding Operations	April 29, 1990, 6:49:57 a.m. PDT, Runway 22, Edwards Air Force Base. Orbiter returned to Kennedy Space Center on May 7, 1990.
Rollout Distance (ft/m)	8,874/2,705
Rollout Time (seconds)	61
Mission Duration	121 hr, 16 min, 5 sec
Landed Revolution No.	79
Mission Support	STDN
Primary Objective	Deployment of Hubble Space Telescope
Deployed Satellites	Hubble Space Telescope
Experiments	IMAX Cargo Bay Camera (ICBC): The ICBC documented operations outside the crew cabin. The crew used the ICBC to film several Earth observation sites, including the country of Botswana, the San Francisco Bay Area, the Andes Mountains in South America, the Amazon Rainforest, and Japan.
	IMAX Handheld Camera: The crew used the handheld camera for filming inside the crew cabin.

Table 3–59. STS-31 Mission Characteristics (Continued)

Experiments	Ascent Particle Monitor (APM): An automatic system mounted in the payload bay measured particle contamination or particle detachment during the immediate prelaunch period and during ascent. It consisted of a small box with a fixed door and a moving door mounted in a clamshell arrangement atop an aluminum housing.
	PCG III: In a continuing series of experiments, the crew modified Vapor Diffusion Apparatus trays to allow for hand seeding of the protein solution droplets. See STS-29.
	RME-III: See STS-28.
	Investigations into Polymer Membrane Processing (IPMP): The Batelle organization sponsored this experiment through the Advanced Materials Center for the Commercial Development of Space, partially funded by NASA. The experiment determined porosity control in a microgravity environment. The objective was to flash-evaporate mixed solvent systems in the absence of convection to control the porosity of a polymer membrane.
	SSIP–Investigation of Arc and Ion Behavior in Microgravity: This experiment studied the effects of weightlessness on electrical arcs and observed the effects of microgravity on an electric arc in a sealed aluminum arc chamber box.[a]
	AMOS Calibration Test: See STS-29.
Get Away Specials	None
Mission Results	Successful

[a] Called the Student Science Investigation Project in the STS-31 Press Kit.

Table 3–60. STS-41 Mission Characteristics

Vehicle	OV-103
	Discovery
Crew	CDR: Richard N. Richards
	PLT: Robert D. Cabana
	MS: Bruce E. Melnick, Thomas D. Akers, William M. Shepherd
Launch	October 6, 1990, 7:47:15 a.m. EDT, Kennedy Space Center, Pad 39-B. Liftoff occurred 12 minutes after a 2 1/2 hour launch window opened at 7:35 a.m. EDT, October 6; a brief delay was due to weather concerns. An additional 11-second hold occurred at T-5 minutes due to a Ground Launch Sequencer glitch, and at T-31 seconds, the count halted for 22 seconds to correct an orbiter purge, vent, and drain system glitch.
Orbital Altitude & Inclination	160 nmi (296 km), 28.45 deg
Launch Weight (lb/kg)	259,593/117,749
Landing & Postlanding Operations	October 10, 1990, 6:57:18 a.m. PDT, Runway 22, Edwards Air Force Base. Orbiter returned to Kennedy Space Center October 16, 1990.
Rollout Distance (ft/m)	8,478/2,584
Rollout Time (seconds)	49
Mission Duration	98 hr, 10 min, 3 sec
Landed Revolution No.	65
Mission Support	STDN
Primary Objective	Launch of Ulysses
Deployed Satellites	Ulysses/IUS/PAM-S
Experiments	SSBUV: See STS-34
	INTELSAT Solar Array Coupon (ISAC): The ISAC obtained data from the interaction of atomic oxygen with the solar array silver interconnects to assess the condition of the INTELSAT spacecraft stranded in lower Earth orbit (see STS-49).
	CHROMEX: See STS-29.
	Voice Command System (VCS): The VCS collected data on voice command recognition accuracy and operated the orbiter's closed-circuit TV system.
	Solid Surface Combustion Experiment (SSCE): The SSCE studied flame spread in microgravity and improved fire safety aspects of space travel.
	IPMP: See STS-31.

Table 3–60. STS-41 Mission Characteristics (Continued)

Experiments	PSE: The PSE studied the effects of a proprietary protein molecule on animal physiological systems in microgravity.
	RME-III: See STS-28.
	SSIP–Convection in Zero Gravity (SE 81-9): This experiment studied surface tension induced flows in microgravity.
	AMOS Calibration Test: See STS-29.
Get Away Specials	None
Mission Results	Successful

Table 3–61. STS-38 Mission Characteristics

Vehicle	OV-104
	Atlantis
Crew	CDR: Richard O. Covey
	PLT: Frank L. Culbertson, Jr.
	MS: Carle J. Meade, Robert C. Springer, Charles D. Gemar
Launch	November 15, 1990, 6:48:15 p.m. EST, Kennedy Space Center, Pad 39-A. The launch was originally scheduled for July 1990. However, a liquid hydrogen leak found on *Columbia* during the STS-35 countdown prompted three precautionary mini-tanking tests on *Atlantis* at the pad June 29, July 13, and July 25. Tests confirmed a hydrogen fuel leak on the external tank side of the external tank/orbiter 17-in (43.1-cm) quick disconnect umbilical. With a leak that could not be repaired at the pad; *Atlantis* was rolled back to the Vehicle Assembly Building (VAB) August 9, demated, and transferred to the Orbiter Processing Facility (OPF). During the rollback, the vehicle was parked outside the VAB about a day while the *Columbia*/STS-35 stack transferred to the pad for launch. Outside, *Atlantis* suffered minor hail damage to its tiles during a thunderstorm. After repairs were made in the OPF, *Atlantis* transferred to the VAB for mating on October 2. During hoisting operations, a platform beam that should have been removed from the aft compartment fell and caused minor damage that was repaired. The vehicle rolled out to Pad A on October 12. The fourth mini-tanking test was performed October 24 with no excessive hydrogen or oxygen leakage detected. At the Flight Readiness Review, the launch date was set for November 9. The launch was reset for November 15 due to payload problems. Liftoff occurred during a classified launch window lying within a launch period extending from 6:30 p.m. to 10:30 p.m. EST, November 15.
Orbital Altitude & Inclination	142 nmi (262 km), 28.5 deg
Launch Weight (lb/kg)	Classified
Landing & Postlanding Operations	November 20, 1990, 4:42:42 p.m. EST, Runway 33, Kennedy Space Center. Mission was extended one day due to unacceptable crosswinds at the original planned landing site, Edwards Air Force Base. Continued adverse conditions led to a decision to shift landing to Kennedy Space Center. It was the first Kennedy Space Center landing for *Atlantis* and the first end-of-mission landing at Kennedy Space Center since April 1985.
Rollout Distance (ft/m)	9,003/2,744

Table 3–61. STS-38 Mission Characteristics (Continued)

Rollout Time (seconds)	57
Mission Duration	117 hr, 54 min, 22 sec
Landed Revolution No.	78
Mission Support	STDN
Primary Objective	DOD mission
Deployed Satellites	USA 67 electronics intelligence satellite[a]
Experiments	None
Get Away Specials	None
Mission Results	Successful

[a] "USA 67," *http://nssdc.gsfc.nasa.gov/database/MasterCatalog?sc=1990-097B* (accessed November 30, 2005). The Federation of American Scientists lists this payload as USA 67–SDS B-2. It is listed as USA 67 QUASAR 2 in the Launch Log of *Jonathan's Space Report*.

Table 3–62. STS-35 Mission Characteristics

Vehicle	OV-102
	Columbia
Crew	CDR: Vance D. Brand
	PLT: Guy S. Gardner
	MS: John M. Lounge, Jeffrey A. Hoffman, Robert A.R. Parker
	PS: Ronald A. Parise, Samuel T. Durrance
Launch	December 2, 1990, 1:49:01 a.m. EST, Kennedy Space Center, Pad 39-B. Launch was first scheduled for May 16, 1990. Following the Flight Readiness Review, the announcement of a firm launch date was delayed to change out a faulty Freon coolant loop proportional valve in the orbiter's coolant system. At a subsequent Flight Readiness Review, the date was set for May 30. Launch on May 30 was scrubbed during tanking due to a minor hydrogen leak in the tail service mast on the mobile launcher platform and a major leak in the external tank/orbiter 17-in (43.1-cm) quick disconnect assembly. Hydrogen also was detected in the orbiter's aft compartment believed associated with a leak involving the 17-in (43.1-cm) umbilical assembly.
	A mini-tanking test on June 6 confirmed the leak at the 17-in (43.1-cm) umbilical. The umbilical could not be repaired at the pad and the orbiter was returned to the VAB June 12, demated, and transferred to the OPF. Changeout of the orbiter-side 17-in (43.1-cm) umbilical assembly was made with one borrowed from *Endeavour*; the external tank was fitted with new umbilical hardware. The Astro-1 payload was reserviced regularly and remained in *Columbia*'s cargo bay during orbiter repairs and reprocessing.

Table 3–62. STS-35 Mission Characteristics (Continued)

Launch	*Columbia* rolled out to Pad A a second time on August 9 to support a September 1 launch date. Two days before launch, the avionics box on the Broad Band X-Ray Telescope portion of the Astro-1 payload malfunctioned and had to be changed out and retested. Launch was rescheduled for September 6. During tanking, high concentrations of hydrogen were detected in the orbiter's aft compartment, forcing another postponement. NASA managers concluded that *Columbia* had experienced separate hydrogen leaks from the beginning: one of umbilical assembly (now replaced) and one or more in the aft compartment that had resurfaced. Suspicion focused on a package of three hydrogen recirculation pumps in the aft compartment. These were replaced and retested. A damaged Teflon cover seal in the main engine No. 3 hydrogen prevalve was replaced. Launch was rescheduled for September 18. A fuel leak in the aft compartment resurfaced during tanking and the mission was scrubbed again. The STS-35 mission was put on hold until problems were resolved by a special tiger team assigned by the Space Shuttle director.
	Columbia transferred to Pad B October 8, to make room for *Atlantis* on STS-36. Tropical Storm Klaus forced a rollback to the VAB on October 9. The vehicle transferred to Pad B again October 14. A mini-tanking test was conducted on October 30 using special sensors and video cameras and employing a see-through Plexiglas aft compartment door. No excessive hydrogen leakage was detected. Liftoff December 2 was delayed 21 minutes to allow Air Force range time to observe low-level clouds that might impede tracking of Shuttle ascent.
Orbital Altitude & Inclination	190 nmi (252 km), 28.45 deg
Launch Weight (lb/kg)	256,385/116,294
Landing & Postlanding Operations	December 10, 1990, 9:54:08 p.m. PST, Runway 22, Edwards Air Force Base. Mission was cut short one day due to impending bad weather at the primary landing site, Edwards Air Force Base. The orbiter returned to Kennedy Space Center on December 20, 1990.
Rollout Distance (ft/m)	10,450/3,185
Rollout Time (seconds)	58
Mission Duration	215 hr, 5 min, 8 sec
Landed Revolution No.	143
Mission Support	STDN
Primary Objective	Astrophysics observations using Astro-1
Deployed Satellites	None

Table 3–62. STS-35 Mission Characteristics (Continued)

Experiments	Astro-1: An observatory consisting of four telescopes: Hopkins Ultraviolet Telescope (HUT), Wisconsin Ultraviolet Photo-Polarimeter Experiment (WUPPE), Ultraviolet Imaging Telescope (UIT), and Broad Band X-Ray Telescope (BBXRT)—all designed for round-the-clock observations of the celestial sphere in the ultraviolet and x-ray ranges.[a] Ultraviolet telescopes were mounted on the Spacelab instrument pointing system carried on pallets in the cargo bay. Loss of both data display units (used for pointing telescopes and operating experiments) during the mission impacted crew-aiming procedures and forced ground teams at Marshall Space Flight Center to aim the ultraviolet telescopes with fine-tuning by the flight crew.

• HUT: This telescope studied faint astronomical objects such as quasars, active galactic nuclei, and supernova remnants in the little-explored ultraviolet range below 1200 angstroms. The telescope Observed the outer planets of the solar system to investigate auroras and gained insight into the interaction of each planet's magnetosphere with the solar wind.

• WUPPE: This experiment measured the polarization of ultraviolet light from celestial objects such as hot stars, galactic nuclei, and quasars.

• UIT: This telescope investigated the present stellar content and history of star formation in galaxies, the nature of spiral structure, and non-thermal sources in galaxies.

• BBXRT: This telescope studied various targets, including active galaxies, clusters of galaxies, supernova remnants, and stars. The BBXRT directly measured the amount of energy in electron volts of each x-ray detected.

Orbiter Experiments Program (OEX): The OEX was developed to perform flight experiments on a full-scale, lifting vehicle:

• SEADS. See STS-32.

Table 3–62. STS-35 Mission Characteristics (Continued)

Experiments	• Shuttle Upper Atmosphere Mass Spectrometer (SUMS): The SUMS complemented SEADS by enabling measurement of atmospheric density above 300,000 ft (91,440 m). The SUMS sampled air through a small hole on the lower surface of the vehicle aft of the nosecap. It used a mass spectrometer operating as a pressure-sensing device to measure atmospheric density in the high altitude and rarefied flow regime where the pressure was too low to use ordinary pressure sensors. The mass spectrometer, incorporated in the SUMS experiment, was spare equipment originally developed for the Viking Mars Lander.
	• ACIP. See STS-28.
	• High Resolution Accelerometer Package (HiRAP): This instrument was a three-axis set of highly sensitive accelerometers that measured vehicle motions during the high altitude portion (above 300,000 ft) (91,440 m) of entry. This instrument provided the companion vehicle motion data to be used with the SUMS results. HiRAP had flown on previous missions of the orbiters *Columbia* and *Challenger*.
	• SILTS: This experiment used a scanning infrared radiometer located atop the vertical tail to collect infrared images of the orbiter's leeside (upper) surfaces during entry, for the purpose of measuring the temperature distribution and the aerodynamic heating environment. On STS-32, the experiment obtained images of the left wing. For STS-35 and STS-40, the experiment was configured to obtain images of the upper fuselage. SILTS had flown on four *Columbia* flights.
	Shuttle Amateur Radio Experiment-2 (SAREX-2): The SAREX-2 communicated with amateur radio stations within line-of-sight of the orbiter in voice mode or data mode.
	AMOS Calibration Test: See STS-29.
	The Shuttle crew conducted a Space Classroom Program called "Assignment–The Stars," to spark student interest in science, math, and technology.

Table 3–62. STS-35 Mission Characteristics (Continued)

Experiments	• Aerothermal Instrumentation Package (AIP): The AIP Comprised about 125 measurements of aerodynamic surface temperature and pressure at discrete locations on the upper surface of the orbiter's left wing, fuselage, and vertical tail. These sensors were originally part of the development flight instrumentation system that flew aboard *Columbia* during its Orbital Flight Test missions (STS-1 through STS-5). They were reactivated through an AIP-unique data handling system. Among other applications, the AIP data provided "ground-truth" information for the SILTS experiment. The AIP had flown on previous *Columbia* flights.
Get Away Specials	None
Mission Results	Marshall and Goddard Space Flight Centers estimated that 70 percent of the planned science data was acquired. Other mission objectives were achieved.
	The crew experienced trouble dumping waste water due to a clogged drain, they were able to use spare containers.

[a]   See chapter 4, Space Science, of this volume for further discussion of the Astro-1 mission.

Table 3–63. STS-37 Mission Characteristics

Vehicle	OV-104 *Atlantis*
Crew	CDR: Steven R. Nagel
	PLT: Kenneth D. Cameron
	MS: Linda M. Godwin, Jerry L. Ross, Jay Apt
Launch	April 5, 1991, 9:22:44 a.m. EST, Kennedy Space Center, Pad 39-B. Launch set for April 5, 9:18 a.m. was briefly delayed due to low-level clouds in the area.
Orbital Altitude & Inclination	248 nmi (459 km), 28.453 deg
Launch Weight (lb/kg)	255,824/116,040
Landing & Postlanding Operations	April 11, 1991, 6:55:29 a.m. PDT, Runway 33, Edwards Air Force Base. Landing was originally scheduled for April 10 but was delayed one day due to weather conditions at Edwards Air Force Base and Kennedy Space Center.
Rollout Distance (ft/m)	6,364/1,940
Rollout Time (seconds)	56
Mission Duration	143 hr, 32 min, 44 sec
Landed Revolution No.	93
Mission Support	STDN
Primary Objective	Deployment of the Gamma Ray Observatory
Deployed Satellites	Gamma Ray Observatory
Experiments	APM: See STS-31.
	SAREX II: See STS-35.
	PCG: A continuing series of experiments, the STS-37 set of PCG experiments used the batch process and flew in a new hardware configuration, the Protein Crystallization Facility, developed by the PCG investigators. See STS-29.
	Bioserve/Instrumentation Technology Associates Materials Dispersion Apparatus (BIMDA): The BIMDA gathered data by mixing fluids in the microgravity of space.
	RME-III: See STS-28.
	AMOS Calibration Test: See STS-29.
Get Away Specials	None
Mission Results	Successful

Table 3–63. STS-37 Mission Characteristics (Continued)

Remarks	Astronauts Ross and Apt performed an unscheduled contingency spacewalk to manually deploy the GRO high-gain antenna. Although the GRO was designed for servicing by the Shuttle, an early mishap with its propulsion system and the later failure of one of its gyroscopes made this impossible. It was brought down in a controlled reentry on June 4, 2000.
	Crew and Equipment Translation Aids (CETA): Ross and Apt performed a scheduled 6-hour spacewalk to test a method for astronauts to move themselves and equipment while maintaining Space Station *Freedom*.

Table 3–64. STS-39 Mission Characteristics

Vehicle	OV-103
	Discovery
Crew	CDR: Michael L. Coats
	PLT: L. Blaine Hammond, Jr.
	MS: Gregory J. Harbaugh, Donald McMonagle, Guion S. Bluford, Jr., Charles Lacy Veach, Richard J. Hieb
Launch	April 28, 1991, 7:33:14 a.m. EDT, Kennedy Space Center, Pad 39-A. The launch was originally scheduled for March 9, but, during processing work at Pad A, significant cracks were found on all four lug hinges on the two external tank umbilical door drive mechanisms. NASA managers opted to roll back the vehicle to the VAB on March 7 and then to the OPF for repair. Hinges were replaced and reinforced with units taken from the orbiter *Columbia*. *Discovery* returned to the launch pad April 1, and launch was reset for April 23. The mission was again postponed when, during prelaunch external tank loading, a transducer on the high-pressure oxidizer turbopump for main engine No. 3 showed readings out of specification. The transducer and its cable harness were replaced and tested. Launch was rescheduled for April 28.
Orbital Altitude & Inclination	140 nmi (259 km), 57.007 deg
Launch Weight (lb/kg)	247,373
Landing & Postlanding Operations	May 6, 1991, 2:55:35 p.m. EDT, Runway 15, Kennedy Space Center. Landing diverted to Kennedy Space Center because of unacceptably high winds at the planned landing site, Edwards Air Force Base.
Rollout Distance (ft/m)	9,234/2,815
Rollout Time (seconds)	56
Mission Duration	199 hr, 22 min, 22 sec
Landed Revolution No.	133
Mission Support	STDN
Primary Objective	Unclassified DOD mission with multiple payloads
Deployment Satellites	Released and retrieved the SPAS-II

Table 3–64. STS-39 Mission Characteristics (Continued)

Experiments	Air Force Program-675 (AFP-675): AFP-675 was a collection of scientific instruments observing targets such as the atmosphere and the aurora and stars in infrared, far ultraviolet, ultraviolet, and x-ray wavelengths. AFP-675 instruments also analyzed the spectrum of various targets and gases released from or around the Shuttle.
	Cryogenic Infrared Radiance Instrumentation for Shuttle (CIRRIS): The CIRRUS measured infrared characteristics of atmospheric emissions.
	Critical Ionization Velocity (CIV): This experiment released four pressure vessels mounted in the payload bay. Each vessel contained a non-hazardous gas. Instruments on SPAS-II observed the resultant gas plume effect.
	Chemical Release Observation (CRO): The CRO deployed three subsatellites. Each subsatellite released a different chemical. SPAS-II, ground, and airborne scientific instruments observed the resulting cloud.
	MPEC: See STS-28.[a]
	RME-III: See STS-28.
	CLOUDS-I: See STS-28.
	Space Test Payload-1 (STP-1): This payload consisted of a varied collection of scientific instruments. One instrument observed the luminous "airglow" effect of atomic oxygen on *Discovery*; a second instrument tested a new method of flowing rocket propellants in weightlessness to assist in the design of future engines; and a third instrument observed the fringes of Earth's atmosphere at various times, including sunrise and sunset, in ultraviolet wavelengths.
Get Away Specials	None
Mission Results	Successful
Remarks	Unclassified DOD mission. Work with payloads during the flight involved extensive maneuvering, rendezvous, and proximity operations by *Discovery*.

[a] The STS-39 Mission Chronology called this experiment the "Multi-Purpose Release Canister." That seems to be an error in that the STS-39 Press Kit and the biography of astronaut Guion Bluford, who released the experiment, called it the "Multi-Purpose Experiment Canister." MPEC also stood for "Multi-Purpose Experiment Canister" in other missions where it was deployed.

Table 3–65. STS-40 Mission Characteristics

Vehicle	OV-102
	Columbia
Crew	CDR: Bryan D. O'Connor
	PLT: Sidney M. Gutierrez
	MS: James P. Bagian, Tamara E. Jernigan, Margaret Rhea Seddon
	PS: F. Drew Gaffney, Millie Hughes-Fulford
Launch	June 5, 1991, 9:24:51 a.m. EDT, Kennedy Space Center, Pad 39-B. Launch was originally set for May 22, 1991. The mission was postponed less than 48 hours before launch when it became known that a leaking liquid hydrogen transducer in the orbiter main propulsion system, which had been removed and replaced during leak testing in 1990, had failed an analysis by the vendor. Engineers feared that one or more of the nine liquid hydrogen and liquid oxygen transducers protruding into fuel and oxidizer lines could break off and be ingested by the engine turbopumps, causing engine failure.
	In addition, one of the orbiter's five general purpose computers failed completely, along with one of the multiplexer demultiplexers that controlled orbiter hydraulics ordinance and orbiter maneuvering system/ reaction control system functions in the aft compartment.
	A new general purpose computer and multiplexer demultiplexers were installed and tested. One liquid hydrogen and two liquid oxygen transducers were replaced upstream in the propellant flow system near the 17-in (43.1-cm) disconnect area, which was protected by an internal screen. Three liquid oxygen transducers were replaced at the engine manifold area, while three liquid hydrogen transducers were removed and the openings plugged. The launch was reset for 8 a.m. EDT, June 1, but it was postponed again after several failed attempts to calibrate inertial measurement unit No. 2. The unit was replaced and retested, and launch was rescheduled for June 5.
Orbital Altitude & Inclination	157 nmi (291 km), 39.0156 deg
Launch Weight (lb/kg)	251,970/114,292
Landing & Postlanding Operations	June 14, 1991, 8:39:11 a.m. PDT, Runway 22, Edwards Air Force Base. Orbiter returned to Kennedy Space Center June 21.
Rollout Distance (ft/m)	9,438/2,877
Rollout Time (seconds)	55

Table 3–65. STS-40 Mission Characteristics (Continued)

Mission Duration	218 hr, 14 min, 20 sec
Landed Revolution No.	145
Mission Support	STDN
Primary Objective	SLS-1 mission
Deployed Satellites	None
Experiments	SLS-1 using the Spacelab long module: The tests subjects were humans, 30 rodents, and thousands of jellyfish. The primary SLS-1 experiments studied six body systems. The body systems investigated were: 1) cardiovascular/ cardiopulmonary (heart, lungs and blood vessels); 2) renal/endocrine (kidneys and hormone-secreting organs and glands); 3) blood (blood plasma); 4) immune system (white blood cells); 5) musculoskeletal (muscles and bones); and 6) neurovestibular (brains and nerves, eyes, and inner ear). Of the 18 investigations, 10 involved humans, 7 involved rodents, and 1 used jellyfish. The experiments were:

- Influence of Weightlessness Upon Human Autonomic Cardiovascular Controls

- In-flight Study of Cardiovascular Deconditioning

- Vestibular Experiments in Spacelab

- Protein Metabolism During Spaceflight

- Fluid-Electrolyte Regulation During Spaceflight

- Pulmonary Function During Weightlessness

- Lymphocyte Proliferation in Weightlessness

- Influence of Spaceflight on Erythrokinetics in Man

- Cardiovascular Adaptation to Microgravity

- Pathophysiology of Mineral Loss During Spaceflight

- Regulation of Erythropoiesis During Spaceflight

- Regulation of Blood Volume During Spaceflight

- Bone, Calcium, and Spaceflight

- A Study of the Effects of Space Travel on Mammalian Gravity Receptors

- Effects of Microgravity-Induced Weightlessness on Aurelia Ephyra Differentiation and Statolith Synthesis

- Skeletal Myosin Isoenzymes in Rats Exposed to Microgravity

- Effects of Microgravity on Biochemical and Metabolic Properties of Skeletal Muscle in Rats

- The Effects of Microgravity on the Electron Microscopy, Histochemistry, and Protease Activities of Rat Hind-limb Muscles

Table 3–65. STS-40 Mission Characteristics (Continued)

Experiments	Orbital Acceleration Research Experiment (OARE): This experiment was designed to accurately measure aerodynamic acceleration rates in zero gravity to expand the database of knowledge in predicting orbital drag in the design of future space systems like Space Station *Freedom*.
	Middeck Zero-Gravity Dynamics Experiment (MODE): This experiment studied the behavior of space structures and contained fluids in microgravity. Scale models of truss beams for large space structures of the future were attached to a vibrating device to analyze the stresses that developed. Fluid slosh forces, in a partially filled container, were measured during vibration. The results yielded insight into developing efficient techniques for fluid transfer in space.
	OEX: The OEX program provided a mechanism for flight research experiments to be developed and flown aboard a Space Shuttle orbiter. Since the program's inception, 13 experiments were developed for flight. Principal investigators for these experiments represented Langley and Ames Research Centers, Johnson Space Center, and Goddard Space Flight Center. Seven OEX experiments flew on STS-40. Included among this group were six experiments conceived by Langley researchers and one experiment developed by Johnson.
	• SEADS: See STS-32.
	• SUMS: See STS-35.
	Both SEADS and SUMS provided entry atmospheric environmental (density) information. This data, when combined with vehicle motion data, was used to determine in-flight aerodynamic performance characteristics of the orbiter.
	• ACIP: See STS-28.
	• HiRAP: See STS-35.
	• SILTS: This experiment used a scanning infrared radiometer located atop the vertical tail to collect infrared images of the orbiter's leeside (upper) surfaces during entry, for the purpose of measuring the temperature distribution and the aerodynamic heating environment. On STS-32, the experiment obtained images of the left wing. For STS-35 and STS-40, the experiment was configured to obtain images of the upper fuselage. SILTS had flown on four *Columbia* flights.
	• AIP: See STS-35.

Table 3–65. STS-40 Mission Characteristics (Continued)

Get Away Specials	Twelve GAS canisters were installed on the GAS bridge in the cargo bay for experiments in materials science, plant biology, and cosmic radiation.
	G-021
	Customer: ESA
	Solid State Microaccelerometer Experiment: This experiment tested a new kind of very sensitive, highly miniaturized accelerometers intended for applications on a number of ESA space missions.
	G-052
	Customer: GTE Laboratories, Inc.
	Experiment in Crystal Growth: This experiment grew two crystals of gallium arsenide (GaAs). Growth of the two crystals in space was part of a comprehensive research program to systematically investigate the effect of gravity-driven fluid flow on GaA crystal growth.
	G-091
	Customer: CSUN Aerospace Group
	Orbital Ball Bearing Experiment: A team of researchers from California State University, Northridge (CSUN) built an apparatus called the Orbital Ball Bearing Experiment (OBBEX) to test the effects of melting cylindrical metal pellets in microgravity. If successful, this experiment might produce a new type of ball bearing.
	G-105
	Customer: Alabama Space & Rocket Center
	In-Space Commercial Processing: Scientists at the University of Alabama in Huntsville (UAH) used five experiments to study possible commercial in-space processing opportunities.
	Two experiment packages in the canister processed organic films and crystals that might be used in optical communications and computers. A third investigated electroplated metals to study special catalytic or reactive properties, or resistance to corrosion. A fourth experiment studied technology used to refine and process organic materials such as medical samples. The fifth experiment collected cosmic ray interactions on film emulsion while helping scientists assess materials that might be used in future massive cosmic ray detectors to be flown on the Shuttle or Space Station *Freedom* or to determine exposure to energetic particles on Earth.
	The U.S. Space and Rocket Center provided the sixth experiment which studied the effects of cosmic radiation on the chromosomes and genes of a common yeast.

Table 3–65. STS-40 Mission Characteristics (Continued)

Get Away Specials	G-286
	Customer: OMNI International, Ltd., and Duke University
	Foamed Ultralight Metals: This experiment demonstrated the feasibility of producing, in orbit, foams of ultralight metals for possible application as shock-absorbing panel-backing to improve the shielding of both crewed and uncrewed vehicles and satellites, including Space Station *Freedom*, against hypervelocity impacts either from micrometeroids or orbiting debris.
	G-405
	Customer: Frontiers of Science Foundation
	Chemical Precipitate Formation: This experiment returned data on the formation of six insoluble inorganic chemical precipitates. The experiment investigated the rate of formation and terminal size of precipitate particles when the growth was unimpaired by settling due to gravity.
	G-408
	Customer: The Mitre Corporation
	Five Microgravity Experiments: One GAS can contained five student experiments from the Worcester Polytechnic Institute. One attempted to grow large zeolite crystals. Another studied the behavior of fluids in microgravity. A third, the Environmental Data Acquisition System, recorded information about sound, light, temperature, and pressure within the GAS can. The fourth measured Shuttle acceleration along three axes with a high degree of precision. A fifth experiment studied the fogging of film in space.
	G-451
	Customer: Nissho Iwai American
	Flower and Vegetable Seeds Exposure to Space: Sakana Seeds Corp. in Yokohama, Japan, and the Nissho Iwai American Corp. in New York, New York, jointly sent 19 varieties of flower and vegetable seeds into space to determine how the unknown variables of microgravity affected seed growth. After the Shuttle landed and the seeds were recovered, the companies distributed the seeds to amateur growers.

Table 3–65. STS-40 Mission Characteristics (Continued)

Get Away Specials	G-455 Customer: Nissho Iwai American Semiconductor Crystal Growth Experiment: Investigated the potential advantages of crystal growth under microgravity. There were two experiments: PbSnTe crystal growth from vapor and GaAs crystal growth from metallic solution. This payload was sponsored by Fujitsu Ltd. in Kawasaki, Japan, and Nissho Iwai Corp. in Tokyo, Japan. G-507 Customer: Goddard Space Flight Center Orbiter Stability Experiment: This experiment measured the Space Shuttle's spectrum of small angular motions (or "jitter") produced by the operation of mechanical systems, thruster firings, and human motions during normal crew activity. In addition to the vibration measurements made, Goddard's GAS can also carried a passive experiment to test the effects of radiation on photographic film. G-616 Customer: Thomas Hancock The Effect of Cosmic Radiation on Floppy Disks and Plant Seeds Exposure to Microgravity: This payload consisted of two experiments. The first investigated static computer memory (floppy disks) to determine if cosmically charged particles produced changes in data integrity or structure. The second looked for changes in the physiology or growth of 38 different types of plant seeds. Each cultivator was examined postflight and compared with samples from the same seed lot that remained on Earth for a wide variety of possible effects or changes. Several of the floppy disks contained programs developed by elementary school students. In addition, a large number of plant seeds were distributed to every elementary and junior high school student in the Redlands, California, Unified School District, the sponsor of the experiment. G-486 Sponsor: EDSYN, Inc., Van Nuys, California Six Active Soldering Experiments
Mission Results	Successful
Remarks	This was the first mission dedicated solely to life sciences using the Spacelab habitable module.

Table 3–66. STS-43 Mission Characteristics

Vehicle	OV-104 *Atlantis*
Crew	CDR: John E. Blaha PLT: Michael A. Baker MS: Shannon W. Lucid, James C. Adamson, G. David Low
Launch	August 2, 1991, 11:01:59 a.m. EDT, Kennedy Space Center, Pad 39-A. Launch was originally set for July 23, but was moved to July 24 to allow time to replace a faulty integrated electronics assembly that controlled orbiter/external tank separation. The mission was postponed again about 5 hours before liftoff on July 24 due to a faulty main engine controller on the No. 3 main engine. The controller was replaced and retested; launch was reset for August 1. Liftoff set for 11:01 a.m. was delayed and postponed because of a cabin pressure vent valve reading at 12:28 p.m. due to unacceptable weather at the return-to-launch-site location. Launch was reset for August 2.
Orbital Altitude & Inclination	174 nmi (322 km), 28.45 deg
Launch Weight (lb/kg)	259,374/117,650
Landing & Postlanding Operations	August 11, 1991, 8:23:25 a.m. EDT, Runway 15, Kennedy Space Center.
Rollout Distance (ft/m)	9,890/3,014
Rollout Time (seconds)	60
Mission Duration	213 hr, 21 min, 22 sec
Landed Revolution No.	142
Mission Support	STDN
Primary Objective	Deployment of TDRS-E
Deployed Satellites	TDRS-5/IUS
Experiments	SHARE II: This experiment demonstrated microgravity thermal vacuum performance of a heat pipe radiator for heat rejection. SSBUV Instrument: See STS-34. Tank Pressure Control Equipment (TPCE): This experiment determined the effectiveness of jet mixing for controlling tank pressures and equilibrating fluid temperatures. Optical Communications Through Windows (OCTW): This experiment demonstrated the optical transmission of data from the crew cabin to the payload bay. APE-B: See STS-33.

Table 3–66. STS-43 Mission Characteristics (Continued)

Experiments	PCG III: Part of a continuing series of experiments, this mission's experiments were conducted using bovine insulin. See STS-29.
	BIMDA: See STS-37.
	IPMP: See STS-31.
	Space Acceleration Measurement System (SAMS): The SAMS provided acceleration data to characterize the middeck and/or middeck-mounted experiments acceleration environment.
	SSCE: See STS-41.
	Ultraviolet Plume Imager (UVPI): No flight hardware; the orbiter was used as a calibration target for space-based ultraviolet sensors.
	AMOS Calibration Test: See STS-29.
Get Away Specials	None
Mission Results	Successful

Table 3–67. STS-48 Mission Characteristics

Vehicle	OV-103
	Discovery
Crew	CDR: John O. Creighton
	PLT: Kenneth S. Reightler, Jr.
	MS: James F. Buchli, Charles D. Gemar, Mark N. Brown
Launch	September 12, 1991, 7:11:04 p.m. EDT, Kennedy Space Center, Pad 39-A. Launch was delayed 14 minutes by a faulty communication link between Kennedy Space Center and Mission Control in Houston.
Orbital Altitude & Inclination	313 nmi (580 km), 57 deg
Launch Weight (lb/kg)	240,062/108,890
Landing & Postlanding Operations	September 18, 1991, 12:38:42 a.m. PDT, Runway 22, Edwards Air Force Base. Landing was scheduled for Kennedy Space Center but was diverted to Edwards due to bad weather. Orbiter returned to Kennedy Space Center September 26, 1991.
Rollout Distance (ft/m)	9,384/2,860
Rollout Time (seconds)	50
Mission Duration	128 hr, 27 min, 34 sec
Landed Revolution No.	80
Mission Support	STDN
Primary Objective	Deployment of the UARS
Deployed Satellites	UARS
Experiments	APM: See STS-31.
	MODE: See STS-40.
	SAM: See STS-28.
	Cosmic Ray Effects and Activation Monitor (CREAM): The monitor collected data on cosmic ray energy loss spectra, neutron fluxes, and induced radioactivity. The data was obtained from the same locations used to gather data for the SAM experiment in an attempt to correlate data between the two.

Table 3–67. STS-48 Mission Characteristics (Continued)

Experiments	Physiological and Anatomical Rodent Experiment (PARE): First in a series of planned experiments, on physiological and developmental adaptation to microgravity, the PARE-01 experiment examined changes caused by exposure to microgravity in anti-gravity muscles (used for movement) and in tissues not involved in movement. Eight young, healthy rats flew on the Shuttle. After flight, full ground studies housing an identical group of animals under identical conditions (except for the presence of gravity) were conducted. Both groups were housed in self-contained animal enclosure modules that provided food, water, and environmental control throughout the flight. The experiment's design and intent received the review and approval of the animal care and use committees at both NASA and the University of Arizona. Laboratory animal veterinarians oversaw the selection, care, and handling of the rats. Following the flight, the principle investigator thoroughly evaluated the rat tissues.
	PCG II-2: A continuation of earlier studies, PCG II-2 investigated processes for growing large protein crystals in space. This experiment consisted of 60 vapor diffusion crystal growth chambers. See STS-29.
	IPMP: See STS-31.
	AMOS Calibration Test: See STS-29.
Get Away Specials	None
Mission Results	Successful

Table 3–68. STS-44 Mission Characteristics

Vehicle	OV-104
	Atlantis
Crew	CDR: Frederick D. Gregory
	PLT: Terence T. Henricks
	MS: F. Story Musgrave, Mario Runco, Jr., James S. Voss
	PS: Thomas J. Hennen
Launch	November 24, 1991, 6:44:00 p.m. EST, Kennedy Space Center, Pad 39-A. Launch set for November 19 was delayed due to a malfunctioning redundant inertial measurement unit on the IUS booster attached to the DSP satellite. The unit was replaced and tested. Launch was reset for November 24 but was delayed for 13 minutes to allow an orbiting spacecraft to pass and to allow external tank liquid oxygen replenishment after minor repairs to the valve in the liquid oxygen replenishment system in the mobile launcher platform.
Orbital Altitude & Inclination	197 nmi (365 km), 28.5 deg
Launch Weight (lb/kg)	259,629/117,766
Landing & Postlanding Operations	December 1, 1991, 2:34:44 p.m. PST, Runway 5, Edwards Air Force Base. Landing was originally scheduled for Kennedy Space Center on December 4, but the 10-day mission was shortened and landing rescheduled following the November 30 on-orbit failure of one of the three orbiter inertial measurement units. Lengthy rollout was due to minimal braking for test.
Rollout Distance (ft/m)	11,191/3,411
Rollout Time (seconds)	107
Mission Duration	166 hr, 50 min, 42 sec
Landed Revolution No.	109
Mission Support	STDN
Primary Objective	Unclassified DOD mission; deployment of DSP satellite
Deployed Satellites	DSP F16/IUS (USA 75)[a]
Experiments	IOCM: See STS-28.
	Terra Scout: To evaluate the effectiveness of real-time visual observation of terrestrial and oceanic targets, the observational and analytical skills of a photointerpretation specialist were compared with earthbound observation technology of designated targets. Results aided in the development of autonomous sensors.

Table 3–68. STS-44 Mission Characteristics (Continued)

Experiments	Military Man in Space (M88-1): To evaluate the effectiveness of real-time visual observations of terrestrial and oceanic targets, a crew member used optical camera systems to attempt to identify various military-related activities such as ship wakes; truck convoys; armored formations; aircraft operations; dust clouds; and smoke.
	AMOS Calibration Test: See STS-29.
	CREAM: See STS-48.
	SAM: See STS-28.
	RME-III: See STS-28.
	VFT-1: See STS-33.
	Extended Duration Orbiter Medical Project: To investigate countermeasures to orthostatic intolerance problems, this experiment used fluid loading, in which crew members ingested salt tablets and water and used the Lower Body Negative Pressure (LBNP) device. The LBNP created a partial vacuum around the lower body, returning some of the fluids to the legs.
	UVPI: See STS-43.
	Bioreactor Flow and Particle Trajectory (BFPT) in Microgravity: This fluid dynamics experiment validated Earth-based predictions for the action of cell cultures in the NASA-developed Slow-Turning Lateral Vessel (STLV) bioreactor. Researchers were interested in the benefits of flying a bioreactor in space because of the expected increased capabilities for cell culturing. The STLV bioreactor, developed as a tool for Space Station *Freedom*, grew cell cultures in a horizontal cylindrical container that slowly rotated, emulating microgravity and keeping the cells continuously suspended while bathing them in nutrients and oxygen. Components from the NASA bioreactor occupied two middeck lockers.
Get Away Specials	None
Mission Results	Successful
Remarks	Unclassified DOD mission
	Ten-day mission was shortened to seven days because of the on-orbit failure of one of the three orbiter inertial measurement units. Despite the early return, most mission objectives were achieved.[b]

[a]   "USA 75," *http://nssdc.gsfc.nasa.gov/database/MasterCatalog?sc=1991-080B* (accessed November 30, 2005). Also listed in the Federation of American Scientists list of military satellites and the Launch Log of *Jonathan's Space Report*.

[b]   "STS-44 Mission Summary," Spacelink Cached Web site, *http://spacelink.nasa.gov/NASA.Projects/ Human.Exploration.and.Development.of.Space/Human.Space.Flight/Shuttle/Shuttle.Missions/ Flight.044.STS-44/Mission.Summary* (cached site accessed December 14, 2005).

Table 3–69. STS-42 Mission Characteristics

Vehicle	OV-103
	Discovery
Crew	CDR: Ronald J. Grabe
	PLT: Stephen S. Oswald
	MS: Norman E. Thagard, David C. Hilmers, William F. Readdy
	PS: Roberta L. Bondar (Canadian Space Agency/CSA), Ulf D. Merbold (ESA)
Launch	January 22, 1992, 9:52:33 a.m. EST, Kennedy Space Center, Pad 39-A. Launch was delayed 1 hour due to weather conditions.
Orbital Altitude & Inclination	163 nmi (302 km), 57 deg
Launch Weight	243,396/110,403
Landing & Postlanding Operations	January 30, 1992, 8:07:17 a.m. PST, Runway 22, Edwards Air Force Base. The mission was extended one day for continued scientific experimentation. Orbiter returned to Kennedy Space Center on February 16, 1992.
Rollout Distance (ft/m)	9,841/3,000
Rollout Time (seconds)	58
Mission Duration	193 hr, 14 min, 44 sec
Landed Revolution No.	128
Mission Support	STDN
Primary Objective	Conduct life sciences research with the IML-1
Deployed Satellites	None
Experiments	IML-1: Working in a pressurized Spacelab long module, the international crew, divided into red and blue teams, conducted experiments on the human nervous system's adaptation to low gravity and the effects of microgravity on other life forms such as shrimp eggs, lentil seedlings, fruit fly eggs, and bacteria. Low gravity materials processing experiments included crystal growth from a variety of substances such as enzymes, mercury iodine, and a virus.
	IML-1 Life Science Experiments
	• Biorack
	– Leukemia Virus Transformed Cells to Microgravity in the Presence of Dimethylsulfoxide (DMSO)
	– Proliferation and Performance of Hybridoma Cells in Microgravity (HYBRID)
	– Dynamic Cell Culture System (CULTURE)

Table 3–69. STS-42 Mission Characteristics (Continued)

Experiments	– Chondrogenesis in Micromass Cultures of Mouse Limb Mesenchyme Exposed to Microgravity (CELLS)
	– Effects of Microgravity and Mechanical Stimulation on the In-Vitro Mineralization and Resorption of Fetal Mouse Bones (BONES)
	– Why Microgravity Might Interfere With Amphibian Egg Fertilization and the Role of Gravity in Determination of the Dorsal/Ventral Axis in Developing Amphibian Embryos (EGGS)
	– Effects of Space Environment on the Development of Drosophila Melanogaster (FLY)
	– Genetic and Molecular Dosimetry of HZE Radiation (RADIAT)
	– Dosimetric Mapping Inside Biorack (DOSIMTR)
	– Embryogenesis and Organogenesis of Carausius (MOROSUS)
	– Gravity-Related Behavior of the Acellular Slime Mold Physarum Polycephalum (SLIME)
	– Microgravitational Effects on Chromosome Behavior (YEAST)
	– Growth and Sporulation in Bacillus Subtilis Under Microgravity (SPORES)
	– Studies on Penetration of Antibiotics in Bacterial Cells in Space Conditions (ANTIBIO)
	– Transmission of the Gravity Stimulus in Statocyte of the Lentil Root (ROOTS)
	– Genotype Control of Graviresponse, Cell Polarity, and Morphological Development of Arabidopsis Thaliana in Microgravity (SHOOTS)
	– Effects of Microgravity Environment on Cell Wall Regeneration, Cell Divisions, Growth and Differentiation of Plants From Protoplasts (PROTO)
	• Gravitational Plant Physiology Facility Experiments
	  – Gravity Threshold (GTHRES)
	  – Response to Light Stimulation: Phototropic Transients (FOTRAN)
	• Microgravity Vestibular Investigations
	• Mental Workload and Performance Experiment
	• Space Physiology Experiments
	  – Space Adaptation Syndrome Experiments (SASE)
	  – Sled Experiment
	  – Rotation Experiment
	  – Visual Stimulator Experiment
	  – Proprioceptive Experiments
	  – Energy Expenditure in Spaceflight (EES)
	  – Position and Spontaneous Nystagmus (PSN)

Table 3–69. STS-42 Mission Characteristics (Continued)

Experiments	– Measurement of Venous Compliance (MVC) and Evaluation of an Experimental Anti-Gravity Suit

– Measurement of Venous Compliance (MVC) and Evaluation of an Experimental Anti-Gravity Suit
– Assessment of Back Pain in Astronauts (BPA)
– Phase Partitioning Experiment (PPE)
• Biostack (four packages)
• Radiation Monitoring Container Device (RMCD)

IML-1 Materials Science Experiments
• PCG
• Cryostat
• Single Crystal Growth of Beta-Galactosidase and Beta-Galactosidase/Inhibiter Complex
• Crystal Growth of the Electrogenic Membrane Protein Bacteriorhodopsin
• Crystallization of Proteins and Viruses in Microgravity by Liquid-Liquid Diffusion
  – Fluids Experiment System
    • Study of Solution Crystal Growth in Low Gravity (TGS)
    • An Optical Study of Grain Formation: Casting and Solidification Technology (CAST)
  – Vapor Crystal Growth System (VCGS)
    • Vapor Crystal Growth Studies of Single Mercury Iodide Crystals
  – Mercury Iodide Crystal Growth (MICG) System
    • Mercury Iodide Nucleations and Crystal Growth in Vapor Phase
• Organic Crystal Growth Facility
• Critical Point Facility (CPF)
  – Study of Density Distribution in a Near-Critical Simple Fluid
  – Heat and Mass Transport in a Pure Fluid in the Vicinity of a Critical Point
  – Phase Separation of an Off-Critical Binary Mixture
  – Critical Fluid Thermal Equilibration Experiment

SAMS: See STS-43.

SSIP (SE 81-09): Convection in Zero Gravity: This experiment studied surface-tension-induced flows in microgravity.

SSIP (SE 83-02): Zero-G Capillary Rise of Liquid Through Granular Porous Media: This experiment studied the flow of liquid through granular porous media.

Gelation of SOLS Applied Microgravity Research (GOSAMR): This experiment involved chemical gelation to form precursors for advanced ceramic materials.

IPMP: See STS-31.

RME-III: See STS-28.

Table 3–69. STS-42 Mission Characteristics (Continued)

Get Away Specials	G-086 Customer: Booker T. Washington Senior High School, Houston, Texas G-086 studied behavioral and physiological effects of microgravity on brine shrimp cysts hatched in space. The experiment also studied thermal conductivity and bubble velocity of air and water in microgravity.
	G-140 Customer: German Space Agency (DARA) G-140 studied the effect that a disturbance of the liquid-liquid interface (due to interfacial tension) had on mass transfer in a liquid-liquid extraction system in a floating zone.
	G-143 Customer: DARA Gas Bubbles in Glass Melts: This experiment researched the process of glass fining, the removal of all visible gaseous inhomogeneities from glass melt.
	G-329 Customer: Swedish Space Corporation G-329 studied solidification phenomena in metal alloys by looking at the dentrite growth in a cadmium-tin alloy.
	G-336 Customer: U.S. Air Force Geophysics Laboratory G-336 measured the visible light reflected by intergalactic dust. Data from those measurements were to be used to validate and update existing data collected in earlier experiments and help provide background measurements of visible light for use in space surveillance.[a]
	G-337 Customer: U.S. Naval Postgraduate School G-337 measured the performance of a thermoacoustic refrigerator under microgravity conditions.
	G-456 Customer: Society of Japanese Aerospace Companies G-456 separated three colored, biologically active enzymes by electrophoresis and compared them to Earth-based patterns.
	G-457 Customer: Society of Japanese Aerospace Companies G-457 cultivated cellular slime mold in microgravity as a preliminary study of a method of gas-liquid separation under conditions of microgravity.

Table 3–69. STS-42 Mission Characteristics (Continued)

Get Away Specials	G-609/G-610 Customer: Australian Space Office Auspace Ltd. designed and built an Australian ultraviolet light telescope for the Australian Space Office. It obtained ultraviolet images of violent events in nearby galaxies. Two interconnected GAS cans housed the payload's components. One contained the optical elements, and the second contained a flight battery and two tape recorders for recording detector data.
	G-614 Customer: Chinese Society of Astronautics and The American Association for Promotion of Space in China G-614 photographed the motion of simulated debris in the Shuttle under microgravity and remelted various low melting point mixtures of paraffin and Wood's metal while in orbit.
Mission Results	Successful

[a] *The First 100 GAS Payloads*, (Greenbelt, MD: NASA Goddard Space Flight Center, no date), p. 148.

Table 3–70. STS-45 Mission Characteristics

Vehicle	OV-104
	Atlantis
Crew	CDR: Charles F. Bolden, Jr.
	PLT: Brian Duffy
	MS: Kathryn D. Sullivan, David C. Leestma, C. Michael Foale
	PS: Byron K. Lichtenberg, Dirk D. Frimout (ESA)
Launch	March 24, 1992, 8:13:40 a.m. EST, Kennedy Space Center, Pad 39-A. Launch was originally scheduled for March 23, but was delayed one day because of higher than allowable concentrations of liquid hydrogen and liquid oxygen in the orbiter's aft compartment during tanking operations. During troubleshooting, the leaks could not be reproduced, leading engineers to believe they were the result of plumbing in the main propulsion system not thermally conditioned to the super cold propellants. Launch was rescheduled for March 24. Liftoff was delayed about 13 minutes due to low-level clouds at the Kennedy Space Center Shuttle runway.
Orbital Altitude & Inclination	160 nmi (296 km), 57 deg
Launch Weight (lb/kg)	233,650/105,982
Landing & Postlanding Operations	April 2, 1992, 6:23:08 a.m. EST, Runway 33, Kennedy Space Center. Mission extended one day to continue science experiments.
Rollout Distance (ft/m)	9,227/2,812
Rollout Time (seconds)	60
Mission Duration	214 hr, 9 min, 28 sec
Landed Revolution No.	142
Mission Support	STDN
Primary Objective	Atmospheric research using the ATLAS-1
Deployed Satellites	None
Experiments	Atmospheric Laboratory for Applications and Science using the Spacelab pallet and igloo. ATLAS-1 investigations in the areas of atmospheric science, plasma physics, and astrophysics.

	• Atmospheric Science
	  – Atmospheric Lyman-Alpha Emission (ALAE)
	  – Atmospheric Trace Molecule Spectroscopy (ATMOS)
	  – Grille Spectrometer
	  – Millimeter Wave Atmospheric Sounder (MAS)
	  – Imaging Spectrometric Observatory (ISO)

Table 3–70. STS-45 Mission Characteristics (Continued)

Experiments	• Solar Science
	– Active Cavity Radiometer Irradiance Monitor (ACRIM)
	– Measurement of Solar Constant (SOLCON)
	– Solar Spectrum (SOLSPEC)
	– Solar Ultraviolet Spectral Irradiance Monitor (SUSIM)
	– Plasma Physics
	– Atmospheric Emissions Photometric Imager (AEPI)
	– Space Experiments with Particle Accelerators (SEPAC)
	– Energetic Neutral Atom Precipitation (ENAP)
	• Astrophysics
	– Far Ultraviolet Space Telescope (FAUST)
	SSBUV: The instrument is housed in two GAS canisters. See STS-34.
	Space Tissue Loss (STL): This experiment was a cooperative effort between NASA's OSSA and the Walter Reed Army Institute of Research with hardware sponsored by the U.S. Army Space Test Program and mission management provided by the U.S. Air Force Space Systems Division. The STL experiment studied the effects of the microgravity environment on the biochemistry and functional activity of muscle, bone, and blood cells. Appropriate cell lines were cultured to develop a cellular model for comparison with whole animal results. The muscle atrophy model, consisting of cultured human myocardial cells in a monitored environment, validated skeletal and cardiac muscle atrophy, collected data on catabolic pathways, and tested candidate pharmaceuticals for efficacy in countering tissue loss.
	IPMP: See STS-31.
	RME-III: See STS-28.
	VFT-2: See STS-28.
	CLOUDS-1A: See STS-28
	SAREX II-B: See STS-35.
Get Away Specials	G-229 Customer: GTE Laboratories Experiment in Crystal Growth: This experiment was designed to grow GaAs crystals. GaAs is a versatile electronic material used in high-speed electronics and optoelectronics. The crystal grown on this mission was 1 inch in diameter by 3.5 inches long and was grown using a gradient freeze growth technique.
Mission Results	Successful

Table 3–71. STS-49 Mission Characteristics

Vehicle	OV-105
	Endeavour
Crew	CDR: Daniel C. Brandenstein
	PLT: Kevin P. Chilton
	MS: Pierre J. Thuot, Kathryn C. Thornton, Richard J. Hieb, Thomas D. Akers, Bruce E. Melnick
Launch	May 7, 1992, 7:40:00 p.m. EDT, Kennedy Space Center, Pad 39-B. Following the Flight Readiness firing of *Endeavour*'s three main engines on April 6, 1992, Shuttle managers decided to replace all three engines due to irregularities detected in two of the high-pressure oxidizer turbopumps; no impact to the launch date was expected. Launch was originally set for May 4 at 8:34 p.m. EDT, but it was moved to May 7 for an earlier launch window opening at 7:06 p.m. EDT to achieve better lighting conditions for photographic documentation of vehicle behavior during the launch phase. Liftoff was delayed 34 minutes due to transoceanic abort landing site weather conditions and a technical glitch with one of the orbiter master events controllers.
Orbital Altitude & Inclination	195 nmi (361 km), 28.34 deg
Launch Weight (lb/kg)	256,597/116,390
Landing & Postlanding Operations	May 16 1992, 1:57:38 p.m. PDT, Runway 22, Edwards Air Force Base. The flight was extended two days to complete mission objectives. The first use of drag chutes during landing, they deployed after nosegear touchdown for data collection only. The orbiter returned to Kennedy Space Center on May 30, 1992.
Rollout Distance (ft/m)	9,490/2,893
Rollout Time (seconds)	58
Mission Duration	213 hr, 17 min, 38 sec
Landed Revolution No.	140
Mission Support	STDN
Primary Objective	Capture, repair, and redeploy INTELSAT VI
Deployed Satellites	INTELSAT VI

Table 3–71. STS-49 Mission Characteristics (Continued)

Experiments	Commercial PCG Experiment: For six years, a variety of hardware configurations were used to conduct PCG experiments aboard 12 Space Shuttle flights. These experiments involved minute quantities of sample materials to be processed. On STS-49, the Protein Crystallization Facility (PCF), developed by the Center for Macromolecular Crystallography (CMC), a NASA Center for the Commercial Development of Space at the University of Alabama-Birmingham, used much larger quantities of materials to grow crystals in batches, using temperature as a means to initiate and control crystallization. The PCF was reconfigured to include cylinders with the same height but varying diameters to obtain different volumes (500 ml, 200 ml, 100 ml, and 20 ml). These cylinders allowed for a relatively minimal temperature gradient and required less protein solution to produce quality crystals. This industry-driven change was brought about by a need to reduce the cost and amount of protein sample needed to grow protein crystals in space, while at the same time increasing the quality and quantity of crystals. The PCF served as the growth chamber for significant quantities of protein crystals.
	Also flying as part of the CPCG payload complement was a newly designed, state-of-the-art Commercial Refrigerator Incubator Module (CRIM) that allowed for a preprogrammed temperature profile. Developed by Space Industries, Inc., for CMC, the CRIM also provided improved thermal capability and had a microprocessor that used "fuzzy logic" to control and monitor the CRIM's thermal environment.
	UVPI Experiment: See STS-43.
	AMOS Calibration Test: See STS-29.
Get Away Specials	None
Mission Results	Successful

Table 3–71. STS-49 Mission Characteristics (Continued)

Remarks	INTELSAT VI had been stranded in an unusable orbit since its launch aboard a Titan rocket in March 1990. The INTELSAT VI capture required three EVAs. The first spacewalk was on flight day four by Thuot, who was unable to attach a capture bar to INTELSAT from his position on the remote manipulator system arm. A second unscheduled but identical attempt by Thuot failed on the following day. After resting on flight day six, an unprecedented three-person EVA was performed on flight day seven. During the longest EVA so far in U.S. space history (8 hours, 29 minutes), Hieb, Thuot, and Akers grasped the rotating INTELSAT by hand while Brandenstein maneuvered the orbiter, ultimately attaching the capture bar to the satellite and attaching INTELSAT to its new upper stage. The day after capture of the satellite, INTELSAT flight controllers ignited the upper stage to send the satellite to its intended geosynchronous orbit.

On flight day eight, Akers and Thornton performed an EVA as part of the Assembly of Station by EVA Methods (ASEM) experiment to demonstrate and verify maintenance and assembly capabilities for Space Station *Freedom*. |

Table 3–72. STS-50 Mission Characteristics

Vehicle	OV-102
	Columbia
Crew	CDR: Richard N. Richards
	PLT: Kenneth D. Bowersox
	PC: Bonnie J. Dunbar
	MS: Ellen S. Baker, Carl J. Meade
	PS: Lawrence J. DeLucas, Eugene H. Trinh
Launch	June 25, 1992, 12:12:23 p.m. EDT, Kennedy Space Center, Pad 39-A. Liftoff was delayed 5 minutes due to weather. This was the first flight of *Columbia* after its scheduled checkout and extensive modification period.
Orbital Altitude & Inclination	160 nmi (296 km), 28.45 deg
Launch Weight (lb/kg)	257,265/116,693
Landing & Postlanding Operations	July 9, 1992, 7:42:27 a.m. EDT, Runway 33, Kennedy Space Center. Landing was delayed one day due to rain at the primary landing site, Edwards Air Force Base. This was the first landing using new synthetic tread tires.
Rollout Distance (ft/m)	10,674/3,253
Rollout Time (seconds)	59
Mission Duration	331 hr, 30 min, 04 sec
Landed Revolution No.	220
Mission Support	STDN
Primary Objective	Microgravity research using USML-1
Deployed Satellites	None
Experiments	USML-1: A pressurized Spacelab long module with connecting tunnel to the orbiter crew compartment. USML-1 was a national effort to advance microgravity research in a broad array of disciplines. Experiments conducted were:
	• Crystal Growth Furnace (CGF)
	• Drop Physics Module (DPM)
	• Surface Tension Driven Convection Experiment (STDCE)
	• Zeolite Crystal Growth (ZCG)
	• PCG
	• Glovebox Facility (GBX)
	• SAMS
	• Generic Bioprocessing Apparatus (GBA)
	• ASTROCULTURE-1 (ASC)
	• Extended Duration Orbiter Medical Project (EDOMP)
	• SSCE

Table 3–72. STS-50 Mission Characteristics (Continued)

Experiments	IPMP: See STS-31.
	SAREX II: See STS-35.
	UVPI: See STS-43.
Get Away Specials	None
Mission Results	Successful

Table 3–73. STS-46 Mission Characteristics

Vehicle	OV-104
	Atlantis
Crew	CDR: Loren J. Shriver
	PLT: Andrew M. Allen
	PC: Jeffrey A. Hoffman
	MS: Franklin R. Chang-Diaz, Claude Nicollier (ESA), Marsha S. Ivins
	PS: Franco Malerba (Italian Space Agency Agenzia Spaziale Italiana/ASI)
Launch	July 31, 1992, 9:56:48 a.m. EDT, Kennedy Space Center, Pad 39-A. Liftoff was delayed 48 seconds at L-5 minutes to allow orbiter computers to verify that the orbiter auxiliary power units were ready to start.
Orbital Altitude & Inclination	230 nmi (426 km), 28.45 deg
Launch Weight (lb/kg)	256,031/116,134
Landing & Postlanding Operations	August 8, 1992, 9:11:51 a.m. EDT, Runway 33, Kennedy Space Center. The mission was extended one day to complete scientific objectives.
Rollout Distance (ft/m)	10,860/13,310
Rollout Time (seconds)	66
Mission Duration	191 hr, 15 min, 03 sec
Landed Revolution No.	126
Mission Support	STDN
Primary Objective	Operation and testing of the TSS and deployment of ESA's EURECA
Deployed Satellites	EURECA Joint NASA/Italian Space Agency TSS-1
Experiments	Evaluation of Oxygen Integration with Materials/ Thermal Energy Management Processes (EOIM-III/ TEMP 2A-3): This experiment gathered accurate data on the reaction rate of atomic oxygen, present in low orbit, on Space Shuttle materials. Collisions with atomic oxygen in orbit cause erosion of many materials. This experiment attempted to identify materials resistant to erosion to be used on future spacecraft.

Table 3–73. STS-46 Mission Characteristics (Continued)

Experiments	Consortium for Materials Development in Space Complex Autonomous Payload (CONCAP II and CONCAP III): This experiment investigated reactions occurring on the surface of materials when exposed to the atomic oxygen flow in Earth orbit on high-temperature, super-conducting films, and on materials degradation/reaction samples. The payloads were flown in 5-ft (1.5-m)-high cylindrical GAS canisters. CONCAP-II studied the changes materials underwent in low-Earth orbit. The payload experiments studied the surface reactions resulting from exposing materials to the atomic oxygen flow experienced by the Space Shuttle in orbit. CONCAP-III measured and recorded absolute accelerations (microgravity levels) in one experiment and electroplated pure nickel metal and recorded the conditions (temperature, voltage, and current) during this process in another experiment.
	ICBC: The ICBC payload documented operations outside the crew cabin, including prerelease and postrelease of EURECA, TSS-1 flyaway, and TSS reel-out and reel-in. The ICBC also observed Typhoon Janis; the Windward Islands; Java the Sahara desert; Madagascar; Brazil; the Andes mountains; the Tuamoto Archipelago; and the area from Indonesia to Australia.
	Limited Duration Space Environment Candidate Materials Exposure (LDCE): This experiment evaluated candidate space structure composite materials for degradation due to exposure in low-Earth orbit (passive systems).
	AMOS: See STS-29.
	Pituitary Growth Hormone Cell Function (PHCF): This experiment was a study to determine if the exposure of cultured rat pituitary cells to microgravity affected their capacity to produce biologically active growth hormone.
	UVPI: See STS-43.
Get Away Specials	None
Mission Results	During TSS deployment, the satellite reached a maximum distance of only 840 ft (256 m) from the orbiter instead of the planned 12.5 mi (201 km) because of a jammed tether line. After numerous attempts spanning several days to free the tether, TSS operations were curtailed, and the satellite was stowed for return to Earth. Other mission objectives were accomplished.

Table 3–74. STS-47 Mission Characteristics

Vehicle	OV-105
	Endeavour
Crew	CDR: Robert L. Gibson
	PLT: Curtis L. Brown, Jr.
	PC: Mark C. Lee
	MS: Jerome Apt, N. Jan Davis, Mae C. Jemison
	PS: Mamoru Mohri (Japanese Aerospace Exploration Agency/JAXA)
Launch	September 12, 1992, 10:23:00 a.m. EDT, Kennedy Space Center, Pad 39-B. This launch was the first on-time Shuttle launch since STS-61-B in November 1985.
Orbital Altitude & Inclination	166 nmi (307 km), 57.00 deg
Launch Weight (lb/kg)	258,679/117,335
Landing & Postlanding Operations	September 20, 1992, 8:53:23 a.m. EDT, Runway 33, Kennedy Space Center. The mission was extended one day for further scientific experimentation. This mission was the first time the drag chute was deployed in operational mode, before nosegear touchdown. Postlanding assessment showed that the orbiter veered off the runway centerline, possibly due to the drag chute.
Rollout Distance (ft/m)	8,567/2,611
Rollout Time (seconds)	51
Mission Duration	190 hr, 31 min, 11 sec
Landed Revolution No.	125
Mission Support	STDN
Primary Objective	Materials and life sciences research using Spacelab-J
Deployed Satellites	None

Table 3–74. STS-47 Mission Characteristics (Continued)

Experiments	Spacelab-J: A joint NASA-National Space Development Agency of Japan (NASDA) mission using a Spacelab long module. The international crew was divided into red and blue teams for round-the-clock operations. Spacelab-J included materials science and life sciences experiments, of which NASDA sponsored 37 and NASA sponsored 8. Materials science investigations covered such fields as biotechnology; electronic materials; fluid dynamics and transport phenomena; glasses and ceramics; metals and alloys; and acceleration measurements. Life sciences investigations covered human health; cell separation and biology; development biology; animal and human physiology and behavior; space radiation; and biological rhythms. Test subjects included crew members; Japanese koi fish; cultured animal and plant cells; chicken embryos; fruit flies; fungi and plant seeds; and frogs and frog eggs.

Sponsored by NASA
• Materials Science
  – SAMS
  – Inflight Demonstration of the Space Station
    Freedom Health Maintenance Facility Fluid
    Therapy System
• Life Science
  – PCG
  – Monitoring Astronauts' Functional State
  – Autogenic Responses to Microgravity
  – Bone Cell Research
  – Amphibian Development in Microgravity: The
    STS-47 Frog Embryology Experiment
  – Lower Body Negative Pressure Countermeasure
    Against Orthostatic Intolerance After Space Flight
  – Plant Cell Research Experiment on Spacelab J:
    Mitotic Disturbances in Daylily (Hemerocallis)
    Somatic Embryos After an 8-Day Spaceflight
  – Magnetic Resonance Imaging After Exposure to
    Microgravity

From the National Space Development Agency of Japan
• Materials Science
  – Growth Experiments of Narrow Band-Gap
    Semiconductor Pb-Sn-Te Single Crystal in Space
  – Growth of Pb-Sn-Te Single Crystal by Traveling
    Zone Method
  – Growth of Semiconductor Compound Single
    Crystal InSb by Floating Zone Method

Table 3–74. STS-47 Mission Characteristics (Continued)

Experiments	
	– Casting of Superconducting Composite Materials
	– Formation Mechanism of Deoxidation Products in Iron Ingot Deoxidized With Two or Three Elements
	– Preparation of Particle Dispersion–Alloys
	– Diffusion in Liquid State and Solidification of Binary System
	– High-Temperature Behavior of Glass
	– Growth of Silicon Spherical Crystals and Surface Oxidation
	– Study of Solidification of Immiscible Alloy
	– Fabrication of Ultra-Low-Density, High-Stiffness Carbon Fiber/Aluminum Composites
	– Study on Liquid Phase Sintering
	– Fabrication of Si-As-Te: Semiconductor in Microgravity Environment
	– Gas Evaporation in Low Gravity
	– Drop Dynamics in an Acoustic Resonant Chamber and Interference with the Acoustic Field
	– Bubble Behavior in Thermal Gradient and Stationery Acoustic Wave
	– Preparation of Optical Materials Used in Non-Visible Region
	– Marangoni Effort-Induced Convection in Material Processing Under Microgravity
	– Solidification of Eutectic System Alloys in Space
	– Growth of Samarskite Crystal in Microgravity
	– Crystal-Growth Experiment on Organic Metals in Low Gravity
	– Crystal Growth of Compound Semiconductors in a Low-Gravity Environment
	• Life Science
	– Endocrine and Metabolic Changes in Payload Specialist
	– Neurophysiological Study of Visuo-Vestibular Control of Posture and Movement in Fish During Adaptation to Weightlessness
	– Comparative Measurement of Visual Stability in Earth and Cosmic Space
	– Crystal Growth of Enzymes in Low Gravity
	– Studies on the Effects of Microgravity on the Ultrastructure and Functions of Cultured Mammalian Cells
	– Effect of Low Gravity on Calcium Metabolism and Bone Formation in Chick Embryo
	– Separation of Biogenic Materials by Electrophoresis Under Zero Gravity
	– Genetic Effects of HZE and Cosmic Radiation

Table 3–74. STS-47 Mission Characteristics (Continued)

    – Manual Control in Space Research on Perceptual
      Motor Functions Under Microgravity Condition
    – Study on the Biological Effect of Cosmic Radiation
      and the Development of Radiation Protection
      Technology
    – Circadian Rhythm of Conidiation in Neurospora
      Crassa
    – Electrophoretic Separation of Cellular Materials
      Under Microgravity
    – Study of the Effects of Microgravity on Cell
      Growth of Human Antibody-Producing Cells and
      Their Secretions
    – Organ Differentiation from Cultured Plant Cells
      Under Microgravity
    – Health Monitoring of Japanese Payload Specialist
    – Autonomic Nervous and Cardiovascular Responses
      Under Reduced Gravity

Israeli Space Agency Investigation About Hornets
(ISAIAH): This experiment attempted to gain greater
insight into the ability of hornets to construct a comb in
the direction of the gravitational vector by observing their
comb-building in microgravity. It also investigated the
effects of microgravity on comb integrity, social
interactions, hornet venom toxicity, and the
semiconductive properties of hornet cuticle.

SSCE: See STS-41.

SAREX II: See STS-35.

AMOS Calibration Test: See STS-29.

UVPI: See STS-43.

Table 3–74. STS-47 Mission Characteristics (Continued)

Get Away Specials	G-102 Customer: TRW Defense and Space Systems Group and Explorer Scouts POSTAR: Consisted of seven experiments: 1. Capillary Pumping Experiment: This experiment investigated pumping liquids using capillary feed tubes. 2. Cosmic Ray Experiment: This experiment studied the direction and composition of cosmic rays. 3. Crystal Growth Experiment: This experiment examined high-quality, lattice-structure crystals using nickel sulfate. 4. Emulsions Experiment: This experiment investigated the formation of oil/water emulsions. 5. Fluid Droplet Experiment: This experiment studied the shape of fluid droplets in microgravity. 6. Floppy Disk Experiment: This experiment investigated the effect of low-level radiation on floppy disks. 7. Fiber Optics Experiment: This experiment examined the degradation of fiber optic cables under the space radiation environment. G-255 Customer: University of Kansas Scientific studies contained four experiments: Composite Materials: This experiment compared composites manufactured in microgravity with those made on Earth. Cell Membrane: This experiment investigated the formation of biological membranes in microgravity. Crystal Growth: This experiment studied the effects of microgravity on bond angles and the structure of crystals. Space Seeds: This experiment studied the effects of the space environment on germination rates and health of seeds. G-300 Customer: Matra Marconi Space Thermal Conductivity of Liquids in Microgravity: This experiment measured the thermal conductivity of liquids in microgravity. It was the first GAS payload from France. G-330 Customer: Swedish Space Corporation Material Science Experiments: Crystal Growth and Electromigration: These experiments investigated the breakdown of a solid/liquid interface.

Table 3–74. STS-47 Mission Characteristics (Continued)

Get Away Specials	G-482 Customer: Space Aerospace, Ltd., Quebec, Canada Baking Bread in Space: This experiment investigated the behavior of bread yeast in the absence of gravity and in the presence of normal atmospheric pressure.
	G-520 Customer: The Independent Television News of England Scientific Experiments for Educational Competition: Consisted of two experiments: Chemical Gardens Experiment: This experiment deposited cobalt nitrate crystals into a container of sodium silicate solution. The ensuing crystal growth was compared to crystals grown on Earth. The Liesegang Rings Experiment: This experiment deposited a silver nitrate solution on a compound containing potassium chromate. Silver nitrate diffused through potassium chromate and formed a precipitate in the shape of rings. The investigation attempted to produce rings on a large scale to aid in determining why they were produced. This was the first British school experiment to fly in space.
	G-521 Customer: Canadian Space Agency QUESTS: This experiment involved performing directional solidification and diffusion experiments in microgravity using 3 gradient furnaces and 12 isothermal furnaces.
	G-534 Customer: NASA Headquarters Pool Boiling Experiment: This experiment investigated the effects of heat flux and liquid subcooling on nucleate pool boiling in a long-term, reduced-gravity environment.
	G-613 Customer: University of Washington Liquid Droplet Rotating-Collector Experiment: This was a proof-of-concept experiment testing the ability of a centrifugal collector to recover a free fluid stream in microgravity.
Mission Results	Successful

Table 3–75. STS-52 Mission Characteristics

Vehicle	OV-102
	Columbia
Crew	CDR: James D. Wetherbee
	PLT: Michael A. Baker
	MS: Charles Lacy Veach, William M. Shepherd, Tamara E. Jernigan
	PS: Steven G. MacLean (CSA)
Launch	October 22, 1992, 1:09:39 p.m. EDT, Kennedy Space Center, Pad 39-B. The targeted launch date in mid-October slipped when managers decided to replace the No. 3 engine, prompted by concerns about possible cracks in the liquid hydrogen coolant manifold on the engine nozzle. Liftoff set for 11:16 a.m. was delayed about 2 hours due to crosswinds at the Kennedy Space Center landing strip, violating return-to-launch-site criteria, and clouds at the Banjul transoceanic abort landing site.
Orbital Altitude & Inclination	163 nmi (302 km), 28.45 deg
Launch Weight (lb/kg)	250,130/113,457
Landing & Postlanding Operations	November 1, 1992, 9:05:52 a.m. EST, Runway 33, Kennedy Space Center. The drag chute was again deployed before nosegear touchdown to allow further study of deployment dynamics.
Rollout Distance (ft/m)	10,708/3,263
Rollout Time (seconds)	63
Mission Duration	236 hr, 56 min, 13 sec
Landed Revolution No.	158
Mission Support	STDN
Primary Objective	Deploy LAGEOS II and operation of USMP-1
Deployed Satellites	LAGEOS II/IRIS

Table 3–75. STS-52 Mission Characteristics (Continued)

Experiments	USMP-1: The payload included three experiments mounted on two connected Multipurpose Experiment Support Structures mounted in the cargo bay. The USMP-1 experiments were:

Experiments USMP-1: The payload included three experiments mounted on two connected Multipurpose Experiment Support Structures mounted in the cargo bay. The USMP-1 experiments were:
- Lambda Point Experiment (LPE): This experiment studied fluid behavior in microgravity.
- French-sponsored *Material Pour l'Etude Des Phenomenes Interessant La Solidification Sur Terre Et En Orbite* (Materials for the Study of Interesting Phenomena of Solidification on Earth and in Orbit) (MEPHISTO): This experiment studied metallurgical processes in microgravity.
- SAMS: This was a study of the microgravity environment on board the Space Shuttle.

Canadian Experiments-2 (CANEX-2): These experiments were a complement of space technology, space science, materials processing, and life sciences experiments located in both the cargo bay and middeck, including:
- Space Vision System (SVS)
- Materials Exposure in Low-Earth Orbit (MELEO)
- Queen's University Experiment in Liquid-Metal Diffusion (QUELD)
- Phase Partitioning in Liquids (PARLIQ)
- Sun Orbiter Glow-2 (OGLOW-2)
- Space Adaptation Tests and Observations (SATO)
- Sun Photo Spectrometer Earth Atmosphere Measurement (SPEAM-2)
- Vestibular-Ocular Reflex Check
- Body Water Changes in Microgravity
- Assessment of Back Pain in Astronauts
- Illusions During Movement

Altitude Sensor Package (ASP): The ASP featured three ESA independent sensors mounted on a Hitchhiker plate in the cargo bay:
- Modular Star Sensor
- Yaw Earth Sensor and Low Altitude Conical Earth Sensor
- Tank Pressure Control Experiment/Thermal Phenomena

Commercial Materials Dispersion Apparatus Instrument Technology Associates Experiments (CMIX): These experiments were designed to mix a variety of sample fluids/solids in microgravity and bio-processing modules designed to mix fluids to perform immune cell response in microgravity.

Table 3–75. STS-52 Mission Characteristics (Continued)

Experiments	CPCG: See STS-49.
	Crystal Vapor Transport Experiment (CVTE): The CVTE consisted of two furnaces, which provided a controlled environment for growth of selected materials.
	• Heat Pipe Performance Experiment (HPP): The HPP experiment developed the understanding of heat pipe behavior in microgravity.
	PSE-02: The PSE-02 determined the effects of a proprietary protein molecule on an animal's physiological system.
	Shuttle Plume Impingement Experiment (SPIE): The SPIE consisted of sensing hardware mounted on the remote manipulating system to measure the atomic oxygen flux and contamination.
	Tank Pressure Control Experiment/Thermal Phenomena (TPCE/TP): This experiment determined the effectiveness of jet mixing as a means of controlling tank pressures and equilibrating fluid temperatures, contained in a GAS canister.
Get Away Specials	None
Mission Results	Successful

Table 3–76. STS-53 Mission Characteristics

Vehicle	OV-103
	Discovery
Crew	CDR: David M. Walker
	PLT: Robert D. Cabana
	MS: Guion S. Bluford, Jr., James S. Voss, Michael R. Clifford
Launch	December 2, 1992, 8:24:00 a.m. EST, Kennedy Space Center, Pad 39-A. Liftoff was originally set for 6:59 a.m. but was delayed to allow sunlight to melt ice on the external tank that had formed after tanking due to overnight temperatures in the upper 40s°F and light wind.
Orbital Altitude & Inclination	174 nmi (322 km), 57.00 deg
Launch Weight (lb/kg)	243,952/110,655
Landing & Postlanding Operations	December 9, 1992, 12:43:47 p.m. PST, Runway 22, Edwards Air Force Base. Landing was originally set for Kennedy Space Center but was diverted due to clouds in the landing strip vicinity. The drag chute was deployed before nosegear touchdown. After landing, a small leak was detected in a forward thruster that delayed crew egress before a fan and winds dissipated the leaking gas. The orbiter returned to Kennedy Space Center on December 18, 1992.
Rollout Distance (ft/m)	10,165/3,098
Rollout Time (seconds)	73
Mission Duration	175 hr, 19 min, 47 sec
Landed Revolution No.	115
Mission Support	STDN
Primary Objective	Deploy a classified DOD payload
Deployed Satellites	SDS-2 (USA 89)

Table 3–76. STS-53 Mission Characteristics (Continued)

Experiments	Glow Experiment (GLO)/Cryogenic Heat Pipe Experiment (CRYOHP) Payload (GCP): These payloads were contained in or attached to GAS hardware. The GLO experiment observed orbiter and air glow, primary reaction control system and vernier reaction control system burns, water dumps, and flash evaporator system operations. The CRYOHP experiment measured the performance of liquid oxygen heat pipes in microgravity.
	Battlefield Laser Acquisition Sensor Test (BLAST): This test evaluated the use of a spaceborne laser receiver to detect laser energy and provide laser communication uplink for GPS information from specific ground sites.
	CLOUDS: See STS-28.
	CREAM: See STS-48.
	Fluid Acquisition and Resupply Experiment (FARE): The FARE investigated the fill, refill, and expulsion of fluid tanks and liquid motion in microgravity.
	Handheld, Earth-oriented, Real-time, Cooperative, User-friendly, Location-Targeting and Environmental System (HERCULES): The HERCULES performed geolocating operations over selected ground sites.
	Microcapsules in Space-1 (MIS-1): This experiment demonstrated the capability to produce microencapsulated ampicillin in microgravity to compare with Earth-produced ampicillin.
	RME-III: See STS-28.
	STL: See STS-45.
	VFT-2: See STS-28.
	Orbital Debris Radar Calibration System (ODERACS): This system qas to release six calibration spheres from *Discovery*. The spheres—two with diameters of 6 in (15-cm), two with 4-in (10-cm) diameters, and two with 2-in (5-cm) diameters—were to be placed in a 175-nmi (377-km) orbit when they were ejected from the Shuttle's cargo bay. The primary objective of the experiment was to provide a source for fine-tuning of the Haystack Radar, located in Tyngsboro, Massachusetts, and operated by the Lincoln Laboratory at the Massachusetts Institute of Technology for the U.S. Air Force.
Get Away Specials	None
Mission Results	The ODERACS was not deployed. After attempts to communicate with the experiment without response, it was determined that a battery had been drained before launch. Other mission objectives were successfully met.
Remarks	This was the final Shuttle flight for the DOD.

Table 3–77. STS-54 Mission Characteristics

Vehicle	OV-105
	Endeavour
Crew	CDR: John H. Casper
	PLT: Donald R. McMonagle
	MS: Mario Runco, Jr., Gregory J. Harbaugh, Susan J. Helms
Launch	January 13, 1993, 8:59:30 a.m. EST, Kennedy Space Center, Pad 39-B. Liftoff was delayed about 7 minutes due to concerns associated with upper atmospheric winds.
Orbital Altitude & Inclination	165 nmi (306 km), 29.45 deg
Launch Weight (lb/kg)	259,764[a]/117,827
Landing & Postlanding Operations	January 19, 1993, 8:37:49 a.m. EST, Runway 33, Kennedy Space Center. Landing was delayed one orbit due to ground fog at Kennedy Space Center.
Rollout Distance (ft/m)	8,724/2,659
Rollout Time (seconds)	49
Mission Duration	143 hr, 38 min, 19 sec
Landed Revolution No.	95
Mission Support	STDN
Primary Objective	Deploy TDRS-6
Deployed Satellites	TDRS-6/IUS
Experiments	Diffuse X-ray Spectrometer (DXS): The DXS was a Hitchhiker experiment sponsored by Goddard Space Flight Center. Data was collected on x-ray radiation from diffuse sources in deep space. The DXS determined the wavelength and intensity of the strongest x-ray lines emitted by the hot stellar gases released by supernovas.[b]
	Commercial General Bioprocessing Apparatus (CGBA): The CGBA performed two functions, biological sample processing and stowage. The Generic Bioprocessing Apparatus (GBA) module was a self-contained mixing and incubation module for samples. Temperature-controlled stowage was achieved in the CRIM.
	CHROMEX: See STS-29.
	PARE-02: See STS-48.
	SAMS: See STS-43.
	SSCE: See STS-41.
Get Away Specials	None
Mission Results	Successful

[a] Jenkins, p. 302.

[b] See chapter 4, Space Science, for further discussion of the Diffuse X-ray Spectrometer experiment.

Table 3–78. STS-56 Mission Characteristics

Vehicle	OV-103
	Discovery
Crew	CDR: Kenneth D. Cameron
	PLT: Stephen S. Oswald
	MS: C. Michael Foale, Kenneth D. Cockrell, Ellen Ochoa
Launch	April 8, 1993, 1:29:00 a.m. EDT, Kennedy Space Center, Pad 39-B. The first launch attempt on April 6 was halted at T-11 seconds by orbiter computers when instrumentation on the liquid hydrogen high point bleed valve in the main propulsion system indicated "off" instead of "on." Later analysis indicated that the valve was properly configured; 48-hour scrub turnaround procedures were implemented. The final countdown on April 8 proceeded smoothly.
Orbital Altitude & Inclination	160 nmi (296 km), 57.00 deg
Launch Weight (lb/kg)	236,659/107,347
Landing & Postlanding Operations	April 17, 1993, 7:37:19 a.m. EDT, Runway 33, Kennedy Space Center. Landing, originally set for April 16, at Kennedy Space Center was waved off due to weather. A second reefing line was added to drag chute for greater stability.
Rollout Distance (ft/m)	9,530/2,905
Rollout Time (seconds)	62
Mission Duration	222 hr, 8 min, 24 sec
Landed Revolution No.	147
Mission Support	STDN
Primary Objective	ATLAS-2
Deployed Satellites	SPARTAN-201: Retrieved April 13[a]
Experiments	ATLAS-2: This mission used Spacelab pallet and igloo to collect data on the relationship between the Sun's energy output and Earth's middle atmosphere and their affect on the ozone layer. The atmospheric instruments were: • ATMOS experiment • MAS • SSBUV/A spectrometer: See STS-34. The solar science instruments were: • SOLSPEC instrument • SUSIM • Active Cavity Radiometer (ACR) • SOLCON experiments SAREX II: See STS-35.

Table 3–78. STS-56 Mission Characteristics (Continued)

Experiments	Commercial Materials Dispersion Apparatus Instrumentation Technology Associates Experiment (CMIX): This collection of experiments was housed in a CRIM, replacing a middeck locker. The CRIM contained four MDA mini-lab units designed to mix a variety of fluids and/or fluids and solids in the microgravity environment. The CRIM also contained 10 Bioprocessing Modules (BPM) designed to mix fluids to perform immune cell response experiments in microgravity. The experiments studied protein crystal growth, collagen polymerization, fibrin clot formation, liquid-solid diffusion, and the formation of thin film membranes. See STS-52.

Experiments developed by the University of Alabama Huntsville Consortium for Materials Development in Space and its affiliates included:
• Bone Cell Differentiation (MDA)
• Immune Cell Response (MDA)
• Diatoms (MDA)
• Mouse Bone Marrow Cells (MDA)
• Nerve/Muscle Cell Interactions
• Phagocytosis (MDA)
• Other experiments evaluating fluids mixing, invertebrate and bone development, virus sub-unit assembly and collagen self-assembly, and formation of drug encapsulated liposomes
• Live Cell Investigations (BPM)

Experiments developed by ITA, Inc. and its affiliates include:
• Collagen Reconstitution (MDA)
• Microencapsulation (MDA)
• Urokinase Protein Crystal Growth (MDA)
• Bacterial Aldolase and Rabbit Muscle Aldolase Protein Crystal Growth (MDA)
• HIV Reverse Transcriptase (MDA)
• RNA Protein Crystal Growth (MDA)
• Methylase Protein Crystal Growth (MDA)
• Lysozyme Protein Crystal Growth (MDA)
• DNA-Heme Protein Crystal Growth (MDA)
• Brine Shrimp Development (MDA)
• Cell Research (MDA)
• Other commercial MDA experiments included inorganic assembly (proprietary), myoglobin protein crystal growth, dye and yeast cell diffusion, and engineering tests
• Mustard Seed Germination (MDA-student)

Table 3–78. STS-56 Mission Characteristics (Continued)

Experiments	PARE-03: See STS-48.
	STL-3: See STS-53.
	CREAM: See STS-48.
	HERCULES: See STS-53.
	RME-III: See STS-28.
	AMOS Calibration Test: See STS-29.
Get Away Specials	Solar Ultraviolet Experiment (SUVE) Customer: Colorado Space Grant Consortium Measured extreme ultraviolet and far ultraviolet solar irradiance with two spectrometers.
Mission Results	Successful

[a] See chapter 4, Space Science, for details of the SPARTAN satellites.

Table 3–79. STS-55 Mission Characteristics

Vehicle	OV-102
	Columbia
Crew	CDR: Steven R. Nagel
	PLT: Terence T. Henricks
	MS: Jerry L. Ross, Charles J. Precourt, Bernard A. Harris, Jr.
	PS: Ulrich Walter (Germany), Hans W. Schlegel[a] (Germany)
Launch	April 26, 1993, 10:50:00 a.m. EDT, Kennedy Space Center, Pad 39-A. Launch first set for February slipped to early March after questions arose about turbine blade tip seal retainers in the high-pressure oxidizer turbopumps on the orbiter main engines. When engineers could not verify whether old or new retainers were on *Columbia*, NASA opted to replace all three turbopumps at the pad as a precaution.
	The March 14 launch date slipped again after a hydraulic flex hose burst in the aft compartment during a Flight Readiness Test. All 12 hydraulic lines in the aft compartment were removed and inspected; 9 lines were reinstalled, and 3 new lines were put in.
	Launch set for March 21 was pushed back 24 hours due to range conflicts caused by a Delta II one-day launch delay. Orbiter computers aborted a liftoff attempt on March 22 at T-3 seconds because of incomplete ignition of the No. 3 main engine. The liquid oxygen preburner check valve leaked internally, causing an overpressur-ized purge system that, in turn, prevented full engine ignition. This was the first on-the-pad main engine abort since return to flight, and the third in program history (51-F and 41-D were the other two). The valve leak was later traced to contamination during manufacturing. NASA decided to replace all three main engines on *Columbia* with spares.
	Launch was reset for April 24 but was scrubbed early on launch morning when one of orbiter's three inertial measurement units (IMU) gave a possible faulty reading. Liftoff was postponed for 48 hours to allow removal and replacement of the IMU. The final launch countdown on April 26 proceeded smoothly.
Orbital Altitude & Inclination	163 nmi (302 km), 28.45 deg
Launch Weight (lb/kg)	255,441[b]/115,866
Landing & Postlanding Operations	May 6, 1993, 7:29:59 a.m. PDT, Runway 22, Edwards Air Force Base. Landing originally set for Kennedy Space Center moved to Edwards because of cloud cover.

Table 3–79. STS-55 Mission Characteristics (Continued)

Rollout Distance (ft/m)	10,125/3,086
Rollout Time (seconds)	61
Mission Duration	239 hr, 39 min, 59 sec
Landed Revolution No.	159
Mission Support	STDN
Primary Objective	Microgravity research using the German Spacelab D-2
Deployed Satellites	None
Experiments	Spacelab D-2: The D-2 mission augmented the German microgravity research program started by the D-1 Spacelab mission in 1985. The German Aerospace Research Establishment (DLR) had been tasked by the DARA to conduct the second mission. DLR, NASA, ESA, and agencies in France and Japan contributed to D-2's scientific program. Eleven nations participated in the experiments.
	Of the 88 experiments conducted on the D-2 mission, four were NASA-sponsored. The crew worked in two shifts around-the-clock to complete investigations in the areas of fluid physics; materials sciences; life sciences; biological sciences; technology; Earth observations; atmospheric physics; and astronomy.
	The payloads were: Material Science Experiment Double Rack for Experiment Modules and Apparatus (MEDEA) • Floating Zone Growth of GaAs • Floating Zone Crystal Growth of Gallium-Doped Germanium • Hysteresis of the Specific Heat CV During Heating and Cooling Through the Critical Point • Diffusion of Nickel in Liquid Copper-Aluminum and Copper-Gold Alloys • Directional Solidification of Ge/GaAs Eutectic Composites • Cellular-Dendritic Solidification with Quenching of Aluminum-Lithium Alloys • Thermoconvection at Dendritic-Eutectic Solidification of an Al-Si Alloy • Growth of GaAs from Gallium Solutions
	Werkstofflabor (WL) Material Sciences Laboratory • OSIRIS: Oxide Dispersion Strengthened Single Crystalline Alloys Improved by Resolidification in Space • Impurity Transport and Diffusion in InSb Melt Under Microgravity Environment • Cellular-Dendritic Solidification at Low Rate of Aluminum-Lithium Alloys • Directional Solidification of the LiF-LiBaF3-Eutectic • Separation Behavior of Monotectic Alloys

Table 3–79. STS-55 Mission Characteristics (Continued)

Experiments	
	• Liquid Columns' Resonances
	• Stability of Long Liquid Columns
	• Higher Modes and Their Instabilities of Oscillating Marangoni Convection in a Large Cylindrical Liquid Column
	• Marangoni-Benard Instability
	• Onset of Oscillatory Marangoni Flows
	• Marangoni Convection in a Rectangular Cavity
	• Stationary Interdiffusion in a Non-Isothermal Molten Salt Mixture
	• Transport Kinetics and Structure of Metallic Melts
	• Nucleation and Phase Selection During Solidification of Undercooled Alloys
	• Heating and Remelting of an Allotropic Fe-C-Si Alloy in a Ceramic Skin and the Effect of the Volume Change on the Mold's Stability
	• Immiscible Liquid Metal Systems
	• Convective Effects on the Growth of GaInSb Crystals
	• Vapor Growth of InP-Crystal with Halogen Transport in a Closed Ampoule
	• Solution Growth of GaAs Crystals Under Microgravity
	• Crystallization of Nucleic Acids and Nucleic Acid-Protein Complexes
	• Crystallization of Ribosomal Particles
	Holographic Optics Laboratory (HOLOP)
	• Marangoni Convection in a Rectangular Cavity
	• Interferometric Determination of the Differential Interdiffusion Coefficient of Binary Molten Salts
	• IDILE: Measurements of Diffusion Coefficients in Aqueous Solution
	• NUGRO: Phase Separation in Liquid Mixtures with Miscibility Gap
	Baroreflex (BA)
	• Residual Acceleration in Spacelab D2
	• Transfer Function Experiment
	• Robotics Experiment (ROTEX)
	• Anthrorack (AR)
	• Cardiovascular Regulation at Microgravity
	• Central Venous Pressure During Microgravity
	• Leg Fluid Distribution at Rest and Under Lower Body Negative Pressure
	• Determination of Segmental Fluid Content and Perfusion
	• Left Ventricular Function at Rest and Under Stimulation
	• Peripheral and Central Hemodynamic Adaptation to Microgravity During Rest, Exercise, and Lower Body Negative Pressure in Humans

Table 3–79. STS-55 Mission Characteristics (Continued)

Experiments	
	• Tonometry–Intraocular Pressure in Microgravity
	• Tissue Thickness and Tissue Compliance Along Body Axis Under Microgravity Conditions
	• Changes in the Rate of Whole-Body Nitrogen Turnover, Protein Synthesis, and Protein Breakdown Under Conditions of Microgravity
	• Regulation of Volume Homeostasis in Reduced Gravity Possible Involvement of Atrial Natriuretic Factor Urodilatin and Cyclic GMP
	• Effects of Microgravity on Glucose Tolerance
	• Effects of Spaceflight on Pituitary-Gonad-Adrenal Function in the Human
	• Adaptation to Microgravity and Readaptation to Terrestrial Conditions
	• Pulmonary Perfusion and Ventilation in Microgravity Rest and Exercise
	• Ventilation Distribution in Microgravity
	• Effects of Microgravity on the Dynamics of Gas Exchange, Ventilation, and Heart Rate in Submaximal Dynamic Exercise
	• Cardiovascular Regulation in Microgravity
	• Biolabor (BB)
	• Development of Vestibulocular Reflexes in Amphibia and Fishes with Microgravity Experience
	• Comparative Investigations of Microgravity Effects on Structural Development and Function of the Gravity-Perceiving Organ of Two Water-Living Vertebrates
	• Structure and Function-Related Neuronal Plasticity of the Central Nervous System of Aquatic Vertebrates During Early Ontogenetic Development Under Microgravity Conditions
	• Immunoelectron Microscopic Investigation of Cerebellar Development at Microgravity
	• Gravisensitivity of Cress Roots
	• Influence of Gravity on Fruiting Body Development of Fungi
	• Significance of Gravity and Calcium Ions on the Production of Secondary Metabolites in Cell Suspensions
	• Influence of Conditions in Low-Earth Orbit on Expression and Stability of Genetic Information in Bacteria
	• Productivity of Bacteria
	• Fluctuation Test on Bacterial Cultures
	• Connective Tissue Biosynthesis in Space: Gravity Effects on Collagen Synthesis and Cell Proliferation of Cultured Mesenchymal Cells
	• Antigen-Specific Activation of Regulatory T-Lymphocytes to Lymphokine Production
	• Growth of Lymphocytes Under Microgravity Conditions

Table 3–79. STS-55 Mission Characteristics (Continued)

Experiments	• Enhanced Hybridoma Production Under Microgravity • Culture and Electrofusion of Plant Cell Protoplasts Under Microgravity: Morphological/Biochemical Characterization • Yeast Experiment HB-L29/Yeast: Investigations on Metabolism
	Cosmic Radiation Experiments • Biological Hze-Particle Dosimetry with Biostack • Personal Dosimetry: Measurement of the Astronaut's Ionizing Radiation Exposure • Measurement of the Radiation Environment Inside Spacelab at Locations Which Differ in Shielding Against Cosmic Radiation • Chromosome Aberration • Biological Response to Extraterrestrial Solar UV Radiation and Space Vacuum

	User Support Structure Payloads
	• Module Optoelectronic Multispectral Stereo Scanner (MOMS)
	• Galactic Ultrawide-Angle Schmidt System (GAUSS)
	• Atomic Oxygen Exposure Tray (AOET)
	• Material Science Autonomous Payload (MAUS)
	   – Reaction Kinetics in Glass Melts Payload (RKGM)
	   – Pool Boiling
	   – Gas Bubbles in Glass Melts

	Crew Telesupport Experiment (CTE): Combined an on-board computer-based multimedia documentation file with a real-time, graphical communication between the on-orbit crew members and the ground station.
	SAREX II: See STS-35.
Get Away Specials	None
Mission Results	Successful
Remarks	Spacelab D-2 conducted the first telerobotic capture of a free-floating object by flight controllers in Germany. The crew conducted the first intravenous saline solution injection in space as part of an experiment to study the human body's response to direct fluid replacement as a countermeasure for amounts lost during spaceflight. The crew also successfully completed an in-flight maintenance procedure for collection of orbiter wastewater that allowed the mission to continue.
	Spacelabs D1 and D2 were the only Spacelab missions to date with payload operations controlled from a foreign country.

[a] Integrated into ESA's single European astronaut corp in 1998.
[b] Jenkins, p. 302.

Table 3–80. STS-57 Mission Characteristics

Vehicle	OV-105
	Endeavour
Crew	CDR: Ronald J. Grabe
	PLT: Brian Duffy
	MS: G. David Low, Nancy J. Sherlock (Currie), Peter J.K. Wisoff, Janice E. Voss
Launch	June 21, 1993, 9:07:22 a.m. EDT, Kennedy Space Center, Pad 39-B. The launch originally targeted for mid-May was rescheduled to June to allow both liftoff and landing to occur in daylight. Liftoff set for June 3 slipped when managers decided to replace the high-pressure oxidizer turbopump on main engine No. 2 after concerns arose over a misplaced inspection stamp on a spring in the pump. Additional time also allowed investigation of an inexplicable loud noise heard after the Shuttle arrived at the launch pad, which was eventually attributed to the ball strut tie-rod assembly inside the 17-in (43-cm) liquid hydrogen line. The launch attempt on June 20 was scrubbed at T-5 minutes due to low clouds and rain at the return-to-launch site at Kennedy Space Center, and weather concerns at all three transoceanic abort landing sites. The launch countdown was the longest since return to flight to allow servicing of payloads at the pad.
Orbital Altitude & Inclination	252 nmi (467 km), 28.45 deg
Launch Weight (lb/kg)	252,710[a]/114,627
Landing & Postlanding Operations	July 1, 1993, 8:52:16 a.m. EDT, Runway 33, Kennedy Space Center. Landing attempts on June 29 and June 30 were waved off due to unacceptable cloud cover and rain showers at Kennedy Space Center. After landing, the STS-57 crew on *Endeavour* talked with the STS-51 crew on *Discovery* at Pad 39-B. It was the first orbiter-to-orbiter crew conversation since the orbiting STS-51-D crew talked with the STS 51-B crew at Kennedy Space Center in 1985.
Rollout Distance (ft/m)	9,954/3,034
Rollout Time (seconds)	65
Mission Duration	239 hr, 44, min, 54 sec
Landed Revolution No.	154
Mission Support	STDN
Primary Objective	Retrieval of EURECA and biomedical and materials science experimentation using the SPACEHAB module
Deployed Satellites	None

Table 3–80. STS-57 Mission Characteristics (Continued)

Experiments	SPACEHAB 01 Experiments:
	• Commercial Material Science Experiments
	– Equipment for Controlled Liquid Phase Sintering Experiment-SPACEHAB (ECLiPSE)
	– GPPM
	– IPMP
	– Liquid Encapsulated Melt Zone (LEMZ)
	– Support of Crystal Growth (SCG)
	– Zeolite Crystal Growth (ZCG)
	• Commercial Life Science Experiments
	– ASTROCULTURE™
	– BioServe Pilot Laboratory (BPL)
	– CGBA
	– Organic Separation (ORSEP)
	– PCG
	– Vapor Diffusion Apparatus and Crystallization Facility Experiments
	– Direct-Control Protein Crystal Growth
	– PSE
	• Johnson Space Center Experiments
	– Application-Specific Pre-programmed Experiment Culture (ASPEC)
	– Charged Particle Directional Spectrometer (CPDS)
	– Human Factors Assessment (HFA)
	– Neutral Body Posture (NBP)
	– Tools and Diagnostics System (TDS)
	• Space Station Experiment
	– Environmental Control Life Support System (ECLSS) Flight Experiment
	• Supporting Hardware Overview
	– Three-Dimensional Microgravity Accelerometer (3-DMA)
	– SAMS
	FARE: See STS-53.
	SAREX-II: See STS-35.
	AMOS Calibration Test: See STS-29.
	Consortium for Materials Development in Space Complex Autonomous Payload IV (CONCAP IV): This complex was designed to grow non-linear optical organic thin films and crystals through the physical vapor transport process. The payload was carried on the GAS Bridge Assembly.
	Superfluid Helium On-Orbit Transfer (SHOOT): SHOOT was designed to develop and demonstrate the technology required to resupply liquid helium containers in space. It was carried on the GAS Bridge Assembly.

Table 3–80. STS-57 Mission Characteristics (Continued)

Get Away Specials	G-022
	Customer: ESA, European Space Research and Technology Centre, Noordwijk, the Netherlands Liquid Gauging Technology Experiment: This experiment demonstrated two on-orbit methods of gauging liquids in tanks.

Get Away Specials

G-022
Customer: ESA, European Space Research and
Technology Centre, Noordwijk, the Netherlands
Liquid Gauging Technology Experiment: This
experiment demonstrated two on-orbit methods of
gauging liquids in tanks.

G-324
Customer: Charleston County School District,
Charleston, South Carolina
CAN-DO (GEOCAM): The CAN-DO consisted of four
35-mm cameras for Earth photography to compare with
Skylab photographs. The canister also contained 350
small passive student experiments that allowed students
to participate directly in research by testing the effect of
space on various materials.

G-399
Customer: Dr. Ronald S. Nelson, Inc., Fresno, California
Insulin/Artemia/Ion Experiments: These experiments
studied Ferritin-tagged insulin, Artemia growth, and salt
ion transport across a permeable membrane.

G-450
Customer: American Institute of Aeronautics and
Astronautics, Vandenberg Air Force Base, California
Multiple Experiments: G-450 was a multidisciplinary
package of six self-contained modules, each containing
multiple experiments designed and developed by
California elementary, middle, and high school students.

G-452
Customer: Society of Japanese Aerospace Companies,
Tokyo, Japan
Crystal Growth of Gallium Arsenide: G-452 had twelve
small electric furnaces used to carry out the following
experiments in low gravity.
Growth of a single gallium-arsenide crystal from the
liquid phase.
Growth of a gallium-arsenide-based mixed crystal.
Addition of a heavy element to gallium-arsenide.
Addition of a heavy element to indium-antimony crystal.

G-453
Customer: Society of Japanese Aerospace Companies,
Tokyo, Japan
Semiconductor/Superconductor Boiling Experiments:
G-453 consisted of three experiments on semi-
conductors and a superconductor and one experiment on
boiling an organic solvent under weightlessness.

Table 3–80. STS-57 Mission Characteristics (Continued)

Get Away Specials	G-454 Customer: Society of Japanese Aerospace Companies, Tokyo, Japan Crystal Growth: This experiment studied the crystal growth of indium-gallium-arsenic from vapor phase under weightlessness, the crystal growth of 3-selenic-niobium from vapor phase, the crystal growth of an optoelectric crystal by the diffusion method, and the formation of superferromagnetic alloy.
	G-535 Customer: NASA Headquarters, OSSA Pool Boiling Experiment: This was a study of heating vapor and bubble growth/collapse.
	G-601 Customer: San Diego Section, American Institute of Aeronautics and Astronautics, San Diego, California High Frequency Variations of the Sun: This experiment measured and analyzed high-frequency solar output to better determine the physics of the Sun and other stars.
	G-647 Customer: Canadian Space Agency, Ottawa, Ontario, Canada Liquid Phase Electroepitaxy, Configurable Hardware for Multidisciplinary Projects in Space (CHAMPS): G-647 provided a versatile payload for materials science experiments in space. The experiment examined a recently developed technique for crystal growth called Liquid Phase Electro-Epitaxy (LPEE) in a microgravity environment.
	GAS Ballast Payload Customer: Goddard Space Flight Center, Greenbelt, Maryland Ballast payloads were flown for stability when a GAS payload was dropped and no replacement was available. A small accelerometer package recorded accelerations during the mission.
	Sample Return Experiment: This experiment sat on top of the ballast GAS cans. It quantified extraterrestrial particles and other orbital debris present in the orbiter bay.
Mission Results	Successful

[a] Jenkins, p. 302.

Table 3–81. STS-51 Mission Characteristics

Vehicle	OV-103
	Discovery
Crew	CDR: Frank L. Culbertson, Jr.
	PLT: William F. Readdy
	MS: James H. Newman, Daniel W. Bursch, Carl E. Walz
Launch	September 12, 1993, 7:45:00 a.m. EDT, Kennedy Space Center, Pad 39-B. The first launch attempt on July 17 was scrubbed during a hold at T-20 minutes due to premature and unexplained charging of pyrotechnic initiator controllers (PIC), located on the mobile launcher platform (MLP), for T-0 liquid hydrogen vent arm umbilical and solid rocket booster hold-down bolts. The problem was traced to a faulty circuit card in the PIC rack on the MLP.
	An abbreviated countdown began July 23. A second liftoff attempt on July 24 was halted at T-19 seconds due to a problem with the auxiliary power unit (APU) turbine assembly for one of the two hydraulic power units on the right solid rocket booster. The APU was removed and replaced at pad.
	Launch was rescheduled for August 4 to August 12 because of concerns relating to the Perseid meteor shower, which was expected to peak on August 11. A liftoff attempt on August 12 was halted at the T-3-second mark due to a faulty sensor monitoring fuel flow on main engine No. 2. This was the fourth pad abort in Shuttle program history that led to a changeout of all three main engines at the pad.
	The launch was rescheduled for September 10 but then slipped to September 12 to allow time to complete a review of the ACTS design, production, and testing history following the loss of contact with the Mars Observer and NOAA-13 satellite.
	The countdown proceeded smoothly to an on-time liftoff on September 12.
Orbital Altitude & Inclination	160 nmi (296 km), 28.45 deg
Launch Weight (lb/kg)	261,486[a]/118,608

Table 3–81. STS-51 Mission Characteristics (Continued)

Landing & Postlanding Operations	September 22, 1993, 3:56:11 a.m. EDT, Runway 15, Kennedy Space Center. A September 21 landing opportunity was waved off due to the possibility of rain within 30 miles of the Shuttle Landing Facility. It was the first end-of-mission night landing at Kennedy Space Center for the Shuttle program.
Rollout Distance (ft/m)	8,271/2,521
Rollout Time (seconds)	50
Mission Duration	236 hr, 11 min, 11 sec
Landed Revolution No.	156
Mission Support	STDN
Primary Objective	Deploy ACTS; deployment and retrieval of ORFEUS-SPAS
Deployed Satellites	ACTS/TOS ORFEUS-SPAS (See chapter 4, Space Science, for a description of science objectives and payload.)
Experiments	IMAX: The ORFEUS-SPAS payload was recorded from the Shuttle using a handheld IMAX camera; the Shuttle was recorded using the Remote IMAX Camera System (RICS) mounted on the free-flying ORFEUS-SPAS.[b] CPCG: See STS-49. CHROMEX: See STS-29. High Resolution Shuttle Glow Spectroscopy (HRSGS-A): This payload obtained high-resolution spectra in the visible and near visible wavelength range of the Shuttle surface glow. APE-B: See STS-33. RME-III: See STS-28. IPMP: See STS-31. AMOS Calibration Test: See STS-29.
Get Away Specials	None
Mission Results	Successful

[a] Jenkins, p. 302.
[b] See chapter 4, Space Science, of this volume for further discussion of the ORFEUS-SPAS mission.

Table 3–82. STS-58 Mission Characteristics

Vehicle	OV-102
	Columbia
Crew	CDR: John E. Blaha
	PLT: Richard A. Searfoss
	MS: Margaret Rhea Seddon, William S. McArthur, Jr., David A. Wolf, Shannon W. Lucid
	PS: Martin J. Fettman
Launch	October 18, 1993, 10:53:10 a.m. EDT, Kennedy Space Center, Pad 39-B. The first launch attempt on October 14 was scrubbed at T-31 seconds due to a failed Range Safety computer. The second launch attempt on October 15 was scrubbed at T-9 minutes due to failed a S-band transponder on the orbiter. Launch was reset for October 18. Countdown proceeded smoothly to liftoff, delayed only by several seconds because of an aircraft in the launch zone.
Orbital Altitude & Inclination	155 nmi (287 km), 39.00 deg
Launch Weight (lb/kg)	256,097[a]/116,164
Landing & Postlanding Operations	November 1, 1993, 7:05:42 a.m. PST, Runway 22, Edwards Air Force Base. The orbiter returned to Kennedy Space Center on November 9 after a two-day trip.
Rollout Distance (ft/m)	9,640/2,938
Rollout Time (seconds)	61
Mission Duration	336 hr, 12 min, 32 sec
Landed Revolution No.	225
Mission Support	STDN
Primary Objective	Dedicated Spacelab Life Sciences research
Deployed Satellites	None

Table 3–82. STS-58 Mission Characteristics (Continued)

Experiments	SLS-2 mission using the Spacelab long module: The crew conducted 14 experiments in four areas: regulatory physiology, cardiovascular/cardiopulmonary, musculoskeletal, and neurosciences. Eight of the experiments focused on the crew, six on 48 rodents. The crew collected more than 650 different samples from themselves and the rodents, increasing the statistical base for life sciences research.
	Cardiovascular/cardiopulmonary experiments: These experiments focused on understanding and quantifying the changes occurring on orbit and on the acute fluid shift and long-term adaptation of the heart and lungs. • Inflight Study of Cardiovascular Deconditioning • Cardiovascular Adaptation to Zero Gravity • Pulmonary Function During Weightlessness
	Regulatory physiology experiments: These experiments investigated the theory that the kidneys and endocrine glands adjusted the body's fluid-regulating hormones to stimulate an increase in fluid to be excreted. The experiments also investigated the mechanisms surrounding the decrease in red blood cells responsible for carrying oxygen to the tissues that occurred in spaceflight. • Fluid-Electrolyte Regulation During Spaceflight • Regulation of Blood Volume During Spaceflight • Regulation of Erythropoiesis in Rats During Spaceflight • Influence of Spaceflight on Erythrokinetics in Man
	Neuroscience investigations: These investigations documented both physical vestibular (balance) changes and perception changes and investigated the mechanisms involved. The investigators also hoped to identify countermeasures to alleviate the effects of space motion sickness. The mission included an Astronaut Science Advisor (ASA), a computer-based intelligent assistant designed to help astronauts work more efficiently and improve the quality of space science. The ASA supported the Rotating Dome Experiment, which measured how the visual and vestibular systems interact and how this interaction was affected as humans adapt to microgravity. The experiment included: • Study of the Effects of Space Travel on Mammalian Gravity Receptors • Vestibular Experiments in Spacelab

Table 3–82. STS-58 Mission Characteristics (Continued)

Experiments	Musculoskeletal investigations: In microgravity, the body's bones and muscles were used less extensively than on Earth. As a result, researchers saw a decrease in the mass of both during spaceflight. The SLS-2 studies provided more information about the complex musculoskeletal system, including: • Protein Metabolism During Spaceflight • Effects of Zero Gravity on the Functional and Biochemical Properties of Antigravity Skeletal Muscle • Effects of Microgravity on the Electron Microscopy, Histochemistry, and Protease Activities of Rat Hindlimb Muscles • Pathophysiology of Mineral Loss During Spaceflight • Bone, Calcium, and Spaceflight
	OARE: See STS-40.
	SAREX: See STS-35.
	Pilot Inflight Landing Operations Trainer (PILOT): The PILOT, a portable laptop computer simulator, allowed the pilot and commander to maintain proficiency for approach and landing during longer missions.
Get Away Specials	None
Mission Results	Successful

[a] Jenkins, p. 302.

Table 3–83. STS-61 Mission Characteristics

Vehicle	OV-105
	Endeavour
Crew	CDR: Richard O. Covey
	PLT: Kenneth D. Bowersox
	PCF: Story Musgrave
	MS: Kathryn C. Thornton, Claude Nicollier (ESA), Jeffrey A. Hoffman, F. Story Musgrave, Thomas D. Akers
Launch	December 2, 1993, 4:27:00 a.m. EST, Kennedy Space Center, Pad 39-B. The launch was originally scheduled to occur from Launch Pad 39-A, but after rollout on October 28, contamination was found in the Pad 39-A Payload Changeout Room, and a decision was made to move the Shuttle and payloads to Pad 39-B. Rollaround occurred on November 15. The first launch attempt on December 1 was scrubbed due to out-of-limit weather conditions at the Shuttle Landing Facility in the event of a return-to-launch-site contingency. Launch on December 2 occurred on schedule.
Orbital Altitude & Inclination	321 nmi (594 km), 28.45 deg
Launch Weight (lb/kg)	250,314/113,541
Landing & Postlanding Operations	December 13, 1993, 12:25:37 a.m. EST, Runway 33, Kennedy Space Center. Second night landing at Kennedy Space Center. The orbiter returned one orbit earlier than originally planned to allow two landing opportunities at Kennedy Space Center.
Rollout Distance (ft/m)	7,922/2,415
Rollout Time (seconds)	53
Mission Duration	259 hr, 58 min, 37 sec
Landed Revolution No.	162
Mission Support	STDN
Primary Objective	First Hubble Space Telescope Servicing Mission
Deployed Satellites	Hubble Space Telescope retrieved and redeployed after servicing
Experiments	None
Get Away Specials	None
Mission Results	Successful

Table 3–83. STS-61 Mission Characteristics (Continued)

Remarks	The record-setting five back-to-back spacewalks totaled 35 hours, 28 minutes.
	EVA No.1, by Hoffman and Musgrave, lasted 7 hours, 54 minutes. The two spacewalkers replaced two sets of remote sensing units, which contained gyroscopes that helped the Hubble Space Telescope point in the correct direction. They also replaced two electronic control units and eight electrical fuse plugs protecting the telescope's electrical circuits. This was the second longest spacewalk in U.S. history to date.
	EVA No. 2, by Akers and Thornton, lasting 6 hours, 36 minutes, replaced the telescope's two solar arrays.
	EVA No. 3, by Hoffman and Musgrave, lasted 6 hr, 47 minutes. They replaced the original WF/PC with a new WFPC-2 in about 40 minutes rather than the 4 hours that had been anticipated. The new camera had a higher rating than the previous model, especially in the ultraviolet range, and included its own spherical aberration correction system. They also replaced two magnetometers.
	EVA No. 4, by Akers and Thornton, lasting 6 hrs, 50 minutes, replaced the Hubble Space Telescope's High Speed Photometer (HSP) with the COSTAR system. COSTAR corrected the telescope's spherical aberration of the main mirror for all instruments except WFPC-2, which had its own built-in corrective optics. During this spacewalk, Akers set a new individual U.S. spacewalking record of 29 hours, 14 minutes.
	EVA No. 5, by Hoffman and Musgrave, lasted 7 hrs, 21 minutes. This EVA replaced the solar array drive electronics and, after several unsuccessful commands from the Space Telescope Operations Control Center, the astronauts cranked the solar arrays' deployment mechanism by hand, successfully deploying them. They also installed the GHRS Redundancy Kit and protective covers over the original magnetometers. The covers, which were fabricated on board by astronauts Nicollier and Bowersox, would contain any debris caused by the older magnetometers, which showed some signs of ultraviolet decay.

Table 3–84. STS-60 Mission Characteristics

Vehicle	OV-103
	Discovery
Crew	CDR: Charles F. Bolden, Jr.
	PLT: Kenneth S. Reightler, Jr.
	MS: N. Jan Davis, Ronald M. Sega, Franklin R. Chang-Diaz, Sergei K. Krikalev (Russian Space Agency/RSA)
Launch	February 3, 1994, 7:10:00 a.m. EST, Kennedy Space Center, Pad 39-A. Launch was on time.
Orbital Altitude & Inclination	191 nmi (354 km), 57.00 deg
Launch Weight (lb/kg)	245,767[a]/111,478
Landing & Postlanding Operations	February 11, 1994, 2:19:22 p.m. EST, Runway 15, Kennedy Space Center. The first landing attempt was waved off due to unfavorable weather in the Kennedy Space Center area.
Rollout Distance (ft/m)	7,771/2,368
Rollout Time (seconds)	50
Mission Duration	199 hr, 9 min, 22 sec
Landed Revolution No.	129
Mission Support	STDN
Primary Objective	Experimentation using WSF-1 and SPACEHAB 02
Deployed Satellites	WSF-1: There were two unsuccessful attempts to deploy the facility. The WSF-1 was instead operated at the end of the remote manipulator system arm.
Experiments	SPACEHAB 02: SPACEHAB 02 carried 12 payloads conducted under the Commercial Middeck Augmentation Module contract. The experiments represented a wide range of space experimentation including nine commercial-development-of-space experiments in materials processing and biotechnology sponsored by five NASA CCDS; three supporting hardware and technology development payloads, one from a CCDS, one from Lewis Research Center; and one from Johnson Space Center. In addition, a Sample Return Experiment sat on the top of the SPACEHAB module.
	SPACEHAB 02 Payloads:[b] • SPACEHAB (ECLiPSE): This experiment used a rack-mounted, enclosed furnace assembly to investigate controlled liquid-phase sintering of metallic systems in microgravity. • Space Experiment Furnace: This payload allowed up to three separate furnaces in one unit. This flight carried one transparent furnace and one opaque core furnace.

Table 3–84. STS-60 Mission Characteristics (Continued)

Experiments	• ASTROCULTURE™: This payload validated the performance of plant growth technologies in the microgravity environment of space. • BPL: This experiment determined the response of cells to various hormones and stimulating agents in microgravity. • CGBA: The apparatus allowed a wide range of sophisticated biomaterials, life sciences, and biotechnology investigations to be performed in one device in a microgravity environment. It processed biological fluids by mixing components in a microgravity environment. The CGBA also supported 32 separate commercial investigations in the areas of biomedical testing and drug development, controlled ecological life support system development, and agricultural development and manufacture of biological-based materials. • IMMUNE-1: This experiment was designed to reduce or prevent changes seen in the immune system of 12 rats after spaceflight. The drug PEG-IL2 was used in an attempt to alleviate the immunosuppression induced by the microgravity environment. The experiment might provide a new therapy to treat the effects of spaceflight on the human immune system, as well as on physiological systems affected by the immune system. • ORSEP: This experiment explored the use of phase separation techniques in microgravity conditions to separate cells, cell fragments, and heavy molecules. • CPCG: This experiment was designed to produce large, well-ordered crystals of various proteins. See STS-49. • Penn State Biomodule: This was a computer-controlled, fluid-transfer mixing device used to test the hypothesis that exposure to near-zero gravity can alter microbial gene expression in commercially useful ways. • 3-DMA: The accelerometer helped chart the effects of deviations of zero gravity on experiments conducted in space, allowing researchers to review experiment results against deviations from zero gravity. • SAMS: The SAMS measured and recorded low-level accelerations during experiment operations. • Stirling Orbiter Refrigerator Freezer (SOR/F): The experiment was a flight test and characterization relating to operation of advanced refrigerator/freezer technology in microgravity. • Sample Return Experiment: This experiment sat on top of the SPACEHAB module to capture intact cosmic dust particles as they came in contact with 160 capture cells. The capture cells consisted of transparent silica aerogel, the lowest density solid material known with extremely fine structure.

Table 3–84. STS-60 Mission Characteristics (Continued)

Get Away Specials	G-071
	Customer: California State University, Northridge
	OBBEX: This apparatus tested the effects of melting cylindrical metal alloy pellets in microgravity to produce a new kind of ball bearing never before built, a seamless hollow ball bearing.
	G-514
	Customer: Goddard Space Flight Center
	The Orbiter Stability Experiment: The objective was to measure the vibration spectrum of the orbiter structure present during normal orbiter and crew operations and to evaluate fogging of photographic emissions due to energetic particles.
	G-536
	Customer: NASA Headquarters, OSSA, Microgravity Sciences Division
	The Pool Boiling Experiment: This experiment marked the 100th GAS payload to fly since the program's inception. The experiment's objective was to improve understanding of the boiling process in microgravity.
	G-557
	Customer: ESA
	The Netherlands Capillary Pumped Loop (CPL) Experiment: This experiment demonstrated in-orbit the working principle and performances of a two-phase CPL, a two-phase Vapor Quality Sensor, and a two-phase multichannel Condenser Profile. It also compared data on CPL behavior in a low-gravity environment with analytical predictions resulting from modeling and on-Earth performance.
	CAPL/ODERACS/BREMSAT/GAS Bridge Assembly
	• Capillary Pumped Loop (CAPL): This system investigated heat rejection in microgravity as a prototype of the two-phase thermal control system planned for use in the Earth Observing System (EOS) platform. This flew as a Hitchhiker payload.
	• ODERACS: This experiment deployed six spheres that were observed, tracked, and recorded by ground-based radars and optical telescopes.

Table 3–84. STS-60 Mission Characteristics (Continued)

Get Away Specials	• BREMEN Satellite (BREMSAT): The satellite conducted scientific activities at various mission phases. This German-built, ejectable satellite consisted of six scientific experiments operated before and after satellite deployment. The experiments measured heat conductivity; residual acceleration forces; density distribution and dynamics of micrometeorites and dust particles in low-Earth orbit; atomic oxygen; exchange of momentum and energy between the molecular flow and the rotating satellite; and pressure and temperature during reentry.
Mission Results	The WSF was not deployed but was operated at the end of the robot arm. Other mission objectives were successfully achieved.

[a] Jenkins, p. 302.
[b] "Space Shuttle Mission STS-60 Press Kit" (February 1994, with Errata and Updates from January 27, 1994, *http://www.jsc.nasa.gov/history/shuttle_pk/pk/Flight_060_STS-060_Press_Kit.pdf* (accessed December 2, 2005).

Table 3–85. STS-62 Mission Characteristics

Vehicle	OV-102
	Columbia
Crew	CDR: John H. Casper
	PLT: Andrew M. Allen
	MS: Pierre J. Thuot, Charles D. Gemar, Marsha S. Ivins
Launch	March 4, 1994, 8:53:00 a.m. EST, Kennedy Space Center, Pad 39-B. Launch set for March 3 was postponed at the T-11 hour mark due to predicted unfavorable weather in the Kennedy Space Center area. The countdown on March 4 proceeded smoothly. The only deviation to normal operating procedures was a delay in deploying the solid rocket booster recovery ships because of high seas. The recovery ships left port on launch day and recovered the boosters and their parachutes on March 6.
Orbital Altitude & Inclination	162 nmi (300 km) and 140 nmi (259 km), 39.00 deg
Launch Weight (lb/kg)	256,584[a]/116,385
Landing & Postlanding Operations	March 18, 1994, 8:09:41 a.m. EST, Runway 33, Kennedy Space Center.
Rollout Distance (ft/m)	10,151/3,094
Rollout Time (seconds)	55
Mission Duration	335 hr, 16 min, 41 sec
Landed Revolution No.	223
Mission Support	STDN
Primary Objective	Experimentation using USMP-2 and OAST-2
Deployed Satellites	None
Experiments	USMP-2: These experiments investigated materials processing and crystal growth in microgravity. • Advanced Automated Directional Solidification Furnace (AADSF) • MEPHISTO • Isothermal Dendritic Growth Experiment (IDGE) • Critical Fluid Light Scattering Experiment (ZENO) • SAMS

Table 3–85. STS-62 Mission Characteristics (Continued)

Experiments	OAST-2: The objective of this payload was to obtain technology data to support future needs for advanced satellites, sensors, microcircuits, and the ISS. Six In-Space Technology Program (INSTEP) experiments were mounted on a Hitchhiker carrier.

OAST-2: The objective of this payload was to obtain technology data to support future needs for advanced satellites, sensors, microcircuits, and the ISS. Six In-Space Technology Program (INSTEP) experiments were mounted on a Hitchhiker carrier.

- Experimental Investigation of Spacecraft Glow (EISG) and Spacecraft Kinetic Infrared Test (SKIRT): These experiments developed an understanding of the physical processes leading to the spacecraft glow phenomena by studying infrared, visible, and far-ultraviolet light emissions as a function of surface temperature and orbital altitude.
- Cryogenic Two Phase (CRYOTP): This experiment determined the performance of microgravity nitrogen space heat pipe and cryogenically-cooled, vibration-free, phase-change-material thermal storage unit thermal energy control technologies.
- Thermal Energy Storage (TES): This experiment determined the microgravity behavior of two different thermal energy storage salts that underwent repeated melting and freezing.
- Emulsion Chamber Technology (ECT): This experiment measured background cosmic ray radiation as a function of shielding and radiation energy photographic films.
- Solar Array Module Plasma Interaction Experiment (SAMPIE): This experiment determined the arcing and current collection behavior of different types, sizes, and shapes of solar cells, solar modules, and spacecraft materials.

Dexterous End Effector (DEE): The DEE worked with the remote manipulator system and demonstrated a Force Torque Sensor, Magnetic End Effector, Targeting and Reflective Alignment Concept (TRAC) grapple alignment system, and Auto TRAC Vision System.

SSBUV/A: See STS-34.

LDCE: This experiment exposed material samples to atomic oxygen in the space environment.

Advanced Protein Crystal Growth (APCG): This experiment produced high-quality, well-ordered crystals of selected proteins for analysis of molecular structures through x-ray diffraction and computer modeling.

PSE: See STS-52.

CPCG: See STS-49.

CGBA: See STS-54.

MODE: See STS-40.

Table 3–85. STS-62 Mission Characteristics (Continued)

Experiments	Bioreactor Demonstration Systems (BDS): This experiment attempted to determine the threshold mass for the transfer/diffusion of glucose and oxygen into a static cell culture in the microgravity environment. APE-B: See STS-33. AMOS Calibration Test: See STS-29.
Get Away Specials	None
Mission Results	Successful

[a] Jenkins, p. 302.

Table 3–86. STS-59 Mission Characteristics

Vehicle	OV-105
	Endeavour
Crew	CDR: Sidney M. Gutierrez
	PLT: Kevin P. Chilton
	PC: Linda M. Godwin
	MS: Jerome Apt, Michael R. Clifford, Thomas D. Jones
Launch	April 9, 1994, 7:05:00 a.m. EDT, Kennedy Space Center, Pad 39-A. Launch set for April 7 was postponed for one day at the T-27-hour mark to allow for additional inspections of the metallic vanes in the SSME high-pressure oxidizer preburner pumps. Launch on April 8 was scrubbed due to weather, high crosswinds, and low clouds at the Shuttle Landing Facility and clouds at the launch pad. The countdown April 9 proceeded smoothly.
Orbital Altitude & Inclination	121 nmi (224 km), 57.00 deg
Launch Weight (lb/kg)	246,851a/111,970
Landing & Postlanding Operations	April 20, 1994, 9:54:30 a.m. PDT, Runway 22, Edwards Air Force Base. Landing was originally planned for Kennedy Space Center on April 19, but two landing opportunities were waved off due to low clouds and possible thunderstorms in the area. An early landing opportunity on April 20 was also waved off in favor of landing at Edwards Air Force Base. The orbiter returned to Kennedy Space Center from Edwards by Shuttle Carrier Aircraft on May 2, 1994.
Rollout Distance (ft/m)	10,691/3,258
Rollout Time (seconds)	54
Mission Duration	269 hr, 49 min, 30 sec
Landed Revolution No.	182
Mission Support	STDN
Primary Objective	Study Earth's global environment using SRL-1
Deployed Satellites	None

Table 3–86. STS-59 Mission Characteristics (Continued)

Experiments	SRL-1: This was the first simultaneous multifrequency (C, L, and X bands), multipolarization phased-array imaging radar in space for geoscientific studies of Earth in different seasons. Used to image sites for geology, hydrology, vegetation science, and oceanography to study vegetation type, extent, and deforestation; water storage and flux; ocean dynamics; wave fields; wind fields; volcanism; tectonic activity; soil erosion; desertification; and topography. The SRL payload consisted of: • SIR-C • X-SAR • MAPS The DARA and the Italian Space Agency provided the X-SAR instrument.
	CONCAP-IV: See STS-57.
	STL/National Institutes of Health-Cells (NIH)-C: Configuration A and B–These experiments validated models for muscle, bone, and endothelial cell biochemical and functional loss induced by microgravity stress; to evaluate cytoskeleton, metabolism, membrane integrity, and protease activity in target cells; and to test tissue loss pharmaceuticals for efficacy.
	VFT-4: This test measured the near and far point of clear vision, as well as the ability to change focus within the range of clear vision.
	SAREX-II: See STS-35.
Get Away Specials	G-203 Customer: New Mexico State University G-203 observed the crystal forming characteristics of water using zeolite desorption-absorption processing.
	G-300 Customer: Matre/Laboratoire de Genie Electrique de Paris (L.G.E.P.) G-300 performed conductivity measurements on two silicon oils in microgravity.
	G-458 Customer: The Society of Japanese Aerospace Companies, Inc. G-458 cultivated cellular slime molds.
Mission Results	Successful

[a] Jenkins, p. 302.

Table 3–87. STS-65 Mission Characteristics

Vehicle	OV-102
	Columbia
Crew	CDR: Robert D. Cabana
	PLT: James D. Halsell, Jr.
	PC: Richard J. Hieb
	MS: Carl E. Walz, Leroy Chiao, Donald A. Thomas
	PS: Chiaki Naito-Mukai (JAXA).
Launch	July 8, 1994, 12:43:00 p.m. EDT, Kennedy Space Center, Pad 39-A. The launch occurred exactly on time at the beginning of a 2 1/2-hour launch window. The countdown progressed smoothly but was held at T-9 minutes due to a return-to-launch-site weather constraint. The weather constraint was cleared at 12:36 p.m., leading to an on-time liftoff.
Orbital Altitude & Inclination	160 nmi (296 km), 28.45 deg
Launch Weight (lb/kg)	258,585a/117,292
Landing & Postlanding Operations	July 23, 1994, 6:38:00 a.m. EDT, Runway 33, Kennedy Space Center. This was the longest Shuttle flight to date. The landing opportunity on July 22 was waved off due to the possibility of rain showers in the area.
Rollout Distance (ft/m)	10,211/3,112
Rollout Time (seconds)	68
Mission Duration	353 hr, 55 min, 1 sec
Landed Revolution No.	234
Mission Support	STDN
Primary Objective	IML-2
Deployed Satellites	None

Table 3–87. STS-65 Mission Characteristics (Continued)

Experiments	IML-2: Two teams performed around-the-clock research. The space agencies represented were: NASA; ESA; CSA; the Centre National d'Etudes spatiales (CNES); DARA; and the NASDA. The activities were divided into two groups: life sciences and microgravity sciences and used a number of facilities and apparatus.
	Life science apparatus included: • Biorack • Biostack • EDOMP • Spinal Changes in Microgravity (SCM) • NIZEMI • Aquatic Animal Experiment Unit (AAEU) • Free Flow Electrophoresis Unit (FFEU) • Real-time Radiation Monitoring Device (RRMD) • Thermoelectric Incubator (TEI)/Cell Culture Kit (CCK) • Performance Assessment Workstation (PAWS)
	Microgravity science apparatus included: • APCF • Bubble, Drop and Particle Unit (BDPU) • CPF • Large Isothermal Furnace (LIF) • Quasi-Steady Acceleration Measurement System (QSAMS) • Applied Research on Separation Methods Using Space Electrophoresis (RAMSES) • SAMS • TEMPUS • Vibration Isolation Box Experiment System (VIBES)
	OARE: See STS-40.
	CPCG: See STS-49.
	Military Application of Ship Tracks (MAST): This experiment used Linhof and Hasselblad cameras to detect ship movement by detecting ship tracks formed in stratus, stratocumulus, and fog when ship-induced disturbances and emissions altered existing cloud structures.
	SAREX: See STS-35.
	AMOS Calibration Test: See STS-29.
Get Away Specials	None
Mission Results	Successful
Remarks	The crew took time during the mission to honor the 25th anniversary of Apollo 11, noting that Apollo 11 also featured a spacecraft named *Columbia*.
	Columbia was outfitted with extended duration orbiter hardware for the flight.

[a] Jenkins, p. 302.

Table 3–88. STS-64 Mission Characteristics

Vehicle	OV-103
	Discovery
Crew	CDR: Richard N. Richards
	PLT: L. Blaine Hammond, Jr.
	MS: Jerry M. Linenger, Susan J. Helms, Carl J. Meade, Mark C. Lee
Launch	September 9, 1994, 6:22:55 p.m. EDT, Kennedy Space Center, Pad 39-B. A late afternoon launch was scheduled to permit nighttime operation of the LITE laser early in the mission. The launch was delayed due to launch weather violations near the launch complex.
Orbital Altitude & Inclination	140 nmi (259 km), 57.00 deg
Launch Weight (lb/kg)	241,434/109,513
Landing & Postlanding Operations	September 20, 1994, 5:12:52 p.m. EDT, Runway 04, Edwards Air Force Base. Mission already extended one day was extended again after first landing opportunities at Kennedy Space Center on September 19 were waved off due to stormy weather. Two additional opportunities at Kennedy Space Center on September 20 were also waved off, and the orbiter was diverted to California. The orbiter was transported to Kennedy Space Center on September 27.
Rollout Distance (ft/m)	9,656/2,943
Rollout Time (seconds)	60
Mission Duration	262 hr, 49 min, 57 sec
Landed Revolution No.	175
Mission Support	STDN
Primary Objective	Experimentation using the LIDAR LITE: Deployment and retrieval of SPARTAN-201
Deployed Satellites	Deployed and retrieved SPARTAN-201
Experiments	LITE: This experiment measured the vertical profile of certain atmospheric parameters (cloud top height, planetary boundary layer height, tropospheric aerosols, stratospheric aerosols, temperature, and density). These measurements were obtained by emitting laser energy into the atmosphere and measuring the return signals scattered from the atmospheric constituents. Unprecedented views were obtained of cloud structures, storm systems, dust clouds, pollutants, forest burning, and surface reflectance. Sites studied included the atmosphere above northern Europe, Indonesia and the south Pacific, Russia, and Africa.

Table 3–88. STS-64 Mission Characteristics (Continued)

Experiments	Shuttle Plume Impingement Flight Experiment (SPIFEX): The SPIFEX studied the characteristics and behavior of exhaust plumes from *Discovery*'s Reaction Control System thrusters during the mission. The SPIFEX, when picked up by *Discovery*'s mechanical arm, is a 33-ft (10-m)-long extension for the arm with a package of instruments to measure the near-field, transition, and far-field effects of thruster plumes. The SPIFEX plume information gathered would assist planners in understanding the potential effects of thruster plumes on large space structures, such as the Russian *Mir* space station and the ISS, during future Shuttle docking and rendezvous operations.
	SAREX-II: See STS-35.
	SSCE: See STS-41.
	Biological Research in Canister (BRIC): This experiment investigated the effects of spaceflight on small arthropod animals and plant specimens.
	RME-III: See STS-28.
	MAST: See STS-59.
	AMOS Calibration Test: See STS-29. Robot Operated Materials Processing Systems (ROMPS): These systems used the microgravity environment to develop commercially valuable methods of processing semiconductor materials. ROMPS also advanced automation and robotics for material processing in ways that could lower the costs of developing and manufacturing semiconductors. The ROMPS experiment investigated in-space processing of semiconductor materials and consisted of a robot, furnace, samples, and control electronics. It used the robot to transport a variety of semiconductors from the storage racks to halogen lamp furnaces where their crystal structures were reformed in heating and cooling cycles. ROMPS flight hardware was contained in a pair of GAS cans mounted on the Hitchhiker-G Carrier. ROMPS was the first robotics system operated in space.
	The NASA Office of Advanced Concepts and Technology sponsored ROMPS as part of their mission to develop commercially relevant techniques for in-space materials processing. The project was being carried out by Goddard Space Flight Center and two NASA-sponsored Centers for the Commercial Development of Space: the Consortium for Commercial Crystal Growth at Clarkson University in Potsdam, New York, and the Space Automation and Robotics Center in Ann Arbor, Michigan.

Table 3–88. STS-64 Mission Characteristics (Continued)

Get Away Specials	GBA: This assembly held 10 GAS canisters containing experiments to investigate different physical and biological phenomena and two ballast GAS cans containing accelerometers.
	G-178 Customer: Sierra College, Rocklin, California Spectrometer Measurements of the Upper Atmosphere in the UV Range: G-178 took ozone measurements of Earth's upper atmosphere in the ultraviolet 200-nanometer to 400-nanometer spectral range using a Charge Coupled Device (CCD)-based spectrometer. A CCD photographic camera also flew as part of the experiment and provided target verification for the spectrometer.
	G-254 Customer: The Kinkaid School and Utah State University Four experiments were flown in individual spacepaks, including a new aluminum Isogrid construction. The payload contained popcorn kernels and radish seeds in separate Ziploc bags as an experiment by Edith Bowen Elementary School located on the Utah State University campus. After the flight, the students popped and tasted the popcorn. The radishes were grown and compared with a similar sample maintained in 1g. The experiments dealt with fluid distillation, flat zone instability, photosynthesis, and bubble interferometry.
	G-0325 Customer: Norfolk, Virginia Public Schools NORSTAR experiment: Acoustic Wave Study and 60 minor experiments: This experiment recorded visually how sound affected dust particles in near-zero gravity to advance understanding of acoustics.
	G-417 Customer: Beijing Institute of Environmental Engineering Three experiments: Paramecium Reproduction, Oil and Water Mixing, Soldering Examples.

Table 3–88. STS-64 Mission Characteristics (Continued)

Experiments	G-453

Customer: N. Tateyama, The Society of Japanese Aerospace Co., Inc. (SJAC)

Reflight of G-453 on STS-57 due to incomplete results following a battery failure. The experiments investigated the formation of superconducting material and the boiling phenomenon under microgravity and the absence of convection. There were two experiments:

1) Formation of Silicon-Lead (Si-Pb) Alloy: This experiment investigated the formation of superconducting alloy (not mixable on the ground). Each sample, in a platinum crucible located inside a quartz ampoule (small glass container), was heated in a furnace up to 1,450°C (2,640°F) for 25 minutes.

2) Boiling Experiment: This experiment observed the bubble formation when an organic solvent (Freon 113) boiled under microgravity and in the absence of convection. The organic solvent was heated and boiled in a small sealed vessel. The behavior of bubbles formed while boiling was observed and recorded using a video system.

G-454
Customer: Society of Japanese Aerospace Co.
Crystallization via a Temperature Gradient Furnace: G-454 investigated the crystallization or the formation of materials under microgravity and in the absence of convection.

G-456
Customer: Society of Japanese Aerospace Co.
Separation of Biologically Active Materials via Electrophoresis: G-456 observed electrophoresis (the movement of suspended particles through a fluid or gel under the action of an electromotive force applied to electrodes in contact with the suspension) with a video camera above the separation chamber. The experiment was recorded on video cassette recorders. The separation results would be compared to results obtained on Earth.

G-485
Customer: ESA
Material Evaporation/Exposure experiment via a motorized door assembly: This experiment tested the feasibility of depositing different materials in a microgravity and vacuum environment by flying the payload in a GAS canister with a motorized door assembly.

Table 3–88. STS-64 Mission Characteristics (Continued)

Experiments	G-506 Customers: Goddard Space Flight Center and Morgan State University Three experiments: Orbiter Stability Experiment: This experiment was designed to evaluate the Space Shuttle as a platform for imaging the Sun in x-rays and extreme ultraviolet light. Radiation Effects on Photographic Film. Radiation Effects on Seeds: This experiment studied the effects of radiation and zero gravity on germination and growth.
	G-562 Customer: Canadian Space Agency QUESTS: Reflight of G-521: G-562 consisted of 12 isothermal furnaces and three gradient furnaces for materials science, a computer control system, a data acquisition system, and batteries. There were two types of furnaces: temperature-gradient (for directional crystal growth studies) and constant-temperature (for metal diffusion studies).
	Sample Return Experiment: The experiment sat on top of the GAS can in position 4. The primary science objective was to quantify extraterrestrial particles and other orbital debris in the orbiter bay. A secondary objective was a realistic test for comet sample collection concepts.
Mission Results	Successful
Remarks	Astronauts Lee and Meade completed the first untethered EVA in a decade. During the 6-hour, 15-minute EVA, they tested a new SAFER backpack designed for use in the event a crew member became untethered while conducting an EVA.

Table 3–89. STS-68 Mission Characteristics

Vehicle	OV-105
	Endeavour
Crew	CDR: Michael A. Baker
	PLT: Terrence W. Wilcutt
	PC: Thomas D. Jones
	MS: Steven L. Smith, Daniel W. Bursch, Peter J.K. Wisoff
Launch	September 30, 1994, 7:16:00 a.m. EDT, Kennedy Space Center, Pad 39-A. The first launch attempt August 18 was halted at T-1.9 seconds when orbiter computers shut down all three main engines after detecting an unacceptably high discharge temperature in the high-pressure oxidizer turbopump turbine for main engine No. 3. *Endeavour* returned to the VAB and all three engines were replaced. The countdown for the second launch attempt proceeded smoothly to an on-time liftoff September 30.
Orbital Altitude & Inclination	120 nmi (222 km), 57.00 deg
Launch Weight (lb/kg)	247,129/112,096
Landing & Postlanding Operations	October 11, 1994, 10:02:08 a.m. PDT, Runway 22, Edwards Air Force Base. The landing was diverted to Edwards Air Force Base because of unacceptable weather at Kennedy Space Center. The postlanding video showed what appeared to be water dripping from the area of the centerline latch for the orbiter/external tank doors. The source later was found to be a cracked valve in water spray boiler No. 3. The orbiter returned to Kennedy Space Center atop the 747 Shuttle Carrier Aircraft October 2.
Rollout Distance (ft/m)	8,495/2,589
Rollout Time (seconds)	60
Mission Duration	269 hr, 46 min, 8 sec
Landed Revolution No.	181
Mission Support	STDN
Primary Objective	Research using the SRL-2
Deployed Satellites	None

Table 3–89. STS-68 Mission Characteristics (Continued)

Experiments	SRL-2: The SRL-2 was the second simultaneous multfrequency (C, L, and X bands), multipolarization phased array imaging radar in space for geoscientific studies of Earth's different seasons. The crew used the SRL-2 to image sites for geology, hydrology, vegetation science, and oceanography to study vegetation type, extent, and deforestation; water storage and flux; ocean dynamics; wave fields; wind fields; volcanism; tectonic activity; soil erosion; desertification; and topography. Using SIR-C/X-SAR, the crew imaged a volcanic eruption in Russia and the islands of Japan after an earthquake. The SRL-2 payload consisted of the following: • SIR-C • X-SAR • MAPS The DARA and Italian Space Agency provided the X-SAR instrument. CHROMEX: See STS-29. CPCG: See STS-49. BRIC: See STS-64. CREAM: See STS-48. MAST: See STS-65.
Get Away Specials	G-316 Customer: North Carolina A&T State University This experiment determined the effect of microgravity on arthropod development and crystal growth. G-503 Customer: University of Alabama This experiment determined the effect of microgravity on diatoms, the curing of concrete, root growth, and the pitting of metals. G-541 Customer: Swedish Space Corporation This experiment studied the breakdown of a planar solid/liquid interface in space. GAS Postal Payloads: The U.S. Postal Service used GAS hardware to fly 500,000 commemorative stamps to recognize the 25th anniversary of the Apollo 11 Moon Landing. The stamp was a $9.95 Express Mail stamp. Father and son team Paul and Chris Calle, experienced stamp designers and NASA Art Program participants, created the artwork for the stamp.

Table 3–89. STS-68 Mission Characteristics (Continued)

Get Away Specials	Sample Return Experiments Principal Investigator: Dr. Peter Tsou, Jet Propulsion Laboratory Two Sample Return Experiments sat on top the postal payloads. The primary science objective was the quantification of extraterrestrial particles and other orbital debris present in the orbiter bay. A secondary objective was a realistic test for comet sample collection concepts.
Mission Results	Successful

Table 3–90. STS-66 Mission Characteristics

Vehicle	OV-104
	Atlantis
Crew	CDR: Donald R. McMonagle
	PLT: Curtis L. Brown, Jr.
	PC: Ellen Ochoa
	MS: Joseph R. Tanner, Jean-Francois Clervoy (ESA), Scott E. Parazynski
Launch	November 3, 1994, 11:59:43 a.m. EST, Kennedy Space Center, Pad 39-B. The 11:56 a.m. launch was delayed slightly while Shuttle managers assessed the weather at the transoceanic abort landing sites. The liftoff was *Atlantis*'s first since an extended checkout and modification period at the Rocketdyne Rockwell plant in Palmdale, California.
Orbital Altitude & Inclination	164 nmi (304 km), 57.00 deg
Launch Weight (lb/kg)	243,089[a]/110,263
Landing & Postlanding Operations	November 14, 1994, 7:33:45 a.m. PST, Runway 22, Edwards Air Force Base. The landing was diverted to California due to high winds, rain, and clouds in Florida from Tropical Storm Gordon.
Rollout Distance (ft/m)	7,642/2,329
Rollout Time (seconds)	49
Mission Duration	262 hr, 34 min, 2 sec
Landed Revolution No.	173
Mission Support	STDN
Primary Objective	Research using ATLAS-3; deployment and retrieval of the CRISTA-SPAS
Deployed Satellites	Deployed and retrieved CRISTA-SPAS: The spacecraft carried two instruments—the German CRISTA telescope and the U.S. Middle Atmosphere High Resolution Spectrograph Investigation (MAHRSI) instrument.

Table 3–90. STS-66 Mission Characteristics (Continued)

Experiments	ATLAS-3: The ATLAS-3 collected data about the Sun's energy output and chemical makeup of Earth's middle atmosphere and how these factors affected global ozone levels. The experiments included the following:

• ATMOS Experiment: This experiment collected data on trace gases in the atmosphere.

• SSBUV: The spectrometer took ozone measurements to calibrate the NOAA-9 satellite ozone monitor. The SSBUV also took cooperative measurements with other ATLAS-3 instruments. (See STS-34.)

• ACRIM: The ACRIM made extremely precise measurements of the Sun's total radiation for 30 orbits as a calibration reference for a sister instrument on the UARS launched in 1991.

• SOLCON: Provided by Belgium, the SOLCON measured solar radiation as a reference point to track changes over years.

• SOLSPEC: A French instrument. The SOLSPEC measured the Sun's radiation as a function of wavelength.

• SUSIM: The SUSIM measured the fluctuation of the Sun's ultraviolet radiation and determined how much the measured ultraviolet light degraded the accuracy of the measuring instrument.

• MAS: The MAS collected 9 hours of observations, measuring the distribution of water vapor, chlorine monoxide, and ozone at altitudes between 12 mi and 60 mi (20 km and 100 km) before a computer malfunction halted instrument operations.

Experiment of the Sun Complementing the ATLAS Payload and Education-II (ESCAPE-II): This experiment collected solar data with solar imaging and ultraviolet solar irradiance experiments. The data was correlated with the co-manifested ATLAS-3 solar experiments to understand upper atmosphere photochemistry.

Protein Crystal Growth-Thermal Enclosure System (PCG-TES): This experiment investigated the mechanisms of PCG and retrieved high-quality crystals grown during spaceflight using a double locker TES.

Protein Crystal Growth-Single Locker Thermal Enclosure System (PCG-STES): This experiment investigated the mechanisms of PCG and retrieved high-quality crystals grown during spaceflight using a single locker TES.

HPP Experiment-Reflight: See STS-52.

PARE/NIH-R: Both PARE and NIH-R studied the physiological and anatomical changes occurring in mammals under weightless spaceflight conditions.

Table 3–90. STS-66 Mission Characteristics (Continued)

Experiments	SAMS: See STS-43.
	STL-A: See STS-53.
Get Away Specials	None
Mission Results	Successful
Remarks	This mission successfully tested a different method of approaching a spacecraft to retrieve CRISTA-SPAS as a prelude to the upcoming U.S. Shuttle/Russian space station *Mir* docking flights. Called an R-Bar approach, it was expected to save propellant while reducing the risk of contamination to *Mir* systems from orbiter thruster jet firings.

[a] Jenkins, p. 302.

Table 3–91. STS-63 Mission Characteristics

Vehicle	OV-103 *Discovery*
Crew	CDR: James D. Wetherbee PLT: Eileen M. Collins PC: Bernard A. Harris, Jr. MS: C. Michael Foale, Janice Voss, Vladimir G. Titov (RSA)
Launch	February 3, 1995, 12:22:04 a.m. EST, Kennedy Space Center, Pad 39-B. NASA adjusted the countdown sequence to accommodate a short 5-minute window required for rendezvous with *Mir*, including adding more hold time at T-6 hours and T-9 minutes. The launch first scheduled for February 2 was postponed on February 1 when one of the three inertial measurement units on the orbiter failed. The countdown on February 3 proceeded so smoothly that there was extra time left in the T-9-minute hold. The launch marked the first at a 51.6-degree inclination to the equator to put the orbiter into the same orbital plane as *Mir*.
Orbital Altitude & Inclination	213 nmi (394 km), 51.60 deg
Launch Weight (lb/kg)	247,555[a]/112,289
Landing & Postlanding Operations	February 11, 1995, 6:50:19 a.m. EST, Runway 15, Kennedy Space Center. This was the first end-of-mission landing since the runway was resurfaced in fall 1994 to decrease wear on orbiter tires and increase crosswind tolerances. After landing, cosmonauts aboard *Mir* radioed their congratulations to the *Discovery* crew. *Discovery* became the first orbiter in the U.S. fleet to complete 20 missions. The orbiter transferred to the OPF later that day.
Rollout Distance (ft/m)	11,002/3,353
Rollout Time (seconds)	80
Mission Duration	198 hr, 28 min, 15 sec
Landed Revolution No.	128
Mission Support	STDN
Primary Objective	Experimentation using SPACEHAB-3, deployment and retrieval of SPARTAN-204
Deployed Satellites	Deployed and retrieved SPARTAN-204 with Far Ultraviolet Imaging Spectrograph (FUVIS)

Table 3–91. STS-63 Mission Characteristics (Continued)

Experiments	SPACEHAB-3: This commercially developed module carried experiments in biotechnology and advanced materials development, technology demonstrations, and two pieces of supporting hardware measuring on-orbit accelerations.

- Biotechnology experiments:
  - ASC-04
  - BPL-03
  - CGBA-06
  - Fluids Generic Bioprocessing Apparatus (FGBA-01)
  - IMMUNE-02
  - Commercial Protein Crystal Growth-Vapor Diffusion Apparatus (CPCG-VDA)
  - Protein Crystallization Facility-Light Scattering/ Temperature Controlled (PCFLS/T)
- Materials processing experiments were:
  - ECLiPSE-Hab
  - Gas Permeable Polymer Membranes (GPPM-02) Technology experiments:
  - 3-DMA Charlotte™ Robotic Experiment Monitor
- Life and biomedical sciences and applications experiments:
  - BRIC-03
  - CHROMEX-06
  - NIH-C-03
- Microgravity science and applications experiments:
  - PCG-STES-03
  - SAMS-03
- Johnson Space Center Space and Life Sciences Directorate experiment:
  - CPDS
- DOD, U.S. Air Force experiments:
  - CREAM-06
  - RME-III
  - Window Experiment (WINDEX-01)

Cryo Systems Experiment (CSE): This experiment tested cryogenic cooling system and oxygen diode heat pipes for use on future spacecraft designs.

GLO-2: See STS-53.

ODERACS-2: See STS-53.

SSCE: See STS-41.

AMOS Calibration Test: See STS-29.

IMAX: The crew used the IMAX handheld motion picture camera to film inside the crew cabin.

Get Away Specials	None
Mission Results	Successful

Table 3–91. STS-63 Mission Characteristics (Continued)

Remarks	Astronauts Foale and Harris began an EVA suspended at the end of the robot arm, away from the payload bay, to test modifications to their spacesuits to keep spacewalkers warmer in the extreme cold of space. They were then scheduled to practice handling the approximately 2,500-lb (1,134-kg) SPARTAN-204 to rehearse Space Station assembly techniques. However, both astronauts reported they were becoming very cold– this portion of EVA was performed during a night pass and mass-handling was curtailed. The EVA lasted 4 hours, 38 minutes.
	This was the first flight as part of Phase I of the ISS program. The Shuttle performed first approach and fly-around of Russian space station *Mir*.

[a] Jenkins, p. 302.

Table 3–92. STS-67 Mission Characteristics

Vehicle	OV-105
	Endeavour
Crew	CDR: Stephen S. Oswald
	PLT: William G. Gregory
	MS: John M. Grunsfeld, Wendy B. Lawrence
	PC: Tamara E. Jernigan
	PS: Samuel T. Durrance, Ronald A. Parise
Launch	March 2, 1995, 1:38:13 a.m. EST, Kennedy Space Center, Pad 39-A. After a smooth countdown, liftoff was delayed for about a minute due to concerns about a heater system on the flash evaporator system. A backup heater was used, and the countdown proceeded.
Orbital Altitude & Inclination	187 nmi (346 km), 28.45 deg
Launch Weight (lb/kg)	256,293[a]/116,253
Landing & Postlanding Operations	March 18, 1995, 1:47:01 p.m. PST, Runway 22, Edwards Air Force Base. The orbiter was diverted to Edwards Air Force Base after landing opportunities in Florida were waved off on March 17 and in the early day on March 18. The orbiter returned to Florida on March 27 and was taken to the OPF on March 28.
Rollout Distance (ft/m)	9,975/3,040
Rollout Time (seconds)	59
Mission Duration	399 hr, 8 min, 48 sec
Landed Revolution No.	261
Mission Support	STDN
Primary Objective	Research using Astro-2
Deployed Satellites	None
Experiments	Astro-2: The Astro-2 made ultraviolet observations of stars, galaxies, magnetospheres, and quasars.[b] Three experiments were mounted to the SPACELAB instrument pointing system: • HUT: Considered a complement to the Hubble Space Telescope, the HUT completed more than 200 separate observations of more than 100 celestial objects. • WUPPE: The WUPPE greatly expanded the database on ultraviolet spectropolarimetry. • UIT: The UIT cameras imaged about two dozen large spiral galaxies for inclusion in an atlas of such galaxies; it made the first ultraviolet images of the entire Moon.

Table 3–92. STS-67 Mission Characteristics (Continued)

Experiments	Middeck Active Control Experiment (MACE): The MACE measured and controlled the dynamics of complex systems in the microgravity environment. The experiment developed a well verified set of Control Structure Integration/Controlled Structures Technology (CSI/CST) methods and approaches, allowing designers of future CST spacecraft, which cannot be dynamically tested on the ground in a sufficiently realistic zero-gravity simulation, to have confidence in the eventual orbit of such spacecraft.
	PCG-TES: See STS-66.
	PCG-STES: See STS-63.
	CMIX: See STS-56.
	SAREX-II: See STS-35.
Get Away Specials	G-387 and G-388 Customer: Australian Space Office and Auspace Limited G-387 and G-388 made ultraviolet observations of deep space to aid in the study of the structure of galactic supernova remnants, distribution of hot gas in the Magellanic Clouds, hot galactic halo emissions, and emissions associated with galactic cooling flows and jets.
Mission Results	One of the two cameras on the UIT malfunctioned undetected while on orbit, and only 80 percent of UIT's science objectives were met. Other mission objectives were met successfully.
Remarks	*Endeavour* was outfitted with extended duration orbiter hardware for the flight.

a Jenkins, p. 303.
b See chapter 4, Space Science, of this volume for further discussion of the Astro-2 mission.

Table 3–93. STS-71 Mission Characteristics

Vehicle	OV-104
	Atlantis
Crew	CDR: Robert L. Gibson
	PLT: Charles J. Precourt
	PC: Ellen S. Baker
	MS: Gregory J. Harbaugh, Bonnie J. Dunbar
	Anatoly Solovyev (RSA)–*Mir*-19 crew upload
	Nikolai Budarin (RSA)–*Mir*-19 crew upload
	Norman E. Thagard–*Mir*-18 crew download
	Vladimir Dezhurov (RSA)–*Mir*-18 crew download
	Gennadiy Strekalov (RSA)–*Mir*-18 crew download
Launch	June 27, 1995, 3:32:19 p.m. EDT, Kennedy Space Center, Pad 39-A. The launch was originally targeted for late May but slipped into June to accommodate Russian space program activities necessary for the first Space Shuttle–*Mir*–Space Station docking, including a series of spacewalks to reconfigure the Station for docking and launch of the new Spektr module to *Mir* containing U.S. research hardware. The launch set for June 23 was scrubbed when rainy weather and lightning prevented loading of the external tank earlier that day. A second try June 24 was scrubbed at T-9 minutes, again due to persistent stormy weather in central Florida, coupled with a short 10-minute launch window. Liftoff was reset for June 27, and final countdown proceeded smoothly.
Orbital Altitude & Inclination	170 nmi (315 km)/216 nmi (400 km) when docking, 51.60 deg
Launch Weight (lb/kg)	248,857[a]/112,880
Landing & Postlanding Operations	July 7, 1995, 10:54:34 a.m. EDT, Runway 15, Kennedy Space Center. The runway switched from 33 to 15 about 20 minutes before touchdown due to concerns of Chief Astronaut Robert Cabana, flying the Shuttle Training Aircraft, about clouds blocking runway landing aids from view.
Rollout Distance (ft/m)	8,364/2,549
Rollout Time (seconds)	51
Mission Duration	235 hr, 22 min, 17 sec
Landed Revolution No.	152
Mission Support	STDN
Primary Objective	First Shuttle-*Mir* docking
Deployed Satellites	None

Table 3–93. STS-71 Mission Characteristics (Continued)

Experiments	IMAX camera: This 70-mm motion picture camera system photographed rendezvous and spacecraft operations from within the crew compartment.
	SAREX-II: See STS-35.
Get Away Specials	None
Mission Results	Successful
Remarks	Spacelab-*Mir*: This was a combined science and logistical transfer mission. Life and microgravity science investigations were performed jointly with *Mir* to complement investigations on the Shuttle. The crews performed research into biomedical life sciences and microgravity with an emphasis on the effects of long-duration spaceflight on the human body. The logistical transfers included transporting the *Mir*-19 crew to *Mir*; returning the *Mir*-18 crew; transferring water to *Mir*; science specimens and hardware transferred to and from *Mir*; food resupply sent to *Mir*; and Russian hardware returned. The crews conducted fifteen separate biomedical and scientific investigations, covering seven different disciplines: 1) cardiovascular and pulmonary functions; 2) human metabolism; 3) neuroscience; 4) hygiene, sanitation, and radiation; 5) behavioral performance and biology; 6) fundamental biology; and 7) microgravity research. The *Mir*-18 crew served as test subjects for the investigations.
	The joint *Atlantis-Mir* spacecraft was the largest ever in orbit when the two were linked and the first on-orbit changeout of a Shuttle crew. Astronaut Thagard logged the longest U.S. spaceflight with his return from *Mir*. For the return flight, the crews made provisions on the *Atlantis* to accommodate an eight-person crew.
	President William J. Clinton called and congratulated the crews on the successful docking and invited them to visit the White House.

[a] Jenkins, p. 303.

Table 3–94. STS-70 Mission Characteristics

Vehicle	OV-103
	Discovery
Crew	CDR: Terence T. Henricks
	PLT: Kevin R. Kregel
	MS: Nancy J. Currie (Sherlock), Donald A. Thomas, Mary Ellen Weber
Launch	July 13, 1995, 9:41:55 a.m. EDT, Kennedy Space Center, Pad 39-B. The count was held for 55 seconds at T-31 seconds due to fluctuations seen on the external tank automatic gain control external tank range safety system receiver. Launch Commit Criteria contingency procedures were worked and the count then proceeded on schedule.
	STS-70 had originally moved ahead of the launch of STS-71 because of a delay in the launch of the Russian Spektr laboratory module to the Russian space station *Mir*. However, on May 31, NASA Shuttle managers assessed damage to the *Discovery*'s external tank caused by nesting flicker woodpeckers. The damage consisted of about 71 holes (ranging in size from 0.5 in (1.3 cm) to 4 in (10 cm) in diameter) in the external tank's thermal protection foam insulation. Technicians installed safeguards against additional damage. On June 2, NASA managers decided to delay the launch of *Discovery* on STS-70 to repair the foam insulation on the vehicle's external tank. STS-71 was moved ahead of STS-70 and *Discovery* was rolled back to the VAB. It was the quickest turnaround landing (STS-71) to launch (STS-70).
Orbital Altitude & Inclination	160 nmi (296 km), 28.45 deg
Launch Weight (lb/kg)	258,798[a]/117,389
Landing & Postlanding Operations	July 22, 1995, 8:02:00 a.m. EDT, Runway 33, Kennedy Space Center. The first landing opportunities on July 21 at Kennedy Space Center were waved off due to fog and low visibility. The first opportunity on July 22 at Kennedy Space Center was also waved off.
Rollout Distance (ft/m)	8,465/2,580

Table 3–94. STS-70 Mission Characteristics (Continued)

Rollout Time (seconds)	57
Mission Duration	214 hr, 20 min, 7 sec
Landed Revolution No.	143
Mission Support	STDN
Primary Objective	Deployment of TDRS-G
Deployed Satellites	TDRS-G (7)/IUS
Experiments	PARE/NIH-R: See STS-66.
	Bioreactor Demonstration System (BDS): The BDS developed the capability and demonstrated the ability to grow mammalian cells in fluid growth medium in microgravity.
	CPCG: See STS-49.
	STL/NIH-C: See STS-59.
	BRIC: See STS-64.
	SAREX-II: See STS-35.
	VFT-4: See STS-59.
	HERCULES: This system provided an on-orbit capability to geolocate a ground target.
	MIS-B: See STS-53.
	WINDEX: The WINDEX obtained spectrally isolated images of the Shuttle surface glow, thruster plumes, water dumps, aurora, and airglow.
	RME-III: See STS-28.
	MAST: See STS-65.
Get Away Specials	None
Mission Results	Successful. It was the most trouble-free mission to date.[b]
Remarks	This was the first mission in which ground support used the new Mission Control Center at Johnson Space Center.[c]

[a]   Jenkins, p. 303.

[b]   "STS-70/Flight 70 Mission Report," *http://members.aol.com/WSNTWOYOU/STS70MR.HTM* (accessed December 6, 2005).

[c]   "STS-70 Flight Brings New Tools On-Line," *Space News Roundup*, Johnson Space Center 34, no. 28 (July 14, 1995): 1, *http://www.jsc.nasa.gov/history/roundups/issues/95-07-14.pdf* (accessed December 7, 2005).

Table 3–95. STS-69 Mission Characteristics

Vehicle	OV-105
	Endeavour
Crew	CDR: David M. Walker
	PLT: Kenneth D. Cockrell
	PC: James S. Voss
	MS: James H. Newman, Michael L. Gernhardt
Launch	September 7, 1995, 11:09:00 a.m. EDT, Kennedy Space Center, Pad 39-A. The launch originally set for August 5 was postponed indefinitely to allow further review of solid rocket motor nozzle joint hardware from STS-70 and STS-71. An inspection team was formed to assess the significance of the gas path in nozzle internal joint No. 3, extending from insulation in the motor chamber to, but not past, the primary O-ring seal. The team concluded that the nozzle joint design was sound and that gas paths were being created when insulation material, known as Room Temperature Vulcanizing, was applied. Small air pockets were forming in the thermal insulation that could later become pathways for hot gas during motor operation. Attention then focused on developing procedures to allow Non-Destructive Evaluation (NDE) inspection of insulation at the pad, and a new launch date of August 31 was set. The August 31 launch was scrubbed about 5.5 hours before liftoff due to the failure of one of the orbiter's three fuel cells. Fuel cell No. 2 indicated higher than allowable temperatures during activation as countdown proceeded. The fuel cell was removed and replaced. Liftoff on September 7 was preceded by a smooth countdown.
Orbital Altitude & Inclination	165 nmi (306 km), 28.45 deg
Launch Weight (lb/kg)	256,645[a]/116,412
Landing & Postlanding Operations	September 18, 1995, 7:37:56 a.m. EDT, Runway 33, Kennedy Space Center. The landing occurred at the first opportunity.
Rollout Distance (ft/m)	10,230/3,118
Rollout Time (seconds)	56
Mission Duration	260 hr, 28 min, 26 sec
Landed Revolution No.	170
Mission Support	STDN
Primary Objective	Deployment and retrieval of SPARTAN 201-03 and WSF-2
Deployed Satellites	Deployed and retrieved SPARTAN 201-03 and WSF-2

Table 3–95. STS-69 Mission Characteristics (Continued)

Experiments	International Extreme Ultraviolet Hitchhiker-1 (IEH): The IEH consisted of a Hitchhiker carrier with four experiments: • Solar Extreme Ultraviolet Hitchhiker (SEH): The SEH was contained in an extended GAS canister with a Hitchhiker Motorized Door Assembly. The SEH used rare gas ionization cells, photodiodes, and a spectrometer for solar viewing. • Ultraviolet Spectrograph Telescope for Astronomical Research (UVSTAR): This was a spectrograph with internal gimbals, allowing for stellar observations. • GLO-3: This spectrograph measured Shuttle glow phenomena in the 115 nm to 1150 nm spectral range • CONCAP IV-03: See STS-57. STL/NIH-C: See STS-59. Electrolysis Performance Improvement Concept Study (EPICS): This study was a characterization in microgravity of the water electrolysis concepts to be used for metabolic oxygen generation in Space Station *Freedom* and other life support, propulsion, EVA, and space power applications. Commercial Materials Dispersion Apparatus (MDS) CMIX: See STS-56. CGBA Configuration A: See STS-54. BRIC Block 2: See STS-64.
Get Away Specials	Capillary Pumped Loop (CAPL)/Gas Bridge Assembly (GBA): The combined CAPL-2/GBA payload consisted of the CAPL-2 Hitchhiker payload, the TES-2 payload, four GAS payloads, and the Sample Return Experiment. CAPL-2: CAPL-2 was a reflight of the CAPL-1 Hitchhiker payload flown on STS-60 with modifications to enhance the startup of its capillary system. This flight verified the heat transport requirements of the thermal control system under design for the EOS. See STS-60. TES-2: The TES-2 was designed to provide data for understanding the long-duration behavior of TES fluoride salts that undergo repeated melting and freezing in microgravity. It developed a melt/freeze behavior database for TES phase change materials, leading to performance enhancements for solar-dynamic power system heat receivers.

Table 3–95. STS-69 Mission Characteristics (Continued)

Get Away Specials	G-645 Customer: Millcreek Township School District Investigation of Electroheological Fluids: This experiment investigated the performance of electroheological fluid-filled beams used as structural dampers in space.
	G-702 Customer: Lewis Research Center Microgravity Smoldering Combustion: This experiment increased the understanding of smoldering combustion in long-term microgravity.
	G-726 Customer: Langley Research Center Joint Damping Experiment: The experiment measured influence of gravity on the structural damping of a three-bay truss.
	G-515 Customer: ESA G-515 studied active damping control loops using a flexible plate and two piezo (pressure) actuators.
	Sample Return Experiment: The Sample Return Experiment sat on top the ballast GAS can. The primary science objective was the quantification of extraterrestrial particles and other orbital debris present in the orbiter bay. A secondary objective was a realistic test for comet sample collection concepts.
Mission Results	Successful
Remarks	This was the first time that two different payloads were retrieved and deployed during the same mission.

ª Jenkins, p. 303.

Table 3–96. STS-73 Mission Characteristics

Vehicle	OV-102
	Columbia
Crew	CDR: Kenneth D. Bowersox
	PLT: Kent V. Rominger
	MS: Catherine G. Coleman, Michael E. Lopez-Alegria
	PC: Kathryn C. Thornton
	PS: Fred W. Leslie, Albert Sacco, Jr.
Launch	October 20, 1995, 9:53:00 a.m. EDT, Kennedy Space Center, Pad 39-B. The launch was after *Columbia*'s return to the fleet following its scheduled modification and refurbishment. A successful launch took place after six scrubs, which tied STS-61-C for the greatest number of launch scrubs. Liftoff originally set for September 25 was scrubbed shortly after tanking began, when a hydrogen leak was detected in the main engine No. 1 main fuel valve. The valve was replaced at the pad. The launch was reset for October 5, but Hurricane Opal led to an L-1 day decision to postpone launch one day to October 6.
	The October 6 launch attempt was scrubbed before external tank loading when it was determined that the hydraulic fluid had been inadvertently drained from hydraulic system 1 following the main engine No. 1 fuel valve replacement. A compressibility test demonstrated that the system was satisfactory for launch, and liftoff was reset to occur October 7. The launch attempt October 7 was scrubbed at T-20 seconds when master events controller 1 failed to operate properly and mission managers determined that it needed to be replaced. The launch was reset for October 14, and then rescheduled for October 15 to allow additional time to inspect the main engine oxidizer ducts because of crack in a test engine oxidizer duct found at Stennis Space Center. A launch attempt October 15 was postponed at T-5 minutes due to low clouds and rain. The launch was tentatively reset to October 19, pending a successful Atlas launch October 18. However, the Atlas launch was delayed, and STS-73 launch was moved to October 20. Countdown to liftoff on October 20 was delayed 3 minutes due to a range computer glitch.
Orbital Altitude & Inclination	150 nmi (277 km), 39.00 deg
Launch Weight (lb/kg)	257,017[a]/116,581
Landing & Postlanding Operations	November 5, 1995, 6:45:21 a.m. EST, Runway 33, Kennedy Space Center. The landing occurred on the first opportunity.

Table 3–96. STS-73 Mission Characteristics (Continued)

Rollout Distance (ft/m)	9,117/2,779
Rollout Time (seconds)	55
Mission Duration	381 hr, 52 min, 22, sec
Landed Revolution No.	255
Mission Support	STDN
Primary Objective	Research using the USML-2
Deployed Satellites	None

Experiments	USML-2: This was the second U.S. Spacelab mission dedicated to microgravity research. It consisted of 14 facilities performing 18 experiments and 7 investigations. The research was dedicated to fluid dynamics, crystal growth, combustion science, biological science, and technology demonstrations. Government, colleges, and private industry were involved in all facets of the research.
	• Primary experiments:
	  – CGF
	  – DPM
	  – Geophysical Fluid Flow Cell (GFFC)
	  – STDCE
	  – GBX
	• Small Middeck Experiment Rack (SMIDEX)
	  – APCF
	  – LBNP
	  – CGBA
	
	ASTROCULTURE™ facility and experiment
	The Glovebox-enclosed cabinet offered a clean working space and minimized contamination risks for these experiments:
	• Interface Configuration Experiment (ICE)
	• Oscillatory Thermocapillary Flow Experiment (OTFE)
	• Fiber Supported Droplet Combustion (FSDC)
	• Protein Crystal Growth–Glovebox (PCGG)
	• Zeolite Crystal Growth–Glovebox (ZCGG)
	• Colloidal Disorder-Order Transitions (CDOT)
	• Particle Dispersion Experiment (PDE)
	
	OARE: Provided near real-time data for characterizing the low frequency microgravity in the orbiter in support of USML-2. See STS-40.

Get Away Specials	None
Mission Results	Successful
Remarks	*Columbia* was outfitted with extended duration orbiter hardware for the flight.

a Jenkins, p. 303.

Table 3–97. STS-74 Mission Characteristics

Vehicle	OV-104
	Atlantis
Crew	CDR: Kenneth D. Cameron
	PLT: James D. Halsell, Jr.
	MS: Chris A. Hadfield (CSA), Jerry L. Ross, William S. McArthur, Jr.
Launch	November 12, 1995, 7:30:43 a.m. EST, Kennedy Space Center, Pad 39-A. The planned rendezvous with *Mir* necessitated a brief launch window of about 7 minutes. Liftoff originally set for November 11, was scrubbed due to unacceptable weather at the TAL sites.[a] Countdown the following day proceeded smoothly to an on-time liftoff.
Orbital Altitude & Inclination	213 nmi (394 km), 51.60 deg
Launch Weight (lb/kg)	274,560[b]/124,538
Landing & Postlanding Operations	November 20, 1995, 12:01:27 p.m. EST, Runway 33, Kennedy Space Center
Rollout Distance (ft/m)	8,607/2,623
Rollout Time (seconds)	57
Mission Duration	196 hr, 30 min, 46 sec
Landed Revolution No.	128
Mission Support	STDN
Primary Objective	Second Shuttle-*Mir* docking
Deployed Satellites	None
Experiments	ICBC: A 65-mm color camera mounted in the payload bay documented DM installation, the *Mir* rendezvous, docking, flyaround, and separation.
	GLO Experiment/Photogrammetric Appendage Structural Dynamics Experiment (PASDE) Payload (GPP): The GLO experiment obtained data from the Shuttle and *Mir* glow emissions for various conditions. Three PASDE canisters, located throughout the cargo bay, photogrammetrically recorded structural response data of the *Mir* solar arrays during the docked phase of the mission.
	SAREX-II: See STS-35.
Get Away Specials	None
Mission Results	Successful

Table 3–97. STS-74 Mission Characteristics (Continued)

Remarks	Shuttle-*Mir* Mission 2 (S/MM-2): This mission delivered the Russian-built DM with two solar arrays attached. The DM was installed on the ODS to be docked to the *Mir* Kristall module. The DM remained attached the *Mir* to provide for future Shuttle and Soyuz-TM dockings. The port solar array was a Russian-built Reusable Solar Array (RSA) while Lewis Research Center built the starboard Cooperative Solar Array (CSA). The crews retrieved and resupplied microgravity and life science experiments on board *Mir*, performed ISS Risk Mitigation Experiments (RME), and resupplied *Mir*.

[a] NASA used the term "transoceanic abort landing" in Section 6.4 of the 1997 *Shuttle Flight Operations Manual* rather than "transatlantic landing," which it used in the 1988 *NSTS Shuttle Reference Manual*. Although some references to the older abort term continued to appear in mission documents, the Mission Chronology for STS-74 in 1995 specifically used "transoceanic abort landing" in its mission description. See "Space Station Mission Chronology: STS-74," *http://www-pao.ksc.nasa.gov/kscpao/chron/sts-74.htm* (accessed November 28, 2005); E-mail from Kyle Herring, November 30, 2005.

[b] Jenkins, p. 303.

Table 3–98. STS-72 Mission Characteristics

Vehicle	OV-105
	Endeavour
Crew	CDR: Brian Duffy
	PLT: Brent W. Jett, Jr.
	MS: Leroy Chiao, Winston E. Scott, Koichi Wakata (JAXA), Daniel T. Barry
Launch	January 11, 1996, 4:41:00 a.m. EST, Kennedy Space Center, Pad 39-B. The countdown to the first Shuttle launch of the year proceeded smoothly except for a 23-minute delay due to communication glitches between various sites on the ground and to reduce the risk of colliding with space debris.
Orbital Altitude & Inclination	250 nmi (463 km), 28.45 deg
Launch Weight (lb/kg)	258,391[a]/117,204
Landing & Postlanding Operations	January 20, 1996, 2:41:41 a.m. EST, Runway 15, Kennedy Space Center. *Endeavour* landed on its first opportunity.
Rollout Distance (ft/m)	8,770/2,673
Rollout Time (seconds)	66
Mission Duration	218 hr, 0 min, 45 sec
Landed Revolution No.	142
Mission Support	STDN
Primary Objective	Deployment and retrieval of SPARTAN OAST-Flyer; retrieval of Space Flyer Unit
Deployed Satellites	Deployed and retrieved SPARTAN OAST-Flyer Retrieved Japanese Space Flyer Unit
Experiments	SSBUV-A: See STS-34.
	PARE/NIH-R-03: See STS-66.
	STL/NIH-C-05: See STS-59.
	PCG-STES-04: See STS-63.
	CPCG-08: See STS-49.
	EDFT-03: This test evaluated and demonstrated mission-critical EVA hardware for its planned use to support the scheduled EVAs for the Space Station.
	Shuttle Laser Altimeter-01: This experiment was the first of four planned remote sensing flights to precisely measure the distance between Earth's surface and the Space Shuttle. The experiment acquired samples of land topology and vegetation data to demonstrate laser altimeter operation in low Earth orbit and measure cloud top height, structure, and aerosol layering. The experiment also provided an in-space engineering testbed for future spaceflight laser sensors.

Table 3–98. STS-72 Mission Characteristics (Continued)

Get Away Specials	G-342 Customer: U.S. Air Force, Space and Missile Systems Center Flexible Beam Experiment 2 (FLEXBEAM 2): The FLEXBEAM 2 investigated vibrations in space by exciting two Aluminum 6061 T-6 cantilevered beams. Each beam was subjected to different initial conditions resulting in exciting different modes. Electromagnetic sensors measured the vibrations while a recorder stored the data.
	G-459 Customer: Society of Japanese Aerospace Co.'s, Inc. (SJAC)
	PCG This experiment reexamined the effect of the microgravity environment on protein-crystal nucleation. Crystal form and size were recorded on photographic film and analyzed after recovery of the payload. To adapt to a GAS payload canister, researchers developed a hardware system using 16 independent crystallization units. Each of the units could carry out crystallization experiments by one of three crystallization methods, i.e., batch, vapor diffusion, and free-interface diffusion.
	G-740 Customer: Lewis Research Center Pool Boiling Experiment: This experiment was an extension of the study of the fundamentals of nucleate pool boiling heat transfer under the microgravity conditions of space. An improved understanding of the basic processes that constitute boiling was sought by removing the buoyancy effects that mask other phenomena.
Mission Results	Successful
Remarks	Two EVAs were conducted as part of the continuing series to prepare for on-orbit construction of the ISS. During the first EVA, lasting 6 hours, 9 minutes, Chiao and Barry evaluated a new portable work platform and a structure known as the rigid umbilical, which might eventually be used to hold various fluid and electrical lines. During the second EVA, conducted by Chiao and Scott, lasting 6 hours, 53 minutes, a portable work platform was again evaluated. Also tested were a Space Station utility box designed to hold avionics and fluid line connects. Scott also tested a spacesuit's thermal control in severe cold up to -104°F (-75°C).

[a] Jenkins, p. 303.

Table 3–99. STS-75 Mission Characteristics

Vehicle	OV-102
	Columbia
Crew	CDR: Andrew M. Allen
	PLT: Scott J. Horowitz
	MS: Jeffrey A. Hoffman, Maurizio Cheli (ESA), Claude Nicollier (ESA)
	PC: Franklin R. Chang-Diaz
	PS: Umberto Guidoni (ESA)
Launch	February 22, 1996, 3:18:00 p.m. EST, Kennedy Space Center, Pad 39-B. The liftoff occurred on-time following a smooth countdown. Six seconds after liftoff, the crew reported that the left main engine chamber pressure meter was showing only 40 percent thrust instead of 104 percent thrust that was necessary prior to throttle-down. Mission controllers in Houston reported that telemetry showed all three engines were performing nominally, and there was no effect on the ascent phase.
Orbital Altitude & Inclination	160 nmi (296 km), 28.45 deg
Launch Weight (lb/kg)	261,927[a]/118,808
Landing & Postlanding Operations	March 9, 1996, 8:58:21 a.m. EST, Runway 33, Kennedy Space Center. A March 8 landing was waved off due to unfavorable weather conditions.
Rollout Distance (ft/m)	8,459/2,578
Rollout Time (seconds)	64
Mission Duration	377 hr, 40 min, 21 sec
Landed Revolution No.	251
Mission Support	STDN
Primary Objective	Flight of TSS-1R; experimentation using USMP-3
Deployed Satellites	U.S./Italian TSS-1R—The satellite was lost during the mission (see Remarks below).

Experiments	USMP-3: The crew performed microgravity research to advance understanding of materials science and condensed matter physics.
	• Supporting hardware included:
	  – SAMS
	  – OARE
	• USMP-3 experiments were:
	  – AADSF
	  – ZENO
	  – IDGE
	  – MEPHISTO

Table 3–99. STS-75 Mission Characteristics (Continued)

Experiments	Middeck Glovebox Facility (MGBX): This facility provided a safe laboratory workbench for three combustion experiments: • Forced-Flow Flamespreading Test (FFFT) • Radiative Ignition and Transition to Spread Investigation (RITSI) • Comparative Soot Diagnostics (CSD) CPCG: See STS-49.
Get Away Specials	None
Mission Results	The tether snapped on flight day three as TSS-1R was just short of full deployment at about 12.8 mi (20.6 km). The satellite immediately began speeding away from the orbiter because of orbital forces; the crew was never in danger. Other mission objectives were successfully achieved.

[a] Jenkins, p. 303.

Table 3–100. STS-76 Mission Characteristics

Vehicle	OV-104
	Atlantis
Crew	CDR: Kevin P. Chilton
	PLT: Richard A. Searfoss
	MS: Ronald M. Sega, Michael R. Clifford, Linda M. Godwin, Shannon W. Lucid (remained on *Mir*)
Launch	March 22, 1996, 3:13:04 a.m. EST, Kennedy Space Center, Pad 39-B. The first launch attempt set for March 21 was scrubbed before beginning tanking operations March 20 due to concerns about high winds. The launch reset for March 22 proceeded smoothly to an on-time liftoff. During ascent, a leak occurred in the hydraulic system powered by APU No. 3. The leak stopped after hydraulic system shutdown on orbit. Mission managers concluded that the system would remain stable and proceeded with plans for a full-duration mission.
Orbital Altitude & Inclination	160 nmi (296 km), 51.60 deg
Launch Weight (lb/kg)	246,337[a]/111,737
Landing & Postlanding Operations	March 31, 1996, 5:28:57 a.m. PST, Runway 22, Edwards Air Force Base. Mission managers rescheduled the landing from March 31 to March 30 in anticipation of rain and clouds at the Kennedy Space Center landing site, but landing attempts at Kennedy Space Center on both March 30 and March 31 were waved off due to weather. The orbiter was finally diverted to California. More conservative weather criteria were employed for landing due to the leak in the APU No. 3 hydraulic system and special measures were taken during reentry to minimize use of this APU. During landing preparations, 3 of 38 reaction control system thrusters failed, but backup thrusters were available to perform the same functions. It was not considered a night landing because landing occurred 11 minutes before sunrise. Flight rules define night launch/landing as one occurring at least 15 minutes after sunset and at least 15 minutes before sunrise.
Rollout Distance (ft/m)	8,357/2,547
Rollout Time (seconds)	55
Mission Duration	221 hr, 15 min, 33 sec
Landed Revolution No.	145
Mission Support	STDN
Primary Objective	Third Shuttle-*Mir* docking; research and transfer of supplies using SPACEHAB-Single Module
Deployed Satellites	None

Table 3–100. STS-76 Mission Characteristics (Continued)

Experiments	SPACEHAB-SM: A single module configuration carried a mix of supplies and scientific equipment to and from *Mir*.

• Equipment in this module:
 – Russian logistics
 – EVA tools
 – ISS Risk Mitigation Experiment
 – American logistics
 – Science and technology experiments
• RME:
 – *Mir* Electric Field Characterization (MEFC) hardware
 – *Mir* Environmental Effects Payload (MEEP)
• Science and Technology Experiments:
 – ESA's Biorack
 – Life Sciences Laboratory Equipment Refrigerator/ Freezer (LSLE R/F)
 – *Mir* Glovebox Stowage
 – QUELD
 – High Temperature Liquid Phase Sintering (LPS)
• *Mir* Environmental Effects Payload (MEEP):
 – Polished Plate Micrometeoroid Debris (PPMD) experiment
 – Orbital Debris Collector (ODC) experiment
 – Passive Optical Samples (POSA) I and II
• *Mir* Glovebox Stowage:
 – Combustion Experiments Parts Box
 – FFFT
 – Passive Accelerometer
 – PCG
 – PCG-TES Ancillary

Kidsat: This project gave middle school students the opportunity to participate in space exploration by configuring their own payload of digital video and a camera for flight on the Shuttle. They could command the camera from their classrooms and download their images of Earth in near real-time.

SAREX: See STS-35.

Get Away Specials	G-312 Customer: U.S. Air Force Space Test Program Naval Research Laboratory's Trapped Ions In Space (TRIS) Experiment: This experiment measured a recently discovered belt of energetic cosmic ray nuclei trapped in Earth's magnetic field to quantify radiation hazards in space to develop a better theoretical understanding of how these cosmic ray nuclei became trapped in Earth's magnetic field. TRIS flew previously on a Space Shuttle mission in 1984.
Mission Results	Successful

Table 3–100. STS-76 Mission Characteristics (Continued)

Remarks	During this Shuttle-*Mir* docking, astronauts Linda Godwin and Michael Clifford conducted the first U.S. EVA around two mated spacecraft. During the 6-hour, 2-minute EVA, they attached four MEEP experiments to *Mir*'s DM that would characterize the environment around *Mir* over an 18-month period. The two wore SAFER propulsive devices.

[a] Jenkins, p. 310.

Table 3–101. STS-77 Mission Characteristics

Vehicle	OV-105
	Endeavour
Crew	CDR: John H. Casper
	PLT: Curtis L. Brown, Jr.
	MS: Andrew S.W. Thomas, Daniel W. Bursch, Mario Runco, Jr., Marc Garneau (CSA)
Launch	May 19, 1996, 6:30:00 a.m. EDT, Kennedy Space Center, Pad 39-B. The original launch date of May 16 was changed to May 19 due to Eastern Range schedule conflicts. The countdown proceeded smoothly to an on-time liftoff on May 19.
Orbital Altitude & Inclination	153 nmi (283 km), 39.00 deg
Launch Weight (lb/kg)	254,891[a]/115,617
Landing & Postlanding Operations	May 29, 1996, 7:09:18 a.m. EDT, Runway 33, Kennedy Space Center. Favorable weather allowed for a landing at the first opportunity.
Rollout Distance (ft/m)	9,291/2,832
Rollout Time (seconds)	42
Mission Duration	240 hr, 39 min, 24 sec
Landed Revolution No.	160
Mission Support	STDN
Primary Objective	Deployment and retrieval of SPARTAN 207 IAE; experimentation using SPACEHAB-04 and TEAMS
Deployed Satellites	Deployed and retrieved SPARTAN 207 carrying the IAE. The PAMS and Satellite Test Unit (STU) were deployed but not retrieved; however, *Endeavour* did rendezvous three times with the satellite.
Experiments	SPACEHAB 04: This mission carried 12 experiments primarily involved in the materials and life sciences.
	• Advanced Separations (ADSEP)
	• Commercial Float Zone Furnace (CFZF)
	• CGBA
	• Commercial Vapor Diffusion Apparatus (CVDA)
	• FGBA
	• Hand Held-Diffusion Test Cells (HH-DTC)
	• IMMUNE
	• PCF
	• Protein Crystal Facility-Light Scattering and Temperature Controlled (PCF-LST)
	• Plant Growth Bioprocessing Apparatus (PGBA)
	• Space Experiment Facility (SEF)
	• GPPM

Table 3–101. STS-77 Mission Characteristics (Continued)

Experiments	TEAMS: Hitchhiker carrier had four experiments: • GPS Attitude and Navigation Experiment (GANE) • Vented Tank Resupply Experiment (VTRE) • Liquid Metal Thermal Experiment (LMTE) • PAMS
	Brilliant Eyes Ten-Kelvin Sorption Cryocooler Experiment (BETSCE): This experiment tested advanced sorption cooler techniques using hydrogen as a coolant.
	Aquatic Research Facility-01 (ARF-01): This facility supported life science research using a broad range of small aquatic species.
	BRIC-07: See STS-64.
	TPCE/RFL: This experiment obtained data required to develop the technology for pressure control of cryogenic tankage.
Get Away Specials	G-056 Customer: California Institute of Technology, Pasadena, California Gamma-ray Astrophysics Mission (GAMCIT): The GAMCIT studied gamma-ray bursts, an enigmatic source of cosmic radiation.
	G-142 and G-144 Customer: DARA Heat Transfer Phenomena (G-142) and Reaction Kinetics in Glass Melts (G-144): Autonomous Material Science Experiments Under Microgravity (MAUS): Germany offered scientists from disciplines of material research and processing the opportunity to perform material science investigations under microgravity conditions.
	G-063 Customer: Pennsylvania State University G-063 gave students first-hand experience at designing a self-contained space experiment. Once the payload returned from its flight, the students observed its results. The experiments included an accelerometer to measure the impact of orbital debris and a magnetometer to measure the magnetic fields. There was also an experiment to test the effect of a single event upset due to effects of cosmic radiation on semiconductors.
	G-163 Customer: Johnson Space Center Diffusion Coefficient Measurement Facility (DCMF): This facility measured the speed at which mercuric iodide (solid) evaporated and then was transported as a vapor under microgravity conditions.

Table 3–101. STS-77 Mission Characteristics (Continued)

Get Away Specials	G-200
	Customer: Utah State University
	Three experiments were flown in the canister. The payload also contained popcorn kernels in ziplock bags as part of an experiment by elementary school students. After return to Earth, students popped the popcorn and compared it to a similar sample maintained in Earth's gravity.
	G-490
	Customer: British Sugar PLC
	The experiment was designed and constructed by the School of Electronics and Electrical Engineering at the Robert Gordon University, Aberdeen, Scotland. British Sugar PLC sponsored the launch services. The first experiment was to verify a proposal that a low-level gravitational field could be measured by observing its effect on the convection currents present in a heated liquid. The second project was devised by a group of children from Elrick Primary School near Aberdeen, Scotland. A series of controlled experiments were carried out on selected samples of seeds, oats, wheat, barley, and nape-oil to quantify the effects of spaceflight on growth patterns.
	G-564 and G-565
	Customer: CSA
	Nanocrystal Get Away Special (NANO-GAS) and Atlantic Canada Thin Organic Semiconductors (ACTORS): The results of these experiments contributed to the development of new materials with applications in high-performance laser, electronic equipment, and components.
	G-703
	Customer: Lewis Research Center
	Microgravity Smoldering Combustion (MSC): This experiment studied the smolder characteristics of porous combustible materials in a microgravity environment.
	G-741
	Customer: Lewis Research Center
	Pool Boiling Experiment: This experiment was an extension of the study of the fundamentals of nucleate pool boiling heat transfer under the microgravity conditions of space.
Mission Results	Successful

[a] Jenkins, p. 310.

Table 3–102. STS-78 Mission Characteristics

Vehicle	OV-102
	Columbia
Crew	CDR: Terence T. Henricks
	PLT: Kevin R. Kregel
	MS: Richard M. Linnehan, Charles E. Brady, Jr.
	PC: Susan J. Helms
	PS: Jean-Jacques Favier (CNES), Robert Brent Thirsk (CSA)
Launch	June 20, 1996, 10:49:00 a.m. EDT, Kennedy Space Center, Pad 39-B. The liftoff proceeded on time. An in-cabin camera provided the first video images from the flight deck, beginning with crew ingress and continuing through main engine cutoff. Postlaunch assessment of spent solid rocket boosters revealed a hot gas path in motor field joints to, not past, the capture feature O-ring. This marked the first occurrence of combustion product penetration into the J-joint of the redesigned solid rocket motor. Flight safety was not compromised, and motor performance met design specification requirements. The probable cause was a new, more environmentally friendly adhesive and cleaning fluid.
Orbital Altitude & Inclination	150 nmi (278 km), 39.00 deg
Launch Weight (lb/kg)	256,145a/116,185
Landing & Postlanding Operations	July 7, 1996, 8:36:45 a.m. EDT, Runway 33, Kennedy Space Center. This mission had the first live downlink video during an orbiter's descent. After landing, Henricks and Kregel participated in the Olympic Torch Ceremony at Kennedy Space Center Visitors Center.
Rollout Distance (ft/m)	9,339/2,847
Rollout Time (seconds)	45
Mission Duration	405 hr, 47 min, 36 sec
Landed Revolution No.	271
Mission Support	STDN
Primary Objective	Experimentation using the LMS
Deployed Satellites	None

Table 3–102. STS-78 Mission Characteristics (Continued)

Experiments	LMS using the Spacelab long module: Five space agencies represented a complement of multinational experiments representing microgravity science and applications with emphasis on material and life science processing. These space agencies were: NASA, ESA, the French Space Agency, CSA, and the Italian Space Agency. Research scientists from 10 countries worked together on this payload, which made use of the Spacelab long module.

• Human Physiology Experiments: Musculoskeletal Investigations
   – Effects of Weightlessness on Human Single Muscle Fiber Function
   – Relationship of Long-Term Electromyographic Activity and Hormonal Function to Muscle Atrophy and Performance
   – Effects of Microgravity on Skeletal Muscle Contractile Properties
   – Effects of Microgravity on the Biomechanical and Bioenergetic Characteristics of Human Skeletal Muscle
   – Magnetic Resonance Imaging After Exposure to Microgravity (Ground Study)
   – An Approach to Counteract Impairment of Musculoskeletal Function in Space (Ground Study)
• Human Physiology Experiments: Metabolic Investigations
   – Direct Measurement of the Initial Bone Response to Spaceflight in Humans
   – Measurement of Energy Expenditures During Spaceflight Using the Doubly Labeled Water Method
• Human Physiology Experiments: Pulmonary Investigations
   – Extended Studies of Pulmonary Function in Weightlessness
• Human Physiology Experiments: Human Behavior and Performance Investigations
   – Human Sleep, Circadian Rhythms, and Performance in Space
   – Microgravity Effects on Standardized Cognitive Performance Measures
• Human Physiology Experiments: Neuroscience Investigations
   – Torso Rotation Experiment
   – Canal and Otolith Integration Studies (COIS)

Table 3–102. STS-78 Mission Characteristics (Continued)

Experiments	• Microgravity Science: BDPU
	– Bubbles and Drop Interaction with Solidification Front
	– Boiling on Small Plate Heaters Under Microgravity and a Comparison with Earth Gravity
	– A Liquid Electrohydrodynamics Experiment
	– Thermocapillary Convection in Multilayer Systems
	– Nonlinear Surface Tension-Driven Bubble Migration
	– Oscillatory Thermocapillary Instability
	– Thermocapillary Migration and Interactions of Bubbles and Drops
	• Microgravity Science: Advanced Gradient Heating Facility (AGHF)–Materials Processing
	– Directional Solidification of Refined Al–4 wt. % Cu Alloys
	– Coupled Growth in Hypermonotectics
	– Effects of Convection on Interface Curvature During Growth of Concentrated Tenary Compounds
	– Directional Solidification of Refined Al–1.5 wt.% Ni Alloys
	– Interactive Response of Advancing Phase Boundaries to Particles
	– Particle Engulfment and Pushing by Solidifying Interfaces
	• Microgravity Science: APCF–Medical Research
	– Crystallization of EGFR-EGF
	– Crystallization of Apocrustacyanin Cl
	– Crystallization and X-ray Analysis of 5S rRNA and the 5S rRNA Domain A
	– Growth of Lysozyme Crystals at Low Nucleation Density
	– Comparative Analysis of Aspartyl tRA-Synthetase and Thaumatin Crystals Grown on Earth and in Microgravity
	– Crystallization of the Nucleosome Core Particle
	– Crystallization of Photosystem I
	– Mechanism of Membrane Protein Crystal Growth: Bacteriorhodopsin-Mixed Micelle Packing at the Consolution Boundary, Stabilized in Microgravity
	– Crystallization in a Microgravity Environment of CcdB, a Protein Involved in the Control of Cell Death
	– Crystallization of Sulfolobus Solfataricus
	– Lysosome Crystal Growth in the Advanced Protein Crystallization Facility Monitored via Mach-Zehdner Interferometry and CCD Video
	– Analysis of Thaumatin Crystals Grown on Earth and in Microgravity

Table 3–102. STS-78 Mission Characteristics (Continued)

Experiments	• Microgravity Science: Accelerometers–Characterizing the Microgravity Environment – OARE – Microgravity Measurement Assembly (MMA) – SAMS • Space Biology: – Compression Wood Formation in a Microgravity Environment – Development of the Fish Medaka in Microgravity – Role of Corticosteroids in Bone Loss During Spaceflight SAREX-II: See STS-35. BRIC-07: See STS-64.
Get Away Specials	None
Mission Results	Successful
Remarks	*Columbia* was outfitted with Extended Duration Orbiter hardware for this mission.

^a Jenkins, p. 310.

Table 3–103. STS-79 Mission Characteristics

Vehicle	OV-104
	Atlantis
Crew	CDR: William F. Readdy
	PLT: Terence W. Wilcutt
	MS: Jerome Apt, Thomas D. Akers, Carl E. Walz, John E. Blaha (remained on *Mir*), Shannon W. Lucid (departed *Mir* for return to Earth)
Launch	September 16, 1996, 4:54:49 a.m. EDT, Kennedy Space Center, Pad 39-A. The launch, originally set for July 31, slipped when mission managers decided to switch out *Atlantis*'s twin solid rocket boosters. The new launch date of September 12 was targeted and *Atlantis* returned to the launch pad. The launch date was delayed to September 16 when the Shuttle was returned to the VAB due to the threat from Hurricane Fran. The countdown proceeded smoothly to an on-time liftoff September 16. Approximately 13 minutes into flight, auxiliary power unit No. 2 shut down prematurely. After review and analysis, the Mission Management Team concluded the mission could proceed to its nominal end-of-mission as planned.
Orbital Altitude & Inclination	170-213 nmi (315-394 km), 51.60 deg
Launch Weight (lb/kg)	249,328[a]/113,093
Landing & Postlanding Operations	September 26, 1996, 8:13:15 a.m. EDT, Runway 15, Kennedy Space Center. The landing went smoothly at first opportunity.
Rollout Distance (ft/m)	10,981/3,347
Rollout Time (seconds)	62
Mission Duration	243 hr, 18 min, 24 sec
Landed Revolution No.	159
Mission Support	STDN
Primary Objective	S/MM-04; experimentation using SPACEHAB-05
Deployed Satellites	None

Table 3–103. STS-79 Mission Characteristics (Continued)

Experiments	SPACEHAB-05: A double SPACEHAB module carried experiments in its forward portion that were conducted by the crew before, during, and after *Atlantis*'s docking to *Mir*. The aft portion of the module housed primarily the logistics equipment to be transferred to *Mir*.

• Experiments returning from *Mir*:
  – Environmental Radiation Measurements
  – Greenhouse-Integrated Plant Experiments
  – Human Life Sciences
  – Assessment of Humoral Immune Function During Long Duration Space Flight
• Experiments remaining on *Mir* for later retrieval:
  – BTS
  – MIDAS
  – CGBA
• Roundtrip experiments on *Atlantis*:
  – Extreme Temperature Translation Furnace (ETTF)
  – CPCG experiments
  – Mechanics of Granular Materials (MGM)
• Risk Mitigation Experiments:
  – Mated Shuttle and *Mir* Structural Dynamics Test (1301)
  – *Mir* Electric Field Characterization (1302)
  – Shuttle/*Mir* Experiment Kit Transport (1303)
  – Shuttle/*Mir* Alignment Stability Experiment (1310)
  – Real-Time Radiation Monitoring Device (RRMD: 1312)
  – Active Rack Isolation System (ARIS: 1313)
  – Inventory Management System (1319)

IMAX: Large format motion picture system photographed *Mir* during undocking and flyaround. The crews also used the camera to photograph *Mir* interior scenes.

Midcourse Space Experiment (MSX): The MSX obtained ultraviolet, infrared, and visible data of the Shuttle and Shuttle thrusters from an independent, space-based sensor satellite in a 99-degree orbit.

SAREX-II: See STS-35.

Get Away Specials	None
Mission Results	Successful

[a] Jenkins, p. 310.

Table 3–104. STS-80 Mission Characteristics

Vehicle	OV-102
	Columbia
Crew	CDR: Kenneth D. Cockrell
	PLT: Kent V. Rominger
	PC: Tamara E. Jernigan
	MS: Thomas D. Jones, F. Story Musgrave
Launch	November 19, 1996, 2:55:47 p.m. EST, Kennedy Space Center, Pad 39-B. Launch date of October 31 was first threatened by the changeout of the STS-79 boosters with those slated to fly on STS-80 and the delay of the STS-79 liftoff. Hurricane preparations because of Hurricane Fran in early September halted the STS-80 booster stacking operations in the VAB, prompting mission managers to reschedule the launch date to November 8. At the Flight Readiness Review (FRR) on October 28, mission managers declined to formalize the launch date pending an analysis of erosion in the STS-79 booster nozzles. At the FRR on November 4, the launch date was changed to no earlier than November 15 to allow engineers more time to complete their study of nozzle erosion. At the follow-up FRR on November 11, November 15 was set as the official launch date, pending a commercial Atlas launch on November 13, and the launch count began. The launch was postponed to November 19, due to the scrub of the Atlas launch and predicted bad weather in the Kennedy Space Center vicinity for a period of several days, and the count remained in an extended hold. The launch on November 19 occurred about 3 minutes after the scheduled opening of the launch window due to a hold at T-31 seconds to assess hydrogen concentrations in the aft engine compartment.
Orbital Altitude & Inclination	189 nmi (350 km), 28.45 deg
Launch Weight (lb/kg)	260,935[a]/118,358
Landing & Postlanding Operations	December 7, 1996, 6:49:05 a.m. EST, Runway 33, Kennedy Space Center. Landing was originally scheduled for December 5, but *Columbia* was waved off two days in a row due to weather conditions in Florida.
Rollout Distance (ft/m)	8,721/2,658
Rollout Time (seconds)	62
Mission Duration	423 hr, 53 min, 19 sec
Landed Revolution No.	277
Mission Support	STDN

Table 3–104. STS-80 Mission Characteristics (Continued)

Primary Objective	Deployment and retrieval of ORFEUS-SPAS-2 and WSF-3
Deployed Satellites	Deployed and retrieved ORFEUS-SPAS-2 and WSF-3
Experiments	NIH-R4: This experiment studied blood pressure regulation and function in rats fed either a high-calcium or a low-calcium diet before and during spaceflight.
	CCM-A (formerly STL/NIH-C-05): This experiment continued the investigation into how microgravity affected bones at the cellular level.
	BRIC-09: See STS-64.
	CMIX-5: See STS-52.
	Visualization in an Experimental Water Capillary Pumped Loop (VIEW-CPL): This experiment collected and transported excess heat generated by spacecraft instruments to a spacecraft radiator for ejection into space. The experiment was used to develop a more complete understanding of CPL physics in a microgravity environment by viewing the fluid flow inside the evaporator.
Get Away Specials	None

SEM	SEM-01 experiments were:
	• Charleston, South Carolina, school district (CAN-DO):
	  – Gravity and Acceleration Readings
	  – Bacteria-Agar Research Instrument
	  – Crystal Research in Space
	  – Magnetic Attraction Viewed in Space
	• Purdue University, West Lafayette, Indiana:
	  – Fluid Thermal Convection
	  – NADH Oxidase Absorbance in Shrimp
	  – Passive Particle Detector Experiment
	• Hampton Elementary School, Lutherville, Maryland:
	  – Experimented with seeds; soil; chalk; crayon; calcite; Silly Putty; bubble solution; popcorn; mosquito eggs; and other organic compounds
	• Glenbrook North High School, Northbrook, Illinois:
	  – Surface Tension Experiment
	• Albion Jr. High, Strongville, Ohio:
	  – Heat Transfer Experiment–studied heating properties of copper tubes and pennies
	• Poquoson Middle School, Poquoson, Virginia:
	  – Bacteria Inoculation in Space Experiment
	• Norfolk Public Schools Science and Technology Advanced Research (NORSTAR):
	  – Observed the behavior of immiscible fluids

Table 3–104. STS-80 Mission Characteristics (Continued)

Mission Results	Two EVAs by Jernigan and Jones planned to evaluate equipment and procedures to be used during construction and maintenance of the ISS were canceled because the crew could not open the outer airlock hatch. Crew and mission troubleshooting did not reveal the cause, so mission managers concluded it would be unwise to attempt the two EVAs and risk unnecessary damage to the hatch or seals. Postlanding assessment of the hatch indicated that a small screw had become loose from an internal assembly and lodged in an actuator, a gearbox mechanism that operated linkages securing the hatch, preventing the crew from opening the hatch. Other mission objectives were achieved.
Remarks	At age 61, Musgrave became the oldest human to fly in space. He set a new record for the most Shuttle flights (six) and tied astronaut John Young's record for the most total spaceflights.

[a] Jenkins, p. 310.

Table 3–105. STS-81 Mission Characteristics

Vehicle	OV-104
	Atlantis
Crew	CDR: Michael A. Baker
	PLT: Brent W. Jett, Jr.
	MS: Peter J.K. Wisoff, John M. Grunsfeld, Marsha S. Ivins, Jerry M. Linenger (remained on *Mir*), John E. Blaha (departed *Mir* for return to Earth)
Launch	January 12, 1997, 4:27:23 a.m. EST, Kennedy Space Center, Pad 39-B. The liftoff occurred on time following a smooth countdown.
Orbital Altitude & Inclination	160 nmi (296 km), 51.60 deg
Launch Weight (lb/kg)	249,936[a]/113,369
Landing & Postlanding Operations	January 22, 1997, 9:22:44 a.m. EST, Runway 33, Kennedy Space Center. The first landing opportunity was waved off due to weather.
Rollout Distance (ft/m)	9,350/2,850
Rollout Time (seconds)	69
Mission Duration	244 hr, 55 min, 23 sec
Landed Revolution No.	160
Mission Support	STDN
Primary Objective	S/MM-05; experimentation using SPACEHAB Double Module
Deployed Satellites	None
Experiments	SPACEHAB Double Module: The double module carried the following experiments: • Environmental Radiation Measurements • Greenhouse-Integrated Plant Experiments • Human Life Sciences • Assessment of Humoral Immune Function During Long Duration Space Flight • Diffusion-Controlled Crystallization Apparatus for Microgravity • Gaseous Nitrogen Dewar • Liquid Metal Diffusion • Optical Properties Monitor
	CREAM: See STS-48.
	KidSat: Provided students access to real-time images of Earth by uplinking commands to the Electronic Still Camera to photograph specific land areas. The images were downlinked in real time via the Ku-band Communication Adapter.

Table 3–105. STS-81 Mission Characteristics (Continued)

Experiments	SAMS: The SAMS provided acceleration data to characterize the experiments acceleration environment on *Mir.* See STS-43.
	MSX: The objective of the MSX was to fire the orbiter thrusters (orbital maneuvering and primary reaction control systems) in space and use the sophisticated sensors of the orbiting MSX satellite to collect ultraviolet, infrared, and visible light data of the event.
Get Away Specials	None
Mission Results	Successful
Remarks	This mission was the largest transfer to date of logistics between the two spacecraft.

[a] Jenkins, p. 310.

Table 3–106. STS-82 Mission Characteristics

Vehicle	OV-103 *Discovery*
Crew	CDR: Kenneth D. Bowersox PLT: Scott J. Horowitz MS: Joseph R. Tanner, Steven A. Hawley, Gregory J. Harbaugh, Steven L. Smith PC: Mark C. Lee
Launch	February 11, 1997, 3:55:17 a.m. EST, Kennedy Space Center, Pad 39-A. The launch originally targeted for February 13 was moved up to February 11 to provide more range opportunities. Countdown proceeded smoothly to an on-time liftoff on February 11. This was the first flight after *Discovery*'s OMDP.
Orbital Altitude & Inclination	313 nmi (579 km), 28.45 deg
Launch Weight (lb/kg)	251,238[a]/113,960
Landing & Postlanding Operations	February 21, 1997, 3:32:26 a.m. EST, Runway 15, Kennedy Space Center. The orbiter landed on the second opportunity after the first was waved off due to low clouds.
Rollout Distance (ft/m)	7,066/2,154
Rollout Time (seconds)	60
Mission Duration	239 hr, 37 min, 7 sec
Landed Revolution No.	149
Mission Support	STDN
Primary Objective	Second Hubble Space Telescope Servicing Mission
Deployed Satellites	Retrieved, serviced, and redeployed the Hubble Space Telescope
Experiments	MSX: See STS-81.
Get Away Specials	None
Mission Results	Successful
Remarks	There were five EVAs: four scheduled and one unscheduled. EVA No. 1: Performed by Lee and Smith, the first EVA lasted 6 hours, 42 minutes. Lee and Smith removed the GHRS and FOS from the telescope and replaced them with the STIS and NICMOS.

Table 3–106. STS-82 Mission Characteristics (Continued)

Remarks	EVA No. 2: Performed by Harbaugh and Tanner, the second EVA lasted 7 hours, 27 minutes. Harbaugh and Tanner replaced a degraded fine guidance sensor and a failed Engineering and Science Tape Recorder with new spares. They also installed a new unit called the Optical Control Electronics Enhancement Kit, which increased the capability of the fine guidance sensor. The astronauts noted cracking and wear on thermal insulation on the side of the telescope facing the Sun and in the direction of travel.
	EVA No. 3: Performed by Lee and Smith, the third EVA lasted 7 hours, 11 minutes. Lee and Smith removed and replaced a Data Interface Unit on the Hubble Space Telescope as well as an old reel-to-reel Engineering and Science Tape Recorder with a new digital solid state recorder (SSR) that allowed simultaneous recording and playback of data. The astronauts also changed out one of four reaction wheel assembly units that used spin momentum to move the telescope toward a target and maintain it in a stable position.
	EVA No. 4: Performed by Harbaugh and Tanner, the fourth EVA lasted 6 hours, 34 minutes. Harbaugh and Tanner replaced a solar array drive electronics package, which controlled the positioning of the Hubble Space Telescope's solar arrays. The astronauts also replaced covers over the telescope's magnetometers and placed thermal blankets of multilayer material over two areas of degraded insulation around the light shield portion of the telescope just below the top of the observatory.
	EVA No. 5: Performed by Lee and Smith, the fifth EVA lasted 5 hours, 17 minutes. Lee and Smith attached several thermal insulation blankets to three equipment compartments at the top of the Support Systems Module section of the telescope that contained key data processing, electronics, and scientific instrument telemetry packages.

[a] Jenkins, p. 310.

Table 3–107. STS-83 Mission Characteristics

Vehicle	OV-102
	Columbia
Crew	CDR: James D. Halsell, Jr.
	PLT: Susan L. Still
	PC: Janice E. Voss
	MS: Michael L. Gernhardt, Donald A. Thomas
	PS: Roger K. Crouch, Gregory T. Linteris
Launch	April 4, 1997, 2:20:32 p.m. EST, Kennedy Space Center, Pad 39-A. The launch originally was set for April 3. It was initially delayed 24 hours on April 1 due to a requirement to add additional thermal insulation to a water coolant line in the orbiter's payload bay. Mission managers determined that the coolant line, which cooled various electronics on the orbiter, was not properly insulated and could possibly freeze on-orbit. On April 4, liftoff was delayed 20 minutes, 32 seconds due to an orbiter access hatch seal that had to be replaced.
Orbital Altitude & Inclination	160 nmi (296 km), 28.45 deg
Launch Weight (lb/kg)	259,144[a]/117,546
Landing & Postlanding Operations	April 8, 1997, 2:33:11 p.m. EDT, Runway 33, Kennedy Space Center. The landing was originally scheduled for April 19, but the mission was cut short due to problems with *Columbia*'s fuel cell No. 2.
Rollout Distance (ft/m)	8,602/2,622
Rollout Time (seconds)	59
Mission Duration	95 hr, 12 min, 39 sec
Landed Revolution No.	63
Mission Support	STDN
Primary Objective	Research using the MSL-1
Deployed Satellites	None
Experiments	MSL-1: The MSL-1 housed a collection of microgravity experiments inside a Spacelab long module. The laboratory featured facilities for material science investigations. Due to an early return because of problems with fuel cell No. 2, only a few experiments were conducted.
	• TEMPUS:
	– Thermophysical Properties of Undercooled Metallic Melts
	• Large Isothermal Furnace:
	– Liquid-Phase Sintering II
	• Combustion Module-1 (CM-1):
	– Laminar Soot Processes
	– Structure of Flame Balls at Low Lewis-number (SOFBALL)

Table 3–107. STS-83 Mission Characteristics (Continued)

Get Away Specials	None
Mission Results	The MSL-1 mission was cut short due to concerns about one of the three fuel cells. Fuel cell No. 2 had shown some erratic readings during prelaunch startup but was cleared to fly after additional checkout and test. Shortly after on-orbit operations began, the differential voltage in the No. 3 substack of fuel cell No. 2 began to rise. Shuttle flight rules required all three to be functioning properly to ensure crew safety and provide sufficient backup capability during reentry and landing. A decision was made after landing to refly the entire mission on STS-94. It was the first mission to end early since STS-44 in 1991.
Remarks	*Columbia* was outfitted with extended duration orbiter hardware for the flight.

[a] Jenkins, p. 310.

Table 3–108. STS-84 Mission Characteristics

Vehicle	OV-104 *Atlantis*
Crew	CDR: Charles J. Precourt
	PLT: Eileen M. Collins
	PC: Jean-Francois Clervoy (ESA)
	MS: Carlos I. Noriega, Edward Tsang Lu, Elena V. Kondakova (RSA), C. Michael Foale (remained on *Mir*), Jerry M. Linenger (departed *Mir* for return to Earth)
Launch	May 15, 1997, 4:07:48 a.m. EDT, Kennedy Space Center, Pad 39-A. The liftoff occurred on time following a smooth countdown.
Orbital Altitude & Inclination	160 nmi (296 km), 51.60 deg
Launch Weight (lb/kg)	249,462[a]/113,154
Landing & Postlanding Operations	May 24, 1997, 9:27:44 a.m. EDT, Runway 33, Kennedy Space Center. The orbiter landed on the second opportunity after being waved off from the first due to low clouds in the vicinity.
Rollout Distance (ft/m)	8,384/2,555
Rollout Time (seconds)	51
Mission Duration	221 hr, 19 min, 56 sec
Landed Revolution No.	143
Mission Support	STDN
Primary Objective	S/MM-06; experimentation using SPACEHAB
Deployed Satellites	None
Experiments	SPACEHAB Double Module: Double module carrying the following experiments: • Environmental Radiation Measurements • Greenhouse-Integrated Plant Experiments • Human Life Sciences Project • Protein Crystal Growth Experiments • Diffusion-Controlled Crystallization Apparatus for Microgravity (DCAM) • Gaseous Nitrogen Dewar (GND) • VDA-2 • Morphological Transition and Model Substances (MOMO)

Table 3–108. STS-84 Mission Characteristics (Continued)

Experiments	• Biorack: This multipurpose unit contained these experiments:
	   – Cytoskeleton of the Lentil Root Statocyte
	   – Morphology and Physiology of Loxodes After Cultivation in Space
	   – Lymphocyte and Monocyte Intra-Cellular Signal Transduction in Microgravity
	   – Microgravity Effects on Bone Cell Gene Expression
	   – Microgravity and Signal Transduction Pathways in Sea Urchin Sperm
	   – Graviperception in Starch-Deficient Plants
	
	PCG-STES: See STS-67.
	
	Liquid Motion Experiment (LME): The LME investigated inertia wave oscillations of liquids in tanks spinning around an exterior axis that was nutating under microgravity conditions.
	
	CREAM: See STS-48.
	
	EPICS: See STS-69.
	
	RME-III: See STS-28.
	
	Shuttle Ionospheric Modification with Pulsed Local Exhaust (SIMPLEX): The orbiter orbital maneuvering system thruster firings were used to create ionospheric disturbances for observation by the SIMPLEX radar sites.
	
	MSX: See STS-79.

Get Away Specials	None
Mission Results	Successful
Remarks	Linenger's 123-day stay on *Mir* and 132 days in space placed him second behind Shannon Lucid for the most time spent on-orbit by an American. Another milestone reached during his stay was the one-year anniversary of a continuous U.S. presence in space that began with Lucid's arrival at *Mir* on March 22, 1996.

[a] Jenkins, p. 310.

Table 3–109. STS-94 Mission Characteristics

Vehicle	OV-102 *Columbia*
Crew	CDR: James D. Halsell, Jr. PLT: Susan L. Still PC: Janice E. Voss MS: Michael L. Gernhardt, Donald A. Thomas PS: Roger K. Crouch, Gregory T. Linteris
Launch	July 1, 1997, 2:02:00 p.m. EDT, Kennedy Space Center, Pad 39-A. The liftoff was delayed about 12 minutes because of unacceptable weather conditions in the launch area in the event a return-to-launch site abort was necessary. The launch window originally was targeted to open at 2:37 p.m., July 1. On June 20, NASA managers decided to move the launch back 47 minutes to 1:50 p.m. to avoid forecasted afternoon thundershowers.
Orbital Altitude & Inclination	160 nmi (296 km), 28.45 deg
Launch Weight (lb/kg)	260,249[a]/118,047
Landing & Postlanding Operations	July 17, 1997, 6:46:34 a.m. EDT, Runway 33, Kennedy Space Center. The landing occurred at the first opportunity.
Rollout Distance (ft/m)	8,892/2,710
Rollout Time (seconds)	55
Mission Duration	370 hr, 44 min, 36 sec
Landed Revolution No.	250
Mission Support	STDN
Primary Objective	Reflight of the MSL-1
Deployed Satellites	None
Experiments	Reflight of STS-83 MSL-1: MSL-1 housed a collection of microgravity experiments inside a Spacelab long module. The laboratory featured material science investigations. The facilities and their experiments were: • LIF: – Measurement of Diffusion Coefficient by Shear Cell Method – Diffusion of Liquid Metals – Diffusion in Liquid Lead-Tin-Telluride – Impurity Diffusion in Ionic Metals – Liquid Phase Sintering II – Diffusion Processes in Molten Semiconductors

Table 3–109. STS-94 Mission Characteristics (Continued)

Experiments	• Expedite the Processing of Experiments to the Space Station Rack (EXPRESS):
	– Physics of Hard Sphere Experiment
	– Astro/Plant Generic Bioprocessing Apparatus (AstroPGBA)
	• TEMPUS:
	– Thermophysical Properties of Undercooled Metallic Melts
	– Thermophysical Properties of Advance Materials in the Undercooled Liquid State
	– Measurement of the Surface Tension of Liquid and Undercooled Metallic Melts by Oscillating Drop Technique
	– Study of the Morphological Stability of Growing Dendrites by Comparative Dendrite Velocity Measurements on Pure Ni and a Dilute Ni-C Alloy in the Earth and Space Laboratory
	– Undercooled Melts of Alloys with Polytetrahedral Short-Range Order
	– Thermal Expansion of Glass Forming Metallic Alloys in the Undercooled State
	– Experiments on Nucleation in Different Flow Regimes
	– Alloy Undercooling Experiments
	– Measurement of Surface Tension and Viscosity of Undercooled Liquid Metals
	– AC Calorimetry and Thermophysical Properties of Bulk Glass-Forming Metallic Liquids
	• CM-1:
	– Laminar Soot Processes
	– SOFBALL
	• Droplet Combustion Apparatus:
	– Droplet Combustion Experiment
	– Fiber-Supported Droplet Combustion
	• Middeck Glove Box:
	– Coarsening in Solid-Liquid Mixtures
	– Bubble and Drop Nonlinear Dynamics
	– A Study of Fundamental Operation of a Capillary-Driven Heat Transfer (CHT) Device in Microgravity
	– Internal Flows in a Free Drop
	• Protein Crystallization Apparatus:
	– Protein Crystallization Apparatus for Microgravity
	– Second Generation Vapor Diffusion Apparatus
	– Handheld Diffusion Test Cells
	• Measuring Microgravity:
	– SAMS
	– QSAMS
	– OARE
	– MMA

Table 3–109. STS-94 Mission Characteristics (Continued)

Experiments	Cryogenic Flexible Diode Experiment (CRYOFD): This experiment determined the behavior of cryogenic two-phase thermal control components in microgravity; demonstrated oxygen and methane heat pipe startups from a super-critical condition, demonstrated operations; verified analytical performance models; and established the correlation between 1g and microgravity thermal performance. A secondary objective was to validate the performance of an American Loop Heat Pipe with Ammonia (ALPHA).
	SAREX-II: See STS-35.
	MSX: See STS-79.
Get Away Specials	None
Mission Results	Successful
Remarks	*Columbia* was outfitted with extended duration orbiter hardware for the flight.
	The mission was the first reflight of the same vehicle, crew, and payloads. It also was the first reservicing of a primary payload, MSL-1, in the orbiter.

[a]   Jenkins, p. 310.

Table 3–110. STS-85 Mission Characteristics

Vehicle	OV-103
	Discovery
Crew	CDR: Curtis L. Brown, Jr.
	PLT: Kent V. Rominger
	PC: N. Jan Davis
	MS: Robert L. Curbeam, Jr., Stephen K. Robinson
	PS: Bjarni V. Tryggvason (CSA)
Launch	August 7, 1997, 10:41:00 a.m. EDT, Kennedy Space Center, Pad 39-A. The liftoff was on time following a smooth countdown.
Orbital Altitude & Inclination	173 nmi (320 km), 51.6 deg
Launch Weight (lb/kg)	249,696[a]/113,260
Landing & Postlanding Operations	August 19, 1997, 7:07:59 a.m. EDT, Runway 33, Kennedy Space Center. A landing opportunity on August 18 was waved off due to threat of ground fog in the local area.
Rollout Distance (ft/m)	8,792/2,680
Rollout Time (seconds)	68
Mission Duration	284 hr, 27 min, 00 sec
Landed Revolution No.	189
Mission Support	STDN
Primary Objective	Deployment and retrieval of CRISTA-SPAS-2
Deployed Satellites	Deployed and retrieved CRISTA-SPAS-2
Experiments	Technology Applications and Science (TAS-1): The overall objective was to fly more science experiments using better, faster, and cheaper avionics and processes. This Hitchhiker payload carried the following experiments:
	• SOLCON
	• Infrared Spectral Imaging Radiometer (ISIR)
	• Shuttle Laser Altimeter (SLA)
	• Critical Viscosity of Xenon (CVX)
	• SEM
	• Two Phase Flow (TPF)
	• Cryogenic Flight Experiment (CFE)
	• Stand Alone Acceleration Measurement Device and the Wide Band Stand Alone Acceleration Measurement Device (SAAMD/WBSAAMD)

Table 3–110. STS-85 Mission Characteristics (Continued)

Experiments	Manipulator Flight Demonstration (MFD): Sponsored by NASDA, this experiment evaluated the use of the Small Fine Arm planned to be part of the Japanese Experiment Module's remote manipulator system on the ISS. It also included two other experiments: • Two-Phase Fluid Loop Experiment (TPFLEX) • Evaluation of Space Environment and Effects on Materials (ESEM)
	EH-2: The IEH-2 consisted of four experiments with the common objective of investigating the uncertainty and long-term variation in the absolute solar extreme ultraviolet (EUV) flux and EUV emissions of the Jupiter Io plasma torus system. The experiments were: • SEH-2 • UVSTAR • Distribution and Automation Technology Advancement–Colorado Hitchhiker and Student Experiment of Solar Radiation (DATA-CHASER) • GLO-5 and GLO-6
	BDS-3: See STS-70.
	PCG-STES: See STS-67.
	MSX: See STS-79.
	SIMPLEX: See STS-84.
	Southwest Ultraviolet Imaging System (SWUIS): This imaging system was used primarily to view the Hale-Bopp comet. The SWUIS also performed ultraviolet astronomy; planetary and cometary imaging; terrestrial airglow and atmospheric background imaging; auroral imaging; and studied Shuttle glow and vehicle plume evaluations.
	BRIC-10: See STS-64.
	SSCE: See STS-41.
Get Away Specials	G-572 Customer: Bellarmine College, University of Utah, Utah State University Hearts in Space: This experiment investigated the effect of weightlessness on physical factors contributing to cardiac function.
	G-745 Customer: Students and Teachers of Mayo High School in Rochester, Minnesota This experiment investigated root growth during a Shuttle mission.

Table 3–110. STS-85 Mission Characteristics (Continued)

Space Experiment Module	CAN-DO, Charleston, South Carolina Several active experiments flew within CAN-DO's single module, including the following: a study of a revival of a Polypodium Polypodioides plant; the measurement of radiation of the internal environment of the module; the recording of "sounds" within the module with a cassette recorder and vibration sensor; and the observation of the dispersion of paint in microgravity. Also, the module included passive and active experiments from other school districts.
Mission Results	Successful

[a] Jenkins, p. 310.

Table 3–111. STS-86 Mission Characteristics

Vehicle	OV-104
	Atlantis
Crew	CDR: James D. Wetherbee
	PLT: Michael J. Bloomfield
	MS: Vladimir G. Titov (RSA), Scott E. Parazynski, Jean-Loup J.M. Chretien (French Air Force), Wendy B. Lawrence, David A. Wolf (remained on *Mir*), C. Michael Foale (departed *Mir* for return to Earth)
Launch	September 25, 1997, 10:34:19 p.m. EDT, Kennedy Space Center, Pad 39-A. On-time liftoff occurred after final approval for flight to *Mir* given earlier in day by NASA Administrator Goldin following his review of independent and internal safety assessments regarding safety of *Mir* and Shuttle-*Mir* docking and two independent studies that were prompted by numerous problems on the station, including a fire and a collision (see discussion of the *Mir* program later in this chapter).
Orbital Altitude & Inclination	160 nmi (296 km), 51.60 deg
Launch Weight (lb/kg)	252,035[a]/114,321
Landing & Postlanding Operations	October 6, 1997, 5:55:09 p.m. EDT, Runway 15, Kennedy Space Center. The Shuttle landed on the first opportunity after two opportunities on October 5 were waved off due to low clouds.
Rollout Distance (ft/m)	11,947/3,641
Rollout Time (seconds)	82
Mission Duration	259 hr, 20 min, 53 sec
Landed Revolution No.	169
Mission Support	STDN
Primary Objective	Seventh Shuttle-*Mir* docking; experimentation using SPACEHAB
Deployed Satellites	None
Experiments	SPACEHAB: This double module carried the following experiments: • Human Life Sciences • ISS Risk Mitigation • Interferometer To Study Protein Crystal Growth (IPCG) • Canadian Protein Crystallization Experiment (CAPE)
	MEEP: This *Mir* payload gathered data on human-made and natural space debris, capturing some debris for later study.

Table 3–111. STS-86 Mission Characteristics (Continued)

Experiments	SEEDS-II: This experiment passively exposed a group of tomato seeds in hand-sewn Dacron bags to the vacuum of space. Seeds flown in the payload were compared with a control group of seeds and an experimental group of seeds in an underwater habitat in Key Largo, Florida. After completion of the mission, the seeds were distributed to schools for education and outreach purposes.
	Kidsat: The Kidsat used an electronic still camera aboard the Shuttle to bring the frontiers of space exploration to a growing number of U.S. middle school classrooms via the Internet.
	CPCG: See STS-49.
	CREAM: See STS-48.
	CCM-A: See STS-80.
	SIMPLEX: See STS-84.
	ESA's European Laser Docking System: This system monitored the Shuttle's approach and departure from *Mir* using GPS receivers and optical rendezvous sensors.
Get Away Specials	None
Mission Results	Successful

[a] Jenkins, p. 310.

Table 3–112. STS-87 Mission Characteristics

Vehicle	OV-102
	Columbia
Crew	CDR: Kevin R. Kregel
	PLT: Steven W. Lindsey
	MS: Kalpana Chawla, Winston E. Scott, Takao Doi (JAXA)
	PS: Leonid K. Kadenyuk (National Space Agency of Ukraine)
Launch	November 19, 1997, 2:46:00 p.m. EST, Kennedy Space Center, Pad 39-B. The liftoff proceeded on time. This was the first use of Pad 39-B since January following extensive modifications to the pad structure.
Orbital Altitude & Inclination	150 nmi (278 km), 28.45 deg
Launch Weight (lb/kg)	260,799[a]/118,296
Landing & Postlanding Operations	December 5, 1997, 7:20:04 a.m. EST, Runway 33, Kennedy Space Center. The Shuttle landed on the first landing opportunity.
Rollout Distance (ft/m)	8,004/2,440
Rollout Time (seconds)	57
Mission Duration	376 hr, 34 min, 2 sec
Landed Revolution No.	251
Mission Support	STDN
Primary Objective	Deployment and retrieval of SPARTAN 201-04; experimentation using the USMP-4
Deployed Satellites	Deployed and retrieved SPARTAN 201-04
Experiments	USMP-4: This payload conducted research in the areas of materials science, combustion science, and fundamental physics.
	USMP experiments operating without crew involvement included: • AADSF • Confined Helium Experiment (CheX) • IDGE • MEPHISTO • SAMS • OARE
	Experiments housed in the MGBX requiring crew involvement: • Enclosed Laminar Flames (ELF) • Wetting Characteristics of Immiscibles (WCI) • Particle Engulfment and Pushing by a Solid/Liquid Interface (PEP)

Table 3–112. STS-87 Mission Characteristics (Continued)

Experiments	Collaborative Ukrainian Experiment (CUE): The CUE was a collection of 10 plant space biology experiments that evaluated the effects of microgravity on pollination and fertilization of *Brassica rapa* (Wisconsin Fast Plants). The experiment also compared change in ultrastructure, biochemical composition, and function induced by the spaceflight environment on the photosynthetic apparatus of *Brassica rapa* seedlings at different stages of vegetative development.
	Teachers and Students Investigating Plants in Space (CUE-TSIPS): High school students in the United States and Ukraine performed special plant biology science experiments while viewing interactive downlinks of Payload Specialist Kadenyuk and U.S. astronauts conducting the same experiments in microgravity.
	Shuttle Ozone Limb Sounding Experiment/Limb Ozone Retrieval Experiment (SOLSE/LORE): A Hitchhiker payload, this experiment generated overall ozone coverage images and cross sections of the atmosphere showing ozone concentrations at different altitudes.
	Loop Heat Pipe/Sodium Sulfur Battery Experiment (LHP/NaSBE): A Hitchhiker payload, this experiment investigated a unique thermal energy management system using a loop heat pipe and studied the microgravity operation of sodium and sulfur liquid electrodes.
	Turbulent Gas-Jet Diffusion Flames (TGDF): A Hitchhiker payload, this experiment used a GAS canister to gain further knowledge of the characteristics of transitional turbulent gas-jet diffusion flames.
Get Away Specials	G-036 Customer: El Paso (Texas) Community College and Goddard Space Flight Center G-036 contained four experiments: • Cement Mixing Experiment (CME): Cement samples were mixed with water and then compared with others produced on Earth to analyze the effects of microgravity on the combination of cement and water. • Configuration Stability of Fluid Experiment (CSFE): The CSFE investigated the effects of microgravity on the configuration stability of a two-phase fluid system. • Computer (Compact) Disc Evaluation Experiment (CDEE): The CDEE investigated the effects of the exosphere, the outer fringe region of the atmosphere of a planet, on the ability of discs to retain their information. • Asphalt Evaluation Experiment (AEE): The AEE explored the effects of exposure to the exosphere on asphalt.

Table 3–112. STS-87 Mission Characteristics (Continued)

Mission Results	A malfunction of SPARTAN's attitude control system caused the rotational spin of about two degrees per second after attempts to regrapple the satellite. Planned research on SPARTAN was not performed. Other mission objectives were successfully achieved.
Remarks	To retrieve SPARTAN, Winston Scott and Takao Doi began a 7-hour, 43-minute EVA. They captured SPARTAN by hand and then completed a series of activities continuing preparations for on-orbit assembly of the ISS. Doi became the first Japanese citizen to walk in space.

[a] Jenkins, p. 310.

Table 3–113. STS-89 Mission Characteristics

Vehicle	OV-105
	Endeavour
Crew	CDR: Terrence W. Wilcutt
	PLT: Joe Frank Edwards, Jr.
	MS: James F. Reilly, Michael P. Anderson, Salizhan Shakirovich Sharipov (RSA), Andrew S.W. Thomas (remained on *Mir*), David A. Wolf (departed *Mir* for return to Earth)
	PC: Bonnie J. Dunbar
Launch	January 22, 1998, 9:48:15 p.m. EST, Kennedy Space Center, Pad 39-A. The launch originally targeted for January 15, 1998, was changed first to no earlier than January 20 and then to January 22 per request from the Russian Space Agency (RSA) to allow completion of activities on *Mir*. *Endeavour* returned to the Shuttle fleet after completing its first OMDP. *Endeavour* was the second orbiter to dock with *Mir*.
Orbital Altitude & Inclination	150 nmi (279 km), 51.60 deg
Launch Weight (lb/kg)	252,316[a]/114,449
Landing & Postlanding Operations	January 31, 1998, 5:35:09 p.m. EST, Runway 15, Kennedy Space Center. The Shuttle landed on the first opportunity.
Rollout Distance (ft/m)	9,790/2,983
Rollout Time (seconds)	70
Mission Duration	211 hr, 46 min, 55 sec
Landed Revolution No.	138
Mission Support	STDN
Primary Objective	S/MM-08; experimentation using SPACEHAB
Deployed Satellites	None
Experiments	The SPACEHAB double module carried the following experiments: • Mechanics of Granular Materials • ASTROCULTURE™ • X-Ray Detector Test • DCAM • Gaseous Nitrogen Dewar
	Closed Equilibrated Biological Aquatic System (CEBAS): The CEBAS mini-module was a habitat for aquatic organisms. The CEBAS conducted various gravity-related experiments in zoology, botany, and developmental biology, and interdisciplinary areas such as scientific research on artificial ecosystems.

Table 3–113. STS-89 Mission Characteristics (Continued)

Experiments	Microgravity Plant Nutrient Experiment (MPNE): This experiment tested nutrient delivery technology that would support plant growth in space.
	EarthKAM: Students from 51 middle schools in three nations operated a digital camera mounted in the overhead window of the Shuttle, selecting sites around the world to photograph during the Shuttle flight.
Get Away Specials	G-093 Customer: University of Michigan Vortex Ring Transit Experiment (VORTEX): The VORTEX investigated the propagation of a vortex ring through a liquid-gas interface in microgravity.
	G-141 Customer: German Aerospace Center and the University of Giessen, Germany Structure of Marangoni Convection in Floating Zones: Marangoni convection was studied without disturbances in microgravity.
	G-145 Customer: German Aerospace Center and Technical University of Clausthal, Germany Glass Fining: G-145 studied the process of glass fining or the removal of all visible gaseous bubbles from a glass melt.
	G-432 Customer: Chinese Academy of Sciences, Beijing, China G-432 consisted of five experiments: • Super Cooling • Processing of High Critical Test • Growth of Gallium Antimony Experiment • Liquid Phase Epitaxy • Wetability Test
Mission Results	Successful

[a] Jenkins, p. 310.

Table 3–114. STS-90 Mission Characteristics

Vehicle	OV-102
	Columbia
Crew	CDR: Richard A. Searfoss
	PLT: Scott D. Altman
	PC: Richard M. Linnehan
	MS: Kathryn P. Hire, Dafydd (Dave) Rhys Williams (CSA)
	PS: Jay C. Buckey, James A. Pawelczyk
Launch	April 17, 1998, 2:19:00 p.m. EDT, Kennedy Space Center, Pad 39-B. The launch was postponed on April 16 for 24 hours due to difficulty with one of *Columbia*'s two network signal processors that format data and voice communications between the ground and the Space Shuttle. Network signal processor No. 2 was replaced, and the liftoff on April 17 occurred on time.
Orbital Altitude & Inclination	150 nmi (279 km), 39.00 deg
Launch Weight (lb/kg)	262,357[a]/119,003
Landing & Postlanding Operations	May 3, 1998, 12:08:59 p.m. EDT, Runway 33, Kennedy Space Center. The Shuttle landed on the first opportunity.
Rollout Distance (ft/m)	9,998/3,047
Rollout Time (seconds)	58
Mission Duration	381 hr, 49 min, 58 sec
Landed Revolution No.	255
Mission Support	STDN
Primary Objective	Conduct final Spacelab mission: Neurolab
Deployed Satellites	None
Experiments	Neurolab: The Neurolab, dedicated to study of life sciences, focused on the most complex and least understood part of the human body: the nervous system. The crew served as both experiment subjects and operators. Other subjects included rats, mice, crickets, snails, and two kinds of fish. The Neurolab teams performed the following experiments:
	• Autonomic Nervous System Team:
	– Artificial Neural Networks and Cardiovascular Regulation
	– Integration of Neural Cardiovascular Control in Space
	– Autonomic Neuroplasticity in Weightlessness
	– Autonomic Neurophysiology in Microgravity

Table 3–114. STS-90 Mission Characteristics (Continued)

Experiments	• Sensory Motor Performance Team:
	– Frames of Reference and Internal Models
	– Visuo-Motor Coordination During Spaceflight
	– Role of Visual Cues in Spatial Orientation
	• Vestibular Team:
	– Visual-Otolithic Interactions in Microgravity
	– Spatial Orientation of the Vestibulo-Ocular Reflex
	• Sleep Team:
	– Sleep and Respiration in Microgravity
	– Clinical Trial of Melatonin as Hypnotic for Neurolab Crew
	• Mammalian Development Team:
	– Neuro-Thyroid Interaction on Skeletal Isomyosin Expression in Zero Gravity
	– Neuronal Development Under Conditions of Spaceflight
	– Reduced Gravity: Effects in the Developing Nervous System
	– Microgravity and Development of Vestibular Circuits
	– Effects of Microgravity on Neuromuscular Development
	– Postnatal Development of Aortic Nerves in Space
	– Effects of Gravity on Postnatal Motor Development
	• Adult Neuronal Plasticity Team:
	– Central Nervous System Control of Rhythms and Homeostasis During Spaceflight
	– Anatomical Studies of Central Vestibular Adaptation
	– Multidisciplinary Studies of Neural Plasticity in Space
	– Ensemble Neural Coding of Place and Direction in Zero-G
	– Effects of Microgravity on Gene Expression in the Brain
	• Aquatic Team:
	– Chronic Recording of Otolith Nerves in Microgravity
	– Development of Vestibular Organs in Microgravity
	• Neurobiology Team:
	– Development of an Insect Gravity Sensory System in Space

Table 3–114. STS-90 Mission Characteristics (Continued)

Experiments	Shuttle Vibration Forces (SVF): Measured dynamic forces acting between the Space Shuttle and a canister attached to the Shuttle sidewall during the mission.
	BDS-04: The crew performed the following two cell biology experiments under controlled conditions on small samples of material: • Human Renal Cell Experiment • Microgravity Induced Differentiation of HL-60 Promyelocytic Leukemia Cell
Get Away Specials	G-197 Customer: Lockheed Martin Astronautics, Denver, National Institute of Standards and Technology (NIST), and Ames Research Center, Mountain View, California This experiment demonstrated pulse tube cooling technology in the zero gravity environment of space to gain operational experience with the smallest such cryocooler yet built.
	G-772 Customer: University of Colorado, Boulder Collisions into Dust Experiment (COLLIDE): The COLLIDE analyzed the gentle collisions of dust particles in space to learn more about the sources of dust in planetary rings.
	G-744 Customer: Sierra College, Rocklin, California This experiment took ozone measurements of Earth's upper atmosphere in the ultraviolet 200-nanometer to 400-nanometer spectral range using a charge coupled device-based spectrometer.
Mission Results	The mission was successful except for the results from the Mammalian Development Team, which had to reprioritize its science activities because of the unexpected high mortality rate of neonatal rats on board.
Remarks	*Columbia* was outfitted with extended duration orbiter hardware for the flight.
	Astronaut Kathryn Hires was the first Kennedy Space Center employee to be chosen as an astronaut candidate.

[a] Jenkins, p. 310.

Table 3–115. STS-91 Mission Characteristics

Vehicle	OV-103 *Discovery*
Crew	CDR: Charles J. Precourt
	PLT: Dominic L. Pudwill-Gorie
	MS: Franklin R. Chang-Diaz, Wendy B. Lawrence, Janet Lynn Kavandi, Valery Victorovitch Ryumin (RSA)
Launch	June 2, 1998, 6:06:24 p.m. EDT, Kennedy Space Center, Pad 39-A. The countdown proceeded smoothly except for a slight delay in operations to load the external tank with cryogenic propellant to evaluate a few technical issues. As planned, launch managers determined the exact orbital location of the *Mir* space station during the countdown's T-9-minute built-in hold. The decision was then made to launch *Discovery* at 6:06 p.m. to achieve optimum Shuttle system performance and to accommodate Shuttle-*Mir* rendezvous activities.
Orbital Altitude & Inclination	204 nmi[a] (379 km), 51.60 deg
Launch Weight (lb/kg)	259,653[b]/117,777
Landing & Postlanding Operations	June 12, 1998, 2:00:18 p.m. EDT, Runway 15, Kennedy Space Center. The Shuttle landed on the first landing opportunity.
Rollout Distance (ft/m)	11,730/3,575
Rollout Time (seconds)	64
Mission Duration	235 hr, 54 min, 00 sec
Landed Revolution No.	154
Mission Support	STDN
Primary Objective	S/MM-09; experimentation using SPACEHAB
Deployed Satellites	None
Experiments	AMS: This experiment was a collaboration between NASA and the U.S. Department of Energy. This was the first time a high-energy particle magnetic spectrometer was placed in orbit. The spectrometer was designed to detect and catalogue, with a high degree of precision, high-energy charged particles (including antimatter) outside Earth's atmosphere. During its time aboard the Shuttle, a complete system check was performed to ensure it would function properly on the Space Station. The spectrometer also carried out a search for anti-helium and anti-carbon nuclei and measured the spectrum of antiprotons.

Table 3–115. STS-91 Mission Characteristics (Continued)

Experiments	Shuttle-*Mir* Science:
	• Advanced Technology–Commercially initiated research to evaluate new technologies and techniques using the *Mir* space station and the Shuttle as a testbed.
	  – ASTROCULTURE™
	  – X-Ray Detector Test (XDT)
	  – Optizon Liquid Phase Sintering Experiment (OLiPSE)
	• Earth Sciences–Visual observations and photography of sites of interest,
	• Human Life Sciences–Investigations focusing on crew members' adaptation to weightlessness in terms of skeletal muscle and bone changes, cardiovascular acclimatization, and psychological interactions. The investigations continued to characterize the integrated human response to a prolonged presence in space.
	  – Crew member and Crew-Ground Interactions During NASA-*Mir*
	  – Magnetic Resonance Imaging (MRI)
	  – Autonomic Investigations (Cardio)
	  – Bone Mineral Loss and Recovery After Shuttle/*Mir* Flights (Bone)
	  – Assessment of Humoral Immune Function During Long-Duration Spaceflight (Immunity)
	  – Renal Stone Risk Assessment During Long-Duration Spaceflight (Renal-2)
	• ISS Risk Mitigation
	  – CREAM
	  – Space Portable Spectroreflectometer (SPSR)
	  – Test of Portable Computer System (TCPS) Hardware
	  – RME
	• Microgravity–Materials science research
	  – Microgravity Isolation Mount (MIM) Facility Operations PCG-Dewar
	  – SAMS
	  – QUELD
	  – Biotechnology System Diagnostic Experiment (BTSDE) Reflight
	  – Biotechnology System Coculture (COCULT)
	  – DCAM
	
	CPCG: See STS-49.
	
	SSCE: See STS-41.
	
	Growth and Morphology, Boiling and Critical Fluctuations in Phase Separating Supercritical Fluids (GMSF): This experiment increased knowledge in the fundamental science of critical fluids.
	
	SIMPLEX: See STS-84.

Table 3–115. STS-91 Mission Characteristics (Continued)

Get Away Specials	G-648
	Customer: Canadian Space Agency's Microgravity Sciences Program and University of Moncton, New Brunswick ACTORS: The ACTORS processed organic materials in space where the gravitational forces were minimal to compare thin films.
	G-765
	Customer: Canadian Space Agency and several other partners • Microgravity Industry Related Research for Oil Recovery (MIRROR): The MIRROR conducted research to develop new technologies to extract oil from Earth and clean up accidental oil spills.
	G-090
	Customer: Utah State University designed this GAS payload to carry the following four experiments for high school students: • Chemical Unit Process (CUP)–Shoshone-Ba Junior/Senior High School, Fort Hall Reservation, Idaho • Nucleic Boiling–Box Elder High School, Brigham City, Utah • Crystal Growth Experiment–Moscow (Idaho) High School and Moscow University, Idaho • Popcorn/Radish Experiment–St. Vincent Elementary School, Salt Lake City, Utah
	G-743
	Customer: Broward (Davie, Florida) and Brevard (Cocoa, Florida) Community Colleges and Belen Jesuit Preparatory School (Miami, Florida) A genotoxicology experiment determined the degree to which DNA was damaged by exposure to cosmic radiation in a space environment.

Table 3–115. STS-91 Mission Characteristics (Continued)

Space Experiment Module	SEM-03:
	• Effect of Microgravity on Crossing-Over in Sordaria Fimicola–Shoreham, New York Wading River High School
	• Crystal Growth in Microgravity–Tomasita Young Astronauts Club, Albuquerque, New Mexico
	• Norfolk, Virginia Public Schools Science and Technology Advanced Research (NORSTAR) experiments:
	  – Effect of microgravity on development of Daphnia, Eubranchipus, and Triops eggs
	  – Separation of immiscible fluids in microgravity
	• Boy Scouts Troop 177 and Four Rivers District, Gambrills, Maryland–Merit Badge Madness
	• CAN-DO Project, Charleston, South Carolina, experiments:
	• MAVIS–Magnetic Attraction Viewed In Space
	• BEST–Big Experiment in Small Tubes
	• Cosmic Radiation Effects on Programmable Logic Devices (CREPLD)–Purdue University, West Lafayette, Indiana
	• Woodmore Elementary School, Mitchellville, Maryland (WESTAR)
	
	SEM-05:
	• Effect of Spaceflight on Food Yield–Chesapeake Bay Girl Scout Council, Salisbury, Maryland
	• Exposure of the Space Experiment Module to the Space Environment–Excel Interactive Science Museum, Salisbury, Maryland
	• Comparative Microgravity Response of Fungi and Mold–Grand Coulee, Washington Elementary School
	• Effect of Microgravity on Plant Seeds–Olin-Sang-Ruby Union Institute, Oconomowoc, Wisconsin
	• Flower Garden in Space–Virginia Parent Teachers and Students Association, Accomac, Virginia
	• Effects of Microgravity on Sordaria Fimicola–Wicomico High School, Salisbury, Maryland

Mission Results	Successful
Remarks	After undocking of *Discovery* from *Mir*, the crew carried out a gas release procedure consisting of the release of a tracer gas composed of acetone and biacetyl into the depressurized Spektr module on *Mir*. The procedure was designed to enable Shuttle astronauts to document the ionization glow from the gas through any hole in Spektr's hull before sunrise and any fluorescent glow from the gas after sunrise.

a Altitude not found in NASA Mission Archives. Source of altitude is Jenkins, p. 311.
b Jenkins, p. 311.

Table 3–116. STS-95 Mission Characteristics

Vehicle	OV-103 *Discovery*
Crew	CDR: Curtis L. Brown, Jr. PLT: Steven W. Lindsey PC: Stephen K. Robinson MS: Scott E. Parazynski, Pedro Duque (ESA) PS: Chiaki Mukai (JAXA), Senator John H. Glenn, Jr.
Launch	October 29, 1998, 2:19:34 p.m. EST, Kennedy Space Center, Pad 39-B. The countdown proceeded to T-9 minutes but was held an additional 8.5 minutes while the launch team discussed the status of a master alarm heard during cabin leak checks after hatch closure. When the count picked up and the Orbiter Access Arm was retracted, the Range Safety Officer (RSO) requested a hold at T-5 minutes due to an aircraft in the restricted air space around Kennedy Space Center. When the aircraft cleared the area, the RSO gave the all-clear signal and the countdown proceeded. Following main engine start, but before booster ignition, the drag chute compartment door fell off but, according to the NASA Space Shuttle Mission Chronology for STS-95, "posed no problem for the mission."[a] Managers decided not to deploy the chute upon landing. This was the first flight of the Space Shuttle Main Engine–Block II.
Orbital Altitude & Inclination	310 nmi (574 km), 28.45 deg
Launch Weight (lb/kg)	263,987[b]/119,743
Landing & Postlanding Operations	November 7, 1998, 12:04:00 p.m. EST, Runway 33, Kennedy Space Center. Landed on first opportunity.
Rollout Distance (ft/m)	9,508/2,898
Rollout Time (seconds)	59
Mission Duration	213 hr, 14 min, 57 sec
Landed Revolution No.	134
Mission Support	STDN
Primary Objective	Experimentation using SPACEHAB; deployment and retrieval of SPARTAN 201; operation of HOST: return of John Glenn to flight
Deployed Satellites	Deployed and retrieved SPARTAN 201

Table 3–116. STS-95 Mission Characteristics (Continued)

Experiments	The SPACEHAB Single Module experiments included:
	• Vestibular Function Experiment Unit (VFEU)
	• BRIC
	• Oceaneering SPACEHAB Refrigerator Freezer (OSRF)
	• OCC 3-DMA
	• AGHF
	• Facility for Adsorption and Surface Tension (FAST)
	• APCF 3-DMA
	• BIOBOX
	• Self-Standing Drawer-Morphological Transition and Model Substances (SSD-MOMO)
	• Osteoporosis Experiment in Orbit (OSTEO)
	• NIH-C8
	• Clinical Trial of Melatonin as Hypnotic for Space Crew (SLEEP-2)
	• Protein Turnover During Space Flight (PTO)
	• CPCG-PCF
	• CPCG-CVDA
	• MGBX
	• MGBX–Colloidal Disorder Order Transition (MGBX-CDOT)
	• MGBX–Colloidal Gelation Experiment Transition (MGBX-CGEL)
	• Commercial Instrumentation Technology Associates (ITA) Biomedical Experiments (CIBX)
	• CGBA 1
	• ASC-8
	• ADSEP
	• Protein Crystallization Apparatus for Microgravity-1 (PCAM-1)
	• Biotechnology Dynamics-A (BIODYN-A)
	• Aerogel
	• MBGX-Internal Flows in a Free Drop (MGBX-IFFD)
	• Microencapsulation Electrostatic Processing System (MEPS)
	The HOST Platform experiments:
	• Tested flight of NICMOS cooler, planned for installation into the Hubble Space Telescope
	• Verified the zero-gravity operation of the Reverse Turbo Brayton Cycle Cooler/CPL system
	• Tested flight of the Hubble Space Telescope 486 computer, planned for installation into the Hubble Space Telescope
	• Tested flight of the solid state recorder and correlated to known Hubble Space Telescope flight performance
	• Verified the operation of the Fiber-Optic Flight Experiment, a fiber-optic data link between the crew cabin and the payload bay

Table 3–116. STS-95 Mission Characteristics (Continued)

Experiments	The IEH-03 experiments consisted of: • SEH • UVSTAR • STAR-LITE • CONCAP IV (see STS-57) • Petite Amateur Navy Satellite (PANSAT) • SOLCON Cryogenic Thermal Storage Unit Flight Experiment (CRYOTSU): Fifth in a series of Cryogenic Test Bed flights. These experiments consisted of: • 60 K Thermal Storage Unity (TSU) • Cryogenic Capillary Pump Loop (CCPL) • Cryogenic Thermal Switch (CTSW) • Phase Change Upper End Plate (PCUEP) E-Nose: The E-Nose was an environmental monitoring instrument that detected and identified a wide range of organic and inorganic molecules down to the parts-per-million level. PCG-STES: See STS-49. BRIC: See STS-64.
Get Away Specials	G-467 Customer: ESA This experiment investigated the performance of a two-phase CPL with two advanced evaporators, a two-phase Vapor Quality Sensor (VQS), and a control reservoir. G-779 Customer: Bellarmine College Louisville, Kentucky This experiment examined the role of gravitationally dependent hydrostatic pressure effects on the adaptation of the cardiovascular system to the microgravity environment of spaceflight. GAS canisters that were carried on the IEH-03 Hitchhiker: G-764 Customer: University of Bremen, Germany and Zentrum fur Angewandte Raumfahrttechnologie und Mikrogravitation (ZARM) Cosmic Dust Aggregation (CODAG): The CODAG experiment simulated the aggregation of dust particles and dynamics of dust clouds that occurred in the early stages of the formation of our solar system. G-238 Customer: American Institute of Aeronautics and Astronautics and DuVal High School, Lanham, Maryland Roach Experiment: This experiment studied the effects of space on the life cycle of the American cockroach.

Table 3–116. STS-95 Mission Characteristics (Continued)

Space Experiment Module	SEM-04: This canister contained the following eight student experiment modules, part of an educational initiative to increase student access to space.

- The Effect of Microgravity and Temperature on Human Tissue and Human Used and Consumed Items–Blue Mountain School, Floyd, Virginia.
- Effects of Microgravity on an Object's Physical Characteristics–Dowell Elementary School, Marietta, Georgia
- The Effect of Cosmic Radiation on Wisconsin Fast Plants and the Development of Brine Shrimp Eggs and Chia Seeds–Fort Couch Middle School, Upper Saint Clair, Pennsylvania Monrovia Elementary School, Madison, Alabama
- The Effects of Microgravity on Surface Tension–Glenbrook North High School, Northbrook, Illinois
- Growing "Montello" Transglobally–Montello High School, Montello, Wisconsin; Istituto Technico Commerciale Riccatl, Treviso, Italy
- Analysis of Three-Dimensional Sprag Performance in a Microgravity Environment–University of Maryland, College Park, Maryland
- The Effect of Microgravity and Temperature on Mold Growth–West Richland Elementary School, Noble, Illinois
- Woodmore Elementary School, Teaching And Researching–2 (WESTAR-2)/ GERMINAcion ARgentina (GERMINAR)–The Effect of Microgravity on Seed Growth and Survival Woodmore Elementary School, Mitchellville, Maryland; Colegio Santa Hilda, Buenos Aires, Argentina

Mission Results	Successful
Remarks	This was the first time a U.S. President (President William J. Clinton) attended a Shuttle launch.[c]

[a] "Space Shuttle Mission Chronology: STS-95," *http://www-pao.ksc.nasa.gov/kscpao/chron/sts-95.htm* (accessed November 30, 2005).

[b] Jenkins, p. 311.

[c] "The First U.S. Launch for the International Space Station: Wrapping Up a Successful Year of Space Shuttle Missions," *Spaceport News* 37, no. 25 (December 18, 1998): 1, 4.

Table 3–117. STS-88 Mission Characteristics

Vehicle	OV-105
	Endeavour
Crew	CDR: Robert D. Cabana
	PLT: Frederick R. Sturckow
	MS: Jerry L. Ross, Nancy J. Currie, James H. Newman, Sergei K. Krikalev (RSA)
Launch	December 4, 1998, 3:35:34 a.m. EST, Kennedy Space Center, Pad 39-A. The originally scheduled launch of *Endeavour* on December 3 was postponed for 24 hours when time ran out on the launch window. About T-4 minutes in the launch countdown, after orbiter hydraulic systems were powered on, a master alarm associated with hydraulic system No. 1 in the crew cabin was noted. The countdown was held at T-31 seconds to further assess the situation. Shuttle system engineers attempted to quickly complete an assessment of the suspect hydraulic system and eventually gave an initial "go" to resume the countdown. With only seconds to respond, launch controllers were unable to resume the countdown in time to launch within the remaining window. The launch was completed on time on December 4.
Orbital Altitude & Inclination	173 nmi (320 km), 51.60/31.363 deg
Launch Weight (lb/kg)	239,059[a]/108,435
Landing & Postlanding Operations	December 15, 1998, 10:53:29 p.m. EST, Runway 15, Kennedy Space Center. Landing made on the first opportunity.
Rollout Distance (ft/m)	8,343/2,543
Rollout Time (seconds)	44
Mission Duration	283 hr, 17 min, 3 sec
Landed Revolution No.	185
Mission Support	STDN
Primary Objective	To deliver the first U.S. ISS module Unity and assemble with the first Russian ISS module Zarya, already in space
Deployed Satellites	Argentinean National Commission of Space Activities' Satelite de Aplicaciones/Cientifico-A (SAC-A)

Table 3–117. STS-88 Mission Characteristics (Continued)

Experiments	MightySat 1: The MightySat 1 was a non-recoverable all-composite spacecraft structure and experiments integrated with a Hitchhiker Ejection System. The program was dedicated to providing frequent, inexpensive, on-orbit demonstrations of space system technologies. The experiments were: • Advanced Composite Structure • Advanced Solar Cell • Microsystem and Packaging for Low Power Electronics
	Assessment of Human Factors Configuration A: The experiment analyzed human-machine, human-environment, and human-human interfaces.
	Effects of Microgravity on Cell-Mediated Immunity and Reactivation of Latent Viral Infections: This experiment assessed the immune system function using the immune cells from the standard Flight Medicine blood draw. The objective was to examine the mechanisms of spaceflight-induced alterations in the human immune function and latent virus shedding.
	Individual Susceptibility to Post-Spaceflight Orthostatic Intolerance: This experiment investigated mechanisms responsible for differences in post-spaceflight orthostatic intolerance to customize countermeasure protocols.
	Interaction of the Space Shuttle Launch and Entry Suit and Sustained Weightlessness on Egress: This experiment identified the impact of the Launch Entry Suit/Advanced Crew Escape Suit (LES/ACES) and sustained weightlessness on egress locomotion mechanical efficiency as measured by oxygen consumption and gait change.
	Low Iodine Residual System: This system used a newly developed technology that replaced the Galley Iodine Removal Assembly (GIRA) to reduce the concentration of iodine in the Shuttle potable water system, demonstrating that iodine concentrations in Shuttle drinking water can be reduced to medically acceptable levels while maintaining microbial control in the water distribution system.
	Single String Global Positioning System: This system demonstrated GPS performance and operation during orbiter ascent, on-orbit operations, entry, and landing phases using a modified military GPS receiver processor and existing orbiter GPS antennas.

Table 3–117. STS-88 Mission Characteristics (Continued)

Experiments	Space Integrated Global Positioning System/Inertial Navigation System (SIGI): The SIGI mitigated the technical and schedule risks of applying this new technology to the Shuttle navigation systems by evaluating the systems' performance in spaceflight.
	Structural Dynamics Model Validation: This test excited the structural dynamics of the joined Shuttle and ISS to acquire several critical natural frequencies and their corresponding structural damping to allow confirmation of the acceptability of the Shuttle primary jet control algorithm tuning before attitude control using the algorithm.
	USA SAFER Flight Demonstration: This demonstration showed, through an end-to-end on-orbit functional checkout, that the USA SAFER design performed as expected.
	ICBC: The ICBC was a 65-mm color motion picture camera system used to film the Unity installation onto the orbiter docking system; the Zarya rendezvous; docking; EVA tasks; separation burn; and flyaround.
Get Away Specials	G-093R Customer: University of Michigan (Ann Arbor) Students for the Exploration and Development of Space
	Vortex Ring Transit Experiment: Investigated the propagation of a vortex ring through a liquid-gas interface in microgravity.
Space Experiment Module	SEM-07: This module contained 11 experiments as part of an educational initiative to increase student access to space.
Mission Results	Successful

[a] Jenkins, p. 311.

Table 3–118. *Space Station* Freedom *Prime Contractor*[a]

Prime Contractor/ NASA Center	Work Package Description	Phase I Value	Phase II Value	Partner Companies
Boeing Aerospace/ Marshall Space Flight Center	Work Package 1: Laboratory and habitation modules; resource node structures; airlock systems; environmental control/life support; thermal/video/ audio systems; logistics elements	$1.6 billion	$25 million	Teledyne Brown; Lockheed; Hamilton Standard; Garrett AiResearch; Grumman; ILC; Fairchild-Weston
McDonnell Douglas Astronautics/Johnson Space Center	Work Package 2: Truss structure; mobile servicing transporter; airlocks; resource node outfitting; data management; communications and tracking; guidance; navigation and control; EVA systems; propulsion; thermal control	$2.6 billion	$140 million (Dual-keel structure)	IBM; Lockheed; GE/RCA; Honeywell; Spar Astro
General Electric Astro- Space/ Goddard Space Flight Center	Work Package 3: Polar platform; two attach points on crewed base; integrated telerobotic servicer; defined satellite servicing facility	$895 million	$570 million (Co-orbit platform; three more attach points; satellite servicing facility)	TRW
Rockwell International Rocketdyne Division/ Lewis Research Center	Photovoltaic power generation system	$1.6 billion	$740 million (Solar dynamic power system)	Ford Aerospace; Harris; Garrett; General Dynamics; Lockheed

[a] "Space Station *Freedom* Contract Negotiations Concluded," *NASA News* Release 88-132, September 28, 1988. Also "Space Station Prime Contracts Awarded at Last," *Interavia* (September 1988): 18.

Table 3–119. Space Station Freedom *Characteristics*
(May 1992[a])

Element	Shape	Characteristic
Station end-to-end length		108 m (353 ft)
Station total weight		281,430 kg (309.6 tons)
Truss assembly and equipment size	Hexagonal	Length: 65.9 m (216.0 ft) Width: 3.7 m by 4.9 m (12 ft by 16 ft)
Truss assembly and equipment weight		146,000 kg (160.6 tons)
U.S. laboratory size	Cylindrical	8.4 m by 4.4 m (27.4 ft by 14.5 ft)
U.S. laboratory weight		15,545 kg (17.1 tons)
U.S. habitation module size	Cylindrical	8.4 m by 4.4 m (27.4 ft by 14.5 ft)
U.S. habitation module weight		16,182 kg (17.8 tons)
Japanese laboratory	Cylindrical	10.6 m by 4.0 m (34.8 ft by 13.1 ft)
Japanese Exposed Facility	Cylindrical	5.0 m by 5.6 m (16.4 ft by 18.4 ft)
Japanese module weight		32,818 kg (36.1 tons) (both laboratory and exposed facility)
International standard payload racks		1 m (3.5 ft) wide
U.S. resource nodes size	Cylindrical	Three each 5.2 m by 4.4 m (17 ft by 14.5 ft); centrifuge in third resource node
U.S. resource node weight		Two nodes and cupola: 23,545 kg (25.9 tons)
Canadian Mobile Servicing System		
Space System Remote Manipulator		55 ft (16.8 m)
Space System Remote Manipulator capacity		113,398 kg (250,000 lb)
Special Purpose Dexterous Manipulator		12 ft (3.6 m) (two 6-ft (1.8-m) arms)
ESA Columbus Free-Flying laboratory	Cylindrical	11.8 m (38.7 ft) by 4.5 m (14.7 ft)
Solar arrays size	Rectangular	12 m (39 ft) and 34 m (112 ft)
Solar arrays weight		7,909 kg (8.7 tons) (does not include truss)
Number of cells per array wing		32,800
Number of solar cells for six array wings		196,800

Table 3–119. Space Station Freedom *Characteristics*
(May 1992[a]) (Continued)

Element	Shape	Characteristic
Power generation		120 volts DC, 18.75 kW at man-tended capability with at least 11 kW available to researchers. Would grow to 56.25 kW in the follow-on phase (permanently tended capability), with nominal 30 kW available for users, 26.25 kW available for housekeeping[b]
Crew		Four persons (two dedicated for payload operations and two researchers), growing to eight crew in the follow-on phase
Altitude		335 km–460 km (208 nmi–285 nmi)
Inclination		28.5 degrees

[a] "Space Station *Freedom*, Gateway to the Future," National Aeronautics and Space Administration, May 1992 (NASA History Office Folder 009554). Also *Space Station Freedom Media Handbook,* 1992, p. 25.

[b] *Space Station Freedom User's Guide,* August 1992, p. 3-3 (NASA History Office Folder 009554). Also Launius, p. 234. Power was reduced to 56.25 kW from 75 kW for permanently tended capability in a 1992 redesign.

Table 3–120. Shuttle-Mir Flights

Mission	Dates	Orbiter	Type of Mission	Astronaut to/from *Mir*
STS-63	February 3 – February 11, 1995	*Discovery*	Rendezvous	
STS-71	June 27 – July 6, 1995	*Atlantis*	Docked	Returned Norman Thagard to Earth after 115 days in space. (Thagard was delivered to *Mir* on the Soyuz-TM 21 mission)
STS-74	November 12 – November 20, 1995	*Atlantis*	Docked	No astronaut. Delivered the docking module.
STS-76	March 22 – March 30, 1996	*Atlantis*	Docked	Delivered Shannon Lucid.
STS-79	September 19 – September 26, 1996	*Atlantis*	Docked	Returned Lucid to Earth after 188 days. Delivered John Blaha.
STS-81	January 12 – January 22, 1997	*Atlantis*	Docked	Returned Blaha to Earth after 128 days. Delivered Jerry Linenger.
STS-84	May 15 – May 24, 1997	*Atlantis*	Docked	Returned Linenger to Earth after 132 days. Delivered Michael Foale.
STS-86	September 25 – October 6, 1997	*Atlantis*	Docked	Returned Foale to Earth after 145 days. Delivered David Wolf.
STS-89	January 22 – January 31, 1998	*Endeavour*	Docked	Returned Wolf to Earth after 128 days. Delivered Andrew Thomas.
STS-91	June 2 – June 12, 1998	*Discovery*	Docked	Returned Thomas to Earth after 143 days.

Table 3–121. ISS Contributor[a]

Country/Space Agency	Component
Canada	Mobile servicing system
European Space Agency	Columbus laboratory module
Japan	Experiment module with centrifuge facility Exposed facility
Russia	Power platform Service module Functional cargo block (FGB)[b] Two research modules
United States	Integrated truss Habitation module Laboratory module Docking modules and passageway (Node 1) Joint airlock
Italy	Nodes 2 and 3 built for NASA

[a] "International Space Station Builds on 'Freedom,'" *HQ Bulletin* (April 4, 1994): 1 (NASA History Office Folder 009577); *Press Information Book,* The Boeing Company (NASA History Office Folder 16482).

[b] The first Russian module is referred to in NASA and Russian documents and Web sites as both the Functional Energy Block and the Functional Cargo Block. For uniformity, it is called Functional Cargo Block in this document. The acronym FGB comes from the Russian translation written in the Cyrillic alphabet.

Table 3–122. ISS Major Milestones (as of April 1994[a])

Date	Event
November 1997	Russian FGB vehicle launch
December 1997	First U.S. launch
January 1998	Russian service module added, followed by the addition of the universal docking module and science power platform
May 1998	U.S. laboratory module attached (marks the beginning of human-tended science operations)
June 1998	Canadian-built robotic arm added
August 1998	Addition of the Soyuz transfer vehicle
Early 2000	ESA laboratory module added
June 2002	Assembly complete

[a] "International Space Station Builds on 'Freedom,'" *HQ Bulletin* (April 4, 1994): 1 (NASA History Office Folder 009577).

Table 3–123. ISS Assembly Schedule (June 1994[a])

Schedule	Date	Payload
First Russian element launch	November 1997	FGB
First U.S. element launch	December 1997	Node 1 (with four racks)
Human-tended capability	June 1998	U.S. lab outfitting
Three-person on-orbit capability	August 1998	Soyuz crew transfer vehicle
Japanese lab launch	March 2000	Japanese Experiment Module pressurized lab
European Space Agency lab launch	March 2001	Attached pressurized module
Habitation module launch	February 2002	U.S. habitation module
Permanent human capability	June 2002	Soyuz crew transfer vehicle

[a] *International Space Station Fact Book*, National Aeronautics and Space Administration, June 1994.

Table 3–124. Columbus Characteristics [a]

Item	Description
Total module length	687 cm (22.5 ft)
Largest diameter	448 cm (14.7 ft)
Total internal volume	75 cu meters (2,649 cu ft)
Total volume of payload racks	25 cu meters (883 cu ft)
Mass without payload	10,300 kg (22,708 lb)
Launch mass	12,800 kg, including 2,500-kg payload (28,219 lb, including 5,512-lb payload)
Maximum payload mass	8,000 kg (17,637 lb)
Maximum on-orbit mass	19,300 kg (42,549 lb)
Communications	Downlink via Artemis; downlink and uplink via TDRS
Crew size	Three
Cabin temperature	Between 16°C and 30°C (61°F and 86°F)
Air pressure	Between 959 and 1013 hPa
Total electrical power	20 kW provided by the Station
Payload power	13.5 kW
Main contractor	Daimler Benz Aerospace leading a consortium of subcontractors[b]

[a] "Columbus: European Laboratory," ESA, *http://www.esa.int/esaHS/ESAFRG0VMOC_iss_0.html* (accessed June 14, 2005).

[b] Daimler Benz Aerospace merged with Matra Marconi Space in May 2000 to form Astrium GmbH.

Table 3–125. Partial Revised Manifest, Revision C,
Through 1999 (as of May 15, 1997[a])

Date	Flight	Element
June 1998	1A/R	Functional Cargo Block
July 1998	2A	STS-88/U.S. Node 1 and two pressurized mating adapters (PMA)
December 1998	1R	Service module
December 1998	2A.1	Either service module logistics or the Interim Control Module
January 1999	3A	Integrated truss structure Z1, PMA-3, Ku-band communications system, control moment gyros
January 1999	2R	Three-person crew begins permanent presence on ISS: Soyuz provides assured crew return capability
March 1999	4A	Integrated truss structure Port 6, photovoltaic module, S-band antenna system
May 1999	5A	Lab provides initial U.S. user capability
June 1999	6A	Adds U.S. multipurpose logistics module, ultra-high frequency antenna, ISS remote manipulating system; carries second ISS crew
August 1999	7A	Joint airlock provides U.S. and Russian EVA capability, high pressure gas assembly
Phase II Complete		
October 1999	7A.1	Multipurpose logistics module, additional battery sets
December 1999	4R	Docking compartment 1 provides egress, ingress for Russia-based EVA and a Soyuz docking port

[a] "Assembly Sequence, 5/15/97 Rev C," National Aeronautics and Space Administration, International Space Station (NASA History Office Folder 11613). Also "Station Buildup Sequence Combines Complex Hardware," *Aviation Week & Space Technology* (December 8, 1997): pp. 52–53 (History Office Folder 16949).

Table 3–126. ISS Assembly Sequence Revision D, Through 1999 (as of May 31, 1998)

Date	Flight	Launch Vehicle	Activity
November 1998	1A/R	Proton	Control Module (Functional Cargo Block)
December 1998	2A	Space Shuttle (STS-88)	Unity node, two PMAs
April 1999	1R	Proton	Service module
May 1999	2A.1	Space Shuttle (STS-96)	SPACEHAB Double Cargo Module
June 1999	3A	Space Shuttle (STS-92)	Integrated Truss Structure Z, PMA-3, Control Moment Gyros
July 1999	2R	Soyuz	Soyuz. Station begins permanent human presence
August 1999	4A	Space Shuttle (STS-97)	Integrated Truss Structure P6, photovoltaic module, radiators
October 1999	5A	Space Shuttle (STS-98)	U.S. Laboratory Module
December 1999	6A	Space Shuttle (STS-99)	Multipurpose Logistics Module, ultra-high frequency antenna, SSRMS

Table 3–127. Functional Cargo Block (Zarya) Specifications[a]

Item	Specification
Length (end-to-end)	12.6 m (41.2 ft)
Diameter at widest point	4.10 m (13.5 ft)
Solar arrays	10.7 m (35 ft) by 3.4 m (11 ft)
Gross launching mass	23,500 kg (53,020 lb)
Orbital mass	19,323 kg (42,600 lb)
Orbital operation lifetime	No less than 15 years
Orbital inclination	51.6 degrees to the equator
Preliminary orbit	220.4 km (137 mi) by 339.6 km (211 mi)
Docking orbit altitude	386.2 km (240 mi)[b]
Propellant	16 tanks together holding more than 5.4 metric tons (6 tons)
Power supply	Two solar arrays and six nickel-cadmium batteries
Power to be supplied to the U.S. segment	Can supply an average of 3 kW. Daily average before docking with service module: 0.8 kW. Daily average after docking with service module: 1.2 kW

[a] "Functional Energy Block," Khrunichev State Research and Production Space Center (History Office Folder 17083). Also "Zarya," ISS Element, *http://www.shuttlepresskit.com/ISS_OVR/element1.htm* (accessed June 13, 2005); "Space Station Assembly: Elements: Zarya Control Module," National Aeronautics and Space Administration, *http://spaceflight.nasa.gov/station/assembly/elements/fgb/index.html* (accessed June 17, 2005).

[b] The docking orbit altitude is the altitude at which *Endeavour* made rendezvous and captured the spacecraft to attach it to the U.S.-built Unity module. "Space Station Assembly: Elements: Zarya Control Module," *http://spaceflight.nasa.gov/station/assembly/elements/fgb/index.html* (accessed December 2, 2005).

Table 3–128. Unity Characteristics[a]

Item	Characteristic
Module length	18 ft (5.5 m)
Module diameter	15 ft (4.6 m)
Shape	Six-sided
Pressurized mating adapters length	8 ft (2.4 m)
Launch weight	25,600 lb (11,612 kg)
Ports	Six 50-in (127-cm) ports (4 radial and 2 axial)
Material	Aluminum
Lines	216 lines to carry fluids and gases
Cables	121 internal/external electrical cables using 6 mi (9.7 km) of wire
Stowage space	Four 27-cu-ft (9.8-cu-m) racks

[a] *Press Information Book, Mission Modules, Station Overview*, NASA, Boeing, pp. 5–6, 12.

Table 3–129. Zarya-Unity Orbital Events Summary[a]

Day	Date	Event
1	November 20, 1998	Proton launch, ascent, and orbit insertion of Zarya. Begin multi-axis spin for thermal control and to reduce fuel consumption.
2	November 21, 1998	Engine test burn (10 seconds duration, single engine). Television camera test. Perform perigee raising burn (single engine). Resulting orbit is 153 statute mi (246 km) by 215 statute mi (346 km).
4	November 23, 1998	Perform two burns to raise orbit; resulting orbit: 190 mi (306 km) by 238 mi (383 km).
5	November 24, 1998	Russian flight controllers place the module in its final orbit to achieve *Endeavour* rendezvous. Resulting orbit is 240 miles (386.2 km) circular.[b]
6	November 25, 1998	Perform on-board computer system test. Maneuver to test *Endeavour* capture, docking orientation. Maneuver to assess solar array performance.
8	November 27, 1998	Maneuver to test *Endeavour* capture, docking orientation. Assess solar array, battery charging performance. Begin multi-axis spin.
14	December 3, 1998	*Endeavour* launches on STS-88. Astronauts activate Shuttle Orbiter Docking System.
15	December 4, 1998	Perform Shuttle remote manipulator system, Orbiter Space Vision System, spacesuit checkout.
16	December 5, 1998	Attach Unity PMAs to the Orbiter Docking System using the Shuttle remote manipulator system.
17	December 6, 1998	Rendezvous with and capture Zarya with *Endeavour* remote manipulator system. Zarya berths to Unity PMA-1.
18	December 7, 1998	First spacewalk to connect utilities between PMA-2. Unity PMA-1, and Zarya. Activate Unity computerized control units and PMA shell heaters.
19	December 8, 1998	Pressurize PMA-1 via Zarya.
20	December 9, 1998	Second spacewalk to install early communications system antennas and cable, install computerized control sunshade, EVA aids, and remove common berthing mechanism hatch launch restraint pins.
21	December 10, 1998	ISS entered for the first time. Install and activate communications system; remove shear panels; transfer spare equipment to ISS.
22	December 11, 1998	ISS entered; complete remaining tasks; doors closed to ISS at end of day.
23	December 12, 1998	Third spacewalk to install EVA node stowage bag, perform photo documentation survey, and other tasks. ISS first element completed.

Table 3–129. Zarya-Unity Orbital Events Summary[a] (Continued)

Day	Date	Event
24	December 13, 1998	*Endeavour*/ISS undock, flyaround.
25	December 14, 1998	Perform *Endeavour* secondary payload operations.
25–34	December 14–23, 1998	Systems checkout by ground controllers through S-band communications system installed on Unity.
26	December 15, 1998	*Endeavour* returns home.

[a] "Summary Flight Plan, The International Space Station," *http://www.shuttlepresskit.com/ISS_OVR/assembly1_summary_timeline.htm* (NASA, Boeing, and United Space Alliance Web site) (accessed June 13, 2005). Also *Press Information Book, Mission Modules, Station Overview*, NASA, Boeing, p. 4.

[b] "Space Station Assembly: Elements: Zarya Control Module," *http://spaceflight.nasa.gov/station/assembly/elements/fgb/index.html* (accessed June 17, 2005).

Table 3–130. Science Laboratories Accommodations[a]

Type of Accommodation	Feature
Overall	• 30 kW average power available for payloads • 75 Mb/sec data downlink • Teleoperations • Multipayload modular environment • Standardized service interfaces for payloads
Internal payload	• 37 rack locations • Vibration-free environments • Microgravity environment: 10 locations at 1g level • 3 kW, 6 kW, and 12 kW power options • Vacuum, nitrogen, argon, helium, and carbon dioxide service options • Ethernet, video, high-rate data download • Payload volumes: more than 40 cu ft (1.1 cu m) per rack • Many racks support multiple modular sub-rack payloads • One location with an Earth-facing science-quality window
External payload	• 14 payload sites • 10 locations with active thermal control • 3 kW and 6 kW power options • Video, high-rate data downlink • Earth and stellar viewing sites • Robotic payload manipulation

[a] *Press Information Book*, p. 53.

Table 3–131. Space Station Chronology

Date	Event
January 25, 1984	President Ronald Reagan delivers State of the Union address calling for a "permanently manned space station" to be built within a decade.
February 20, 1986	*Mir* space station core sent into space on a Proton booster rocket.
March 1986	First Russian crew arrives on *Mir*.
1987	Space Station development split into two phases: the revised baseline configuration and the enhanced configuration.
1987	NASA Administrator Fletcher requests $3 million to study crew emergency return vehicles.
July 1988	President Ronald Reagan names the Space Station "Freedom."
September 28, 1988	Negotiations concluded for four 10-year contracts with Boeing Aerospace, McDonnell Douglas Astronautics, GE Astro-Space Division, and Rocketdyne Division of Rockwell International to correspond with four work packages.
June 1989	Assembly sequence is revised to allow only the Shuttle for lifting and assembling components.
July 1989	NASA forms Configuration Budget Review team headed by W. Ray Hook to develop preliminary options for ways for the program to exist within severe budget constraints threatened by Congress.
October 1989	Congress funds program at $1.8 billion, $250 million less than the administration's request.
October 1989	NASA releases a request for proposal for the Assured Crew Return Vehicle.
Late 1989	"Rephasing" of program announced to reduce risk and meet anticipated budget cut of nearly $300 million for FY 1990. First element launch remains at March 1995.
January 1990	NASA forms the External Maintenance Task Team to address EVAs needed for Station maintenance.
June 1990	NASA forms the External Maintenance Solutions Team to address problems raised by the External Maintenance Task Team and to recommend ways to reduce the number of spacewalks.
Fall 1990	White House forms the Advisory Committee on the Future of the U.S. Space Program, chaired by Norman Augustine, to assess alternative approaches and make recommendations for future civil space goals. Committee recommends reducing the Station's size and complexity.
March 1991	NASA delivers restructuring report to Congress for a smaller and simpler Station with a $30 billion price tag. Work Package 2 with GE is eliminated. A rephased assembly sequence moved first element launch to early 1996, human-tended capability to mid-1997, and permanent occupation to 2000.

Table 3–131. Space Station Chronology (Continued)

Date	Event
March 21, 1991	Vice President Dan Quayle and the National Space Council endorse report and redesign.
July 1991	Vice President Dan Quayle meets with Oleg Shishkin, minister of General Machine Building in the Soviet Union, to discuss a cooperative effort using the *Mir* space station for human missions.
July 31, 1991	President George H. W. Bush and Soviet President Mikhail Gorbachev sign an agreement for an astronaut to visit *Mir* and a cosmonaut to fly on the Space Shuttle.
September 27, 1991	The House Appropriations Committee recommends cutting off all funds to the Station, but the Senate agrees to a House funding bill and grants NASA its full FY 1992 funding request of $2,028,900,000 for the Space Station.
December 9, 1991	President George H. W. Bush signs NASA's funding bill.
April 1, 1992	Daniel Goldin replaces Richard Truly as NASA Administrator.
April 1992	Russian President Boris Yeltsin creates the civilian Russian Space Agency, headed by Yuri Koptev. Goldin and Koptev meet informally in Washington, DC to discuss possibilities for cooperation.
June 17, 1992	President George H. W. Bush and Russian President Yeltsin hold a summit in which the two agree to consider a joint space mission. They sign the "Agreement Between the United States of America and the Russian Federation Concerning Cooperation in the Exploration and Use of Outer Space for Peaceful Purposes" that includes a Shuttle-*Mir* mission.
June 19, 1992	Russia and the United States formally sign a new U.S.-Russian Space Cooperation Agreement and ratify a $1 million contract between NASA and Russian aerospace firm, NPO-Energia.
October 5, 1992	NASA and the Russian Space Agency sign an "Implementing Agreement Between the National Aeronautics and Space Administration of the United States of America and the Russian Space Agency of the Russian Federation on Human Space Flight Cooperation" that details cooperation called for in the June 1992 agreement.
March 9, 1993	President William J. Clinton orders NASA to undertake a redesign of the Station to reduce costs and complexity. The administration goal was $9 billion.
March 10, 1993	First meeting of the Station Redesign Team, led by Dr. Joseph Shea.
April 1993	An Advisory Committee, also known as the Vest Panel, is formed to assess redesign options.
April 3–4, 1993	President William J. Clinton and Vice President Albert A. Gore meet with Russian leaders at a summit in Vancouver, Canada, on further cooperation in space. President William J. Clinton invites Russia to participate in the new Station, and Russian President Yeltsin agrees.

Table 3–131. Space Station Chronology (Continued)

Date	Event
June 10, 1993	Advisory Committee presents final report to President William J. Clinton.
June 17, 1993	President William J. Clinton announces his selection of a reduced cost, scaled-down version of the original Space Station *Freedom*, called Alpha, with a $10.5 billion cost spanning five years. The President also directs NASA to develop an implementation plan by September 1993.
August 17, 1993	Goldin names Johnson Space Center as the host Center for the new Space Station program and Boeing as the single prime contractor.
September 1–2, 1993	United States-Russian Commission on technological cooperation in the areas of energy and space, headed by Vice President Albert A. Gore and Russian Prime Minister Viktor Chernomyrdin, meets and agrees on a three-phase structure to complete the Space Station.
September 7, 1993	President William J. Clinton formally chooses the small, four-person Alpha Station, a merger of Space Station *Freedom* and the Russian *Mir*. Congress and the administration agree to a total cost cap of $17.4 billion and a fixed annual budget of $2.1 billion. NASA slips the date for permanent habitability to September 2003.
October 16, 1993	The United States meets with international partners in Paris, France to formally inform them of the intent to invite Russia to join the Space Station program.
November 1, 1993	Goldin and Koptev sign an "Addendum to Program Implementation Plan" for Space Station Alpha describing the overall concept of the relationship between NASA and the Russian Space Agency.
November 7, 1993	Space Station partners jointly meet with the Russian Space Agency to review details of the November 1 addendum.
November 15, 1993	NASA signs a letter contract with Boeing making the company the Space Station prime contractor.
November 29, 1993	An agreement is reached for Russia to be "the primary partner" in the Space Station program, which would be designated the International Space Station. Russia agrees to cancel the planned sale of missile technology to India and would receive $100 million annually from NASA as compensation.
December 6, 1993	Space Station partners decide to formally invite Russia to join the partnership.

Table 3–131. Space Station Chronology (Continued)

Date	Event
December 16–17, 1993	The Gore-Chernomyrdin Commission meets in Moscow. Russia to announce the Commission's acceptance of the invitation to join the Space Station program. Goldin and Koptev sign a protocol expanding the terms of the 1992 HSF Cooperation agreement and agree on Shuttle-*Mir* flights during 1995–1997. Albert A. Gore and Chernomyrdin sign a "Joint Statement on Space Station Cooperation" describing the steps needed to formally bring Russia into the Station partnership. The two parties note that NASA and the Russian Space Agency have agreed to a $400 million contract through 1997 for the Shuttle-*Mir* program and other Station development.
February 1, 1994	Space Station *Freedom* is formally terminated when contracts ending the work package contracts are ended and responsibility is consolidated in a contract with Boeing.
February 3, 1994	Phase I of the ISS begins when Cosmonaut Sergei Krikalev becomes the first Russian to fly on a U.S. spacecraft on the STS-63 Shuttle mission, inaugurating the Shuttle-*Mir* program.
March 1994	The ISS assembly schedule is revised with the first Shuttle launch moved from July 1997 to December 1997, and the completion date slipped from October 2001 to June 2002.
March 1994	The successful System Design Review marks a major technical milestone; it confirms the validity of the baseline configuration, schedule, and cost. Planned assembly is scheduled to begin in November 1997 with the Russian FGB. Assembly planned to be complete in 2002.
April 1994	Heads of the Space Station agencies meet in Washington, DC, to endorse the successful review and reaffirm Russia's part in the program.
June 23, 1994	The U.S.-Russian Joint Commission on Economic and Technological Cooperation signs a new "Joint Statement on Space Station Cooperation" reiterating the two governments' commitment to develop an integrated Space Station and to expedite Russia's involvement as a full partner in the program.
June 29, 1994	A bipartisan House coalition defeats a motion to cancel the Station.
July 1994	The Space Station Control Board approves a revised assembly sequence moving launch of Russia's service module from January to May 1998. The Board also agrees to purchase the FGB from Khrunichev to assure its availability when Station assembly begins.
August 3, 1994	The Senate rejects motion to cancel the Station.
August 31, 1994	NASA and Boeing agree on the key elements of the ISS prime contract.
September 1994	Space Station managers release another updated assembly plan incorporating a centrifuge to augment the Station's science capabilities and provide more power.

Table 3–131. Space Station Chronology (Continued)

Date	Event
1995	X-38 project begins at Johnson Space Center.
January 13, 1995	NASA and Boeing sign a $5.63 billion contract to manage the building of the core Station, including two nodes, an airlock, and laboratory and habitation modules and their integration. The contract also calls for the design and development of the Station.
February 5, 1995	NASA and the Russian Space Agency sign a protocol reflecting the contract negotiated between Boeing subcontractor Lockheed Missiles & Space and Khrunichev for the FGB.
May 1995	ISS completes tests to evaluate the Water Recovery System.
May 20, 1995	Spektr module is launched toward *Mir*.
July 1995	The orbiter *Atlantis* permanently attaches a new docking module to the *Mir* Kristall androgynous docking unit.
July 1995	The House defeats an attempt to cut off Station funding.
September 1995	The Senate defeats a motion to cut off Station funding.
Mid-September 1995	By this date, the United States has produced 54,000 pounds of ISS hardware; international partners have manufactured a total of more than 60,000 pounds of hardware. Boeing completes the main structure, Node 2, of the U.S. laboratory module.
October 18, 1995	The ESA Council meets in Toulouse, France, and approves the program "European Participation in the International Space Station Alpha," providing for the Columbus laboratory module, ATV, and studies of a European CTV.
December 1995	The Russian Space Agency announces that the Russian government owes Khrunichev money for work performed in 1995 and, if the government does not release the funds needed for the FGB and service module, it would be unable to meet the FGB's launch date and unable to build the service module.
January 1996	Exterior of the U.S. Station's modules is completed.
March 27, 1996	NASA Administrator Goldin states that he is giving Russia one month or six weeks to "get [the] Station moving again" and that he is "cautiously optimistic" that Russia will be able to "meet its commitment to deliver the critical service module on time..."[a]
May 1996	The ISS air purification system passes a major test of ability to control carbon dioxide, oxygen, and air pressure inside the Station's living and laboratory quarters.
November 1996	Node 1, the U.S. module, successfully completes the module's final pressure test.
December 1996	Russian FGB is assembled and ready for testing. The Russian Space Agency acknowledges that the service module will have to be delayed until December 1998 because of lack of funds.
January 1997	NASA allocates $100 million to Lockheed to develop the Interim Control Module as a backup to Russia's service module.

Table 3–131. Space Station Chronology (Continued)

Date	Event
February 24, 1997	A fire on *Mir* endangers the crew.
April 1997	NASA and the Russian Space Agency formally agree to slip launch of the FGB from November 1997 to mid-1998.
April 9, 1997	NASA announces a slip in the Station's on-orbit assembly to "no later than" October 1998.
April 11,1997	Russia arranges for bank loans to Energia by the end of May, allowing work to resume on the service module.
May 15, 1997	The Space Station Control Board releases a new assembly schedule, Revision C. The FGB is scheduled to launch in June 1998, the U.S. node in July 1998, and the service module in December 1998.
May 31, 1997	The heads of space agencies accept Revision C at a meeting in Tokyo. Japan.
June 25, 1997	Collision between Progress cargo ship and *Mir* causes air leak and extensive damage.
June 1997	Node 1 (Unity) is shipped to Kennedy Space Center from Boeing plant in Alabama.
July 1997	Unpiloted, captive-carry flight tests of X-38 test airframes attached to B-52 aircraft begin at Dryden Flight Research Center.
September 1997	NASA and Boeing reveal that Boeing's prime contract will have at least a $600 million overrun at completion and that NASA needs $430 million more than expected for FY 1998.
September 1997	Goldin requests that the NASA Advisory Council establish a cost control task force.
September 1997	Revision C of the assembly and launch schedule is formally approved by all partners. The first U.S.-built element, Node 1, is scheduled to launch in July 1998, the Russian service module in December 1998, and the ESA's Columbus is scheduled to launch in October 2002.
September 18, 1997	The GAO releases a report describing worsening cost overruns.
October 14, 1997	NASA and the Brazilian Space Agency sign an implementing arrangement providing for the design, development, operation, and use of Brazilian-developed flight equipment and payloads for the ISS in exchange for Brazil's access to ISS facilities on orbit and a flight opportunity for a Brazilian astronaut.
November 1997	Boeing admits to a House panel that its costs are millions of dollars over its contract amount.
January 29, 1998	The United States and international partners sign a multilateral agreement formalizing the framework for cooperation among the ISS partners. Goldin also signs bilateral agreements with the heads of the ESA, the Canadian Space Agency, and the Russian Space Agency describing their roles and responsibilities. A similar agreement with the government of Japan is signed on February 24, 1998.

Table 3–131. Space Station Chronology (Continued)

Date	Event
March 12, 1998	First free-flight X-38 drop tests take place at Dryden Flight Research Center.
April 15, 1998	The Cost Control Task Force, chaired by Jay Chabrow, delivers its report to the Advisory Council. The report states that NASA will need an estimated $7 billion extra and up to three additional years to complete the program.
May 31, 1998	NASA's partners agree to officially target a November 1998 launch for the first Station component and to revise remaining launch dates. The partners set an April 1999 launch date for the service module and a summer 1999 date for transport of the first crew by Soyuz to the ISS.
June 15, 1998	The NASA response to the Cost Control Task Force report is released. NASA identifies $1.4 billion in additional costs. The schedule has been changed to accommodate a four-month service module schedule slip. The first element launch was moved to November 1998, and the ISS assembly complete date is scheduled for January 2004.
November 20, 1998	Launch of Zarya (FGB) takes place from the Baikonur Cosmodrome in Kazakhstan.
December 3, 1998	First U.S. component, Unity, is launched on STS-88.
December 6, 1998	Unity and Zarya dock.

[a] "Goldin Gives Russia Six Weeks To Get Station Moving Again," *Aerospace Daily* (March 27, 1996): Article 28066 (NASA History Office Folder 17083).

www.ingramcontent.com/pod-product-compliance
Lightning Source LLC
Chambersburg PA
CBHW081253170526

45165CB00011B/3300